Nonequilibrium Statistical Mechanics

Nonequilibrium statistical mechanics (NESM), practically synonymous with time-dependent statistical mechanics (TDSM), is a *beautiful and profound subject*, vast in scope, diverse in applications, and indispensable in understanding the changing natural phenomena we encounter in the physical, chemical and biological world. Although time dependent phenomena have been studied from antiquity, the modern subject, the nonequilibrium statistical mechanics, has its genesis in Boltzmann's 1872 classic paper that aimed at extending Maxwell's kinetic theory of gases by including intermolecular interactions. Subsequent development of the subject drew upon the seminal work of Einstein and Langevin on Brownian motion, Rayleigh and Stokes on hydrodynamics, and on the works of Onsager, Prigogine, Kramers, Kubo, Mori, and Zwanzig.

One major goal of this book is to develop and present NESM in an organized fashion so that students can appreciate and understand the flow of the subject from postulates to practical uses. This book takes the students on a journey from fundamentals to applications, mostly using simple mathematics, and fundamental concepts. With the advent of computers and computational packages and techniques, a deep intuitive understanding can allow the students to tackle fairly complex problems, like proteins in lipid membranes or solvation of ions in electrolytes used in batteries. The subject is still evolving rapidly, with forays into complex biological events, and materials science.

Nonequilibrium Statistical Mechanics: An Introduction with Applications is, thus, an introductory text that aims to provide students with a background and skill essential to study and understand time-dependent (relaxation) phenomena. It will allow students to calculate transport properties like diffusion and conductivity. The book also teaches the methods to calculate reaction rate on a multidimensional energy surface, in another such application.

For a beginner in the field, especially for one with an aim to study chemistry and biology, and also physics, one major difficulty faced is a lack of organization of the available study material. Since NESM is a vast subject with many different theoretical tools, the above poses a problem. This book lays the foundations towards understanding time-dependent phenomena in a simple and systematic fashion. It is accessible to students and researchers who have basic training in physics and mathematics. The book can be used to teach advanced undergraduates. Some involved topics, like the projection operator technique and mode coupling theory, are more suitable for Ph.D. level.

Nonequilibrium Statistical Mechanics

An Introduction with Applications

Biman Bagchi

CRC Press
Taylor & Francis Group
CHAPMAN & HALL

First edition published 2024
by CRC Press
2385 NW Executive Center Drive, Suite 320, Boca Raton FL 33431

and by CRC Press
4 Park Square, Milton Park, Abingdon, Oxon, OX14 4RN

CRC Press is an imprint of Taylor & Francis Group, LLC

© 2024 Biman Bagchi

ISBN: 9780367743956 (hbk)
ISBN: 9780367743994 (pbk)
ISBN: 9781003157601 (ebk)

DOI: 10.1201/ 9781003157601

Typeset in Times New Roman
by Newgen Publishing UK

This book is dedicated to the memory of my scientific mentor,
Professor Robert Zwanzig (1928–2014).

Contents

PART IV   Advanced Topics

Preface

Nonequilibrium statistical mechanics (NESM), practically synonymous with time-dependent statistical mechanics (TDSM), is a beautiful and profound subject. It is vast in scope and wide-ranging in practical use. Unfortunately, however, this subject has remained not only less appreciated among students but also insufficiently organized as a subject than, say, quantum mechanics or classical mechanics, or even equilibrium statistical mechanics. As a result, this subject is hardly taught at the B.Sc. or the M.Sc. levels, and often not even to first-year PhD students, particularly in India. The reasons for this disparity can be multifaceted. Common perception is that NESM involves complex mathematical formalisms and requires a solid understanding of advanced concepts based on statistical theorems, dynamical systems, and stochastic processes. However, it is important to note, especially for students, and this must be emphasized by the teachers, that a large part of the subject and its applications is simple and phenomenological in nature and relies heavily on intuitive understanding. This aspect is not often stressed enough, that NESM, although mathematical, does not always need to invoke complex mathematical tools. Understanding and intuition serve as powerful weapons. In fact, with the advent of computers and computational packages and techniques, a deep intuitive understanding can help the students to tackle fairly complex problems, like, for example, proteins in lipid membranes or the role of solvation of ions in batteries.

As will be made evident throughout this book, nonequilibrium statistical mechanics is essential in the understanding of relaxation functions routinely measured in spectroscopy experiments, in the calculation of transport properties like diffusion and conductivity, also in the calculation of the rate of a reaction on a multidimensional surface, in a viscous medium, or in the formulation of theories of phase transitions. It is used to understand the catalytic cycle of enzyme kinetics or the nucleation of a new (daughter) phase from the old parent phase, and indeed is needed in understanding endless collective phenomena. Many of the topics mentioned above are missed in most of the physics-directed books and reviews, where the emphasis often lies on deciphering general behavior, such as finding the scaling properties.

The present book was conceived from the very outset as a textbook of nonequilibrium statistical mechanics directed toward the students of physical and biophysical chemistry or biophysics, and additionally, materials science. Most of the available textbooks are written by physicists for the physicists. Moreover, the concepts and tools required for students of these *other* disciplines are often quite different, and are not available in standard statistical mechanics books, like the ones written by Kerson Huang or Haken.

A second objective of writing this book is to present a sequel to our earlier text on equilibrium statistical mechanics, also published by Francis-Taylor/CRC Press (2018). The ideas of both books were conceived when I was teaching a two-semester course on statistical mechanics, where I taught a first-semester course on equilibrium statistical mechanics that was followed by a second-semester course on time-dependent phenomena. Much of the material that forms the core of the present book comes from this latter course.

In the area of nonequilibrium statistical mechanics, students and researchers of chemistry and biology (such as ourselves) routinely use three books. The first one is the wonderful book *Simple Liquids* (Academic Press) by Hansen and McDonald. The second one is entitled *Molecular Hydrodynamics* (Dover) by Boon and Yip. The third one is *Dynamic Light Scattering* by Berne and Pecora. There is, of course, the elegant and masterly book by Zwanzig (Oxford) that caters to a certain audience. That is, Zwanzig's book requires pre-exposure to the field. The great book by van Kampen again caters to a selective audience. The last two can be called advanced texts. I must also mention the pedagogical books by H. Haken in the series named Synergetics.

But the main difficulty of not having an organized textbook on time-dependent statistical mechanics is that students might not get the chance to learn a connected and/or coherent approach toward relaxation phenomena, chemical dynamics, spectroscopy, transport properties, to name a few. For example, in order to study solvation dynamics and electron transfer reactions, we make use of the time correlation formalism of TDSM and of linear response theory. The last two are needed to explain solvation dynamics in terms of dielectric relaxation. The connection is often left unclear. Of course, advanced students could always turn to advanced books like that of Zwanzig and van Kampen, but prior exposure to certain basic concepts and methods serves students in the long run.

To stress this point, a topic often ignored is *the relationship between theory and experiment.* A quantitative study of this relationship requires the use of both quantum and statistical mechanics. We have covered this topic in **Chapter 3.**

Our next example is hydrodynamics. This is a subject that is entirely neglected in chemistry and biology courses. This may create difficulties because much of the developments, even the language often employed, are derived from hydrodynamics. While the kinetic theory of gases is discussed, albeit at a preliminary level, hydrodynamics is left out.

With the above lacunae in mind, we have tried to develop a course that serves as an introduction to the main concepts and techniques of nonequilibrium statistical mechanics. In order to create a coherent exposition of the diverse concepts involved, the entire book is divided into **five parts.**

Part I addresses preliminary aspects. The first chapter of Part I of the book addresses the need to study nonequilibrium statistical mechanics, while the second chapter introduces the theory of probability catered to nonequilibrium statistical mechanics. The third chapter describes in detail the relationship between theory and experiment. This is followed by a pedagogical introduction to the essential concepts of nonequilibrium thermodynamics. Subsequently, we discuss two important chapters – hydrodynamics and the kinetic theory of gases. Our treatment of these two chapters differs from conventional treatments in the sense that we bring out the role of these two subjects to the future growth of NESM/TDSM in the years that followed.

Part II deals with fundamental aspects of time dependent statistical mechanics. It starts with Liouville's equation, and covers important topics like time correlation functions, linear response theory, projection operator technique, and also the method of moments and cumulants. All these topics are of great importance, and our goal in bringing them together is to drive home the point that these are closely related to each other.

Part III covers phenomenological aspects of time-dependent statistical mechanics and consists of eight chapters. This part describes, in detail, equations and techniques related to the Langevin equation and Brownian motion, random walks, Fokker-Planck and Smoluchowski equation. *Thus, this part contains topics that find maximum use in the treatment of nonequilibrium phenomena, even at a molecular level, in modeling various transport and relaxation phenomena.* An important point often not discussed in classes is that most conventional applications include only single particle variables, like single-particle density and momentum. This involves a huge reduction in degrees of freedom. There are of course, many-body versions of the phenomenological equations which can be solved by computer simulations. We have discussed such solution methods in this part of Chapter 19. This part also includes three chapters related to general relaxation phenomena. These are (i) the rate theory, (ii) translational diffusion, and (iii) rotational diffusion. The rate theory is a topic that has seen great advancement in the last three decades, although usually not included in the textbooks of statistical mechanics. We pay special attention to the calculation of rate in a multidimensional potential energy surface in a dissipative medium. The chapter on translational diffusion includes many currently popular subjects like the relation between diffusion and entropy. The chapter on rotational diffusion is closely connected to many experimental techniques as rotational diffusion can be measured with relative ease but reveals important information about the interactions of a system.

Part IV we discuss three advanced topics that could be of use to students in their research work. They include coupling theory, advanced tools on irreversible thermodynamics and the computational approach to rare events.

Thus, throughout the book attempts have been made to bring together subjects and topics that are used to understand relaxation phenomena, not just in chemistry and biology but also in physics, and materials science.

However, a book like this that attempts to cover a wide range of topics tends to get disorganized and out-of-focus. We made special efforts to keep the chapters coherent, including the division of chapters into parts that was meant to keep the focus on the evolution of the subject. Of course, this could not simply be done by following the chronology because phenomenology was developed over many years and long before Kubo's theory of linear response and Zwanzig's contribution to the development of the projection operator technique. However, the idea was to formulate a course that could be taught even if the instructor picks up a few topics from each part. For example, *one can weave together a course with the first four chapters of Part I, the first three chapters of Part II, the first three chapters and chemical reaction theory chapter of Part III, and the last chapter of Part IV. This would make a course of twelve chapters, which is doable in a semester.*

A word on the requirement of certain knowledge and expertise in mathematics and physics needed for the book. These are however minimal. A working knowledge of solving ordinary and simple partial differential equations, mostly linear, by use of the Laplace and Fourier transforms is all that is required. We do use a small amount of vector algebra but that can be picked up along the way.

As before, the students of the Bagchi group contributed significantly toward the writing, editing, correcting, and completion of this book. Ms. Sangita Mondal diligently read through most of the chapters, prepared the table of contents. Mr. Ved Mahajan was heroic in his efforts to understand and correct each chapter, and modified and improved many figures. Dr. Saumyak Mukherjee, Dr. Sayantan Mondal, Mr. Subhajit Acharya, Dr. Neeta Bidhoodi, Dr. Sarmistha Sarkar, and Dr. Anjali Nair helped in many of the chapters. This book owes a lot to Dr. Saumyak Mukherjee, Dr. Sayantan Mondal, and Dr. Sarmistha Sarkar, in particular, for help in the early stages with writing of chapters and in the planning of the book – special thanks to Saumyak and Shubham Kumar for drawing and redrawing many of the figures.

Many of my faculty colleagues helped by reading various chapters and offering suggestions. Professor Awadhesh Narayan (IISc), Professor Govardhan Reddy (IISc), Professor Shinji Saito (IMS, Japan), Professor K. Seki (AIST, Japan), Professor Srabani Tarafdar (IITKgp), Professor Sarika M Bhattacharyya (NCL), Professor Madhav Ranganathan (IITK), Professor Mantu Santra (IITG), and Professor Rakesh S. Singh (IISERT) helped in various ways.

As is evident from above, writing of this book was a formidable effort that lasted over several years. During this rather long period, my students and also my two sons, Kaushik and Kushal provided motivation, encouragement, and hope. I also thank my wife Sukla for encouragement and for being of help when needed.

Finally, nonequilibrium statistical mechanics is perceived to be a difficult subject. This is because a unified approach is often not provided. We hope that the emphasis of the present book on concepts and ideas, and the present effort toward creating an order in the teaching of the course, prove helpful to students and young researchers.

10-05-2023
Bengaluru

About the Author

Biman Bagchi is currently holding an India National Science Chair (DST-SERB) & Honorary Professorship at the Solid State and Structural Chemistry Unit (SSCU), Indian Institute of Science (IISc), Bangalore. He obtained his B.Sc. degree from the Presidential College, Calcutta and M.Sc. from Science College, Calcutta University. He received a Ph.D. from Brown University, RI, in 1981 and carried out postdoctoral work at the University of Chicago and University of Maryland, before returning to India in 1984 to join the Indian Institute of Science. He is a Fellow of all three National Science Academies of India. He is an elected Fellow of the Third World Academy of Sciences (TWAS) and Elected Foreign Member of the American Academy of Arts and Sciences (2020) [AAA&S]. He has received a fair number of Awards in India, including an early Bhatnagar (at 36 years of age), GD Birla, Goyal Prizes. He received International TWAS Prize in 1998. He is the recipient of the 2021 Joel Henry Hildebrand ACS National Award from the American Chemical Society in Theoretical and Experimental Chemistry of Liquids and also the prestigious Humboldt Science Research Award of Alexander von Humboldt Foundation (2019). J. Physical Chemistry brought out a Festschrift special issue J. Phys. Chem. B, 2015, Vol.: 119, in his honor. Bagchi has published more than 500 papers including 26 reviews, authored three books on different aspects of statistical mechanics and is currently writing a fourth book on nonequilibrium statistical mechanics. He has also published two nontechnical books (mostly for students) available on Amazon Kindle.

Part I

Preliminaries

1 Preliminaries
The Scope of Nonequilibrium Statistical Mechanics

OVERVIEW

Nonequilibrium statistical mechanics [often described as time-dependent statistical mechanics (TDSM)] is an ever-expanding discipline that has its origin in our need and desire to understand many natural phenomena occurring around us and also to explore their temporal evolution. While equilibrium statistical mechanics aims at explaining the reason "why a certain phenomenon happens as observed" based on an average picture, nonequilibrium statistical mechanics aim at explaining "how it happens as observed" as a function of time. Nonequilibrium statistical mechanics aims at obtaining the values of transport properties like diffusion, viscosity, conductivity, relaxation time, and kinetics of phase change, etc. This subject has its origin in the works of Maxwell and Boltzmann in the second half of the nineteenth century and got a real boost from the work of Einstein on Brownian motion. It has grown tremendously in recent times, partly due to the availability of computer simulations and our increasing desire to understand biological systems from a molecular perspective. In this chapter, we introduce and define the subject of nonequilibrium statistical mechanics, outline its scope and directions, articulate the history of its developments, and a summary of the goal, and expectations from, the subject. However, before we start, it should be stressed that TDSM remains a fairly mathematical subject, with extensive use of partial differential equations even at the simplest level. Nevertheless, progress and rewards await when physical insight is developed and employed in understanding dynamical phenomena. So, let us embark on the journey.

1.1 INTRODUCTION

Man has remained fascinated with time from the beginning of civilization, largely because of the necessities in daily life and also to have a predictive ability over phenomena, like tides in rivers, movement of the stars, Sun and Moon, change of seasons, floods, rain and forest fire, to name a few. The flow of heat from hot to cold, flow of matter from high concentration to the low concentration region, time-dependent changes in our body also drew attention. Time-dependent statistical mechanics is an ever-expanding discipline that has its origin in our need and desire to understand such time-dependent phenomena occurring around us in nature that affect our life and occupation.

The above-mentioned examples all constitute what is called large-scale and long-time phenomena. This commonly used jargon means that these phenomena are happening over a length scale that is thousands of molecular diameters large and over a time scale that is many orders of magnitudes longer than, say, the time taken between two sequential collisions between molecules.

The theory that was first developed to explain such large-length and long-time phenomena is hydrodynamics. Hydrodynamics is a powerful theoretical discipline that allows us to make

DOI: 10.1201/9781003157601-2

quantitative predictions, such as the velocity of a rigid body immersed in a liquid and driven by a force. *Hydrodynamics should be rightly considered as the beginning of time-dependent statistical mechanics.* This might appear surprising because these two subjects stand widely separated in our course work and are seldom taught together. In fact, these two appear together in the work on Einstein on Brownian motion of large particles (like colloids, pollens) suspended or immersed in liquids of small molecules, like water.

Unlike hydrodynamics, statistical mechanics aims at explaining the observed world in terms of atoms and molecules that comprise it. The positions and orientations of these particles define the microscopic states of the system. At any given time, our macroscopic system (like air in a room or water in a glass) inhabits a specific microscopic state. The thermal motions of the molecules drive the system from one microscopic state to another. As a macroscopic state comprises of a huge number of microscopic states, this passage through microscopic states can go on forever. The equilibrium state of a macroscopic system is defined by the average over these microscopic states. That is, the equilibrium, time-independent properties, like specific heat and compressibility, are obtained by averaging over the system's journey over many microscopic states. The temporal evolution of the system through its microscopic states is called a time trajectory. We can then say that the time-independent properties we observe are the average over this trajectory.

The study of equilibrium properties of a system, however, does not require any detailed consideration of the time trajectory. There are theoretical methods that invoke techniques borrowed from probability theory that avoids the time-dependent approach. This is the method of ensembles devised by J.W. Gibbs that leads to the definition of partition functions. The partition functions, in turn, give us the thermodynamic functions, like entropy, free energy, specific heat, etc. This well-developed branch of science is called equilibrium statistical mechanics. However, this method does not provide interpretation or information about time-dependent processes. This is provided by time-dependent statistical mechanics (TDSM). The objective of this book is to provide an introduction to time-dependent statistical mechanics with applications to chemical and biological systems.

As discussed later in Chapter 4, the important dynamical variables are position, orientation, momentum of a molecule, the total polarization and dipole moment of the system, the energy, to name a few. Different correlations between these molecules are measured experimentally and studied theoretically. In a condensed phase, such as liquids, the variables mentioned above are coupled to many degrees of freedom. As many of our experimental probes are dynamical in nature, as for example fluorescence depolarization that measures the time-dependent orientation of molecule, we need a detailed dynamical theory to understand such processes.

Time-dependent statistical mechanics (TDSM) allows us to calculate the rate of chemical reactions, like the cis-trans isomerization of stilbene and the electron transfer involved in redox reactions. It allows us to understand viscosity and temperature effects on an activated barrier crossing reaction. It slows us to understand the logistics behind a protein folding problem, the dimerization, and dissociation of proteins, such as insulin. The intercalations of a chemotherapeutic drug like daunomycin into DNA, so important to the drug industry and drug discovery research, are treated with methods of TDSM. TDSM also allows us to explain the line shapes of infra-red and NMR spectra. Furthermore, TDSM allows us to understand electrochemical phenomena, electrolyte dynamics, and kinetics of phase transformations which are important to the industry.

Therefore, the scope of time-dependent statistical mechanics (TDSM) is enormous. It is studied with different goals and perspectives, depending on the problems at hand and the disciplines involved.

In Figure 1.1, we depict the multitude of physical and chemical processes that are routinely studied by using the methods of TDSM.

The vast multitude of processes studied over the wide time scales – ranging from femtoseconds to seconds and hours. On the shortest time scales, we have electronic and vibrational motions, while on the longest time scales, we address biological processes and also motions of colloids

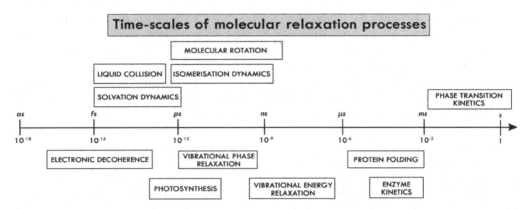

FIGURE 1.1 Schematic illustration of a range of time scales for different kinds of physical and chemical processes that are regularly studied with the help of several methods of TDSM processes. The time scales range from fs to s.

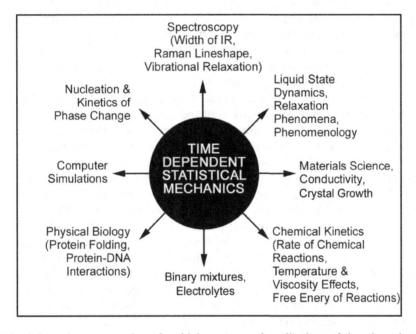

FIGURE 1.2 Schematic representation of multiple scopes and applications of time-dependent statistical mechanics. In this diagram, different processes that are being studied routinely by using several methods of TDSM are shown.

and proteins. Understanding phenomena at different time scales often involves the development of different theories because physics is often different at different time and length scales, although a general methodology is available.

In Figure 1.2, we summarize the processes being studied by TDSM.

1.2 EVOLUTION OF TDSM: HISTORICAL PERSPECTIVE

As already mentioned, the time-dependent theories that were developed initially addressed the large-scale phenomena, like the flow of water in rivers. *Impressively, scientists recognized very early the importance of the constraints due to the conservation of mass, momentum, and energy.* It was also realized that dissipation occurs through momentum and energy as energy gets converted from one form to another. These beautiful concepts guided much of the subsequent evolution of the field.

The conservation laws (which are exact over appropriate length scales), when combined with the phenomenological but robust laws of dissipation (termed the constitutive relations), form the equations of hydrodynamics. This description remains an integral part of time-dependent statistical mechanics. Hydrodynamics plays an extremely important role in describing many aspects of macroscopic processes with quantitative accuracy. Additionally, it often finds use in the description of molecular processes too. The reason for the surprising validity of hydrodynamics at a molecular length scale has been extensively studied.

The modern study of hydrodynamics starts with the Navier-Stokes equations that provide far-reaching predictions, including expression for the Rayleigh-Brillouin spectrum that in turn allows us to obtain values of thermal conductivity, sound velocity, and bulk viscosity. There is hardly any other experiment that provides so many important physical quantities. We have described hydrodynamics and Rayleigh-Brillouin spectroscopy in Chapter 5.

Kinetic theory of gases constitutes the second, and more ambitious, approach in time-dependent statistical mechanics. It is a completely different approach where one starts with atoms and molecules, and interactions among them. These features are absent in hydrodynamics, that deals with densities of conserved variables but not with atoms and molecules. The kinetic theory of gases gave rise to many fundamental ideas, concepts, and quantities, like ideal gas law, Avogadro number. This is also called the atomic theory of heat, a name that has a historical origin. We have discussed a few aspects of the kinetic theory of gases in Chapter 6.

The kinetic theory of gases that we study in the undergraduate first- or second-year courses was developed in its present form mostly by Maxwell and Boltzmann. The ideal gas picture has also been extended greatly by including the effects of inter-particle interactions. The first notable development was made by Boltzmann, whose well-known kinetic equation presented in 1872 included the effects of intermolecular interactions through binary collisions that conserve momentum and energy. The studies of Boltzmann in an effort to develop a molecular theory of heat led to important concepts and discoveries, including a probabilistic definition of entropy that played a seminal role in the development of equilibrium statistical mechanics, developed by J.W. Gibbs.

Boltzmann's reduction of the two-particle position, momentum, and time-dependent distribution into a product of two single-particle distribution through the assumption of "molecular chaos" (Stosszahlansatz) also constituted the first application of probabilistic theory in the mechanical description of molecular dynamics that insists on conservation laws. The student should note the difference between this approach and hydrodynamics. In hydrodynamics, we make approximations (in the form of constitutive relations) at the level of densities involving many particles, while in Boltzmann's treatment, the approximation was made at the level of two atoms.

In the century that followed Boltzmann, time-dependent statistical mechanics was developed into a rigorous discipline, helped by computer simulations that allowed tests of the predictions. The hydrodynamic theory was extended to take into account, albeit approximately, the effects of correlations at the molecular level. Kinetic theory of gases, on the other hand, was generalized to include local structure by Enskog. Subsequently, the kinetic theory was extended to include repeated collisional events. Remember that Boltzmann considered only binary collisions.

Parallel to the developments of hydrodynamics and the formal structure of time-dependent statistical mechanics, a third approach was developed almost spontaneously due to the need to explain experimental results that started accumulating from the early twentieth century. This third approach should be termed "*phenomenological time-dependent statistical mechanics*". This is quite distinct from the formal structure initiated by Boltzmann but is mostly applicable to single-particle dynamics. This was developed initially to describe the diffusive motion of a solute (colloid particle, large molecule) in liquids. This approach has a glorious beginning with the 1905 work of Einstein on Brownian motion that established the relation between diffusion and friction. The latter could be obtained from hydrodynamics and related to shear viscosity. It was almost immediately generalized by Langevin, who formulated a stochastic description using the ideas of Einstein.

In the next step of the development of this phenomenological approach to TDSM, Smoluchowski presented his famous algebraic equation of the probability distribution of the motion of a particle in the presence of an external potential. The formulation of the Smoluchowski equation allowed many applications, which were not possible with the Langevin equation, which was stochastic. The most important result from the Smoluchowski equation was the prediction of the inverse viscosity dependence of the barrier crossing rate, which finds use even today after a hundred years. The phenomenological approach was further developed by Fokker-Planck, who derived a more general probabilistic equation for the case when we need to consider the dependence on both position and momentum. This is also known as Kramers' equation. This has also found wide use in chemical reaction kinetics and is particularly useful at low viscosity where the Smoluchowski equation breaks down.

The phenomenological approach benefitted and was enriched tremendously by the development of the linear response theory and the fluctuation-dissipation theorem, pioneered by Onsager, Green, and Kubo, among others. The phenomenological approach blended seamlessly with the later advances in time correlation function formalism and aided the advances of the formal theory as well.

Much of the discussions presented above pertain to dynamics at equilibrium, although the experimental studies employ small perturbations to displace the system from its equilibrium state. There are many systems that are time independent but not at equilibrium. They are usually steady states maintained by the supply of energy and/or matter. The prime example is our body. There are also many chemical reactions that exhibit interesting features and are away from equilibrium. The theoretical discipline that investigates such time-dependent processes are usually termed "irreversible processes". This field was essentially created by the landmark paper of Onsager that established the reciprocal relations and also proposed the principle of microscopic reversibility. It was further developed by the work of Prigogine and coworkers in the 1950s and 1960s. We have discussed this area later in Chapter 4.

The latest entrance in the field of time-dependent statistical mechanics is computer simulation methods. Of late, this has become the method of choice for most theoretical studies on dynamics, especially for complex systems so much so that many analytical methods have become almost obsolete. For the study of dynamical processes, one employs mostly molecular dynamics simulations. With rapid advances in computer technologies, one can now comfortably simulate fairly large and complex systems beyond 100 ns and even up to several µs. In fact, the combination of phenomenology and MD simulations form a powerful synergy to attack real problems. Simulations are particularly useful to understand processes that occur on short to intermediate time scales. They still falter to study really slow processes, like glassy dynamics and kinetics of phase transitions. But simulations seem to have an ever-expanding reach.

In Figure 1.3, we show the evolution of time-dependent statistical mechanics.

1.3 SUMMARY

Despite its great importance, the nonequilibrium statistical mechanics has remained somewhat less structured than the equilibrium statistical mechanics. There have also been fewer books, or rather systematic books, on TDSM to help students and researchers. A systematic development starting from basic equations like the Liouville equation and BBGKY's hierarchy is not just nontrivial but often does not lead to a fruitful end. The approach is too difficult to implement for interacting many-body systems that we need to deal with routinely in TDSM. Thus, the quantitative approach of Boltzmann was fraught with difficulties that never really got solved in its entirety. On the other

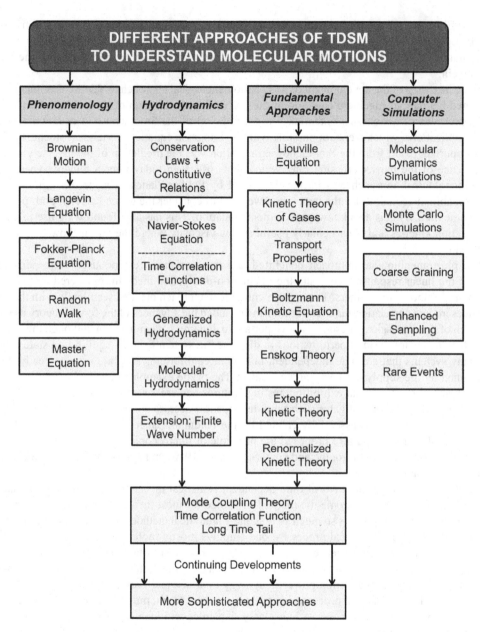

FIGURE 1.3 Schematic illustration depicting the multiple and diverse evolution of time-dependent statistical mechanics shown partially chronologically here.

hand, the phenomenological theory of Brownian motion advanced by Einstein motivated great developments. It is remarkable how phenomenology played a critical role. Also, modern computer simulation methods have proven to be useful. Rapid progress in recent times in TDSM is fueled by the great developments in theoretical understanding that started from the 1950s, initiated by Green-Kubo's formalism of observables in terms of time correlation functions.

Time correlation functions are natural generalizations of the fluctuation formulae for the response functions that we use in equilibrium statistical mechanics, and the time integral over the TCFs give the transport coefficients, like diffusion, viscosity, and the rest. These expressions for the transport coefficients are one of the most fruitful results of TDSM and shall be described in detail in this book.

We shall discuss in detail the relationship between experiment and theory in Chapter 3, where we shall present several expressions for the spectral linewidths in terms of Fourier transforms of appropriate TCFs. These expressions are immensely important results in the study of dynamics and liquids through spectroscopy.

The area of nonequilibrium statistical mechanics is huge. However, theoretical evaluation of time correlation functions is difficult. As described above, computer simulations are now widely used. However, the theoretical formulation is still required in chemical reaction dynamics, spectroscopy, enzyme kinetics, protein folding, and in a large number of problems.

The purpose of this textbook is to introduce to students and researchers the beautiful, vast, and highly useful area of time-dependent statistical mechanics in a systematic fashion by keeping the historical developments in perspective so that students can appreciate the reason for the developments.

As mentioned earlier, this area suffers from the absence of any systematic exposition. This is partly because formal development is difficult and lacking. So, phenomenology and intuition play a key role. Nevertheless, a formal approach that keeps a focus on the time correlation function formalism, linear response theory, and experimental observables is possible. Such an approach is adopted here.

We believe that this practical approach shall be useful for the students to master this subject, and allow them in applications.

BIBLIOGRAPHY

1. *Haken, H. 1977. *Synergetics: An Introduction.* New York: Springer-Verlag.
2. *Kampen, N. G. V. 1992. *Stochastic Processes in Physics and Chemistry.* Amsterdam: Elsevier.
3. Zwanzig, R. W. 2001. *Non-Equilibrium Statistical Mechanics.* Oxford: Oxford University Press.
4. Balescu, R. 1975. *Equilibrium and Nonequilibrium Statistical Mechanics.* New York: John Wiley & Sons.
5. *Prigogine, I. 1977. Time Structure and Fluctuations. *Nobel Prize Lecture* www.nobelprize.org/prizes/chemistry/1977/prigogine/lecture/
6. Nicolis, G. and I. Prigogine. 1977. *Self-Organization in Nonequilibrium Systems: From Dissipative Structures to Order through Fluctuations.* New York: John Wiley & Sons.
7. *Berne, B. and R. Pecora. 1976. *Dynamic Light Scattering.* New York: John Wiley & Sons.
8. Fleming, G. R. 1986. *Chemical Applications of Ultrafast Spectroscopy,* Oxford: Oxford University Press.
9. Bagchi, B. 2012. *Molecular Relaxation in Liquids.* Oxford: Oxford University Press.
10. Bagchi, B. 2013. *Water in Biological and Chemical Processes: From Structure and Dynamics to Function.* New York: Cambridge University Press.

* Highly recommended.

2 Time-Dependent Probability Distribution Functions

OVERVIEW

The subject of statistical mechanics, both equilibrium and nonequilibrium, is built on probabilistic concepts and employs certain basic techniques that were well-established in the mathematical theory of probability even in the nineteenth century. Statistical mechanics borrows ideas from probability theory literature, such as the distribution functions, conditional probability, moments and cumulants, master equations, etc. These form the very basis of statistical mechanics. In this chapter, we shall review many of these concepts. There exist excellent texts on probability theory that we refer to, like the ones by Feller and by Kai Lai Chung. We recommend the monograph entitled "*Synergetics*" by H. Haken (details are given in the Bibliography). In this chapter, we discuss a few elementary concepts and methods that are essential to understand the formulation of statistical mechanics. Several concepts essential to nonequilibrium statistical mechanics are introduced in this chapter. They include random walks, conditional probability, time-dependent distribution functions.

2.1 INTRODUCTION

Statistical mechanics deals with large macroscopic systems comprising many molecules and many degrees of freedom. Typically, 1 cm^3 of liquid water contains $\sim 10^{22}$ water molecules. Therefore, it is impossible and futile to write down the equation of motion for each constituent and calculate the macroscopic quantity of interest by solving these equations. The reason we need statistical mechanics is much deeper than dealing with such a large number of degrees of freedom. In the classical system, in general, the dynamical systems are chaotic. Hence, we need some kind of statistical or probabilistic approach. In order to describe the properties of such macroscopic systems, one needs to adopt a probabilistic description not only because of convenience but also because this is the correct way to approach the understanding of macroscopic phenomena. As explained in our earlier text on "*Equilibrium Statistical Mechanics*", the subject is built from two postulates and one hypothesis. The postulates are (i) *equality of time average with ensemble average, and (ii) all states in the NVE ensemble are equally probable (the postulate of equal a priori probability)*. These two postulates are connected by the ergodic hypothesis, which assumes that all the microscopic states are visited an equal number of times. While the second postulate asserts equal probability of the microscopic states, the ergodic hypothesis asserts that they indeed are visited an equal number of times. The ergodic hypothesis is clearly required for the validity of the first postulate.

Equilibrium statistical mechanics is concerned with thermodynamics, structure, and phase transitions. It does not have the notion of time, and dealing with an average of macroscopic variables gives us thermodynamics. Time-dependent statistical mechanics is concerned with time-dependent

DOI: 10.1201/9781003157601-3

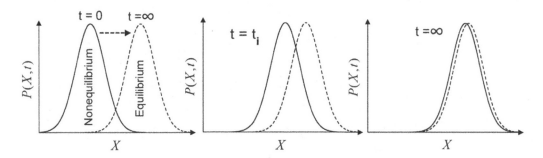

FIGURE 2.1 Representative diagram of the relaxation or progression of a nonequilibrium distribution functions $P(X, t)$ from an initial ($t = 0$) state that has been perturbed from the equilibrium distribution that shall be achieved at a long time when the system relaxes back to its usual state.

changes in the system due to time-dependent perturbation, either external or internal. By putting an external field, we derive the system out of equilibrium. The next question that comes to mind is what happens to physical quantities in the presence of external perturbation. The time-dependent change is described in terms of the motion of atoms and molecules that constitute the system.

There is an intimate connection between equilibrium and nonequilibrium statistical mechanics. For most of the studies of nonequilibrium statistical mechanics, we consider the relaxation of distribution functions from a state disturbed from the equilibrium distribution. The situation is depicted in Figure 2.1.

Here we consider that at time $t = 0$, we created a perturbation or distortion in the distribution of a system variable X, which could be population in a harmonic potential, such as vibration that can be perturbed by excitation to create a nonequilibrium distribution; or, X can be polarization if an external electric field is employed. The main idea is that the perturbation disturbs the distribution of X to produce a nonequilibrium distribution at time $t = 0$. Below we discuss several other examples too.

Let us consider two specific examples. First, let us place at time $t = 0$, a drop of blue ink in the middle of a bucket of water. We can see that with time the ink spreads out and then disappears into the bulk water, leaving no trace. We understand that molecules of the ink, the dye molecules, have become dispersed into the bulk. How do we describe this process? The spread happens because of interactions between the molecules of the dye with the surrounding water molecules. The initial localization of the dye molecules means that there is a high density of the dye molecules at the center of the beaker where the ink drop is placed. For the purpose of understanding like the time needed for the ink to disappear or be dispersed, it is neither necessary nor meaningful to track the motion of each dye molecule. Rather, it makes sense to discuss it in terms of the time-dependent density distribution of the ink as a function of position. In Figure 2.2, we show the evolution of the density distribution, as a function of the distance (X) from the center. An analytical expression of the time-dependent distribution, $P(X, t)$ is shown in Figure 2.2.

Our second specific example comes from spectroscopy. A common application involves the excitation of the molecule from the ground state potential energy surface to the excited state surface by optical excitation. We depict this common scenario in Figure 2.3. We further note that the photochemical process being discussed occurs in the liquid state. That is, the dye being excited (like malachite green or crystal violet) is in the solution with liquid water. Subsequent to the optical excitation, the population distribution in the excited state potential energy surface is in the nonequilibrium state, and relaxes back to the minimum of the excited state surface. We show this relaxation in Figure 2.3. As the population distribution evolves on the excited state surface, the energy of the molecule decreases, which can be measured optically. As mentioned, this is a common scenario. If the

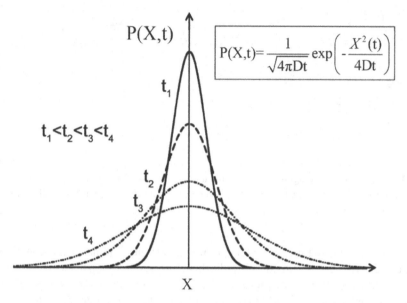

FIGURE 2.2 The evolution of the probability density distribution, $P(X, t)$, as a function of the distance (X) from the center. This is a representative diagram where the particle was initially localized at the origin, $X = 0$, at time $t = 0$. The density spreads outward at the subsequent times. Here D is the diffusion constant.

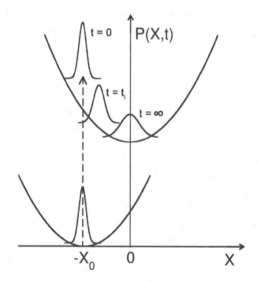

FIGURE 2.3 The representative diagram depicts the relaxation of an initially excited population. Initially, the fluorescent molecule is excited from the ground state potential energy surface to the upper state potential surface by optical excitation. Subsequent to the optical excitation, the population distribution in the excited state potential energy surface is in a nonequilibrium state and begins to relax to the minimum of the excited state surface. This relaxation is depicted above.

excited potential energy surface is assumed to be harmonic of frequency ω, then the time-dependent population distribution can be given by,

$$P\left(X,t|X_0\right)=\frac{\sqrt{B}}{\sqrt{2\pi D\left(1-e^{-2Bt}\right)}}\exp\left(-B\frac{\left(X-X_0 e^{-Bt}\right)^2}{2D\left(1-e^{-2Bt}\right)}\right) \qquad (2.1)$$

where $B=\dfrac{\mu\omega^2}{\zeta}, D=\dfrac{k_B T}{\zeta}$. Here ω is the harmonic frequency, and D is the diffusion constant, ζ is the friction constant.

B is a constant proportional to frequency and diffusion. x_0 is the initial position of the system on the excited state surface. Here B is determined by a combination of the frequency (ω) of the surface, friction (ζ) experienced by the particle, and its effective mass, μ.

In both the examples discussed above, we have used a highly reduced one-dimensional description. In a complete description of the microscopic process in atomic or molecular details, we would need to consider the interaction of the dye molecules with a huge number of solvent molecules. For example, one cc of liquid water (that approximately weighs 1 gm, at 298 K) contains $(6.023 \times 10^{23})/18 = 3.34 \times 10^{22}$ water molecules. We know that each water molecule has nine degrees of freedom. Even if we consider a fraction, we have to consider a very large number of degrees of freedom for a detailed microscopic description of the dynamics (i.e., time evolution). In general, both in the fields of chemistry and biology, the system sizes may vary from the nanoscale to macroscale. Even in nanoscale, except for very small-sized systems, the systems of interest comprising millions of atoms and molecules.

The foundation of the first probabilistic approach with intermolecular interactions was laid down by Boltzmann. He attempted to solve the position (r), momentum (p), and time (t) dependent single particle joint probability distribution function $f_1(\mathbf{r}, \mathbf{p}, t)$. Here Boltzmann encountered a problem. He found that an accurate description singlet distribution requires knowledge of two-particle distribution. And then, a description of two-particle phase space distribution requires knowledge of three-particle distribution, and so on. This difficulty afterward became known as the problem of hierarchy. This hierarchical coupling makes the problem mathematically intractable. This means that in order to obtain a description, a single-particle distribution, we need to truncate the hierarchy. Here Boltzmann introduced a statistical assumption, popularly known as "*molecular chaos*", which says that the velocities of colliding particles are independent of positions. Hence, joint two-particle probability distribution function $f_2(\mathbf{r},\mathbf{p}_1,\mathbf{p}_2,t)$ can be written as the product *of two singlet distribution function, $f_1(\mathbf{r},\mathbf{p}_1,t) f_1(\mathbf{r},\mathbf{p}_2,t)$,* in order to truncate the hierarchy mentioned above.

Therefore, in statistical mechanics, *we make the transition from a deterministic approach or language to a probability.* The latter is developed in terms of distribution functions. Our experimental observables are averages over such distributions. The distributions themselves are not directly measured, except maybe via recent single-molecule spectroscopy and imaging techniques that allow us in a limited way to find the distribution function in a few specific cases.

The basic idea behind a distribution function is that our observable (say, X) is a random quantity (which can be discrete or continuous). In the examples above, this X is the position of the ink dye suspended in water inside the beaker or the position of the system on the excited state potential energy surface. In real experiments, a large number of dye molecules are suspended in the water. They are within a small area at the initial time ($t = 0$), but afterward, each molecule follows a different trajectory due to interactions with different water molecules. Thus, the position of each dye molecule is different. Its value at a given moment of time can be calculated by the positions and velocities of the constituent atoms and molecules of the system. If we can follow the instantaneous value of X with respect to time, we would observe that its value fluctuates with time, around an average. As we shall discuss later, both the distribution function of instantaneous values of X and correlations between fluctuations contain essential information about the system. In the forthcoming section, we are going to discuss the probability distribution, denoted by $P(X, t)$.

2.2 BASIC PREREQUISITES

As mentioned above, Boltzmann was perhaps the first to explicitly introduce a probabilistic description in the description of motions of atoms and molecules. The random variables in his description were the position and velocity of the atoms that constituted the system. This approach introduced a time-dependent distribution function explicitly.

In statistical mechanical formulation of a phenomenon, we first need to determine the relevant variables. In many applications, these are positions, orientations, momenta, and specific functions of these variables. After the specification of the relevant random variables, we need to determine the values that the variable can take. The latter is called the sample space of the random variable. In the following, we further elaborate on these concepts, with examples.

2.2.1 CONCEPT OF RANDOM VARIABLES

In classical probability theory, a *random variable* (X) is the outcome of an experiment where the value is not predetermined, not known *a priori* and is subject to variations. This implies that the same experiment repeated many times can give rise to different values. The values can be discrete or continuous. Let us take the example of throwing a dice. Here the random variable is the "number of spots" on the side facing upward. The possible outcomes are the numbers between 1 and 6, each with a probability of 1/6 if all other conditions while throwing the dice remain unchanged. Similarly, when we toss an unbiased coin, we have two outcomes – heads and tails, and the probability of each event is 1/2.

In the following, we denote the random variable by X and the possible outcomes by x. In many cases, x can be continuous, taking up all the values in a range, say $a \leq x \leq b$. While examples of discrete distribution are the throwing of the dice or coin, that of a continuous distribution is the distance traveled by different walkers within a given time interval.

In problems of interest in physics and chemistry, the random variable is often a thermo-dynamic property (such as pressure or enthalpy). Although the property is describable by an average value, it can, and often does, exhibit fluctuations, which makes it a random variable, with an average and a standard deviation. We give several examples. If we measure accurately (with minute resolution) the instantaneous pressure of a liquid in equilibrium with its vapor, we see that the pressure continuously fluctuates around an average value. The same holds for the energy of a system kept at a constant temperature, or the volume of a system kept at constant pressure. *Thus, the energy, the pressure, and/or the volume of a system (of molecules) are all be considered as random variables.*

Let us give an important example from the microscopic world. The number of nearest neighbors that an atom or a molecule, can have within its first shell in its liquid state is a random number. For a one-component system of spherical atoms, like argon, this number varies between 9 and 13. In addition, this number varies with time. We can define an average number for the nearest neighbor as a time average, but the real number at a given time is a fluctuation around the average.

In all the above examples, there exists an average, *but the individual outcome of an experiment can be different from the average and not predictable by an observer. One needs a probability distribution to describe the behaviors of the random variable. Observable macroscopic properties are most of the time determined by this average.*

2.2.2 CONCEPT OF SAMPLE SPACE

Sample space is the collection of all the outcomes/values of the observable obtainable in experiments repeated many times. A simple example is provided by the experiment of tossing a coin or throwing dice. When a coin is tossed once, the total numbers of outcomes (which are two in this case, either heads or tails) are different when the same coin is tossed more than once. For example, if we toss the same coin twice, we have four possible outcomes: HH, HT, TH, and TT (where H and T stand for heads and tails, respectively). If the coin is tossed n times, the possible outcomes are 2^n. Thus, our sample space contains all the possible sequences of events.

Next, we consider the experiment of throwing a dice. We have six outcomes, each of which is equally probable. If we now toss the dice twice, we get $6^2 = 36$ possible outcomes. Then the probability of finding three in the first toss and five in the second will be 1/36. But if we ask what is the probability of getting three in one and five in another toss separately, then the probability is 2/36, as you may have the possibility to get three and five on both of the dice. Thus, the sample space determines all possible outcomes. Thus, the measure of the sample space is related to the entropy of the system.

Therefore, in probability theoretic formulation, an event could be a series of actions, and we need to consider all possible outcomes. *This total set of the outcome of an event constitutes the sample space.* The sample place allows us to define a probability distribution that we are going to discuss in the forthcoming subsection.

2.2.3 PROBABILITY DISTRIBUTION – BOTH EQUILIBRIUM AND TIME-DEPENDENT

While characterizing a random variable, we need to obtain information about all the possible outcomes of elementary events of the random variable, and subsequently, we need to construct the probability of a given value as a possible outcome. Hence, not just the sample space but also the occupation probability of any given region of the sample space is necessary. This occupation probability is known as the *probability distribution*. When the possible realizations (outcomes) do not form a discrete set but a continuum in real number space, then we construct a continuous probability distribution function $P(X)$ or a probability density function $p(x)$. In the latter case, we describe the probability of an outcome to have a value between x and $x+dx$ as $p(x)dx$.

The variables and the related questions are different in the case of nonequilibrium, time-dependent processes from the equilibrium ones. In the former case, the random variables are time-dependent. So, outcomes or values of measurements are tabulated as a function of time. The difference from the time-independent case is that in the latter, we can measure the variable at any given time without measuring time intervals, but in the time-dependent case, we need to carry the additional time variable. Thus, the probability distribution is time-dependent, $P(X,t)$.

The next question that arises is the following: *why do we need time-dependent values, even at equilibrium?* The simple answer is that the relaxation of a system initially displaced from equilibrium (as shown in Figure 2.1) relaxes back to the equilibrium state by using processes that are essentially unaltered from the equilibrium state can be well-described by using natural dynamics at equilibrium. We shall address this question in more detail in the next chapter, but the reader should keep this question in mind.

We are interested in the time-dependent probability distribution, as discussed in Section 2.3. For the time evolution of the time-dependent density distribution, we need $P(X, t)$. For the dye diffusion problem, the probability distribution can be solved exactly and is given in the following calculation.

We assume that the dye molecules are located at the origin of a three-dimensional coordinate system. The position vector $\boldsymbol{R}$ denotes the location of a dye molecule at time t. In order to obtain the time-dependent solution for the dye problem, we solve the given diffusion equation

$$\frac{\partial P(\boldsymbol{R},t)}{\partial t} = D\nabla^2 P(R,t). \tag{2.2}$$

The boundary condition is $P(\boldsymbol{R},t) \to 0$ as $\boldsymbol{R} \to \infty$ in all directions, i.e., we do not find any ink molecule at a very far distance. The initial condition is $P(\boldsymbol{R},0) = \delta^{(d)}(\boldsymbol{R})$, i.e., at $t = 0$ the concentration of ink molecules is at the starting point, d is the dimension, and everywhere the concentration is zero like a delta function. The fundamental Gaussian solution to the diffusion Eq.(2.2) is as below

$$P(\boldsymbol{R},t) = \left(\frac{1}{4\pi Dt}\right)^{\frac{d}{2}} \exp\left(-\frac{\boldsymbol{R}(t)^2}{4Dt}\right). \tag{2.3}$$

Next, to illustrate, we calculate the variance in one dimension ($d = 1$). Towards this goal, we need two quantities $\langle X(t) \rangle$ and $\langle X^2(t) \rangle$. The average value of $X(t)$ is

$$\begin{aligned}
\langle X(t) \rangle &= \int_{-\infty}^{\infty} dX\, X(t) P(X,t) \\
&= \frac{1}{\sqrt{4\pi Dt}} \int_{-\infty}^{\infty} dX X(t) \exp\left(-\frac{X(t)^2}{4Dt}\right) \\
&= 0.
\end{aligned} \tag{2.4}$$

The average displacement of the molecule is zero because diffusion is symmetric in all directions. That is, the molecule moves both in the positive and negative directions by an equal amount. However, the spread of the ink given by the mean square displacement $\langle X(t)^2 \rangle$ is

$$\begin{aligned}
\langle X(t)^2 \rangle &= \int_{-\infty}^{\infty} dX\, X(t)^2 P(X,t) \\
&= \frac{1}{\sqrt{4\pi Dt}} \int_{-\infty}^{\infty} dX X(t)^2 \exp\left(-\frac{X(t)^2}{4Dt}\right) \\
&= 2Dt.
\end{aligned} \tag{2.5}$$

Therefore, the variance of the distribution in 1-d is defined as

$$\begin{aligned}
\sigma^2 &= \langle X^2 \rangle - \langle X \rangle^2 = \langle X^2 \rangle \\
&= 2Dt.
\end{aligned} \tag{2.6}$$

It is important to note that the same logic can be applied to any arbitrary dimension d. In a d dimensional system,

$$\begin{aligned}
\sigma^2 &= \langle X_1^2 \rangle + \langle X_2^2 \rangle + \cdots + \langle X_d^2 \rangle \\
&= 2Dt + 2Dt + \cdots + 2Dt \\
&= 2dDt.
\end{aligned} \tag{2.7}$$

The average position remains zero as the dye spreads uniformly across the liquid. The variance of distribution thus provides a measure of the fluctuation from the mean value. We leave the calculation of the variance of the distribution $P(\boldsymbol{R},t)$ and $P(X, t)$ as two exercises.

2.2.4 Joint Probability Distribution

In several cases, the outcomes of two events can be correlated. That is, the outcome of one event depends on the outcome of another event. In these types of cases, the probability of simultaneous occurrence of two events, namely A and B, can be expressed by a joint probability distribution. The number of events can be generalized to any number. *If two events A and B, are completely uncorrelated, then the probability of both occurring will be a simple multiplication of the probability of those events occurring individually.*

The joint probability of two events, A and B, is written as $P(AB)$. If both A and B are uncorrelated, that is, independent of each other, then the joint probability of occurrence of A and B will be a product of individual probabilities,

$$P(AB) = P(A)P(B). \tag{2.8}$$

We can easily understand this result from the example of throwing a dice twice. Here getting a three in one dice throw is not at all affected by the event of getting a two in the next one.

However, Eq. (2.8) does not hold when the two events A and B are correlated. Here correlation means that results of one experiment (A) influence the outcome of the other (B). For example, we call a dense liquid a strongly correlated system because the position of one molecule affects the probability of finding another molecule next to it. The idea of correlation among different events leads to *conditional probability. This is particularly important in dynamics, where we discuss time correlation functions. This function measures the correlation in the values of a dynamical variable as a function of time. This function is of primary concern in this book.*

2.2.5 Conditional Probability in Time-Dependent Statistical Mechanics

This is an important quantity. In order to understand the concept and the need of conditional probability, let us ask the following question. If we locate a tagged molecule at position $r = 0$ at time $t = 0$, then what is the probability of finding the same molecule at position r at time t? The probability of the second event is clearly related to the first event. The joint probability of the two events is denoted by $P(r, t;0,0)$. Clearly, at short times, the value of the joint probability is different from just the product of the two probabilities. Conditional probability is introduced to describe this correlation. We write,

$$P(r,t;0,0) = P(r,t \mid 0,0)P(0,0). \tag{2.9}$$

Thus, $P(r,t \mid 0,0)$ is the conditional probability, In general, for any two events A and B, this is represented by

$$P(A,B) = P(A \mid B)P(B) \tag{2.10}$$

where, as before, $P(A,B)$ is the joint probability and $P(A|B)$ is the conditional probability. Thus, the conditional probability $P(A|B)$ is defined as

$$P(A \mid B) = \frac{P(A,B)}{P(B)}. \tag{2.11}$$

Eq. (2.11) makes the reason for introducing the conditional probability clear. When two events are independent, then $P(A|B)=P(A)P(B)$, and the conditional probability is just $P(A)$. That is, probability of event A is not affected by B.

In the presence of correlation between two events, the volume of the available sample space becomes less (than in the absence of any correlation) because the outcome of one experiment partly determines that of the other. We discuss the utility of conditional probability further below.

2.2.6 CONDITIONAL PROBABILITY AND CORRELATION

The study of statistical mechanics is the study of the correlation between events, particles, fluctuations, etc. Conditional probability provides a quantitative measure of such a correlation. Many important functions, like the radial distribution function (RDF), are essentially conditional probabilities. The velocity time correlation function (discussed later) is another example of an important correlation, which is a conditional probability.

In fact, the study of statistical mechanics is a study of different types of correlations between atoms and molecules, as will be evident from the subsequent chapters of the book. Actually, what we often mean by "ideal behavior" (like ideal gas or ideal solution) is really meant to be a total absence of correlation among molecules (the noninteracting limit).

2.3 CENTRAL LIMIT THEOREM (CLT)

The central limit theorem, as the name suggests, is central and essential to the entire field of probability theory. It states *that if S is a sum of N number of random variable {x_i} and if x_i are uncorrelated or weakly correlated then the probability distribution P(S) is a Gaussian distribution.* This theorem is attributed to the French mathematician Laplace. In statistical mechanics frequently the above condition of weak correlation among random variables is satisfied. For example, the total energy of the system, E is the sum of kinetic and potential energies of individual particles, and the individual energies are weakly correlated among themselves. Hence, the total energy E is almost invariably Gaussian with the standard deviation being represented by the specific heat.

Another significant example of a probability distribution function is provided by the end-to-end distribution of a long polymer chain that indicates the probability of its two ends being separated by a certain distance R. Because of connectivity along the polymer chain, this distribution can be a formidable function. One finds that the end-to-end distribution is a Gaussian or a nearly Gaussian function of separation R. In fact, the probability distribution of many macroscopic variables (like the energy of a system in contact with a heat bath, the volume of a system kept at constant pressure) is found to be accurately described by a Gaussian function. This ubiquity of the Gaussian nature of the distribution can be understood by using the central limit theorem.

The theorem can be illustrated in a straightforward way as follows. If x_i ($i=1,N$) are mutually independent random variables, then the sum of {x_i}, S, is also a random variable that obeys a Gaussian distribution when the number N is sufficiently large. The plot of $P_{\mu,\sigma}(S)$ against S is shown in Figure 2.4.

If,

$$S = \sum_{i=1}^{N} x_i,$$
$$\langle S \rangle = \mu$$

(2.12)

then,

$$P_{\mu,\sigma}(S) = \frac{1}{\sqrt{2\pi}\sigma} e^{-\frac{(S-\mu)^2}{2\sigma^2}}$$

(2.13)

where $P_{\mu,\sigma}(S)$ is the probability distribution function of S and σ is the standard deviation, given by $\left(\sqrt{(S)^2 - (S)^2}\right)$.

The only restrictive conditions are as follows: (i) {x_i} needs to be mutually weakly correlated, and (ii) N needs to be adequately large. Typically, we get a robust Gaussian distribution when N is

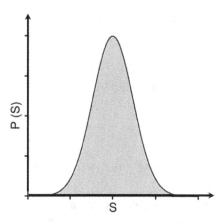

FIGURE 2.4 A Gaussian probability distribution for the sum (S) is generated by adding n ($n > 12$) random numbers. We find a nearly perfect Gaussian, which is in agreement with the central limit theorem.

greater than 12, as shown in Figure 2.4. However, it is advisable to take a larger value of N while using it in computer simulations.

The central limit theorem is successful in explaining why energy and volume distributions are Gaussian functions. We can write,

$$P(E) = A \exp\left(-\left(E - \langle E \rangle\right)^2\right)/\langle \delta E^2 \rangle,$$
$$P(V) = B \exp\left(-\left(V - \langle V \rangle\right)^2\right)/\langle \delta V^2 \rangle. \tag{2.14}$$

Here A and B denote normalization constants and $\delta E = E - \langle E \rangle$ where $\langle E \rangle$ denotes the average value of energy E. Similar considerations have been made for δV. The standard deviations are important material properties, as for example, fluctuation in energy represents the specific heat while that of volume gives isothermal compressibility.

The central limit theorem (CLT) is one of the basic theorems of probability theory. CLT has found its wide use in the statistical mechanical description of matter. We recommend interested students go through the books by Feller (Volume 1), van Kampen, and Kai Lai Chung for a detailed explanation. See the Bibliography for comprehensive information.

2.4 TIME CORRELATION FUNCTION (TCF) AS A CONDITIONAL PROBABILITY

TCFs are the most effective tools of time-dependent statistical mechanics. They play a role analogous to the partition function in equilibrium statistical mechanics, in the sense that both allow us to calculate the response functions.

Let us suppose $X(t)$ and $Y(t')$ are two random dynamical variables defined on the sample space S_x and S_y, respectively, with mean μ_x and μ_y and with positive variances σ_x^2 and σ_y^2. The mean of these random variables at any given instant of time gives us information about the expected values. It does not say anything about how these two variables are related to each other. We want to establish the statistical connection between these two variables, and for that purpose, we use the joint probability distribution, which determines the distribution of X and Y, defined by Eq. (2.11).

The TCF for two random variables is given below,

$$C_{XY}(t,t') = \langle X(t)Y(t') \rangle. \tag{2.15}$$

If we define the fluctuation in the random variables X and Y as $\delta X(t)$ and $\delta Y(t')$ respectively, then the TCF in terms of the fluctuation has the following form:

$$
\begin{aligned}
\langle \delta X(t)\delta Y(t')\rangle &= \langle \left(X(t) - \langle X(t)\rangle \right)\left(Y(t') - \langle Y(t')\rangle \right)\rangle \\
&= \langle X(t)Y(t')\rangle - \langle X(t)\rangle\langle Y(t')\rangle \\
&= C_{XY}(t,t') - \mu_x(t)\mu_Y(t').
\end{aligned} \tag{2.16}
$$

If we define a new function $R_{XY}(t,t')$ by

$$
R_{XY}(t,t') = C_{XY}(t,t') - \mu_x(t)\mu_Y(t'). \tag{2.17}
$$

Then this quantity $R_{XY}(t,t')$ is a measure of the correlation between the variables $X(t)$, and $Y(t')$. If these variables have some relationship through another common internal variable, then they are correlated, i.e., the covariance function $R(t,t')$ will have finite nonzero value.

Thus, when two random variables are dependent, they influence the output values of each other. Observed values can be used to find the relation between them. On the other hand, if there exists no statistical relationship between them, then $X(t)$ and $Y(t')$ are uncorrelated, i.e., $\langle X(t)Y(t')\rangle = \langle X(t)\rangle\langle Y(t')\rangle$ or $C_{XY}(t,t') = \mu_x(t)\mu_Y(t')$. *Therefore, the covariance function $R_{XY}(t,t')$ will be zero for two completely independent variables.* It means that the knowledge of the behavior of one variable with time does not affect the values associated with another variable. These variables can be some dynamical quantity. In that case, $C_{XY}(t,t')$ is called the autocorrelation function and $R_{XY}(t,t')$ is called the autocovariance function.

2.5   REDUCED PROBABILITY DISTRIBUTION

Much of statistical mechanics deals with the reduction of degrees of freedom. This is an essential step because experimental observables are most often an average over a reduced probability distribution (examples given later). This provides an important simplification because although we deal with a many-body system, the measurable properties can be expressed as the average over much simpler few-body distribution functions. We give two examples below.

First is the example of the radial distribution function. This is a conditional probability distribution function that a particle (molecule, atom) can be found at a distance given that one particle is at the origin. Thus, the radial distribution function does not consider all the particles of the system, but their influence is contained in the average.

Our second example is from the study of chemical kinetics. Here we study the motion of the reaction coordinate, and do not explicitly consider other (like solvent) degrees of freedom. Thus, the equation of motion of the time-dependent probability distribution of the reaction coordinate is often all that we need to consider. For example, in an isomerization reaction, like in cis-trans isomerization, the reaction coordinate is the twisting angle, we attempt to write an equation of motion for this coordinate where the effects of the surrounding solvent is contained through a friction term to take into account the collisions.

Mathematically, a reduced description can be understood in the following fashion. If we have a distribution in two degrees of freedom, like $P(x,y,t)$, then the reduced distribution shall be just an average over y-coordinate.

$$
P(x,t) = \int dy P(x,y,t). \tag{2.18}
$$

The method of eliminating certain degrees of freedom is a standard procedure in statistical mechanics, popularized in nonequilibrium phenomena by Kubo.

2.6 GAUSSIAN MARKOV PROCESS AND DOOB'S THEOREM

We discussed above that the observed values of many macroscopic or collective properties are distributed as a Gaussian or normal distribution function. This can be rationalized in most cases by the use of the central limit theorem, discussed above. Let us consider a random variable $X(t)$. While the Gaussian distribution describes the probability of observing a given value, *it does not consider the correlation between any two values of X(t) observed in a succession of events*. Let us assume that the values of X observed in a succession are denoted by $x_1, x_2, x_3\dots$ observed at times t_1, $t_2, t_3,\dots$. We now introduce certain definitions. *A stochastic process is called a Markov process if and only if the values at a given time (say, t_3) depend only on the previous value, and not on the earlier ones.* That is, the probabilities obey the following relation:

$$p\left(x_3,t_3 \mid x_2,t_2,x_1,t_1\right)=p\left(x_3,t_3 \mid x_2,t_2\right). \tag{2.19}$$

Two things are to be noticed. First, we have a sequence of events where values are observed at distinct intervals. A Markov process or a Markov chain is thus defined by a sequence of two-point probabilities.

Markov process is a mathematical entity and is an idealization, and is to be regarded as such. In practical applications, the time gap between two events becomes an issue. The power of a Gauss Markov process is demonstrated by Doob's elegant theorem articulated below. The theorem proved by Doob plays an extremely important role in theoretical developments in nonequilibrium statistical mechanics.

Doob's theorem states that if a random variable x obeys Gaussian distribution, and is, in addition, Markovian, then the time correlation between the values observed at two times separated by τ, decays exponentially, as

$$\boxed{C_x\left(\tau\right) =\sigma_x^2 exp\left(-a\tau\right)} \tag{2.20}$$

where "a" is a decay constant, and σ_x^2 is the standard deviation.

This theorem is not just beautiful and elegant but also a wonderful addition in the nonequilibrium statistical mechanics. Since the process is a Gaussian Markov, we have in addition the following relation for the joint probability distribution:

$$p\left(x_0,t = 0; x_1,t = t\right)=\frac{1}{\sqrt{2\pi.\left(1-exp\left(-2at\right)\sigma_x^2\right)}} exp\left(\frac{\left(x_1 -\overline{x}\right)^2 -\left(x_0 -\overline{x}\right)^2 e^{-at}}{\sqrt{2\left(1-exp(-2at)\sigma_x^2\right)}}\right). \tag{2.21}$$

Note that as the time difference t increases to infinity, the above function reduces to a normal distribution

$$p\left(x_0,t = 0; x_1,t = \infty\right)=\frac{1}{\sqrt{2\pi.\sigma_x^2}} exp\left(-\frac{\left(x_1 -\overline{x}\right)^2}{\sqrt{2\sigma_x^2}}\right), \tag{2.22}$$

which is of course the expected limiting distribution.

2.7 MARKOV CHAIN

We described above that when a dynamical variable in correlated only with the value observed at the previous time, then it is a Markov process. That is, the value at a given time is not dependent on its previous history. This allows us to describe evolution of a random variable as a chain of events where the entire chain consists of intermediate components interconnected to its nearest neighbors.

The Markov process is a powerful idealization. Thus, if we consider an equation such as an ordinary Langevin equation (to be discussed later), then the velocity of a tagged particle in a given time is related only to the previous time because the change that occurs is a delta correlated white noise. Here velocity describes the Markov process.

2.8   MEMORY EFFECTS AND HIERARCHY OF DYNAMICAL EVENTS

The shortest time that we encounter in liquids is the collision time of a molecule, which is the average time interval between two successive collisions. Even in gases, this is short. The experimental observables on the other hand vary over many orders of magnitude longer than the collision time. As discussed in Chapter 1, there is hierarchy of time scales, and different experiments probe different time scales. An event uncorrelated over a long time scale, can be correlated at short time scales.

The temporal correlation between values of a dynamical quantity, like local density or current, is often measured in experiments and is a much sought-after quantity in the study of relaxation phenomena. Just as the radial distribution function, $g(r)$, is introduced to describe spatial correlation, we introduce a time correlation function to quantify the temporal correlation/association of the dynamical quantity between two times. Just as the spatial correlation of a quantity like density becomes uncorrelated as the separation between the two positions of interest increases, the temporal correlation also decays with time.

In a later chapter, we shall discuss the concept of the time correlation function that has been introduced and developed to describe temporal correlation in dynamical quantities. But in this section, we would like to introduce certain basic concepts and ideas that are essential to understand the temporal correlation.

The first such concept is the hierarchy of dynamical events. In the study of dynamics in the condensed matter, the microscopic shortest time dynamical events that form the basis are the collisional events. In dense liquids, for example, a molecule has to move only one tenth of an angstrom to suffer a collision. These are thermal events and have a time gap of about 10–$20\,fs$ only. In solid states these are replaced by vibrations and hence phonons. At the next level, we find the momentum relaxation. These are mediated by collisions but require multiple collisions. At a still higher level comes the number density relaxation. Momentum and density relaxation are determined by conservation laws. Further slowing down happens for energy relaxation.

As a major thrust of time-dependent statistical mechanics is the study of chemical and biological reactions, we already discussed the time scales in Chapter 1. Students should note the diverse time scales, and compare them with the time scales of elementary events mentioned above.

An important concept in the study of dynamical phenomena is the concept of memory. The value of a dynamical quantity, like density in a given region, is correlated with its value a short time earlier but not with its value a long time back. These qualitative statements, "short time", "long time" can be made quantitative by specifying a function that tracks the memory of the past.

When the memory is lost in a short time, the process is called Markovian. If the memory is finite, then the process is termed non-Markovian. This is a relative classification, depending on the choice of time scales. Usually, we take the shortest time as a Markovian, like the duration between any two collisions that a molecule suffers.

A good example is provided by the Brownian motion to be discussed later. Here one considers the diffusion of a molecule much larger than the solvent molecules. The large molecules suffer many collisions with solvent molecules. On the time scale of the movement of the large molecules, two separate collisions are uncorrelated in time and called Markovian. Thus, the concept of memory in nonequilibrium statistical mechanics is intimately connected with the hierarchy of time scales involved in the processes that contribute to the observed dynamics.

2.9 SUMMARY

The comprehensive understanding of statistical mechanics remains incomplete without a good understanding of probability theory. Starting from the basic postulates of statistical mechanics to the formulation of phase space density and theory of density distribution, probability theory plays a crucial role in the development of concepts and basic mathematical techniques of this subject.

In this chapter, a brief description of probability theory is provided. Students are encouraged to invest more time in this important subject to gain an in-depth understanding of a few important concepts like conditional probability, the central limit theorem, Doob's theorem, random walks, stochastic processes, to name a few. As students would learn later in the book, statistical mechanics depends heavily on probabilistic and statistical descriptions. This is because, as we discuss in the next chapter, one cannot hope for a deterministic description to be a viable microscopic approach to the phenomena and also because the systems are inherently stochastic.

The subject of probability theory not only trains a student to think correctly in the area of statistical mechanics, but it is such a fundamental subject that it helps a student to acquire a deeper understanding of natural phenomena.

BIBLIOGRAPHY

1. *Feller, W. 1968. *An Introduction to Probability Theory and Applications.* Vol. 1 and 2. New York: John & Wiley Sons.
2. Chung, K. L. 1974. *A Course in Probability Theory.* New York: Academic Press.
3. *Haken, H. 1977. *Synergetics: An Introduction.* New York: Springer-Verlag.
4. Kac, M. 1959. *Probability and Related Topics in Physical Sciences.* London: Interscience Publishers.
5. *Kampen, N. G. V. 1992. *Stochastic Processes in Physics and Chemistry.* Amsterdam: Elsevier.

* Highly recommended.

3 Relationship Between Theory and Experiment

OVERVIEW

From the very beginning, the study of time-dependent statistical mechanics has been intimately connected with experiments so much so that most of the theoretical methods, starting with Einstein's famous theory of Brownian motion, have been developed in response to experimental results. In experiments, we often subject a system of interest to a small perturbation and measure the resultant change in its properties. The measured properties are called response functions. In equilibrium statistical mechanics, these perturbations are time-independent and the response is given, in the appropriate cases, by quantities such as specific heat, isothermal compressibility, dielectric constant. So, when we want to learn about dynamic response functions of the system and transport properties, we impose time-dependent perturbations, which might be in the form of a beam of neutrons, exciting light, fluctuating electric field, etc. There are also experimental methods that employ an incident radiation and measure absorption or scattering as a function of frequency and wavenumber. In general, it is fair to state that the experiments can be broadly divided into these two categories – time domain and frequency domain. In this chapter, we shall learn about the theoretical analyses and approaches that allow us to build the relationship between theory and experiments. The experimental observables are often related to time correlation functions (TCF), which in turn contain information about the dynamics of the system. For example, infra-red (IR) spectroscopy provides information about orientational TCF, and the Rayleigh-Brillouin spectrum is given by density-density TCF. The present chapter describes these relationships between experimental observables and theoretical counterparts.

3.1 INTRODUCTION

Time-dependent statistical mechanics (TDSM) is intimately connected with time and frequency domain spectroscopic and scattering experiments. It is essential to understand this relationship to appreciate the course of development and the scope of the subject, especially in the applications and understanding of the practical side of TDSM. This relationship between theory and experiments occurs in three steps.

(a) In the first step, a relation is to be derived between the experimental observables, like the scattering cross-section, and the underlying system property, such as dielectric function, or polarizability.

(b) In the second step, the macroscopic system property is related to an average of amicroscopic quantity, as a time correlation function.

DOI: 10.1201/9781003157601-4

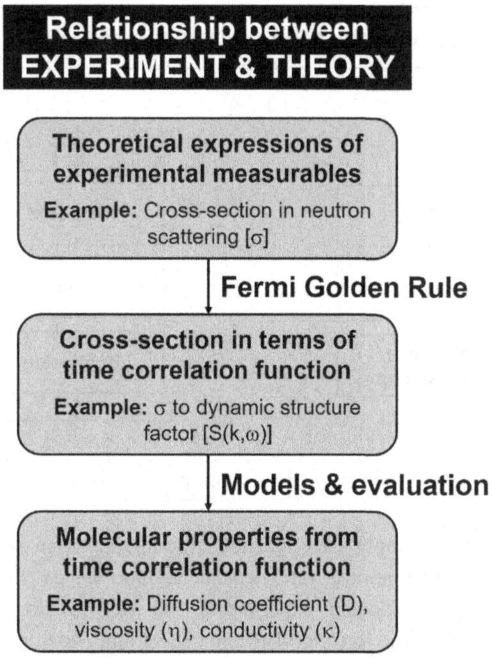

FIGURE 3.1 Schematic illustration of the three distinct steps involved in establishing the relationship between theory and experiment.

(c) The third step is the calculation of the time correlation function, leading to the connection of the time correlation function to desirable quantities like relaxation times and/or viscosity dependence of rates.

In Figure 3.1, we have illustrated this sequence of steps needed to establish such a relationship. Note that these three steps are distinct and often unrelated. The first step involves a theoretical analysis that takes into account the interaction of radiation with matter, as in electromagnetic theory, and considers the response to a perturbation imposed. In the second step, we use the Fermi golden rule or any other expression for the transition probability to reduce the cross-section to a time correlation function (TCF). The third step uses information about molecular interactions, and system properties to evaluate the time correlation function. In this last step, we use time-dependent statistical mechanics to evaluate the time correlation function and correlate observed results with the structure and dynamics of the system; in the process, we learn about the transport properties (diffusion, viscosity, thermal diffusivity).

As described in Chapter 1, relevant physical, chemical, and biological processes together cover a vast time scale, ranging from fs to ms and seconds. Usually, the experiments can be divided into two broad categories: (i) frequency domain experiments and (ii) time domain experiments. The examples of frequency domain experiments are many, infra-red and Raman, NMR, optical emission and absorption, incoherent neutron scattering, sound absorption, to name a few. In the frequency domain experiments, energy (and sometimes momentum) is/are exchanged with the molecules and/or collective modes of the system. In the time domain experiments whose scopes have grown tremendously after the advancement of the ultrashort laser pulses, measurements are made directly in the time domain. This method has proven to be particularly useful to probe short-time dynamical processes that often get obscured in the high-frequency tail of an absorption experiment. Inelastic and quasi-elastic neutron scattering techniques provide important information about collective

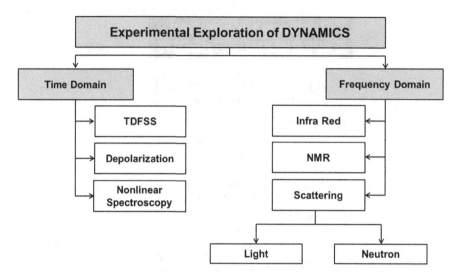

FIGURE 3.2 Classification of commonly used dynamic experiments based on time domain and frequency domain measurements. Here, TDFSS implies time-dependent fluorescence Stokes shift (used in solvation dynamics experiments). In frequency domain experiments, the width of the line shape reveals important information. Light scattering leads to the Rayleigh-Brillouin spectrum, which reveals a wealth of information (thermal diffusion, viscosity, etc.), whereas neutron scattering gives us information about structural arrangements at the microscopic level through wavenumber dependence.

dynamics in the condensed phases. As these techniques depend on the exchange of energy and momentum of the incoming neutrons with many molecules (the nuclei) of the system being probed, we can extract the information about collective dynamics. *Here and hereafter, collective dynamics means the motion of many molecules determines the response. This could be different from a single-particle response, like the diffusion coefficient of a tagged molecule.* Most experiments measure this collective response.

In Figure 3.2, we have made the important distinction/classification between time and frequency domain explicit. Often one uses both the methods together to obtain a better understanding. The different time scales are often intimately connected.

In this chapter, we discuss the intimate relationship between theory and experiments in this area. We discuss how the experiments related to the appropriate time correlation function. Subsequent chapters are devoted to the evaluation of the time correlation functions.

3.2 TIME CORRELATION FUNCTIONS (TCF): DEFINITION WITH MOTIVATION

Time correlation functions are natural generalizations of the fluctuation formulae for the response functions that we use in equilibrium statistical mechanics. The time integral over the TCFs provides the transport coefficients, like diffusion, viscosity, etc., that are the most powerful results of TDSM. The establishment of the relationship between theory and experiment happened over ages starting in the 1960s, initially formulated by Kubo.

We shall introduce the concept of time correlation functions in detail later in Chapters 8, 9, and 10. Here we mention the bare essentials.

As mentioned in Figure 3.1, an intermediate level outcome of experimental studies often is a time correlation function (TCF). This is because the experimentally measurable quantities like infra-red absorption spectrum or light scattering spectrum are expressed in terms of Fourier or

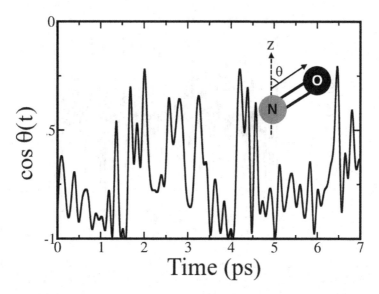

FIGURE 3.3 Time evolution of the orientation of a tagged nitric oxide molecule (NO) in water at 300 K obtained from molecular dynamics simulation. The orientation is measured by the cosine of the angle (θ) between a fixed vector and the dipole moment of the tagged NO molecule.

Laplace-Fourier transformation of the appropriate time correlation function. Let us consider a dynamical quantity $X(t)$ where t is the time. This can be the density of molecules in a local region, the position of a tagged molecule, the orientation of a tagged molecule, like water, or the energy of the system immersed in a bath. Let us consider the orientation of a tagged nitric oxide, NO, molecule for explicit purposes. Because our NO molecule is immersed in water and continuously interacting with surrounding water molecules, its orientation is changing as a function of time. If we denote the orientation with respect to the body fixed frame by θ, we can define a function cosine (θ) or $\cos(\theta)$. Then in our notation $X = \cos(\theta)$. If we plot X against time, we shall find that X executes a random walk between 1 and –1 when θ varies between 0 and π (Figure 3.3)

In an additional example, we can also imagine $X(t)$ as the energy of a system immersed in a temperature bath. Now, the energy of such a system fluctuates continuously around the mean value. Such an energy trajectory is shown in Figure 3.4. We would like to remind the student that these fluctuations determine the specific heat of the system.

Moreover, we can imagine that the orientation of the molecule, or the energy of the system, should be correlated over a short time interval. This is because while the changes are infinitesimal as time changes continuously, there is a memory of the immediate past. The loss of this memory is described in terms of time correlation function (TCF). How this memory becomes lost to give rise to decay of the TCF is the question whose answer is sought after by TDSM. This correlation can be described by using the methods of probability theory, which tells us that the simplest way to describe any correlation is to form a product. Here that should be $X(0)X(t)$.

This gives us the idea that we can then define a correlation function by considering a long time trajectory of $X(t)$ in the following fashion:

$$\langle X(0)X(t) \rangle = \lim_{T \to \infty} \frac{1}{T} \int_0^T ds X(s) X(t+s) \tag{3.1}$$

This is called a time average or the average over time of the time correlation of the dynamical quantity, $X(t)$. There are other ways to define an average that we shall discuss in a later chapter.

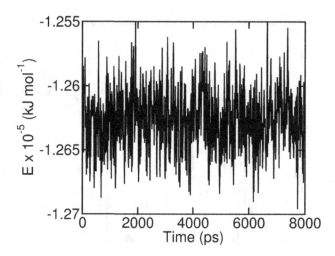

FIGURE 3.4 Variation of energy (E) along a trajectory of a system of ~2000 water molecules simulated under NVT condition (T = 300 K). The energy of the system fluctuates about a mean value of -1.2625×10^{-5} kJ/mol. As discussed in our equilibrium statistical mechanics book, the fluctuation in energy gives rise to specific heat.

But for the time being, it is vital for students to understand that a time correlation of a dynamical quantity measures how the variation or changes are correlated over time. *This is the simplest time correlation function that one can define between two objects, and rather, fortunately, it turned out to be effective in describing a large amount of experimental data.*

In experiments, we measure the time correlation function over different dynamical quantities. In theoretical studies of TDSM, a major effort is devoted to calculate and understand these time correlation functions. We shall discuss TCFs in detail in Chapters 8, 9, and 10.

In the remaining part of this chapter, we shall present several expressions for the spectral line widths in terms of Fourier transforms of appropriate TCFs. These expressions are of enormous importance in studying dynamics and liquids through spectroscopy. In the forthcoming sections, we shall discuss how different experiments provide different TCFs. These TCFs are the windows to the motion of atoms and molecules in a system.

3.3   INELASTIC AND QUASI-ELASTIC NEUTRON SCATTERING (INS AND QENS)

As mentioned above (see Figure 3.1), we follow certain logical steps to establish a quantitative relation between the experimental results and the molecular properties in a quest. For neutron scattering, the first step is the derivation of the following expression for the differential scattering cross-section, which is obtained by using the Fermi golden rule (FGR) for the transition $|1, \mathbf{k}_1\rangle \rightarrow |2, \mathbf{k}_2\rangle$ given by Eq. (3.2). We provide a schematic of the neutron scattering process in Figure 3.5.

In order to quantify the experimental observation, we need to calculate the partial differential cross-section for scattering into the solid angle $d\Omega$ in a range of energy transfer $d\omega$ by averaging over all initial states $|1\rangle$ with their statistical weights $p_1 \propto \exp(-\beta E_1)$ and summing over all final states $|2\rangle$. In continuation with Figure 3.1, we outline the first two steps connecting measurable to the appropriate TCF

$$\frac{d^2\sigma}{d\Omega d\omega} = \frac{k_2}{k_1}\left(\frac{m}{2\pi\hbar^2}\right)\sum_{\{1\}}\sum_{\{2\}} p_1 \left|\langle 2, \mathbf{k}_2 | V | 1, \mathbf{k}_1 \rangle\right|^2 \delta\left(\omega - \omega_{12}\right) \qquad (3.2)$$

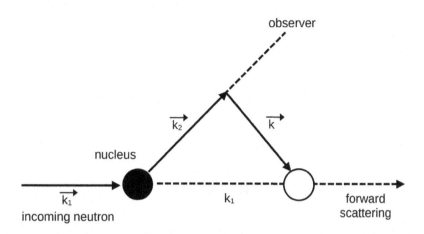

FIGURE 3.5 Schematic representation of the geometry of a neutron scattering experiment. k_1 is the wave vector for an incoming neutron, and k_2 is the wave vector for the scattered neutron. The momentum transfer from the sample to the neutron is measured by $\vec{\mathbf{k}} = \vec{\mathbf{k}}_2 - \vec{\mathbf{k}}_1$ (in units of $\hbar$). For elastic scattering, we have $\left|\vec{\mathbf{k}}_1\right| = \left|\vec{\mathbf{k}}_2\right|$. The scattered intensity is observed at various angles, allowing a variation of k.

Here m is the mass of the neutron. Other variables are explained in Figure 3.6. The coupling between the system and the incoming neutrons is given by

$$\left\langle 2, \mathbf{k}_2 \mid V \mid 1, \mathbf{k}_1 \right\rangle = \frac{2\pi\hbar^2}{m} \left\langle 2 \left| \sum_{i=1}^{N} b_i \exp\left(i\mathbf{k}.\mathbf{r}_i\right) \right| 1 \right\rangle. \tag{3.3}$$

In the next step, one combines the two equations above and utilizes the definition of the dynamic structure factor to obtain the following expression for scattering cross-section:

$$\frac{d^2\sigma}{d\Omega d\omega} = b^2 \left(\frac{k_2}{k_1}\right) N S(\mathbf{k}, \omega) \tag{3.4}$$

where, b is the scattering length of each nucleus. k_1 and k_2 are the magnitude of wave vectors corresponding to the incident and the scattered neutrons, respectively. $S(\mathbf{k}, \omega)$ is the dynamic structure factor, which can be obtained from the density autocorrelation function as,

$$S(\mathbf{k}, \omega) = \frac{1}{2\pi N} \int_{-\infty}^{\infty} \exp\left(i\omega t\right) \left\langle \rho_{-\mathbf{k}} \rho_{\mathbf{k}}(t) \right\rangle dt \tag{3.5}$$

where, $\rho_{-\mathbf{k}} = \rho_{-\mathbf{k}}(t = 0)$.

A large amount of effort is still directed to the experimental measurement and theoretical calculation of the dynamic structure factor. Students should note that the combination $(-\mathbf{k}, \mathbf{k})$ is due to momentum conservation.

To summarize, inelastic and quasi-elastic neutron scattering experiments give us the dynamic structure factor $S(\mathbf{k}, \omega)$, which is the Fourier-Laplace transform of the density-density time correlation function. We shall devote a full chapter on $S(\mathbf{k}, \omega)$ later but let us state here that the dynamic structure factor is the primary interest in our study of complex systems.

The difference between INS and QENS lies in the length and time scales probed. QENS is restricted to somewhat shorter time scales (of the order of 100 ps or so) and contains less information about the local structure than INS, as it is restricted to smaller $\mathbf{k}$ values.

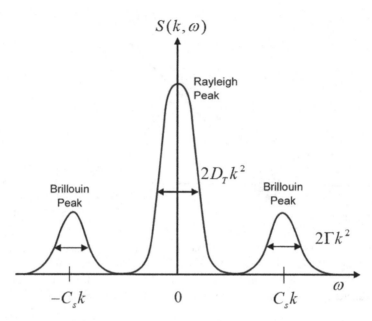

FIGURE 3.6 A Rayleigh-Brillouin spectrum, showing the dynamic structure factor $S(k, \omega)$ in the hydrodynamic limit. The Rayleigh peak describes thermal diffusion while Brillouin doublets are due to sound waves. The width of the central Rayleigh peak gives thermal diffusivity while that of the Brillouin peaks gives sound attenuation coefficient, which is related to shear and bulk viscosity.

Source: Figure taken from [13].

3.4 DYNAMIC LIGHT SCATTERING (DLS)

Dynamic light scattering (DLS) technique, sometimes also known as the Rayleigh-Brillouin spectroscopy (RBS), is a method similar to neutron scattering but works at large length and long time scales. The primary result is again the dynamic structure factor $S(\mathbf{k}, \omega)$. However, the purpose is different from neutron scattering. The latter probes structure and dynamics at molecular length scales. But light scattering is tuned to processes occurring at length scales of several thousand molecular diameters.

Because of the long wavelengths involved, light interacts with the hydrodynamic modes, like sound and heat or thermal diffusion modes. The mechanism is as follows. Light is scattered by *fluctuations in the local dielectric constant (which is proportional to the refractive index) of the liquid, making the dielectric constant of the liquid different at different locations.* The fluctuations in the dielectric constant or refractive index can be related to fluctuations in the density of the liquid. The latter can be expressed in terms of hydrodynamic modes. Therefore, the incoming light gets scattered differently from the incident direction. The wavenumber and frequency-resolved intensity of scattered light can be related to the liquid density-density correlation function, or the dynamic structure factor, in a fashion not too different from what we discussed for neutron scattering except that here the wavelength of the phenomena is several (at least three) orders of magnitude longer and the times scales are also much longer. The relationship between the intensity of scattered light, $I(\mathbf{k}, \omega)$ and the dynamic structure factor, $S(\mathbf{k}, \omega)$ for pure monatomic fluid is given by

$$I(\mathbf{k}, \omega) = \left(\mathbf{n_i}.\mathbf{n_f}\right)^2 \alpha^2 S(\mathbf{k}, \omega) \tag{3.6}$$

where the molecular polarizability along $\mathbf{n_i}$ and $\mathbf{n_f}$ for spherical molecules is $\left(\mathbf{n_i}.\mathbf{n_f}\right)\alpha$. The details of the derivation and the related discussion have been given in the classic book on the subject by

Berne and Pecora [1], and also covered briefly in Chapter 5 on "Hydrodynamics". The line shape function $I(\mathbf{k},\omega)$ is the three-peak structure where the central peak is called the Rayleigh peak, and the two side branches are called Brillouin peaks. The location and widths of these three peaks provide us with valuable information, like values of thermal diffusivity, sound velocity, sound attenuation coefficient.

Dynamic light scattering cannot provide information on the structure and dynamics at small length scales. However, being sensitive to long-length scale phenomena, DLS provides information about phase transition dynamics, and relaxation at long length and time scales. It is a highly useful technique.

3.5 INFRARED (IR) SPECTROSCOPY

Infrared (IR) spectroscopy is one of the oldest and popular tools to study molecular relaxation in liquids. It is a frequency domain technique where one is interested primarily in the lineshape of absorption of light in the range of a few hundred to a few thousand cm^{-1}, which is typical of the vibration of a chemical bond.

Typically, IR and far-infra red (FIR) spectroscopy can probe time scales of the order of ~ns or higher. One of the techniques to study rotational relaxation and vibrational dephasing experimentally is by determining spectral line shapes. The IR absorption line shape $I_{IR}(\omega)$ is given by Eq. (3.7).

$$I_{IR}(\omega) = 3\sum_{i,f} \left|\left\langle f \mid \hat{\varepsilon}.\mu \mid i \right\rangle\right|^2 \rho_i \delta(\omega_{fi} - \omega).$$ (3.7)

Here $\hat{\varepsilon}$ is a unit vector along the direction of polarization of the incident radiation and $\propto$ is the dipole moment operator. This expression evaluates the line shape occurring due to transitions from the initial state i to the final state f. The delta function has a special significance. It ensures the exclusive transitions from i to f only, with an infinite rate. The frequency ω_{fi} signifies the difference of energy between these two states, given by

$$\hbar\omega_{fi} = E_f - E_i.$$ (3.8)

In Eq. (3.7), ρ_i is the probability that the system is initially in state i. This equation is a direct consequence of the Fermi golden rule. Taking the Fourier transform of the delta function and invoking the Heisenberg operator representation, the spectral line can be represented as a Fourier transform of the total dipole moment time correlation function of the system.

$$I_{IR}(\omega) = \frac{1}{2\pi}\int_{-\infty}^{+\infty} dt e^{i\omega t} \left\langle \mu(t)\cdot\mu(0)\right\rangle.$$ (3.9)

If the system dipole moment operator is approximated as the sum of the molecular dipole moment of individual molecules, then the absorption intensity near the vibrational frequency ω_0 can be expressed as,

$$I_{IR}(\omega) \approx \int_{-\infty}^{+\infty} dt e^{i\omega t} \left\langle \mathbf{u}_i(t)\cdot\mathbf{u}_i(0)Q_i(t)Q_i(0)\right\rangle$$ (3.10)

where, $\mathbf{u}_i$ is the direction of the threefold or higher axis, and Q_i is the vibrational coordinate of intereston molecule i. Thus, the line shape derives contributions from both rotational and vibrational relaxations. This aspect has been discussed in the monograph "Molecular Relaxation in Liquids" in detail.

One relatively new entrant in this area of experimental research is the terahertz (THz) spectroscopy, which measures the polarizability-polarizability time correlation function in the time domain, using terahertz pulses became available around the year 2000. This technique is sensitive to low frequency collective motions, although the related theory is still evolving.

3.6 RAMAN SCATTERING

Raman scattering, like IR spectroscopy, is also used to study the phase relaxation of vibration, such as a chemical bond. It also allows the study of rotational relaxation. It is used both as a diagnostic tool and also interrogator of dynamics. The advantage is that it is an off-resonant technique, that is, no absorption is involved. This gives it some unique advantage, for example, working with visible light.

Unlike IR, this technique is also directly coupled to the molecular vibrations via polarizability. Placzek derived a quantum mechanical equation for the scattering cross-section (σ) for Raman scattering into a solid angle $d\Omega$ and frequency range $d\omega$.

$$\frac{d^2\sigma}{d\Omega d\omega} = \left(\frac{2\pi}{\lambda}\right)^4 \sum_{i,f} \left|\langle i|\hat{\varepsilon}_I.\underline{\alpha}.\hat{\varepsilon}_s|f\rangle\right|^2 \rho_i \delta\left(\omega - \omega_{fi}\right). \tag{3.11}$$

Here, $\hat{\varepsilon}_I$ and $\hat{\varepsilon}_s$ are the polarization directions of the incident and scattered lights and $\underline{\alpha}$ is the polarizability tensor, λ is the wavelength of the scattered light. i and f denote the initial and the final states of the system with a probability ρ_i denoting that the system is initially in state i. The scattering intensity $\left(I(\omega)\right)$ is proportional to this cross-section. Depending on whether $\hat{\varepsilon}_I$ and $\hat{\varepsilon}_s$ are parallel or perpendicular to each other, isotropic and anisotropic Raman line shapes are defined as follows:

$$
\begin{aligned}
I_{iso}\left(\omega\right) &= I_{\parallel}\left(\omega\right) - \frac{4}{3}I_{\perp}\left(\omega\right) \\
I_{aniso}\left(\omega\right) &= I_{\perp}\left(\omega\right)
\end{aligned}
\tag{3.12}
$$

Isotropic and anisotropic scattering intensities are related to mean polarizability and polarizability anisotropies by a time correlation function formalism.

$$\boxed{I_{iso}(\omega) \propto \left(\frac{\partial \alpha}{\partial Q}\right)^2 \int_{-\infty}^{\infty} dt e^{i\omega t} \left\langle Q_i(0)Q_i(t)\right\rangle} \tag{3.13}$$

$$\boxed{I_{aniso}(\omega) \propto \int_{-\infty}^{\infty} dt e^{i\omega t} \left\langle P_2(\mathbf{u}_i(t)\cdot\mathbf{u}_i(0))Q_i(t)Q_i(0)\right\rangle} \tag{3.14}$$

where $\mathbf{u}_i$ is the direction of the threefold or higher axis and Q_i is the vibrational coordinate of interest on molecule i and P_2 is the second Legendre polynomial.

3.7 FÖRSTER RESONANCE ENERGY TRANSFER (FRET)

Förster resonance energy transfer (FRET) has recently become immensely popular, for reasons we discuss below. This technique allows us to measure distances between two groups as a function of time.

FRET is a process where radiationless excitation energy transfer occurs from an optically excited state of a donor (D^*) molecule to the ground state of an acceptor (A) molecule. This is a long-range process where electronic excitation energy in some suitable cases can be efficiently transferred between a donor and an acceptor over distances as large as 10 nm. The fundamental advantage of

FRET is that the rate of such energy transfer strongly depends on the donor-acceptor separation distance. This energy transfer occurs via long-range coulombic (for example, dipole-dipole). As a result, the donor molecule proceeds to the ground state, and simultaneously the acceptor molecule gets excited. Förster energy transfer typically happens when the energy differences between the ground and the excited states of the donor and acceptor molecules are nearly degenerate.

In 1948, Förster proposed a theory for this radiationless energy transfer process, postulating that the transfer rate *depends on the inverse sixth power of the distance between the donor and acceptor*. This was further validated by fluorescence studies of model systems having a known donor acceptor pairs' separation. Since then, FRET has evolved into a major experimental procedure, routinely employed in studies in physics, chemistry, biology, and materials science. According to Förster, the rate expression for the nonradiative energy transfer (k_{DA}) is

$$k_{DA} = \frac{1}{\tau_D}\left(\frac{R_0}{R}\right)^6 \tag{3.15}$$

where R denotes the distance between the centers of the donor (D) and acceptor (A) molecules, τ_D represents the excited state lifetime of the donor molecule in the absence of an acceptor, R_0 is termed as Förster critical radius and symbolizes the distance at which the efficiency of energy transfer decreases by 50%.

It is the expression for R_0 that has tremendous spectroscopic significance. Förster expressed the critical radius (R_0) in terms of spectral overlap between the emission spectrum of D and the absorption spectrum of A as

$$R_0^6 = \frac{9000\ln(10)\kappa^2 Q_D c^4}{128\pi^5 \eta^4 N_A}\int dv\, \frac{\Phi_D(v)\varepsilon_A(v)}{v^4} \tag{3.16}$$

where $\Phi_D(v)$ denotes the normalized fluorescence spectrum of the donor, $\varepsilon_A(v)$ is the extinction coefficient of the acceptor molecule. The spectral overlap integral is represented by $\int dv\, \dfrac{\Phi_D(v)\varepsilon_A(v)}{v^4}$. Q_D denotes the quantum yield of the donor molecule, η is the refractive index of the solvent, N_A is Avogadro's number, κ^2 is an orientation factor for the dipole-dipole interaction of the transition dipoles associated with D and A, and is typically set to 2/3 as per the result of isotropic orientational averaging. A more elaborate discussion can be found in the earlier book entitled "*Molecular Relaxation in Liquids*".

3.8 SINGLE MOLECULE SPECTROSCOPY (SMS) AND FRET

Single-molecule spectroscopy has emerged as a powerful technique that allows interrogating both structure and dynamics at mesoscopic length and time scales. It combines, partially, the length scale resolution of microscopy with the time scale resolution of spectroscopy. At a static level, microscopy can provide structural information at a microscopic atomic length scale. Similarly, spectroscopy can provide time resolution even down to the fs time scale. However, traditional spectroscopic methods do not provide spatial resolution, just as microscopy does not provide information about dynamics. SMS allows us to combine the two, albeit with certain sacrifices.

The development of SMS owes its origin to the developments of single photo counting and confocal microscopy that allows us to capture every photon being emitted from a small region of micrometer length scale. One still uses the basic principles of spectroscopy like FRET or fluorescence correlation spectroscopy (FCS) to measure dynamics. The microscopy allows us to probe a local region where certain change could occur; this allows a spatiotemporal map of a dynamical process, not possible by microscopy and spectroscopy alone, separately.

SMS has proven to be exceptionally successful in studying enzyme kinetics, protein and DNA dynamics, and dynamics inside a biological cell. It is being increasingly used in diagnostics. While conventional spectroscopic measurements provide an average ensemble picture, SMS provides us with detailed information about the distribution itself. The distribution, in turn, allows us to understand the details of dynamics otherwise hidden in an ensemble measurement. It is here that TDSM becomes useful.

3.9   OPTICAL KERR EFFECT SPECTROSCOPY (OKE)

Optical Kerr effect (OKE) spectroscopy is an extensively used technique for investigating the ultrafast dynamics in transparent fluids. OKE spectroscopy probes the collective orientational motion, diffusion as well as the dynamics of Raman-active intramolecular and intermolecular modes in the time domain. Thus, this technique has provided several landmark understandings into the microscopic characteristics of simple liquids.

Initially, John Kerr discovered that applied DC electric field in a transparent medium induces birefringence. Subsequently, it was found that the oscillating light field of a laser pulse can also generate induced dipoles in the liquid molecules, which can interact with the laser field and eventually construct a partial preference for molecules to align with their axis of maximum polarizability parallel to the laser polarization. This alignment creates a transient birefringence in the liquid that can be probed with a second pulse of light. Typically, the pump pulse is polarized at 45°, the probe pulse is polarized vertically and has a variable time delay t relative to the pump pulse.

The OKE basically determines the laser pulse induced nonlinear polarization (P^{NL}). This nonlinear polarization depends on the polarizability-polarizability time correlation function defined by

$$C_{ijkl}(t) = \left\langle \alpha_{ij}(0)\alpha_{kl}(t) \right\rangle \tag{3.17}$$

It is essential to realize that Eq. (3.17) is a *collective* orientational correlation function because $\alpha_{ij}(t)$ is a sum over the polarizabilities of all the molecules of the system. In fact, the OKE decay is proportional to the negative time derivative of the collective orientational correlation function. *Note that dielectric, Kerr, and solvation energy relaxations all measure essentially the collective orientational relaxation of a dipolar liquid. In fact, OKE data can be used to explain the solvation dynamics with reasonable accuracy.*

In a landmark work, McMorrow and Lotshaw investigated the Kerr relaxation of neat liquid acetonitrile and found that the OKE signal has an initial relaxation, which is dominated by an ultrafast, Gaussian component with the time constant ~110 fs. Followed by a nonexponential (apparently also non-Gaussian) phase with a time constant ~ 400 fs. The last 20% is exponential decay with a time constant ~1.4 ps.

The long-time part of the OKE signal arises from orientational diffusion, and any high-frequency oscillations originate from intramolecular Raman modes. The source of the rest of the signal, which is associated with intermolecular motions, is rather more complex. The ultrashort component comprises information from inter-molecular vibrations due to interaction-induced effects that can substantially contribute to the magnitude of the polarizability tensor. Libration is the simplest mechanism through which intermolecular modes contribute to the OKE signal. Libration is typically a rocking motion of a molecule that appears from the hindered rotational motion enforced by the surrounding cage of nearest neighbors. Although librations may comprise complex motions of multiple molecules, librational scattering can be expressed in terms of motions of the polarizability tensors of individual molecules and is considered a molecular effect.

3.10 TIME-DEPENDENT FLUORESCENCE STOKES SHIFT (TDFSS) AND SOLVATION DYNAMICS

Solvation dynamics is a tool that has developed over the years into a useful technique to study dynamics, particularly at a local scale. The solvation of a species profoundly influences the way it participates in a chemical reaction. Solvation dynamics explores the time-dependent collective polarization response of a dipolar liquid to a changing charge distribution in a solute probe. For this purpose, the solute probe is first excited to achieve a new dielectric behavior. The neighboring solvent thus reorients to attain a new equilibrium. Subsequently, the time-dependent fluorescence Stokes shift (TDFSS) of the emission spectrum is monitored. This is accompanied by the relaxation of the solvent environment. Hence, the TDFSS indirectly probes the solvent dynamics.

If the shape of the emission spectrum remains unchanged with time, we can define a solvation time correlation function $S(t)$ to study this dynamic.

$$S(t) = \frac{\bar{v}(t) - \bar{v}(\infty)}{\bar{v}(0) - \bar{v}(\infty)} = \frac{\bar{E}(t) - \bar{E}(\infty)}{\bar{E}(0) - \bar{E}(\infty)}. \tag{3.18}$$

Here $\bar{v}(t)$ is the average time-dependent frequency of the emission spectrum and $\bar{E}(t)$ is the corresponding energy. This is basically the first moment of the spectrum. Eq. (3.18) defines a nonequilibrium Stokes shift response $S(t)$ that estimates the time dependence of the progress of solvation, as depicted in Figure 3.7A.

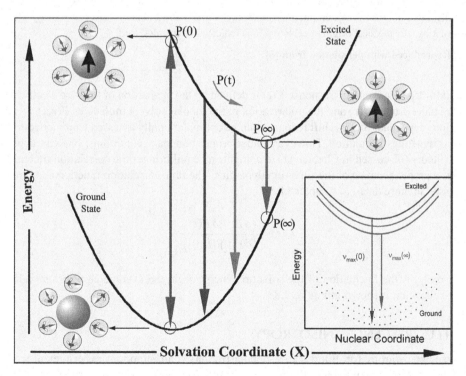

FIGURE 3.7 Schematic representation of the potential energy surfaces involved in solvation dynamics showing the orientational water motions along the solvation coordinate together with instantaneous polarization $P(t)$. The change in the potential energy along the intramolecular nuclear coordinate is shown in the inset. As solvation proceeds, the energy of the solute comes down, giving rise to a redshift in the fluorescence spectrum.

Source: Reproduced with permission from [6].

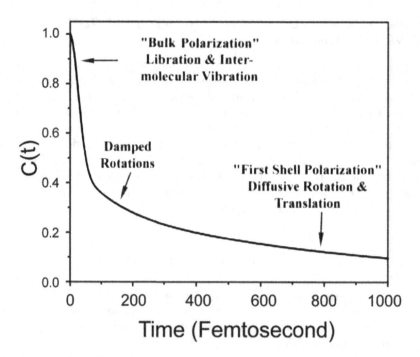

FIGURE 3.8 A typical equilibrium solvation time correlation function for water. It exhibits three distinct regions: the initial ultrafast decay, an intermediate decay of about 200 fs, and the last slow decay with a time constant of 1 ps. The physical origin of each region is indicated on the plot itself.

Source: Reproduced with permission from [6].

The time-dependent energy response $E(t)$ is defined by the interaction of the bare electric field of the polar solute probe $\mathbf{E}_0(\mathbf{r})$ and the polarization of the dipolar solvent molecules $\mathbf{P}(\mathbf{r}, t)$.

The time-dependent Stokes shift response can be computationally assessed from nonequilibrium molecular dynamics simulation. However, under proper boundary conditions, one can apply liner response theory (discussed in Chapter 11) to compute an equilibrium time correlation function $C(t)$, that gives a proper estimate of the solvation dynamics. The time correlation function approach shall be discussed in more detail in Chapter 11.

$$C(t) = \frac{\langle \delta E(0) \delta E(t) \rangle}{\langle \delta E(0) \delta E(0) \rangle}. \tag{3.19}$$

Here, $\delta E(t)$ is the fluctuation of the solvation energy from the average equilibrium value. The resultant relaxation is shown in Figure 3.8.

3.11 FLUORESCENCE ANISOTROPY

Fluorescence anisotropy (or, fluorescence depolarization) is an important experimental technique to study the rotational relaxation of fluorophores in solution. If one excites the sample with a plane polarized light, the emission becomes anisotropic. That is, the fluorescence intensity is different in the parallel ($I_\parallel$) and perpendicular ($I_\perp$) directions to the electric field vector used for excitation. In experiments, one is interested in the time-dependent anisotropy [$r(t)$] defined as

$$r(t) = \frac{I_{\parallel}(t) - I_{\perp}(t)}{I_{\parallel}(t) + 2I_{\perp}(t)}. \tag{3.20}$$

The decay of $r(t)$ indicates the decreasing difference between $I_{\parallel}$ and $I_{\perp}$ with time. Rotational diffusion of the fluorescent probe is the microscopic origin of the observed decay. The denominator (net emission) in Eq. (3.20) is used to normalize the output.

The origin of the observed anisotropy can be understood as follows. The fluorophore molecules are randomly distributed in a homogeneous solution in the ground state. The incoming polarized electric field creates an orientational anisotropy in the angular distribution of excited fluorophores.

Now, we aim to develop a theoretical formalism for fluorescence anisotropy decay. Suppose at a particular moment, the absorption dipole of a single molecule makes θ angle with the Z-axis (Figure 3.9). We assume Z to be the direction of the polarized electric field ($Z \equiv \parallel$). The emission intensity along a particular axis is proportional to the square of the component of transition dipole along that axis.

Hence one can write,

$$I_{\parallel}(\theta) = \cos^2\theta.$$
$$I_{\perp}(\theta, \phi) = \sin^2\theta \sin^2\phi \tag{3.21}$$

If the excitation occurs along the Z-axis, we need to integrate out ϕ as the excitation is isotropic throughout the XY plane. For a particular value of θ an average must be performed over all possible values of ϕ. Therefore,

$$\frac{1}{2\pi}\int_0^{2\pi} d\phi I_{\parallel}(\theta) = I_{\parallel}(\theta) = \cos^2\theta.$$
$$\frac{1}{2\pi}\int_0^{2\pi} d\phi I_{\perp}(\theta, \phi) = I_{\perp}(\theta) = \frac{1}{2}\sin^2\theta \tag{3.22}$$

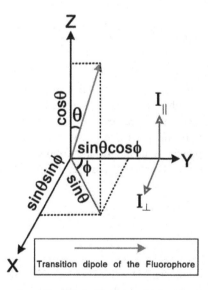

FIGURE 3.9 Schematic representation of parallel and perpendicular components of the emission intensities for a single fluorophore in the Cartesian coordinate system. For simplicity, the magnitude of the transition dipole is scaled to unity.

Now, let us define a probability function $f(\theta)$. Here $f(\theta)d\theta$ is the probability of finding a fluorophore molecule that produces an angle between θ to $\theta + d\theta$ with the Z-axis. As in the experiments, one obtains an averaged output from all the fluorophores, and we perform the average over $f(\theta)$.

$$I_{\parallel} = \int_0^{\frac{\pi}{2}} d\theta \, f(\theta) I_{\parallel}(\theta) = \left\langle \cos^2 \theta \right\rangle$$

$$I_{\perp} = \int_0^{\frac{\pi}{2}} d\theta \, f(\theta) I_{\perp}(\theta) = \frac{1}{2} \left\langle \sin^2 \theta \right\rangle. \tag{3.23}$$

We substitute the expressions of parallel and perpendicular intensity from Eq. (3.23) into the expression of $r(t)$ in Eq. (3.20) to obtain

$$\boxed{r(t) = \frac{3\left\langle \cos^2 \theta(t) \right\rangle - 1}{2} = \left\langle P_2 \left(\cos\theta(t) \right) \right\rangle}. \tag{3.24}$$

However, the derivation of Eq. (3.24) does not take into account the loss of anisotropy due to rotational diffusion. That is, the co-linearity of the absorption and emission dipoles is assumed.

Now, we theoretically calculate the maximum and minimum possible values of $r(t)$. For that, we need to explicitly evaluate the first integral in Eq. (3.23). In the case of an isotropic distribution of fluorophores, $f(\theta)d\theta = \sin\theta \, d\theta$. Hence, $\left\langle \cos^2 \theta \right\rangle = \int_0^{\pi/2} d\theta \sin\theta \cos^2\theta = \frac{1}{3}$ and $r(t) = 0$.

This can be considered the long-time limit and a complete loss of anisotropy, the minimum possible value. On the other hand, the probability of absorption of the plane-polarized light (along Z) is proportional to the square of the z-component of the transition dipole $\cos^2\theta$. If a molecule's transition dipole lies in the XY plane, it does not absorb. Because of the introduction of this anisotropy in the system, the probability function gets modified as $f(\theta)d\theta = \cos^2\theta \sin\theta d\theta$ at $t = 0$. Hence, one can write

$$\left\langle \cos^2 \theta \right\rangle = \frac{\int_0^{\pi/2} d\theta \cos^4\theta \sin\theta}{\int_0^{\pi/2} d\theta \cos^2\theta \sin\theta} = \frac{3}{5}. \tag{3.25}$$

From Eq. (3.24) we get, the initial (maximum) value of $r(t)$ is obtained as 0.4. In practice, one often acquires the initial value slightly less than 0.4, which is attributed to either limited experimental resolution or some excited state processes.

As stated above, depolarization happens due to the rotational diffusion of fluorophores. The decay of $r(t)$ with time is exponential in nature and follows $r(t) = r_0 \exp[-6D_r t]$. For spherical molecules, the ratio to the initial anisotropy to the measured, steady state anisotropy is given by Perrin's equation, Eq. (3.26).

$$\boxed{\frac{r_0}{r} = 1 + 6D_r \tau}. \tag{3.26}$$

Here, τ is the fluorescence lifetime and D_r is the rotational diffusion coefficient. Perrin's equation is particularly useful to study the rotational motions in proteins.

3.12   DIELECTRIC RELAXATION

Dielectric dispersion is a frequency domain experimental technique to study the relaxation of macroscopic electric polarization following the application of an oscillatory electric field. This furnishes information about the collective orientational motion of a dipolar liquid. Therefore, this technique bears certain similarities with Kerr relaxation. Polarization relaxation is described in terms of the frequency-dependent dielectric function, $\varepsilon(\omega)$. In its simplest form $\varepsilon(\omega)$ is given by the Debye expression Eq. (3.27).

$$\varepsilon(\omega) = \varepsilon_\infty + \frac{\varepsilon_0 - \varepsilon_\infty}{1 + i\omega\tau_D}. \tag{3.27}$$

Here ε_0 and ε_∞ stand for dielectric constants at zero (that is, static) and infinite frequency limit. τ_D is known as the Debye relaxation time. The complex dielectric function is separated into real and imaginary components as follows:

$$\varepsilon(\omega) = \varepsilon' - i\varepsilon''$$
$$\text{where, } \varepsilon' = \varepsilon_\infty + \frac{\varepsilon_0 - \varepsilon_\infty}{1 + \omega^2\tau^2} \text{ and } \varepsilon'' = \frac{(\varepsilon_0 - \varepsilon_\infty)\omega\tau}{1 + \omega^2\tau^2}. \tag{3.28}$$

The dielectric loss is defined as the ratio of the imaginary part to the real part as

$$\tan(\delta) = \frac{\varepsilon''}{\varepsilon'} = \frac{(\varepsilon_0 - \varepsilon_\infty)\omega\tau}{\varepsilon_0 + \varepsilon_\infty\omega^2\tau^2}. \tag{3.29}$$

However, in complex systems, the description given by Eq. (3.27) is not sufficient. To incorporate multiple relaxation processes, one often uses the following relations:
Budo formula for bimodal relaxation ($n = 2$ and $\sum_k g_k = 1$),

$$\varepsilon(\omega) = \varepsilon_\infty + (\varepsilon_0 - \varepsilon_\infty)\sum_{k=1}^{n} \frac{g_k}{1 + i\omega\tau_k}. \tag{3.30}$$

Cole-Cole relation,

$$\varepsilon(\omega) = \varepsilon_\infty + \frac{\varepsilon_0 - \varepsilon_\infty}{1 + (i\omega\tau)^\alpha}; \ 0 < \alpha < 1. \tag{3.31}$$

Davidson-Cole relation,

$$\varepsilon(\omega) = \varepsilon_\infty + \frac{\varepsilon_0 - \varepsilon_\infty}{(1 + i\omega\tau)^\beta}; \ 0 < \beta < 1. \tag{3.32}$$

Or, the most general form, Havriliak-Negami relation,

$$\varepsilon(\omega) = \varepsilon_\infty + \frac{\varepsilon_0 - \varepsilon_\infty}{\left[1 + (i\omega\tau)^\alpha\right]^\beta}; \ 0 < (\alpha, \beta) < 1. \tag{3.33}$$

The exponents describe the asymmetry and broadness of the obtained spectra. Now, we aim to relate the frequency-dependent dielectric function to quantities that can be obtained from theoretical or computer simulation studies.

Suppose the system is in equilibrium with a uniform, weak external electric field $\mathbf{E}_{ext}$. Say the external field is turned off at an arbitrary time $t = 0$. Then, according to Kubo's linear response theory (LRT), the decay of the total dipole moment of the system, $\mathbf{M}(t)$ can be written as

$$\mathbf{M}(t) = b(t)\mathbf{E}_{ext}(t = 0). \tag{3.34}$$

Here, $b(t)$ is termed as the *after-effect function*. $b(t)$ becomes exponential for a single relaxation process (Debye relaxation). It can be shown, by using linear response theory (LRT), that $b(t)$ is related to the total dipole moment autocorrelation function as follows:

$$b(t) = \frac{\Phi(t)}{3k_B T} = \frac{\langle \mathbf{M}(0) \cdot \mathbf{M}(t) \rangle}{3k_B T}. \tag{3.35}$$

The frequency-dependent dielectric function of a spherical region that is embedded in a dielectric continuum of the same material is related to the total dipole moment autocorrelation by the following relation:

$$\boxed{\mathscr{L}\left[-\frac{d}{dt}\Phi(t)\right] = \frac{\varepsilon_0 \left[\varepsilon(\omega) - 1\right]\left[2\varepsilon(\omega) + 1\right]}{\varepsilon(\omega)\left[\varepsilon_0 - 1\right]\left[2\varepsilon_0 + 1\right]}.} \tag{3.36}$$

In Eq. (3.36), $\mathscr{L}\left[f(t)\right]$ indicates the Laplace transformation of $f(t)$. This elegant theoretical description, which is extremely useful in studying dielectric spectroscopy, was first developed by Fatuzzo and Mason.

In recent years, intense studies have been initiated to explore the polar response of water under nanoconfinement. It has been found that the dielectric constant of confined water becomes anisotropic. This, in turn, is due to different effects of confinement on different components of the orientational correlations among the dipolar molecules. Thus, the anisotropic dielectric constant may provide valuable information about the structure and dynamics of confined liquids. Theoretical studies are still in progress.

3.13 MULTIDIMENSIONAL SPECTROSCOPY: LOOKING INTO LOCAL STRUCTURE AND DYNAMICS AND HETEROGENEITY OF CHEMICAL ENVIRONMENT

Most of the experimental techniques mentioned above, with the exception of the single-molecule spectroscopy, provide ensemble-averaged information about dynamics. To remind the student, an ensemble averaged picture means that the signal or response is being calculated from a large number of molecules, and the responses are added. In this process, we lose the information about the local environment around each molecule, which affects the details of the response. SMS does allow us to interrogate the local environment but still at mesoscopic length scales. For example, SMS cannot tell us the numbers of nearest neighbors that a water molecule experiences at a given time and how fast that number undergoes fluctuations with time. In fact, to find out such a local, microscopic environment around a given molecule, we need a microscopic technique. Multidimensional IR spectroscopy allows us to find such information. Needless to say that IR spectroscopy studies the dynamics of a chemical bond, like O-D in the molecule HOD in bulk water. Vibrational dynamics of the O-D bond, such as the frequency and the phase relaxation, are sensitive to the local environment.

The similarity with SMS, however, can be continued productively. Ordinary IR spectroscopy gives us an ensemble average picture of vibrational dynamics and a broad IR lineshape. However, we can resolve the spectrum by using multiple pulses by taking advantage of the different frequencies in different environments. This is achieved by using multiple pulses.

TDSM is used to understand this correlation between structure and dynamics in multidimensional spectroscopy, as we shall discuss in more detail later.

3.14   SUMMARY

In this chapter, we have collected a rather large number of useful experimental techniques that are routinely used to explore dynamics in liquids and biological systems. *A unifying feature among them is that understanding the observed results in all of them requires the use of time-dependent statistical mechanics.* This is because all the experimental observables are expressed in terms of relevant time correlation function (TCF). Evaluation of the latter requires the use of time-dependent statistical mechanics. In the process, we get to know about and evaluate molecular transport processes, rates of chemical reactions, even molecular mechanisms, and chemical pathways.

Understanding the relationship between theory and experiment is of utmost importance to gain proper perspective into the motivation and working of the time-dependent statistical mechanics. Broadly speaking, the area can also be named relaxation phenomena, as many books do. However, we often do not pay due attention to the relationship described here.

As outlined in flowcharts Figure 3.1 and Figure 3.2, *there is a unifying approach to connecting the measurables to the molecular properties.* Implementation of this connection involves three steps, which have been illustrated in Figure 3.1 and discussed in the following examples covered in this chapter.

From the examples discussed above, we find that the Rayleigh-Brillouin spectrum gives the value of thermal diffusivity and a part of the viscosity. However, extraction of this information and the relevant expressions required the use of time dependent statistical mechanics.

We now proceed to develop the concepts, methods, and techniques required to evaluate the time correlation functions, by using the method of time-dependent statistical mechanics.

BIBLIOGRAPHY

1. Berne, B. J., and R. Pecora. 2000. *Dynamic Light Scattering: With Applications to Chemistry, Biology, and Physics*. Chelmsford: Courier Corporation.
2. Lakowicz, J. R. 2013. *Principles of Fluorescence Spectroscopy*. New York: Springer Science & Business Media.
3. Fleming, G. R. 1986. *Chemical Applications of Ultrafast Spectroscopy*. Oxford: Oxford University Press.
4. Placzek, G. 1959. *The Rayleigh and Raman Scattering*. Berkeley: Lawrence Radiation Laboratory.
5. Bagchi, B. 2012. *Molecular Relaxation in Liquids*. Oxford: Oxford University Press.
6. Pal, S. K., J. Peon, B. Bagchi, and A. H. Zewail. 2002. Biological water: femtosecond dynamics of macromolecular hydration. *Journal of Physics Chemistry B, 106*, 12376.
7. Zwanzig, R. 2001. *Nonequilibrium Statistical Mechanics*. Oxford: Oxford University Press.
8. Szabo, A., R. Zwanzig, and N. Agmon. 1988. Diffusion-controlled reactions with mobile traps. *Physical Review Letters, 61*(21), 2496.
9. Mantsch, H. H. and D. Naumann. 2010. Terahertz spectroscopy: the renaissance of far infrared spectroscopy. *Journal of Molecular Structure, 964*(1–3), pp.1–4.
10. Fatuzzo, E., and P. R. Mason. 1967. A theory of dielectric relaxation in polar liquids. *Proceedings of the Physical Society, 90*, 741–750.

11. McMorrow, D., and L. T. Lotshaw. 1991. Intermolecular dynamics in acetonitrile probed with femtosecond Fourier-transform Raman spectroscopy, *Journal of Physics Chemistry*, 95, 10395.

12. Basché, T., W. E. Moerner, M. Orrit, and U. P. Wild. 2008. *Single-Molecule Optical Detection, Imaging and Spectroscopy.* New York: John Wiley & Sons.

13. Hansen, J. P. and I. R. McDonald. 2006. *Theory of Simple Liquids.* Cambridge: Academic Press.

4 Force, Flux, and Irreversible Thermodynamics

OVERVIEW

Irreversible thermodynamics often refers to the study of steady states, which are maintained far from equilibrium by supply of energy and are in a state of flux. Thus, study of irreversible thermodynamics deals with force and flux. Historically, the study of irreversible thermodynamics heralded the development of time-dependent statistical mechanics. While the beginning of time-dependent statistical mechanics (TDSM) can be assigned to the Boltzmann era, during the 1870s, the drive towards a systematic development of TDSM could rightfully be attributed to Onsager's landmark work on reciprocal relations that also introduced the concept of "regression hypothesis", which is essentially equivalent to the linear response theory that was later put on a firm basis by Kubo. However, Onsager brilliantly addressed the issue of cross effects and attributed their existence to the coupling between different fluctuations. Kubo's work also formalized the concept of fluctuation-dissipation theorem in terms of the time correlation function at equilibrium, in terms of the *natural motion* of the system. Einstein's theory of Brownian motion, Onsager's regression hypothesis and Kubo's linear response theory can be regarded as the trinity of time-dependent statistical mechanics. Earlier than Kubo's work, Prigogine formulated a theorem on the stability of a steady state far from equilibrium in terms of rate of entropy production. In this chapter, we shall discuss the basic concepts, Onsager's work, linear regression hypothesis, fluctuation dissipation theorem, and Prigogine's minimum rate of entropy production theorem. We shall return to this subject again in Chapter 24 towards the end of the text.

4.1 INTRODUCTION

Historically, our encounter with, and the study of, time-dependent phenomena began with such natural processes as the flow of water in rivers, springs, and oceans, blow of winds and storms, and flow of heat from hot to cold, also melting of ice and fall of rain. Identification of the cause or the origin of the associated changes, the resultant motions and flows as driven by forces, required certain abstraction. *It was aided by the observation that the motion is largely unidirectional.* For example, water flows down the land in hilly terrain or heat flows from the hot end to the cold end in an iron rod. Still, it was quite an achievement to identify the force causing the flow. It took mankind far longer to find the appropriate driving force for a component or a chemical species that causes it to move from a domain of higher to lower concentration till the concentration becomes uniform all over the space. It was a chemical force, termed chemical potential. We later realized that this homogenization is essentially due to entropy.

At the molecular level, motions are known to be reversible, satisfying Newton's equations of motion. The emergence of irreversibility at the *macroscopic level* was an issue of much interest at

DOI: 10.1201/9781003157601-5

one point in the history of time-dependent statistical mechanics and irreversible thermodynamics. The question that perplexed scientists is the origin of irreversibility that we routinely observe in nature, like the molecules of an ink drop placed in a beaker full of water never again aggregate back together to form the initial drop. *It was subsequently realized that this can be explained in terms of entropy – the system with dispersed ink is a state of higher entropy.* The ink molecules cannot go back in any conceivable time. Another (related) issue is the involvement of dissipation of energy. This provides the direction in the flow of water from high level to a lower one.

As evident from the above discussion, *entropy plays an essential role in our understanding of irreversible processes.* The first step towards this understanding was made in Clausius' statement of the second law of thermodynamics that states that entropy of an isolated system increases till the equilibrium is reached. This led to the idea that entropy can be used as an "arrow of time". This same idea found further substance in Boltzmann's entropy formula ($S = k_B \ln \Omega$, where Ω is the number of microscopic states) and in Boltzmann's *H*-theorem. Boltzmann's *H*-function provides a rationale for the direction in irreversibility. We discussed Boltzmann's function and his famous theorem in Chapter 6.

It was perhaps Einstein who first used Boltzmann's formula to define the probability of a fluctuation, $P(S) = \exp (\Delta S/k_B)$, where ΔS is the entropy cost of the fluctuation. We shall use this formula repeatedly in this chapter, in Chapter 24, and several other places.

In the attempt to establish a relation between flow (the effect) and the force (the cause), one first perceives that the magnitude of flow depends on the magnitude of the force. Thus, the flux of the flow, J (of energy or matter), that occurs in a response to a force, X, must be a function of X. Let us now consider the nature of the force-flux relationship in more detail. Flux, by definition, is the number of particles crossing a unit area perpendicular to the flow per unit time. When the force increases, the flux should also increase. Let us first consider the flow of water from a higher plane to a lower plane. Here the force is the gravitational force, which is mgh, where m is the mass, g is the gravitational force, and h is the height of the body of water (in this example). It was an empirical observation that flow was proportional to the force.

But one should note an interesting fact here. Even in the presence of a force, the body of water or the flow of current moves with a constant, steady velocity. In the case of heat transfer, we find that the rate of transfer of heat from the hot side of the rod to the cold side occurs at a steady rate. *Thus, Newton's law of motion that states force generates acceleration, appears to be violated!* However, Newton's laws of motion continue to be obeyed at a microscopic level. The irreversible steady motion in the presence of a driving force is a consequence of dissipation of energy, a subject we shall discuss in greater detail in the chapter on Langevin's equation and Brownian motion. As already mentioned, dissipation leads to an increase of entropy.

In many cases, the rate of motion, termed as the flux, is just proportional to X, and we write down a simple relation

$$J = LX \tag{4.1}$$

This linearity is an assumption, but a good one. This is based on observations, is largely valid, especially when the force is small. *The coefficient "L" is the dynamical counterpart of equilibrium (linear) response functions like specific heat, isothermal compressibility, polarizability* (which is related to the dielectric constant) of a medium. For example, we write the relation between temperature rise (ΔT, T is the temperature) in the system due to a heat (q) given to the system as,

$$\Delta T = \frac{1}{C_P} q \tag{4.2}$$

Here $1/C_p$ plays a role similar to L in Eq. (4.1) should be regarded as the *nonequilibrium counterpart of* Eq. (4.2). The equilibrium response functions are given by mean square fluctuations. Thus, we have the equation $C_V = \dfrac{\left\langle (\delta E)^2 \right\rangle}{k_B T^2}$, where δE is the fluctuation in the value of the energy of the system, the other terms have their usual meaning. We discussed at length in our book on *Equilibrium Statistical Mechanics* (CRC Press, 2018) that the response functions represent the force constant of the "springs" that hold the system together. The harmonic nature of the confining potential is reflected in the Gaussian distribution. We need to develop a similar description and language for the nonequilibrium systems.

Now, we specify J, L, and X for several cases. Let us consider heat flow from high temperature side of a rod to the low temperature side. Here the force is the temperature gradient ($-(\partial T / \partial x)$, a partial derivative) across the length x, J_H is the heat flux. The relation we are looking for is given by

$$J_H = -D_T \frac{\partial T}{\partial x} \tag{4.3}$$

where, D_T is the thermal diffusivity. Here $\dfrac{\partial T}{\partial x}$ is obviously the force. Similarly, for mass transport we have,

$$J_C = -D_S \frac{\partial c}{\partial x} \tag{4.4}$$

where, $c(x)$ is the concentration, and D_S is the self-diffusion coefficient. Similarly, we have the following relation between electric current and conductivity, in an electrolyte solution:

$$J_e = -\kappa \left(V_1 - V_2 \right) \tag{4.5}$$

where κ is the electrical conductivity and $\left(V_1 - V_2 \right)$ is the voltage difference. Another relation of significance in our study here is the one between force, F exerted on a particle being pulled through a dissipative medium, like a liquid. In this case, the relation is

$$v = \frac{1}{\zeta} F \tag{4.6}$$

which defines the friction ζ. Here the velocity v is the response, ζ is the response function. This is Stokes law of hydrodynamics and plays an important role. Note that friction gives rise to dissipation. Dissipation is an essential part of irreversible thermodynamics.

These *linear relations* between the flux and force form the cornerstones of our description of time-dependent phenomena, just as the static response functions form the very basis of equilibrium phenomena. We shall see later that this analogy between equilibrium response functions and dynamic response functions goes further in providing valuable insight.

4.2 RECIPROCAL RELATIONS

In a seminal paper published in 1931, Onsager established important reciprocal relations for irreversible processes. This work essentially gave birth to nonequilibrium (or, irreversible) thermodynamics.

In all the discussions above on the relation between force and flux, we considered direct, one-to-one relation between fluxes and responses. However, in many cases, a force or gradient can give rise to two responses: one direct, as discussed above, and an indirect, or cross response. Let us provide a concrete, almost everyday example of these cross-terms involving J, L, and X. We consider setting up of a temperature gradient across a metal plate that contains a junction of two metals with different conductivities. This can be easily achieved by keeping the two metal plates at the two ends at different temperatures. Surprisingly, we find that the system generates a difference in voltage between the two ends, which can be harnessed to generate electricity. This difference in voltage can be used to light a lamp, for example. This is known as the Seebeck effect. This is quite amazing actually. Many of us have seen this in museums. That is, *a temperature difference not only gives rise to thermal conduction – the transport of heat from high temperature to low temperature – but also a voltage difference.* This is an example of cross-relation. Similarly, when we send electric current across a junction between two metal plates of different metals, we generate heat. Thus, an electrical potential gradient and flow of electric current give rise to a temperature gradient between two conductors. This is called the Peltier effect. Thus, in addition to pure terms, like thermal conduction in the Seebeck effect and electrical conduction in the Peltier effect, we also have reciprocal cross-effects. The importance of such phenomena is that these processes are present almost everywhere.

Such reciprocal relations were known before Onsager. It was actually suggested earlier by J.J. Thomson (Lord Kelvin) but he gave no proof. Onsager proved that these reciprocal cross-terms are exactly equal, that is,

$$\boxed{L_{ij} = L_{ji}} \qquad (4.7)$$

(popularly known as Onsager reciprocal relationship). This is a key result that paved the way for the development of nonequilibrium statistical mechanics, as already mentioned. These relations were ultimately explained by time correlation function formalism, which is the heart of linear response theory. Thus, both in the case of Peltier effects and Seebeck effect, we get $L_{HV} = L_{VH}$, where subscripts "V" and "H" denote voltage and heat, respectively. Simple pictorial descriptions of Seebeck effect and Peltier effect are shown in Figure 4.1.

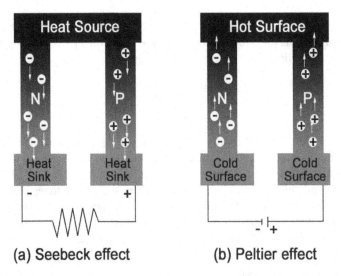

(a) Seebeck effect   (b) Peltier effect

FIGURE 4.1 (a) Seebeck effect – electric current is generated between a hot and a cold surface; (b) Peltier effect – application of potential gradient between two surfaces generates a temperature gradient. P and N represent negative and positive, respectively.

Onsager's paper of 1931 was a significant milestone for two more reasons. It introduced the principle of "microscopic reversibility" that states that at equilibrium, each step must be individually balanced in population change. This is sometimes also referred to as *the principle of detailed balance*. Microscopic reversibility puts constraints on rate constants and is widely used in the study of chemical and biological reactions. The second significant concept that evolved from the 1932 paper was the rate of entropy production. This concept ultimately led to Prigogine's formulation of the principle of minimum value of the rate of entropy production at a steady state maintained at far from equilibrium. Although Prigogine received the Nobel Prize for this work in 1977, it has remained controversial to date.

4.3 ONSAGER'S REGRESSION HYPOTHESIS

Another fundamental concept that can be traced back to Onsager's 1931 paper is "Onsager's regression hypothesis". This hypothesis states that for a small perturbation from equilibrium, the system returns to equilibrium at the same rate as a fluctuation does at equilibrium. The hypothesis can be explained in the following way.

Figure 4.2 shows an energy landscape that represents a system both at equilibrium and away from equilibrium. This picture assumes that the free energy landscape remains the same, and that the system is kept away from the equilibrium, minimum energy configuration by force imposed from outside. A practical example would be a system kept under a steady temperature gradient with the help of heat supplied at one side and removed from the other. Another example could be a liquid under steady flow. There is no reason to believe that static and dynamical properties (such as response functions) of the system in such steady states to be equal to those in equilibrium. Yet, this is assumed in Onsager's treatment. Thus, his theory is valid only near equilibrium

Let the equilibrium fluctuation in a property X at a time t is denoted by $\delta X(t)$. Relaxation of the fluctuation is represented by the normalized time correlation function $f(t)$ defined by

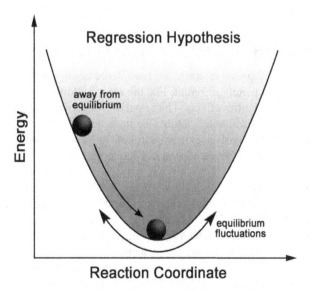

FIGURE 4.2 Representation of a system in equilibrium and away from equilibrium on an arbitrary energy landscape. Onsager's regression hypothesis connects equilibrium fluctuations to the relaxation from a nonequilibrium state.

$$f(t) = \frac{\langle \delta X(0) \delta X(t) \rangle}{\langle \left(\delta X(0) \right)^2 \rangle}. \tag{4.8}$$

Let us now consider a finite nonzero fluctuation in the property X and let $\Delta X(t)$ denote the time evolution of the property subsequent to the perturbation. When perturbed, the system moves away from the equilibrium value. It relaxes back to the equilibrium state with a change in $X(t)$ given by $g(t)$.

$$g(t) = \frac{\Delta X(t)}{\Delta X(0)}. \tag{4.9}$$

According to the regression hypothesis,

$$\boxed{f(t) = g(t)}. \tag{4.10}$$

That is, this hypothesis allows us to relate the decay of a nonequilibrium state in terms of a fluctuation time correlation function at equilibrium. This hypothesis greatly reduces theoretical analysis because calculations at equilibrium are much simpler than those in a nonequilibrium state. The regression hypothesis is closely related to (and often considered the same as) the linear response theory developed by Kubo and others.

However, Onsager's work goes deeper. It answers the question as to the origin of the cross-effects. Onsager showed that this is due to the coupling between different forces at the level of fluctuations, although appear to be absent at the level of the average.

Much appreciated by the community, Onsager was awarded the Nobel Prize in 1968. The citation of the Nobel Committee said, "*The great importance of irreversible thermodynamics becomes apparent if we realize that almost all common processes are irreversible and cannot by themselves go backward. As examples can be mentioned conduction of heat from a hot to a cold body and mixing or diffusion. When we dissolve a cold lump of sugar in a cup of hot tea these processes take place simultaneously*".

4.4 DERIVATION OF THE RECIPROCAL RELATIONS

The reason why Onsager's reciprocal relations are considered nontrivial is that the forces that generate cross-responses are completely different, like the electric voltage and temperature gradient, as in the case of Seebeck and Peltier effects. The popular proof follows the one given in the book on statistical mechanics by Landau and Lifshitz. We have described the detailed proof in a later chapter (Chapter 24) because it involves extensive use of time correlation function formalism and its various properties, which shall be introduced in Chapter 8. However, below we briefly outline the strategy.

As mentioned above, the existence of the cross-effects itself is interesting, and a bit unexpected. Why should an electric filed generate a temperature gradient? The reason, realized by Onsager, that these two are coupled at the level of fluctuations.

The proof uses the fluctuation theory developed by Einstein. Let $\{x_1, x_2, ..., x_n\}$ denote fluctuations from equilibrium values in several thermodynamic quantities and let $S(x_1, x_2, ..., x_n)$ be the entropy cost of fluctuation. The entropy function provides the force to control these fluctuations. The responses are determined the second derivatives of the entropy function. One now uses Boltzmann's entropy formula to obtain the probability distribution function, $w = A \exp(S / k_B)$ where A is a constant. At equilibrium, the first derivatives are zero because of the equilibrium extremum condition.

Therefore, if the fluctuations are small, the probability distribution function can be expressed through the second differential of the entropy

$$S = S_0 + \frac{1}{2} \beta_{ik} x_i x_k$$

$$w = A e^{-\frac{1}{2} \beta_{ik} x_i x_k} \tag{4.11}$$

$$\beta_{ik} = \beta_{ki} = -\frac{1}{k_B} \frac{\partial^2 S}{\partial x_i \partial x_k}$$

where we are using Einstein summation convention and $-\frac{1}{2} \beta_{ik} x_i x_k$ is a positive definite symmetric matrix.

One can understand the reciprocal relations from Eq. (4.11). The proof has been given in Chapter 24. When β_{ik} are nonzero, it gives the coupling between the two fluctuations at the level of entropy. The equality of two cross-correlation coefficients at the level of fluctuation appears to be obvious.

From Eq. (4.11) to the reciprocal relations is still a long road that involves use of several fundamental concepts.

4.5 RESPONSE FUNCTIONS AND TIME CORRELATION FUNCTIONS

As discussed above, the equilibrium response functions are given by the natural fluctuations in appropriate variables, which are present in the system, at equilibrium, in the absence of any perturbation. Thus, the specific heat is given by the mean square energy fluctuation, isothermal compressibility is given by the mean square volume fluctuation, the dielectric constant is given by the mean square total dipole moment fluctuation of the system. That is, the response functions are determined by the fluctuations in the respective conjugate quantities. These fluctuations are the consequence of the natural, thermal motion of the molecules of the system.

Dynamical response functions turn out to be a rather straightforward generalization of the equilibrium response functions. They are given by dynamical correlations present in the system at equilibrium. We shall elaborate on this aspect in detail later. Let us just state here that the dynamical response functions are determined by the time correlation functions. These relations are often known as Green-Kubo relations. Thus, diffusion coefficient is given by the velocity-velocity time correlation function, conductivity by current-current TCF, viscosity by stress-stress TCF.

The cross-correlation functions L_{ij} thus involves the TCFs that involve cross-correlation of dynamic variables at the level of fluctuations.

4.6 RATE OF ENTROPY PRODUCTION AND PRIGOGINE'S THEOREM

In conventional thermodynamics, when a system undergoes a transition from one equilibrium state to another, entropy and enthalpy of the systems change but in such a fashion that the new system is also a free energy minimum with respect to fluctuations or variations in certain parameters like density, the order in molecular arrangements. The equilibrium systems are determined by the principle of free energy minimization. Thermodynamics is applicable for systems at equilibrium. Despite the importance accorded to thermodynamics, its principles are not applicable to nonequilibrium systems.

A large number of systems that we encounter in nature are not in an equilibrium state. Instead, they are maintained in a steady state, which is far from equilibrium. The steady state is maintained by the supply of energy from outside. The temperature-independent steady state involves the continuous dissipation of energy and work done by the system.

However, an extension of thermodynamics to a nonequilibrium system has proven to be extremely difficult. It has proven hard to formulate general principles, such as extremum of free energy or entropy for an open system under flux or change. When a system is kept in a nonequilibrium state by applying a gradient, such as a temperature gradient, then the gradient causes a flow (such as flow of heat, in the example cited). When such a system is near equilibrium, we have Onsager's reciprocal relations, described above. While Onsager's relations are for the cross-coefficients, the use of regression hypothesis and microscopic reversibility are of importance in specifying dynamics of nonequilibrium systems near equilibrium. These principles have been found to be generally valid in the experimental studies of relaxation, like solvation dynamics of a newly created dipole in a molecule, when the system is driven out of equilibrium.

But regression hypothesis presupposes an equilibrium state because we need to have in our possession regression after small fluctuations at equilibrium. However, what happens in states truly far from equilibrium where the regression hypothesis is inapplicable?

The class of commonly encountered nonequilibrium states is time-invariant as it is kept in a steady state away from equilibrium by applying energy. Since the free energy minimization principle is of no use here, there has been a search for a general principle for the stability of the state. Such a nonequilibrium steady state consumes energy and therefore dissipates energy to maintain the steady state. This is a common trait in such systems. The dissipation of energy leads to entropy production.

Irreversible thermodynamics is concerned with entropy production in nonequilibrium systems. The field essentially started with Onsager's reciprocal relations and regression hypothesis, and progressed to a great extent, partly fueled by Prigogine's important work.

In his Nobel Prize-winning work, Prigogine demonstrated that in systems not too far from equilibrium yet in a steady state, the rate of entropy production is a minimum. We now briefly discuss the proof of this important work.

4.7 RATE OF ENTROPY PRODUCTION

While Clausius' second law of thermodynamics states that entropy of an isolated system increases till equilibrium is reached where the entropy is maximum, Prigogine established a theorem that proves that if a system is in a steady state far from equilibrium, it is the rate of entropy production that is minimum. This seductive theorem has drawn a lot of attention.

In the following, we first define the rate of entropy production. We start with Clausius' definition of entropy. Let us consider an isolated system divided into two parts separated by a diathermal wall. The two parts are kept at two different temperatures, as shown in Figure 4.3.

Since the two boxes are separated by a diathermal wall, the heat gets supplied from the higher temperature side and removed from the lower temperature side. That is, there is a continuous supply of heat from the high temperature side to maintain a steady state. The existence of a steady state implies that the flow of heat through the molecules of the matter occurs on a time scale shorter than the observation interval used in the measurement of properties. As a consequence, energy is

FIGURE 4.3 Two boxes at different temperatures (T_1 and T_2) are kept together, separated by a diathermal wall.

dissipated and entropy is produced. If we assume that laws of thermodynamics remain valid in this nonequilibrium state, then the change in entropy can be found out in the following fashion:

$$dS = \delta q \left(\frac{1}{T_1} - \frac{1}{T_2} \right). \tag{4.12}$$

Since T_1 and T_2 are both time-independent under steady state condition, we can write the following expression for the rate of entropy change:

$$\frac{d}{dt} dS = \frac{d}{dt} \delta q \left(\frac{1}{T_1} - \frac{1}{T_2} \right). \tag{4.13}$$

The right-hand side gives the rate of entropy production (in this particular case). If the length of the system is l and the cross-section is given by A, then we can obtain the rate of entropy production per unit volume by dividing the right-hand side of the above equation by lA to obtain which is often denoted by σ.

$$\sigma = \frac{1}{lA} \frac{d}{dt} \delta q \left(\frac{1}{T_1} - \frac{1}{T_2} \right). \tag{4.14}$$

The rate of entropy production σ is a central quantity in irreversible thermodynamics.

We can further generalize the above expression by making the length of the system l tend to zero so that we can take a continuum limit. That is, $(1/T_1 - 1/T_2)$ is replaced by $\frac{d}{dx}\left(\frac{1}{T}\right)$. Subsequent operations lead to the following form:

$$\sigma = -\left(\frac{J_Q}{T^2} \right) \frac{dT}{dx} , \tag{4.15}$$

with $J_Q = \frac{1}{A} \frac{\delta q}{dt}$. This interesting expression for the rate of entropy production is rather general, and

similar expressions can be derived for the transport of molecules where the force is the gradient of the chemical potential.

When several processes are running in parallel, the rate of entropy production per unit volume, σ_k, for each process, k is given by the product of a flux term J_k, associated with the process k or a rate per unit volume and the force,

$$\sigma_k = J_k X_k, \tag{4.16}$$

where X_k is the same force as used earlier in this chapter. This neat form proves the central piece of Prigogine's theorem. In order to obtain the total rate of entropy production we need to sum over all the processes

$$\sigma = \sum_k \sigma_k. \tag{4.17}$$

We recognize that we already have a relation between force and flux through Onsager

$$J_k = \sum_{i=1}^{N} L_{ki} X_i. \tag{4.18}$$

So, the final form of rate of entropy production is given by

$$\sigma = \sum_{k} L_{ki} X_i X_k. \tag{4.19}$$

We next proceed to derive Prigogine's principle.

4.8 PRIGOGINE'S THEOREM ON THE MINIMUM RATE OF ENTROPY PRODUCTION

Prigogine's entropy theorem is to be regarded as the extension of Clausius' well-known entropy statement of the second law of thermodynamics that states that any nonequilibrium process of *an isolated system* increases entropy till equilibrium is reached. Thus, the entropy of an isolated system is maximum. Clausius' statement, often regarded as the second law of thermodynamics, Prigogine's theorem, sometimes regarded as the *principle of minimum entropy production*, states that in a steady state, the rate of entropy production is minimum. Note that the system is in a nonequilibrium steady state, so that thermodynamic properties (if they can be defined) are independent of time. The system can be maintained in such a state only at the expense of energy. This leads to entropy production. The minimum entropy production theorem states that the rate of entropy production is minimum.

While Clausius' second law is always valid, the minimum entropy principle is only approximate but still finds wide use.

Here we shall provide a proof of Prigogine's principle for a nonequilibrium system. As before, the steady state of the system is described by N forces X_k where $k = 1, 2, \cdots, N$. Let J_k be the fluxes corresponding to the force variables X_k. For this system, the entropy production from the formalism of irreversible or nonequilibrium thermodynamics is given by

$$\frac{dS}{dt} = \sum_{k=1}^{N} J_k X_k > 0. \tag{4.20}$$

The relation between forces X_k and fluxes J_k is given by Eq. (4.18), with $L_{ki} = L_{ik}$ as given by Onsagar's reciprocity relation. If we deal with a system that is at equilibrium with respect to one of its state variables (say, k), then the corresponding flux vanishes, i.e., the right-hand side of Eq. (4.18) becomes zero

$$J_k = \sum_{i=1}^{N} L_{ki} X_i = 0. \tag{4.21}$$

Since the entropy production is steady with respect to thermodynamic force X_k generating flux J_k

$$\frac{\partial}{\partial X_k}\left(\frac{dS}{dt}\right) = \sum_{i=1}^{N} (L_{ki} + L_{ik}) X_i = 2\sum_{i=1}^{N} L_{ki} X_i = 0. \tag{4.22}$$

Let us now consider the simple example of a one-component coupled system maintained in nonequilibrium steady state by a temperature gradient and a concentration gradient. These forces are represented by X_{th} and X_m, respectively. In the vicinity of equilibrium, the entropy production per unit time for two fluxes and two force systems is given by

$$\frac{dS}{dt} = J_{th}X_{th} + J_mX_m > 0, \tag{4.23}$$

where J_{th} and J_m are temperature and matter flux, respectively. In the above equation, the entropy production is $\frac{dS}{dt} > 0$ because we are dealing with irreversible processes. The fluxes can be written down as

$$\begin{aligned} J_{th} &= L_{11}X_{th} + L_{12}X_m \\ J_m &= L_{21}X_{th} + L_{22}X_m. \end{aligned} \tag{4.24}$$

Here L_{11} is the coefficient of thermal conductivity, L_{22} is the diffusion coefficient, L_{12} the coefficient of heat flow associated with the concentration gradient, and L_{21} is the coefficient of mass flow associated with the temperature gradient. Substituting Eqs. (4.24) in (4.23) and using Onsagar's reciprocity relation,

$$\frac{dS}{dt} = L_{11}X_{th}^2 + 2L_{12}X_{th}X_m + L_{22}X_m^2 > 0. \tag{4.25}$$

We next write the following expression that follows from above equations:

$$\frac{\partial}{\partial X_m}\left(\frac{dS}{dt}\right)_{X_{th}} = 2(L_{12}X_{th} + L_{22}X_m). \tag{4.26}$$

We now use Eq. (4.24) to obtain

$$\frac{\partial}{\partial X_m}\left(\frac{dS}{dt}\right)_{X_{th}} = 2J_m. \tag{4.27}$$

At steady state $J_m = 0$. Clearly, the two conditions $J_m = 0$ and $\frac{\partial}{\partial X_m}\left(\frac{d_iS}{dt}\right)_{X_{th}} = 0$ are completely equivalent. To show the extremum value, we again differentiate Eq. (4.26).

$$\boxed{\frac{\partial^2}{\partial X_m^2}\left(\frac{dS}{dt}\right)_{X_{th}} = 2L_{22} > 0}, \tag{4.28}$$

since L_{22} is a positive constant, thereby *proving that the extremum is a minimum.*

Thus, Eq. (4.28) is the mathematical statement of Prigogine's principle of minimum entropy production. Prigogine got the Nobel Prize in 1977 for his contribution to irreversible thermodynamics.

4.9 DISSIPATIVE STRUCTURES

Under certain conditions a system driven far from equilibrium by supply of energy and matter can undergo a self-organization to form well-defined structures or structural pattern that exhibits

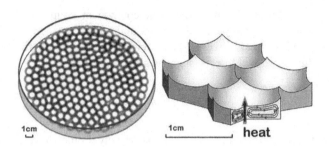

FIGURE 4.4 This figure shows hexagonal Bernard cells in an evenly heated liquid. Due to the density difference between the liquid at the top and at the bottom, gravity pulls the denser, cooler liquid to the bottom. This gravitational pulling is opposed by viscous damping force on fluid. Beyond a certain critical temperature gradient, gravitational force dominates and convection cells appear. Wikimedia CC BY 2.5.

temporal stability. These are called "dissipative structures". In theoretical analyses, such structures are found to appear through bifurcation of stable solutions of differential equations near the equilibrium regime. It has remained a popular area of research that is said to be relevant to understand living organisms. This area of research is attributed to the Brussels school of research led by Prigogine who pointed out that one can generate "order out of chaos" in a system far from equilibrium [Nicolis and Prigogine]. One often observes beautiful spatio-temporal patterns that could oscillate for a long time. The appearance of a temporally stable pattern is also studied in the field, christened *active matter*.

The most famous example is provided by the Belousov-Zhabotinski reaction. This is a class of oscillatory chemical reactions where transition metal ions catalyze the oxidation of metal complexes by bromate ion. In this competition, interesting patterns emerge in certain concentration ranges.

Mathematically, one describes formation of structures through the appearance of a new solution of nonlinear equations of motion. One example is the crossover observed in Rayleigh-Benard convention (RBC) when a fluid confined within a horizontal plate is heated from below. Initially the mechanism of heat transfer occurs via conduction or thermal diffusion. However, beyond a certain temperature gradient, a crossover occurs and one observes vortices that are known as Benrad cells. The emergence of Bernard cells is described pictorially in Figure 4.4. This is perhaps the simplest example. A readable account is available in Chandrasekhar's book on hydrodynamic stability, cited below [10].

This field experienced phenomenal growth after large-scale computing became possible in the late 1970s. Haken coined the name "Synergetics" to describe a broad class of dynamical phenomena under nonequilibrium systems.

4.10 MOLECULAR MOTIONS THAT DRIVE THE FLUX

The above arguments are all macroscopic and begs for a microscopic interpretation. As mentioned above, the equilibrium response functions are determined by the standard deviation of different types of fluctuations at equilibrium. Our dynamic response functions are also determined by the same fluctuations, but by the time correlations between these fluctuations. But what determines the amplitude and the extent of correlations among these fluctuations? Therefore, to understand the force-flux relationship we start to inquire about the details of motions that determine the values of the dynamic response functions.

In fact, the crux of the time-dependent statistical mechanics is to provide a microscopic approach to the dynamic response functions. By the term "microscopic approach" we mean that we start from

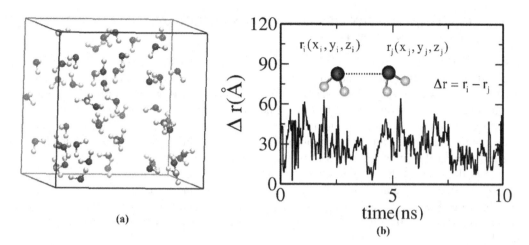

(a)

(b)

FIGURE 4.5 (a) Snapshot of the instantaneous position and orientation of a system of water molecules. (b) Relative displacement with time of two tagged water molecules is exhibited.

an intermolecular potential (or, a set of intermolecular potentials) between atoms and molecules, and the end product is the set of appropriate response functions.

We show in Figure 4.5(a) snapshot of the instantaneous position and orientation of a system of water molecules. In Figure 4.5(b) we show the displacement with time of two tagged water molecules. Both the water molecules show random motions, the displacements usually obey Gaussian distribution. These are the motions that drive the response of the liquid to time-dependent perturbations. These motions determine the rate of heat conduction, the decay of polarization when an electric field acting on the liquid from a long time in the past is removed.

The question then naturally arises: how do we relate these molecular motions to the dynamic response functions? In the case of equilibrium response functions (like specific heat), we employ methods of equilibrium statistical mechanics involving partition functions to translate molecular information (like intermolecular potential) to experimental observables. This in turn required the development of a detailed theoretical formulation. In the case of equilibrium statistical mechanics, the formulation was developed by Maxwell, Boltzmann, and Gibbs. The elegant formulation of Willard Gibbs of equilibrium statistical mechanics was partly to understand phase equilibrium and phase transitions that are driven by fluctuations.

In the case of time-dependent statistical mechanics, the formalism that we are going to study was developed by Boltzmann, Einstein, Onsager, Kubo, and others. The development of time-dependent statistical mechanics and relaxation phenomena were guided, and motivated, by a large number of experimental observations, largely pertaining to spectroscopy, like Rayleigh-Brillouin, nuclear magnetic relaxation (NMR) and infra-red (IR) spectroscopy. Subsequently, time-dependent statistical mechanics was used to understand the results of such experiments as coherent and incoherent neutron scattering, various kinds of Raman scattering, different types of experiments using non-linear optical pulses.

The results of all these experiments can be explained by using the time correlation function formulation of statistical mechanics, which has its origin in the linear response theory, which in turn originated from Onsager's regression hypothesis. The validity of both is tied down to the stability of the system at equilibrium to small perturbations, which can be described by the existence of a set of springs or harmonic force constants. At a molecular level, these springs are given by energy or entropy fluctuations.

However, before we embark on the relationship between theory and experiments, we would do well to learn Einstein's theory of Brownian motion.

4.11   EINSTEIN'S THEORY OF BROWNIAN MOTION AND RELATION WITH REGRESSION HYPOTHESIS

Long before Onsager's work, Einstein in the early twentieth century embarked on an analysis on the origin of the incessant motion of pollens suspended on the surface of the water and observed by botanist Robert Brown by using an optical microscope. This is known as the Brownian motion. Einstein demonstrated that the diffusion constant of the pollen, treated as a large spherical particle, can be described by the viscosity of the solution. This Brownian motion is essentially a random walk well-known in the mathematical theory of probability. The relationship between the regression hypothesis and reciprocal relation is that this Brownian motion gives rise to relaxation through the natural motion of the system. Thus, we need not consider any perturbation in the phase space motion of the systems from its natural equilibrium dynamics at rest. The time correlation function expressions essential to understand regression hypothesis are all evaluated at equilibrium. This simple fact is critical to understand time-dependent statistical mechanics and shall be described later in more detail.

4.12   SUMMARY

One could very well argue that the hydrodynamic relation between the force on a particle and the resulting velocity that introduces the friction coefficient could be regarded as the beginning of TDSM. The work of Einstein on Brownian motion that appeared in 1905 can also be considered as a major turning point in our thinking about random processes. However, the real turning point in our thinking came with the work on Onsager's regression hypothesis and reciprocal relations.

Somewhat paradoxically, the area of study and research of irreversible thermodynamics remained rather separated from the main area of time-dependent statistical mechanics, primarily because it has been hard to carry out detailed first principles calculations of the relaxation rates and response function in the steady state far from equilibrium. In fact, it is still not possible to calculate those properties in systems far from equilibrium, from a microscopic theory. The strength of Onsager's theory arises from the validity of microscopic reversibility.

However, one is uncomfortable with the definition of entropy in a nonequilibrium state.

There have been notable simulation studies on nonequilibrium states under steady flow, such as Couette flow. These studies have explored the relationship between shear viscosity and shear strain rate, with the interesting result that the frequency dependence of viscosity shows a power-law dependence on the strain rate. Zwanzig argued that the behavior is related to long-time tails in the relevant time correlation functions, observed at equilibrium, although the shear strain result was observed in the steady state. The result demonstrates that relations derived at equilibrium are sometimes admissible at not-far-from-equilibrium situations. However, this conclusion should be regarded with caution while applying to realistic and much more complex situations.

The other current area of development is the field of active matter where a coherent state far from equilibrium is created by the supply of energy. This has been applied to explain the observed coherent motion among the flock of birds. This has again been explained by using ideas from time-dependent statistical mechanics and phase transition [Vicsek model]. We shall discuss this in Chapter 24.

Finally, students should note this remains a subject of great importance, as demonstrated by two single Nobel Prizes awarded to this area. However, progress has been slow, the field is hard, and a lot remains to be done.

BIBLIOGRAPHY

1. Onsager, L. 1931. Reciprocal relations in irreversible processes. I. *Physical Review*, *37*(4), 405; Onsager, L. 1931. Reciprocal relations in irreversible processes. II. *Physical Review*, *38*(12), 2265.
2. De Groot, S. R., and P. Mazur. 2013. *Non-Equilibrium Thermodynamics*. Chelmsford: Courier Corporation.
3. Landau, L. D. and E. M. Lifshitz. 1969. *Statistical Physics*. London: Pergamon Press.
4. Zwanzig, R. 2001 *Non-Equilibrium Statistical Mechanics*. Oxford: Oxford University Press.
5. Prigogine, I. 1967. *Introduction to Thermodynamics of Irreversible Processes*, 3rd ed. New York: John Wiley & Sons.
6. Wigner, E P. 1954. Derivations of Onsager's reciprocal relations. *Journal of Chemical Physics, 22*, 191–215.
7. Prigogine, I, 1977. Time. Structure and Fluctuations, *Nobel Prize Lecture* www.nobelprize.org/prizes/chemistry/1977/prigogine/lecture/.
8. Prigogine, I and G. Nicolis. 1985. *Self-Organisation in Nonequilibrium Systems: Towards A Dynamics of Complexity*. London: Wiley.
9. Zwanzig, R. 1981. Nonlinear shear viscosity and long-time tails. *Proceedings of the National Academy of Sciences USA, 78*, 3296.
10. Chandrasekhar, S. 1961. *Hydrodynamic and Hydromagnetic Stability*. Oxford: Oxford University Press; also by New York: Dover Publications 1970.
11. Haken, H., 1983, *Advanced Synergetics: Instability Hierarchies of Self-Organizing Systems and Devices*. Berlin: Springer-Verlag.
12. De Groot, S. R. 1951. *Thermodynamics of Irreversible Processes*. Amsterdam: North-Holland Publishing Comp.

5 Hydrodynamic Approach to Relaxation Phenomena

OVERVIEW

It might come as a surprise to some, but no study of time-dependent statistical mechanics (TDSM) is complete, or can progress, without a basic understanding of hydrodynamics. Hydrodynamics evolved in the early nineteenth century through the efforts of French mathematicians Bernoulli, Navier, and others who realized that large-scale fluid flows could be described by equations based on the conservation laws of mass, momentum, and energy. Hydrodynamics makes the assumption that these globally conserved quantities are also locally conserved and their fluctuations from average values decay slowly, as particles or energy cannot be created or destroyed. In order to obtain these decay rates, the conservation conditions are required to be supplemented by certain constitutive relations that bring in the material transport properties, like viscosity, thermal diffusivity, and so on. When solved, hydrodynamic equations provide a large number of analytical results that can be tested directly against experiments. These results have served the development of TDSM in a very essential way, as described throughout this book.

5.1 INTRODUCTION

The name "hydrodynamics" implies the dynamics of water. Thus, the subject was initially developed to understand the flow of liquid water. This in turn was motivated by flows of water in rivers and oceans as those were the daily companions of mankind from ancient times. Later the more general name fluid mechanics was coined to describe the flow of fluids, which plays an important role in understanding aerodynamics and is routinely used to study flows around aeroplane, missile, wind flows, weather prediction, and many others.

As discussed in greater detail below, strictly speaking, hydrodynamics is valid at large-length (much larger than a molecular diameter) and long-time (much longer than the time between two consecutive collisions that a molecule suffers in liquids or a dense gas) limits. Nevertheless, many hydrodynamic predictions are routinely used to describe phenomena and properties at a molecular level. For example, hydrodynamic estimates of translational and rotational friction of molecular solutes are found to agree rather well with experimental and simulation results.

There are certain essential concepts that need to be understood in this subject.

(i) First is the concept of a fluid point. This assumes a region that is much larger than the size of a molecule but much smaller than any macroscopic length scale, like 1 mm. This is of course a wide window. Fortunately, a detailed prescription of the dimension of the fluid point is not essential, and this measure does not enter in the final result, although present implicitly.

DOI: 10.1201/9781003157601-6

(ii) Hydrodynamics assumes that the globally conserved quantities, like number, momentum, and energy are also conserved locally. *Because of the conservation constraint, fluctuations in these variables relax slowly back to their respective equilibrium values.* If we again imagine this relaxation occurring at a fluid point, then the time-dependent changes occur through fluxes. We then make connections of these fluxes with phenomenological laws. One example of this exercise is Fick's law.

(iii) The third point to remember is the aim of hydrodynamics and its evolving objectives. Let us consider the two main results that we routinely use in TDSM. (a) The relation between friction on a moving sphere and the solvent viscosity. Here viscosity enters through a set of constitutive relations, described below, and friction is calculated by the distortion to the flow field around the sphere that causes energy to dissipate. This leads to the familiar expression $\zeta = 6\pi\eta R$ where ζ is the friction, η is the viscosity, and R is the radius of the sphere. (b) The second major achievement is the expression for the Rayleigh-Brillouin spectrum, which is the frequency dependence of the intensity of light scattered by a liquid. It can be shown that this intensity is proportional to the Laplace-Fourier transform of the density-density space-time correlation function. The spectrum is thus the same as the dynamic structure factor, $S(k, \omega)$. The Rayleigh Brillouin spectrum (RBS) of a liquid has a trimodal structure, with the three peaks providing a large number of important transport and thermodynamic properties like sound attenuation coefficient, thermal diffusivity, bulk viscosity. An accurate expression of RBS was a major success of hydrodynamics. However, our point of contention is that the scattering process involved momentum exchange at a very small wavenumber or large wavelength which is that of visible light, that is wavelength of 5000–10,000 cm^{-1}.

(iv) This brings us to the fourth point. Hydrodynamics does not include any information about intermolecular interactions. It is accurate at large-length and long-time scales. Thus it does include information about the size and the shape of the particle on which the friction due to the solvent is exerted. Some aspects of solute-solvent interaction are also included through the boundary conditions at the surface of the particle.

(v) Thus, hydrodynamics in opposite of the kinetic theory of gases that starts with intermolecular interactions, in the form of binary collisions.

(vi) Dissipation (loss of energy and/or momentum) enters in hydrodynamic description through transport coefficients like viscosity and thermal conductivity but hydrodynamics has no way to calculate them. They are obtained through a kinetic theory approach.

(vii) Hydrodynamics provides expression for friction (in terms of viscosity), and also, of special interest to this book, time correlation functions. In this sense, it bears certain resemblance to thermodynamics.

We shall have occasions to comment on the scope and validity of hydrodynamics throughout this chapter, about the length and time scales involved. Despite any limitations hydrodynamics could have, it is a highly useful, and in many ways, highly elegant subject based on broad principles of science.

Hydrodynamic equations are built in two steps. First is the continuity equation that follows from the local conservation of number, momentum, and energy. This is a set of five equations that have been described below. These equations connect time derivatives of these conserved quantities to respective fluxes. In the second step use certain relations between the fluxes and material and thermodynamic properties. These relations are based on general symmetry properties, and although general, they are approximate in the sense that they are valid under certain conditions, to be discussed below. As discussed elsewhere, both the steps involve approximations but are generally valid at large-length and long-time scales.

5.2 BASIC CONSIDERATIONS: LENGTH AND TIME SCALES

Hydrodynamics provides a systematic way to look into the dynamics of liquid but with a macroscopic approach, unlike the phenomenological theories, which may claim to be microscopic but are often *ad hoc*. Large-scale response in liquids to a given perturbation involve transport of mass, momentum, and energy. Examples are the flow of water due to gravitational forces or tidal waves. Understanding these relaxation processes requires a description of the change in local properties of the liquid. However, the flow of a liquid must obey the conservation laws. As already mentioned above, "local" in hydrodynamics is not microscopically defined – it encompasses a broader region in space.

At such long-length scales, the main hydrodynamic variables (density, momentum, energy) are globally conserved. It should be noted that the dynamics of liquids at short-length scales (over one or two molecular diameters) are different from that of larger-length scales because momentum and energy conservations need not constrain dynamics at small distances. The number however continues to be a conserved quantity at all length scales. This rather unusual observation has a deep impact, as we shall discuss later. Several experiments, such as light scattering, can access the long-wavelength dynamics accurately and easily.

It is important to understand quantitatively the time scales involved here. We can easily estimate the time scales relevant to liquid relaxation dynamics involving number *density* (ρ), *momentum* (g), and *energy* (or *heat*) (H). The respective time scales (τ) are given by the following equations:

$$\tau_\rho = \sigma^2/6D_s$$
$$\tau_g = \mu/\zeta \tag{5.1}$$
$$\tau_H = \sigma^2/D_t$$

Here, D_s and D_t are the self- and heat diffusion coefficients, μ is the effective mass, σ is the molecular diameter, and ζ is the friction coefficient. For typical values of the involved variables, in the case of water, for example, $\tau_\rho \sim 100$ ps, $\tau_g \sim 0.01$ ps, and $\tau_H \sim 0.1$ ps. *Hence density relaxation is often the slowest process creating the bottle-neck for the concerned relaxation.*

5.3 CONSERVATION EQUATIONS AND CONSTITUTIVE RELATIONS

We now derive the hydrodynamic equations of motion for density, momentum, and energy. The first step in this maneuver is the conservation laws. These are given by the continuity theorem, also known as continuity equations. This derivation of this equation is available in elementary textbooks but we repeat the same here for the sake of completeness.

5.3.1 CONTINUITY THEOREM/EQUATION

Continuity theorem, also referred to as continuity equation, is a mathematical statement of the conservation laws, but translated to the local level. It also constitutes the most basic assumption of hydrodynamics: *the globally conserved quantities are also locally conserved.*

Let us consider a property A, which is locally conserved. Let us consider a local variation or fluctuation of this property from the average value. We define a local density of this quantity and denote it by $A(\mathbf{r},t)$. For any arbitrary volume V, we can then write

$$A(t) = \int_V d^3r\, A(\mathbf{r},t) \tag{5.2}$$

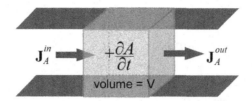

FIGURE 5.1 Schematic representation of the change in the density term $A(r,t)$ within an arbitrary volume V. The rate of change of $A(r,t)$ within V is balanced by the vector sum of the inward and outward fluxes of the property.

where d^3r is an infinitesimal volume element and the total time-dependent property is $A(t)$. The strategy is clear. The rate of change of the total property can be expressed as,

$$\frac{dA(t)}{dt} = \int_V d^3r \frac{dA(\mathbf{r},t)}{dt} \tag{5.3}$$

The total time derivative can consist of only two terms. One is the local increase or decrease. As there is no sink or source (due to conservation), the local change in a small volume element must be balanced by an overall, net, inward-outward flow or flux of the property. The rate of A passing normally through a unit area is known as its flux, $\mathbf{J}_A(t)$. The vector sum of the inward and the outward flux nullifies the rate of change of density to yield the conservation condition.

As shown in Figure 5.1, consider a cube of arbitrary volume V. If the rate of change of the density $A(\mathbf{r},t)$ within this volume is equal to the total flux in and out of the volume so that the total property is conserved. Mathematically, this can be expressed as Eq. (5.4).

$$\boxed{\frac{\partial A(\mathbf{r},t)}{\partial t} + \nabla.\mathbf{J}_A(\mathbf{r},t) = 0} \tag{5.4}$$

This is known as the *equation of continuity* and also as the *continuity theorem*. This equation implies that time-dependent change in A can take place only by inward or outward flow through the surface of volume V. Particles or energy cannot be created or destroyed.

For proper hydrodynamic descriptions, the basic conserved properties are (i) the number (or mass) density (ρ); (ii) the momentum density (g); and (iii) the energy density (e). This discussion becomes clearer by comparing this situation to a nonconserved property. For example, if a chemical reaction is going on inside volume V, the concentration c of a reactant is a nonconserved quantity. In that case, $\sigma_c(\mathbf{r},t) \neq 0$. However, this change can easily be accommodated through the law of mass action.

Note that if $\mathbf{u}(\mathbf{r}, t)$ be the velocity of the fluid at $(\mathbf{r}, t)$, and A is carried only by the convective flow of the fluid, we can write the following *relation*:

$$\mathbf{J}_A(\mathbf{r},t) = A(\mathbf{r},t)\mathbf{u}(\mathbf{r},t). \tag{5.5}$$

Substituting this into Eq. (5.4) gives us the closed relaxation equation for the property. However, the situation is a bit more complex.

The explicit form of the conservation equations are given in the form of Eq. (5.4).

$$\frac{\partial \rho(\mathbf{r},t)}{\partial t} + \nabla . \mathbf{J}_\rho (\mathbf{r},t) = 0 \tag{5.6}$$

$$\frac{\partial \mathbf{g}_i (\mathbf{r},t)}{\partial t} + \nabla_j \tau_{ij} (\mathbf{r},t) = 0 \quad (i = 1,2,3) \tag{5.7}$$

$$\frac{\partial e(\mathbf{r},t)}{\partial t} + \nabla . \mathbf{J}_e (\mathbf{r},t) = 0. \tag{5.8}$$

Here, $\mathbf{J}_\rho(\mathbf{r}, t)$, $\tau_{ij}(\mathbf{r}, t)$, and $\mathbf{J}_e(\mathbf{r}, t)$ are the local fluxes of number, momentum, and energy, respectively. ∇_j is the jth component of the gradient operator and τ_{ij} is the jth component of flux of the ith component of momentum. *Repeated indices denote summation using Einstein's convention.* We note that these equations are assumed to apply to both macroscopic and locally average densities.

In order to derive the relaxation equations for these variables, we need to close the above equations by supplying these with appropriate relations. These relations are called the constitutive relations. We now proceed to discuss these relations.

5.3.2 CONSTITUTIVE RELATIONS

Note that the number of particles in the volume V can change only by the virtue of convection current of the fluid. If the velocity field of the fluid at $(\mathbf{r}, t)$ is $\mathbf{u}(\mathbf{r}, t)$, the constitutive relation for number density is [see Eq. (5.5)]

$$\mathbf{J}_\rho (\mathbf{r},t) = \rho(\mathbf{r},t)\mathbf{u}(\mathbf{r},t). \tag{5.9}$$

For momentum and energy, however, the change in effect is not only due to convection. Besides convection, the change of momentum within the volume V is additionally influenced by the force (or, stress) exerted by the fluid outside the volume. If $\boldsymbol{\sigma}$ is the stress tensor, then the total resultant flux for momentum is given by the following constitutive relation:

$$\tau_{ij} (\mathbf{r},t) = m\rho(\mathbf{r},t)u_i (\mathbf{r},t)u_j (\mathbf{r},t) - \sigma_{ij} (\mathbf{r},t). \tag{5.10}$$

The first term is the convection term and the second term is due to the stress tensor. Note that the first term is at least quadratic in fluctuation. Stress causes flow as it is a force acting on a fluid element. We shall specify it a bit later.

For the change in energy, there are two new terms: (i) work done by the fluid outside V and (ii) thermal diffusion. If $\mathbf{Q}$ be the heat diffusive flux, then the rate of change of heat content inside the fluid element due to thermal diffusion can be written as $\int_V d^3r \nabla . \mathbf{Q}$. Hence, the energy flux becomes

$$\mathbf{J}_e (\mathbf{r},t) = e(\mathbf{r},t)\mathbf{u}(\mathbf{r},t) + \mathbf{Q}(\mathbf{r},t) - \mathbf{u}(\mathbf{r},t).\boldsymbol{\sigma}(\mathbf{r},t). \tag{5.11}$$

Hence, when we combined the continuity equations with the constitutive equations, we obtain the following system of equation:

$$\frac{\partial\rho(\mathbf{r},t)}{\partial t}+\nabla.\rho(\mathbf{r},t)\mathbf{u}(\mathbf{r},t)=0 \tag{5.12}$$

$$\frac{\partial\mathbf{g}_i(\mathbf{r},t)}{\partial t}+\nabla.\left(m\rho(\mathbf{r},t)u_i(\mathbf{r},t)u_j(\mathbf{r},t)-\sigma_{ij}(\mathbf{r},t)\right)=0 \tag{5.13}$$

$$\frac{\partial e(\mathbf{r},t)}{\partial t}+\nabla.\left(e(\mathbf{r},t)\mathbf{u}(\mathbf{r},t)+\mathbf{Q}(\mathbf{r},t)-\mathbf{u}(\mathbf{r},t).\boldsymbol{\sigma}(\mathbf{r},t)\right)=0. \tag{5.14}$$

However, the constitutive relations for momentum and energy densities are incomplete till now as we still need to determine the stress tensor and the thermal diffusive flux. We next proceed to carry out this step.

We know that if the force acting on the surface element $d\mathbf{S}$ is $d\mathbf{F}$, then we can use stress tensor that can be defined as

$$d\mathbf{F} = \boldsymbol{\sigma}.d\mathbf{S}. \tag{5.15}$$

This can be written in the matrix form as

$$\begin{pmatrix} dF_x \\ dF_y \\ dF_z \end{pmatrix} = \begin{pmatrix} \sigma_{xx} & \sigma_{xy} & \sigma_{xz} \\ \sigma_{yx} & \sigma_{yy} & \sigma_{yz} \\ \sigma_{zx} & \sigma_{zy} & \sigma_{zz} \end{pmatrix} \times \begin{pmatrix} dS_x \\ dS_y \\ dS_z \end{pmatrix}. \tag{5.16}$$

The diagonal elements of the stress tensor define forces per unit area acting perpendicularly (or normally) on the respective planes (x, y, or z). Hence these three are called the *normal stresses and they are related to pressure*. The off-diagonal terms are, however, forces per unit area acting parallel to their respective planes of reference. These are the *shear stress* components. For a fluid in equilibrium the shear stresses are zero and the three normal stresses are equal. If p_0 is the equilibrium pressure on the fluid then we may write, for the diagonal terms, the equality $\sigma_{xx} = \sigma_{yy} = \sigma_{zz} = -p_0$. In general, the hydrostatic pressure $p(\mathbf{r},t)$ can be defined as the force per unit area averaged over three mutually orthogonal planes through the point $\mathbf{r}$ at time t, that is

$$p(\mathbf{r},t) = -\frac{1}{3}Tr\boldsymbol{\sigma}(\mathbf{r},t). \tag{5.17}$$

Apart from this pressure part, the symmetric stress tensor has a viscous part σ'_{ij}. Hence the total stress tensor can be written as

$$\sigma_{ij}(\mathbf{r},t) = -p(\mathbf{r},t)\delta_{ij} + \sigma'_{ij}(\mathbf{r},t). \tag{5.18}$$

Determination of $\sigma'_{ij}(\mathbf{r},t)$ is an important and nontrivial part of forming the hydrodynamic equations. First, note that the momentum flux can be induced by the presence of a gradient in the pressure because such a gradient shall induce viscous flow. If we assume that the magnitude of the flux is proportional to the gradient, we then can write the contribution of the gradient term as proportional to $\nabla_i u_j$ and the proportionality constant is the shear viscosity η_s. However, our task is not fully complete yet. The stress tensor should be invariant to a rotation that can exchange the coordinate system. This is accomplished by including the term $\nabla_j u_i$. Next we note that the diagonal stress terms may cause a flow through motion of fluid particles towards each other, resulting in the change of volume, as the liquid is compressible. This brings in a dissipation term through bulk viscosity. As we follow Einstein's convention, we can write it as $\eta_v \nabla.\mathbf{u} \delta_{ij}$. Finally we note that these diagonal compression terms need to be excluded from the shear terms. For this, we note that we have two diagonal terms $\nabla_i u_i$ from the sum of the two symmetrized terms, so that rotational invariance is preserved. In addition, if we make a vector dot product, we would get three terms, which should be identical because of the isotropy of the liquid. Thus, we need to subtract this appropriately.

When we combine all these terms, we obtain the following expression for the stress tensor term:

$$\sigma'_{ij} = \eta_s \left(\nabla_i u_j + \nabla_j u_i - \frac{2}{3} \nabla.\mathbf{u} \delta_{ij} \right) + \eta_v \nabla.\mathbf{u} \delta_{ij}. \tag{5.19}$$

As mentioned, η_s and η_v are the shear and bulk viscosities, respectively. In fact, these expressions can be regarded as providing the definitions of shear and bulk viscosity.

Use of Eqs. (5.18) and (5.19) into the constitutive relation Eq. (5.10), we get

$$\tau_{ij} = m\rho u_i u_j + p \delta_{ij} - \eta_s \left(\nabla_i u_j + \nabla_j u_i - \frac{2}{3} \nabla.\mathbf{u} \delta_{ij} \right) - \eta_v \nabla.\mathbf{u} \delta_{ij}. \tag{5.20}$$

Note that gradient of the stress tensor involves double gradient. This gives rise to both transverse and longitudinal terms in current.

We now turn to the discussion of the constitutive relation for the energy current. According to Fourier's law, heat flows in a direction opposite to the temperature gradient. Hence the energy flux $\mathbf{Q}$ is given by Fourier law,

$$\mathbf{Q} = -\lambda \nabla T \tag{5.21}$$

where λ is the thermal conductivity. We next divide the local energy density into a kinetic part $\left(\frac{1}{2} m\rho u^2 \right)$ and a potential part (e') and substitute all the terms in Eq. (5.11) to obtain the constitutive relation for energy flux as

$$\mathbf{J}_e = \left(\frac{1}{2} m\rho u^2 + e' \right) \mathbf{u} + \mathbf{u}.\sigma - \lambda \nabla T \tag{5.22}$$

where σ is given by Eqs. (5.18) and (5.19). Here the first term is due to convection of energy, and the last term is due to conduction. The middle term on the right-hand side is to account for the work done per unit time due to stress.

5.4 ASSUMPTION OF LOCAL THERMODYNAMIC EQUILIBRIUM AND LINEARIZATION OF HYDRODYNAMIC EQUATIONS

In this section, we perform two more steps required to arrive at the final hydrodynamic equations. They involve (a) the reduction in the number of unknowns in the above equations, and (b) the linearization of the nonlinear equations of momentum and energy.

(a) The above relations in Eqs. (5.12), (5.13) give five fundamental equations of fluid mechanics. These five equations, however, involve seven unknown variables: ρ, $\mathbf{g}$, e, p and T ($\mathbf{g} = \{g_x, g_y, g_z\}$). It is not possible to solve for seven unknowns using five equations. Hence, further mathematical maneuvers are required to solve these relaxation equations. This is accomplished by making the assumption of "*local thermodynamic equilibrium*". This requires us to assume that thermodynamics continues to hold even in the presence of hydrodynamic flows. The flows in turn are assumed to be small fluctuations. *Thus, this step and the subsequent linearization step go hand-in-hand, as described below. In essence, one assumes that for small fluctuations, one can use thermodynamic relations to eliminate two (let us say, entropy and pressure) in favor of density and temperature.* But before that step, let us discuss the linearization.

(b) We define the fluctuation in a property A as

$$A_1^i(\mathbf{r},t) = A^i(\mathbf{r},t) - A_0^i. \tag{5.23}$$

A_0^i is the equilibrium mean value for the ith hydrodynamic mode. Considering the fluctuations of the seven said variables to be small around the equilibrium means ($A_0^i = 0$), we can substitute $A^i = A_1^i + A_0^i$ to obtain the linearized equations of fluid mechanics. Here we need to retain only the first-order terms. We explain this assumption with a few examples.

$$e = \frac{1}{2}m\rho u^2 + e' \approx e' \tag{5.24}$$

since $u = u_1 + u_0$ and $u_0 = 0$ leaves us with u_1^2, which is a second-order term, which we can neglect. Hence we need to consider the fluctuation in internal energy only. Next,

$$\begin{aligned}\rho\mathbf{u} &= (\rho_1 + \rho_0)(\mathbf{u}_1 + \mathbf{u}_0)\\ &= \rho_1\mathbf{u}_1 + \rho_1\mathbf{u}_0 + \rho_0\mathbf{u}_1 + \rho_0\mathbf{u}_0\\ &\approx \rho_0\mathbf{u}_1\end{aligned} \tag{5.25}$$

Since, $\mathbf{u}_0 = 0$ and we can neglect $\rho_1\mathbf{u}_1$ as it is a second-order term. Using these assumptions in Eqs. (5.12), (5.13), and (5.14), we get finally the linearized equations of fluid dynamics as

$$\frac{\partial \rho_1}{\partial t} + \rho_0 \nabla.\mathbf{u}_1 = 0 \tag{5.26}$$

$$m\rho_0 \frac{\partial \mathbf{u}_1}{\partial t} = -\nabla p_1 + \eta_s \nabla^2\mathbf{u}_1 + \left(\eta_v + \frac{1}{3}\eta_s\right)\nabla(\nabla.\mathbf{u}_1) \tag{5.27}$$

$$\frac{\partial e_1}{\partial t} + (e_0 + p_0)\nabla.\mathbf{u}_1 = \lambda\nabla^2 T_1. \tag{5.28}$$

Defining the fluctuation in heat density (or enthalpy density) $h_1 = e_1 - (e_0 + p_0)\dfrac{\rho_1}{\rho_0} = T_0 s_1$

(T_0 = equilibrium temperature and s_1 = entropy density fluctuation), we can rewrite Eq. (5.28) as

$$\frac{\partial h_1}{\partial t} = T_0 \frac{\partial s_1}{\partial t} = \lambda\nabla^2 T_1. \tag{5.29}$$

From Eqs. (5.26), (5.27), and (5.29), we have five linear hydrodynamic equations involving seven fluctuations terms (ρ_1, u_{1x}, u_{1y}, u_{1z}, p_1, s_1, T_1). We proceed by eliminating two of the four scalar variables using equilibrium thermodynamic equations of state. Here, we choose ρ_1 and T_1 to eliminate p_1 and s_1. Although any other pair may be chosen. Considering ρ_1 and T_1 as the independent variables we can express s_1 and p_1 as functions of the former pair. This allows us to use the assumptions of local equilibrium to write

$$s_1 = \left(\frac{\partial s}{\partial \rho}\right)_T \rho_1 + \left(\frac{\partial s}{\partial T}\right)_\rho T_1 \tag{5.30}$$

$$p_1 = \left(\frac{\partial p}{\partial \rho}\right)_T \rho_1 + \left(\frac{\partial p}{\partial T}\right)_\rho T_1. \tag{5.31}$$

These partial derivatives can be written in terms of well-known thermodynamic identities. A thorough description of these identities is omitted here. The details can be found in any book on equilibrium thermodynamics. After some trivial mathematical manipulations, we get

$$s_1 = -\frac{m\rho_0 c_V}{T_0}\left(\frac{\gamma-1}{\alpha\rho_0}\rho_1 - T_1\right) \tag{5.32}$$

$$p_1 = mc_T^2\left(\rho_1 + \alpha\rho_0 T_1\right). \tag{5.33}$$

In the next step we separate the current into longitudinal and transverse components. This is achieved by noting that in Eq. (6.27) we have the double gradient term, which on Fourier transformation give rise to the tensor **kk**. This is decomposed into components longitudinal (which is parallel to the vector **k**) and transverse (perpendicular to **k**). When this procedure is carried out we find that the five hydrodynamic equations decouple into a set of three equations those containing the longitudinal current and the density, and two separate identical equations for the transverse components.

In the following we continue with the set of three equations that contain density and longitudinal current terms.

Substituting these in Eqs. (5.26), (5.27), and (5.29) and expressing

$$\psi_1 = \nabla.\mathbf{u}_1 \tag{5.34}$$

we get the following three linear partial differential equations with three unknowns (ρ_1, ψ_1, T_1).

$$\frac{\partial \rho_1}{\partial t} + \rho_0 \psi_1 = 0 \tag{5.35}$$

$$\left(\frac{\partial}{\partial t} - D_V \nabla^2\right)\psi_1 + \left(\frac{c_T^2}{\rho_0}\nabla^2 + \alpha c_T^2 \nabla^2\right)\rho_1 = 0 \tag{5.36}$$

$$\left(\frac{\partial}{\partial t} - \gamma D_T \nabla^2\right)T_1 - \frac{\gamma-1}{\alpha\rho_0}\frac{\partial}{\partial t}\rho_1 = 0 \tag{5.37}$$

where

$$D_V = \frac{\left(\eta_v + \frac{4}{3}\eta_s\right)}{m\rho_0} = \text{longitudinal kinematic viscosity} \qquad (5.38)$$

$$D_T = \frac{\lambda}{m\rho_0 c_P} = \text{thermal diffusivity.} \qquad (5.39)$$

We recommend the books by Landau and Lifshitz (*Fluid Mechanics*) and Lamb for more details.

5.5 DERIVATION OF NAVIER-STOKES EQUATIONS FROM BOLTZMANN KINETIC EQUATION

We discussed Boltzmann's kinetic equation in Chapter 6. It was also discussed that one can derive the Navier-Stokes equation and higher-order hydrodynamic equations like Burnett and super-Burnett from Boltzmann's equation. As also discussed in Chapter 6, the derivation requires the introduction of two ingredients (at least). First, one introduced the collision invariants, which are essentially the conserved variables: the density, the momenta, and the energy. In the second step, one uses the Hilbert approach of expanding the single-particle phase space (the μ-space) density $f(r,p,t)$ in a series expansion where the first term corresponds to the local equilibrium. This first term leads to the Navier-Stokes equation.

This derivation can be found in advanced texts on the kinetic theory of gases. The references to such texts have been given in Chapter 6. An instructive aspect of this derivation is the clarity that it brings that helps understanding the limitations of Navier-Stokes equations.

5.6 SOLUTION OF LINEARIZED THERMODYNAMICS AND THE RAYLEIGH-BRILLOUIN SPECTRUM

The Rayleigh-Brillouin spectrum is the scattering of light by spontaneous density fluctuations in the liquid. Since the light scattered is of large wavelength, these density fluctuations are also of long wavelength and hence can be treated by hydrodynamics, as described below.

The scattering of light happens due to density fluctuations in dense liquids that introduce varying refractive index. The scattering spectra in such systems give us important information about dynamical, and also thermodynamic properties in the long wavelength regions.

Now Eqs. (5.35), (5.36), and (5.37) form a set of closed linear equations, which can be conveniently solved in the Fourier-Laplace plane. The Fourier-Laplace transform of number density is shown in Eq. (5.40). Similar transforms are applied to momentum and energy densities.

$$\tilde{\rho}_1(\mathbf{q},z) = \int_0^\infty dt\, e^{-zt} \int d^3r\, e^{i\mathbf{q}\cdot\mathbf{r}} \rho_1(\mathbf{r},t) = \int_0^\infty dt\, e^{-zt}\rho_1(\mathbf{q},t). \qquad (5.40)$$

Here, $\tilde{\rho}_1(\mathbf{q},z)$ is the spatio-temporal Fourier-Laplace transform of $\rho_1(\mathbf{r},t)$. We thus transform these equations and then write the corresponding Fourier-Laplace forms as follows:

$$\tilde{\mathbf{M}}(\mathbf{q},z).\tilde{\phi}(\mathbf{q},z) = \mathbf{N}(\mathbf{q}).\phi(\mathbf{q}). \qquad (5.41)$$

Here the matrix elements are given by

$$\tilde{\mathbf{M}}(\mathbf{q},z) = \begin{vmatrix} z & \rho_0 & 0 \\ \dfrac{-\omega^2(q)}{\gamma\rho_0} & (z+D_V q^2) & \dfrac{-\alpha\omega^2(q)}{\gamma} \\ \dfrac{-(\gamma-1)z}{\alpha\rho_0} & 0 & (z+\gamma D_T q^2) \end{vmatrix} \tag{5.42}$$

$$\tilde{\phi}(\mathbf{q},z) = \begin{pmatrix} \tilde{\rho}_1(\mathbf{q},z) \\ \tilde{\psi}_1(\mathbf{q},z) \\ \tilde{T}_1(\mathbf{q},z) \end{pmatrix} \tag{5.43}$$

$$\mathbf{N}(\mathbf{q}) = \begin{vmatrix} 1 & 0 & 0 \\ 0 & 1 & 1 \\ \dfrac{-(\gamma-1)}{\alpha\rho_0} & 0 & 1 \end{vmatrix} \tag{5.44}$$

$$\phi(\mathbf{q}) = \begin{pmatrix} \rho_1(\mathbf{q},0) \\ \psi_1(\mathbf{q},0) \\ T_1(\mathbf{q},0) \end{pmatrix} \tag{5.45}$$

Here $\omega(q) = c_s q$, if $\tilde{\mathbf{M}}(z)$ is the determinant of the matrix $\tilde{\mathbf{M}}(\mathbf{q},z)$, then the solution of Eqs. (5.41) can be obtained (by the standard procedure) in the following form:

$$\begin{pmatrix} \tilde{\rho}_1(\mathbf{q},z) \\ \tilde{\psi}_1(\mathbf{q},z) \\ \tilde{T}_1(\mathbf{q},z) \end{pmatrix} = \frac{1}{\tilde{\mathbf{M}}(z)} \begin{pmatrix} \left\{ \begin{matrix}(z+D_V q^2)(z+\gamma D_T q^2)\\ +(\gamma-1)\omega(q)/\gamma\end{matrix}\right\} & \left\{ -\rho_0(z+\gamma D_T q^2) \right\} & \left\{ -\dfrac{\alpha\rho_0\omega^2(q)}{\gamma} \right\} \\ \left\{ \dfrac{\omega^2(q)}{\gamma\rho_0}(z+\gamma D_T q^2) \right\} & \left\{ z(z+\gamma D_T q^2) \right\} & \left\{ \dfrac{\alpha\omega^2(q)z}{\gamma} \right\} \\ \left\{ -\dfrac{(\gamma-1)\omega^2(q)}{\alpha\gamma\rho_0} \right\} & \left\{ -\dfrac{(\gamma-1)z}{\alpha} \right\} & \left\{ \begin{matrix}z(z+q^2 D_V)\\ +\dfrac{\omega^2(q)}{\gamma}\end{matrix} \right\} \end{pmatrix} \begin{pmatrix} \rho_1(\mathbf{q},0) \\ \psi_1(\mathbf{q},0) \\ T_1(\mathbf{q},0) \end{pmatrix}. \tag{5.46}$$

These three equations are used to obtain the correlation functions of hydrodynamic variables. These correlation functions are obtained from Rayleigh-Brillouin spectroscopy. In Eq. (5.46), the longitudinal modes are all coupled together. These give rise to the dynamic structure factor $S(\mathbf{q},z)$ in the hydrodynamic limit. $S(\mathbf{q},z)$ is the spatio-temporal correlation of density fluctuations in the Laplace-Fourier plane.

$$S(\mathbf{q},z) = \int dt\, e^{izt} \langle \rho(\mathbf{q},t)\rho(-\mathbf{q},0) \rangle. \tag{5.47}$$

The expression for dynamic structure factor can be derived from the hydrodynamic matrix Eq. (5.46) in the following well-known form:

$$S(q,z) = \frac{1}{2\pi} S(q) \left[\left(\frac{\gamma-1}{\gamma} \right) \frac{2D_T q^2}{z^2 + (D_T q^2)^2} \right.$$

$$+ \frac{1}{\gamma} \left\{ \frac{\Gamma q^2}{(z + c_s q)^2 + (\Gamma q^2)^2} \right.$$

$$\left. \left. + \frac{\Gamma q^2}{(z - c_s q)^2 + (\Gamma q^2)^2} \right\} \right]. \quad (5.48)$$

Eq. (5.48) is known as the Landau-Placzek formula. Here, $\gamma = c_P/c_V$, D_T is the thermal diffusivity [Eq. (5.39)], c_s is the adiabatic speed of light and Γ is the acoustic attenuation coefficient given by

$$\Gamma = \frac{1}{2} \left\{ (\gamma - 1) D_T + D_V \right\}. \quad (5.49)$$

D_V is the longitudinal kinematic viscosity given by Eq. (5.38). Eq. (5.48) shows that the R-B spectrum consists of three peaks (Figure 5.2).

The first term in the equation represents the unshifted (elstic) Rayleigh line denoted by R in Figure 5.2. This peak has a Lorentzian nature with a full width at half maxima (FWHM) of $2D_T q^2$. Hence the width of the peak depends on the thermal diffusivity. This peak results from the scattering of light by thermal fluctuations, which are dissipative modes of the fluid. Hence the Rayleigh spectrum probes the fluctuations or incoherent motions in a liquid.

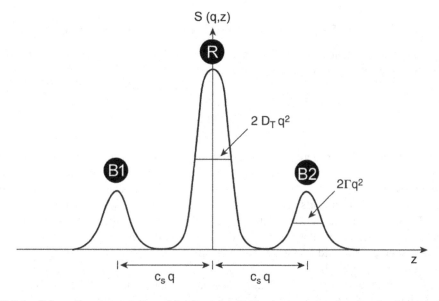

FIGURE 5.2 Schematic representation of the Rayleigh-Brillouin spectrum, showing the dynamic structure factor $S(q,z)$ in the hydrodynamic limit (at long wavelengths or small wavenumbers). D_T is the thermal diffusivity, c_s is sound velocity. "R" denotes the Rayleigh peak and B1 and B2 denote the Brillouin doublets.

The second and the third terms of Eq. (5.48) are known as the Brillouin doublets. These are represented by B1 and B2 in Figure 5.2. These are also Lorentzian in nature with FWHM of $2\Gamma q^2$. They are symmetrical shifted from the origin by $\pm c_s q$. They have their origin in the photon-phonon interactions in the fluid. These are interactions between photons and hydrodynamic fluid modes. The shift in frequency results from the loss of energy due to such inelastic collisions.

We take an example to describe the $R\text{-}B$ spectrum for a real fluid. For liquid argon, at temperature $T = 235$ K and mass density of 1 g cm^{-3}, the following values for the parameters are found.

$$c_s = 6.85 \times 10^4\ cm\ s^{-1}$$
$$D_T = 1.0 \times 10^{-3}\ cm^2\ s^{-1}.$$
$$D_V = 1.3 \times 10^{-3}\ cm^2\ s^{-1}$$

Typically for light scattering, we have

$$\gamma = 1$$
$$q = 2.1 \times 10^5\ cm^{-1}.$$

Hence we can calculate the ratios of FWHM to the frequency shifts as follows:

$$\frac{\gamma D_T q^2}{c_s q} = 1.5 \times 10^{-3}$$
$$\frac{\Gamma q^2}{c_s q} = 1.2 \times 10^{-3}.$$

Note that these ratios are very small (which is true for most fluids) and this is an assumption made in the foregoing discussion.

5.7   DERIVATION OF STOKES FRICTION ON A MOVING SPHERE

When a sphere of radius R, moves through a viscous liquid with viscosity η, the force (F) experienced by it is given by the celebrated Stokes expression, which reads

$$F = \zeta u_\infty$$
$$\zeta = 6\pi\eta R. \tag{5.50}$$

Here, ζ is the friction on the sphere and u_∞ is the far-field velocity vector. In the derivation of this well-known equation for friction on a moving sphere, one starts with the Navier-Stokes equation. The basic philosophy in this derivation is as follows. We consider a steady state flow past a sphere. The presence of the sphere distorts the flow. However, far from the sphere, the flow becomes laminar again. Thus, if we change the coordinate system with the center of the sphere as the origin of the coordinate system, then we find that the flow of the fluid becomes laminar and unperturbed far from the stationary sphere. The fluid flowing past the sphere exerts force on the sphere. This is shown pictorially below in Figure 5.3.

In the first step of the derivation, one obtains the angle (θ) and position (r) dependence of the flow field. The angle is measured in terms of the longitudinal flow direction and the distance is measured from the centre of the sphere. The expression for the stress tensor (σ) is given by

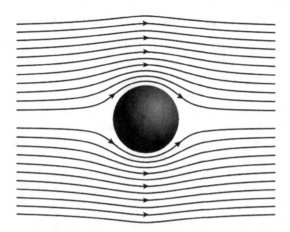

FIGURE 5.3 Schematic representation of the flow of a fluid past a sphere fixed in position. The arrowheads represent the direction of the flow. The flow field around the sphere generates a friction that depends on the boundary condition (slip/stick). Far from the sphere, the flow is laminar.

$$\sigma = -p.\mathbf{I} + \eta\left[(\nabla\mathbf{u}) + (\nabla\mathbf{u})^T\right] \tag{5.51}$$

The flow field is obtained by solving the Navier-Stokes equation, in terms of position-dependent pressure and velocity field. Here, $\mathbf{I}$ is the identity matrix and $\mathbf{u}$ is the distance-dependent velocity field given by (see the Appendix of the book by Lamb for details)

$$\mathbf{u}(\mathbf{r}) = \mathbf{G}(\mathbf{r}).\mathbf{u}_\infty \tag{5.52}$$

$$\mathbf{G}(\mathbf{r}) = \frac{3R^3}{4}\frac{\mathbf{r}\times\mathbf{r}}{|\mathbf{r}|^5} - \frac{R^3}{4}\frac{\mathbf{I}}{|\mathbf{r}|^3} - \frac{3R}{4}\frac{\mathbf{r}\times\mathbf{r}}{|\mathbf{r}|^3} - \frac{3R}{4}\frac{\mathbf{I}}{|\mathbf{r}|} + \mathbf{I}. \tag{5.53}$$

Now, the force acting on the sphere moving through the viscous liquid is given by the surface integral over the stress tensor. Hence we have

$$\mathbf{F} = \oiint d\mathbf{s}.\sigma. \tag{5.54}$$

Introduction of Eqs. (5.51), (5.52), and (5.53) in Eq. (5.54) and subsequent mathematical manipulation gives the friction coefficient as in Eq. (5.55).

$$\boxed{\zeta = 6\pi\eta R}. \tag{5.55}$$

Students should note that the above derivation is for the stick boundary condition. When the fluid slips past the flow, then the expression for the stress tensor is different. Another interesting application is the derivation of Stokes friction on a rotating sphere. This is often referred to as Debye-Stokes friction because Debye extensively used it in his exposition of dielectric relaxation in polar fluids. In spite of the success of the hydrodynamic description, there are certain limitations that we discuss now.

5.8 LIMITATIONS OF NAVIER-STOKES HYDRODYNAMICS

Navier-Stokes hydrodynamics based on three conservation laws, which are globally exact, but assumed (in the Navier-Stokes approach) to be locally valid, with no well-defined length scale. Whenever the length scale becomes smaller, the energy and momentum conservation laws are no longer valid. This, as discussed above, has a serious consequence.

Similarly, the constitutive relation with constant viscosity is applicable in long time. The long-time approach also enters into the assumption of the constitutive equations.

In a different, elegant work, Zwanzig and Bixon had generalized the Navier-Stokes hydrodynamics in order to include the viscoelastic response of the medium (discussed later). This generalization provides an expression for the frequency-dependent friction, which depends on the frequency-dependent shear viscosity, bulk viscosity, and sound velocity.

In an approach of far-reaching consequence, de Schepper and Cohen, and Kirkpatrick developed an approach to extend the hydrodynamical modes to intermediate wavenumbers. They demonstrated that the transport coefficients appearing in the generalized hydrodynamics can be extended in a consistent fashion to include wavenumber dependence. Kirkpatrick used the standard projection operator technique in order to derive the generalized hydrodynamic equations for the equilibrium time correlation functions in the case of a hard-sphere fluid.

As pointed out by these authors, the slow mode at intermediate wavenumbers is important in the theories that include mode coupling effects. Such theories have been utilized to understand quantitatively the anomalous long-time tails of the stress-stress correlation function and the shear dependent viscosity, often observed in computer simulations. The theory of glass transition has also been advanced based on the softening of the heat mode.

5.9 JOURNEY TOWARDS SMALLER WAVELENGTH: SIGNIFICANCE OF DE GENNES' NARROWING AND SLOW DYNAMICS AT LARGE WAVENUMBERS

We have discussed the Rayleigh-Brillouin spectrum observed in light scattering experiments in an earlier section. The wavelengths involved in light scattering experiments are typically of the order of about 5000 Å. The scattering of light occurs from fluctuations in the dielectric constant of the liquid, which occurs due to density fluctuations. Therefore, light scattering probes density fluctuations that occur at length scales ~5000 Å. This means we need to focus on density fluctuations within regions (like cubic boxes) of the order of ~5000 Å. We can calculate density fluctuations, and hence the spectral distribution of scattered light over this large (compared to molecular size) length scales by using the macroscopic equations of hydrodynamics. The neutron scattering (NS) experiments, on the other hand, involve wavelengths of a few angstroms only which are of the order of the nearest-neighbor separation in the liquid. Therefore, NS captures physics completely different from that in light scattering and hence cannot be explained by using the Navier-Stokes hydrodynamics. We shall now discuss several distinct aspects of particle motions at molecular length scales.

The wavenumbers accessible in inelastic neutron scattering experiments are typically within 0.1 and 15 Å^{-1} having the same range as that of the studies of molecular dynamics simulations. As discussed in the earlier section, the dynamic structure factor $S(q, \omega)$ at reduced wavenumbers $q\sigma \ll 1$ (where σ is the atomic diameter), has a three peak structure and the two side peaks correspond to propagating sound waves. At shorter wavelengths, the sound waves seem to be strongly damped and the high-frequency structure disappears when $q\sigma \geq 2$, leaving only the Lorentzian-like central peak. The width of the central peak starts increasing with wavenumber q, but eventually shows a sharp decrease at wavenumbers close to q_m,

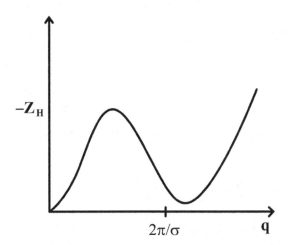

FIGURE 5.4 Nonmonotonic wavenumber (q) dependence of the eigenvalues (z_H) of the extended hydrodynamic modes. Note the slowing down of relaxation at intermediate wave numbers ($q \approx 2\pi / \sigma$).

that corresponds to the wavenumber where the static structure factor has its dominant first peak. Typically, $q_m \approx 2\pi / \sigma$, where σ is the molecular diameter. The spectrum again starts broadening at larger wavenumbers and finally reaches its free particle limit. The wavenumber-dependent behavior is depicted in Figure 5.4.

The sharp decrease in the width near $q = q_m$ is called "de Gennes narrowing". This cannot be explained from the dynamic structure factor obtained using the linearized hydrodynamic theory. de Gennes observed that for intermediate wavenumbers (that is, for intermediate momentum transfer) intermolecular correlation between neighboring atoms plays a crucial role. The nearest-neighbor molecules form a cage around a central molecule. This cage provides an effective field that controls density relaxation, as described in more detail below. As first pointed out explicitly by de Schepper and Cohen, at such small distances or large wavenumbers, momentum and energy conservations no longer pose constrains for the density relaxation, but number conservation must remain valid. *Therefore, the heat mode degenerates into a self-diffusion mode.* This leads to a strong narrowing of the density distribution function.

Since the density relaxation is now controlled by self-diffusion but under the influence of the cage formed by the neighbors, relaxation can be expressed by a Smoluchowski-Vlasov type equation in order to show the number density fluctuations.

It is worth emphasizing this interesting physics that at large wavenumbers, neither the conservation of the momentum nor the energy is relevant constraints to the dynamics at the wavelengths of the order of nearest-neighbor separation. However, number conservation continues to hold, as it has to be conserved at all length scales. Hence, the only slow mode relevant in this regime of wavenumber is the number density.

Thus, the diffusion equation could provide a valid description of density fluctuation. However, a particle has to diffuse in the force field of its neighbors. One can thus write a Smoluchowski-Vlasov type equation for the number density that involves a mean-field force term due to the cage effect as shown below,

$$\dot{\rho}(\mathbf{r},t) = D\nabla . [\nabla - \beta \mathbf{F}(\mathbf{r},t)]\rho(\mathbf{r},t), \qquad (5.56)$$

where $D = k_B T / m\gamma$ represents the self-diffusion coefficient of the liquid, γ be the collisional frequency. $\mathbf{F}$ is the mean field force written as

$$F(\mathbf{r},t) = -\nabla V_{eff}(\mathbf{r},t) \qquad (5.57)$$

$V_{eff}(\mathbf{r}.t)$ represents the effective potential energy, which is determined from mean spherical approximation and can be written in terms of the two particle direct correlation function, $C(\mathbf{r}-\mathbf{r}')$ as given by,

$$\beta V_{eff}(\mathbf{r},t) = -\int d\mathbf{r}' C(\mathbf{r}-\mathbf{r}')\rho(\mathbf{r}',t) \qquad (5.58)$$

ρ_0 is the equilibrium solution of the above equation having the following form:

$$\rho_0(\mathbf{r}) = c' \exp\left[\int d\mathbf{r}' C(\mathbf{r}-\mathbf{r}')\rho_0(\mathbf{r}')\right]. \qquad (5.59)$$

Here c' represents a constant determined from the normalization condition. The above equations can be solved to obtain the intermediate scattering function $F(q,t)$ in the following form:

$$F(q,t) = S(q) e^{-\left(\frac{D}{S(q)}q^2 t\right)} \qquad (5.60)$$

where $S(q)$ denotes the static structure factor.

In general, the value of self-diffusion coefficient is about three orders of magnitude smaller than that of the thermal diffusion coefficient. Additionally, the static structure factor is sharply peaked at the first maximum. These two above mentioned factors combine to give rise to a marked slowing down of the intermediate scattering function, which in turn leads to a substantial narrowing of the Rayleigh peak of the dynamic structure factor. This is known as the de Gennes narrowing.

The Rayleigh peak is often used to explore the stability of the liquid. This de Gennes narrowing forms the foundation for the mode coupling theory explanation of the glass transition.

In a dense liquid, the harsh repulsive part of the inter-molecular potential plays a significant role. This repulsive potential creates a cage around each molecule, which obstructs the movement of the molecule. Thus, the self-diffusion becomes the slowest process. Therefore, the transfer of a molecule becomes the bottle-neck of any relaxation process.

5.10 FREQUENCY-DEPENDENT FRICTION AND GENERALIZED HYDRODYNAMICS

Hydrodynamics provides an elegant expression for the frequency-dependent friction that finds use in many applications, like calculation of rate of barrier crossing, velocity time correlation function, and several others. The most elegant derivation was provided by Zwanzig and Bixon in 1970. They generalized the standard Navier-Stokes treatment by introducing viscoelasticity into the theory, by making the viscosity time-dependent. A detailed discussion of this theory is given in Chapter 10. Once, we get the frequency-dependent friction, we can compute the frequency-dependent diffusion coefficient using the generalized Einstein equation [Eq. (5.61)], which in turn gives the frequency-dependent velocity autocorrelation [Eq. (5.62)] that encapsulates the essence of generalized hydrodynamics.

$$D(z) = \frac{k_B T}{m[z + \zeta(z)]} \tag{5.61}$$

$$C_V(z) = \frac{k_B T}{m[z + D(z)]} \tag{5.62}$$

One should note that the concept was dragged from de Gennes' narrowing slowing down of relaxation at intermediate wavenumbers and generalized hydrodynamics of Zwanzig and bixon and combined to develop the area of molecular hydrodynamics where the transport properties are treated as a wavenumber and frequency-dependent.

5.11 SUMMARY

Hydrodynamic equations, along with their use of conservation laws, which are used to describe slow variables, continue to play an essential role in our understanding of time-dependent statistical mechanics. Much of the large-scale nonequilibrium phenomena, not just flow of water in rivers and oceans, but also atmospheric sciences and even cosmic motions employ hydrodynamics. Even in the understanding of molecular motions, hydrodynamics could play a very important role. An example of such an application at molecular length scale is provided by the Stokes-Einstein law ($D = k_B T / 6\pi\eta r$) to describe diffusion of liquids. Another take-home message is the hydrodynamic description of the Rayleigh-Brillouin spectrum. Rayleigh-Brillouin spectroscopy provides, as discussed above, experimental values of a large number of transport properties (like thermal diffusivity, sound attenuation, speed of light, bulk viscosity, etc.).

As we shall see through the course of this book, hydrodynamic equations and concepts enter again and again in the description of several nonequilibrium phenomena. As emphasized in the very beginning (Chapter 1) hydrodynamics, along with the kinetic theory of gases, form the two pillars of time-dependent statistical mechanics. Therefore, students are encouraged to study hydrodynamics in detail, more than that provided here. Several relevant references are given below.

BIBLIOGRAPHY

1. Landau, L. D., and E. M. Lifshitz. 1987. *Fluid Mechanics*. Oxford: Pergamon Press.
2. Lamb, H. 1993. *Hydrodynamics*. Cambridge: Cambridge University Press.
3. Hansen, J. P., and I. R. McDonald. 2006. *Theory of Simple Liquids*, Cambridge: Academic Press.
4. Berne, B. and R. Pecora. 1976. *Dynamic Light Scattering*, Hoboken: Wiley.
5. Bagchi, B. 2012. *Molecular Relaxation in Liquids*, New York: Oxford University Press.
6. Zwanzig, R., and M. Bixon. 1970. Hydrodynamic theory of the velocity correlation function. *Physical Review A*, 2(5), 2005.
7. Boon, J. P., and S. Yip., 1991. *Molecular Hydrodynamics*. Chelmsford: Courier Corporation, 1991.

6 Kinetic Theory of Gases

Boltzmann's Kinetic Equation and H-Theorem

OVERVIEW

Kinetic theory of gases forms the second pillar of the time-dependent statistical mechanics – the other one being hydrodynamics, discussed in the last chapter. While hydrodynamics is based on a continuum description of fluids, ignoring molecular nature of matter altogether, kinetic theory of gases is based on corpuscular or atomic description following the detailed laws of mechanics. Hydrodynamics is built on laws of conservation of density, momentum, and energy flux across a small volume element that might contain thousands of particles, kinetic theory deals with individual particle kinematics, following Newton's equations. Perhaps the most celebrated expression of kinetic theory of gases is the equipartition theorem that is rooted in Maxwell's velocity distribution. Since the equipartition theorem provides a molecular definition of temperature, kinetic theory of gases is also considered as a theory of heat and theory of motion. Here we first review the simple kinetic theory of gases that was developed by Maxwell and others. Subsequently, we discuss the Boltzmann kinetic equation (BKE), also known as Boltzmann transport equation. The latter can be considered as the starting point of the time-dependent statistical mechanics because it made for the first time the beginning towards a theory of phase space dynamics of interacting atoms. BKE gives us an equation of motion of the phase space density function, $f(r,p,t)$. In the process, Boltzmann introduced important concepts like molecular chaos and Boltzmann's H-theorem. We also discuss subsequent extensions by Enskog, and more modern generalizations by Cohen, Dorfman *et al.* whose work was motivated by computer simulation results, like long-time tails in the velocity-time correlation function of hard sphere particles. However, the latter topics have been discussed minimally, leading the students to useful references.

6.1 INTRODUCTION

The kinetic theory of gas has a chequered history, and played an important role in the evolution of science. That matter consists of small particles was proposed by Indian scientist/physicist Maharishi Kanad, also known as Acharya Kanad, who also does not get much mention in western science literature where credit is usually given to Greek philosopher Democritus. The atomic view of matter was opposed by the philosophers/scientists who believed that matter was continuous. It is interesting to note that this continuum view of matter finds quantification in hydrodynamics while kinetic theory of gases developed independently is based on laws of mechanics.

In modern times, it was Daniel Bernoulli who advanced the idea that gases are formed of elastic molecules, which are moving randomly at large speeds. These elastic spheres collide and rebound following the laws of classical mechanics, which itself was just being developed. Subsequently,

DOI: 10.1201/9781003157601-7

Joule and Clausius led the development of the corpuscular view of matter. The quantitative theory was developed in the mid and late nineteenth century primarily by Maxwell and Boltzmann in order to explain and calculate from first principles the macroscopic physical properties of gases based on the atomistic description of matter. The initial treatments were devoted to dilute gas only that displays close to an ideal behavior. The fond hope was to build a description that would start from microscopic conservation laws of number, momentum, and energy where motions obey Newton's laws of motion. That is a theory built at the microscopic level with intermolecular interaction that would go much beyond a hydrodynamic level description. Such a description was ultimately developed, but took almost a century, starting from Boltzmann to Cohen and de Schepper, to the mode coupling theory (MCT).

An important role was played by the Boltzmann kinetics equation (BKE), also known as the Boltzmann transport equation. It was in the course of the derivation of BKE that the hypothesis of "molecular chaos" was introduced, out of necessity. The same derivation also was involved in Boltzmann's *H*-theorem and, in the process, served to introduce the statistical mechanical definition of entropy. This approach and subsequent attempts to extend by Enskog, Chapman were not too successful. The semi-empirical approach of Enskog leads to an approximate expression of friction by including radial distribution function at contact. This extension offered better agreement with known results but was not fully satisfactory.

The discovery of the long-time tail in the velocity-time correlation function of hard disks by Alder and Wainwright opened a flurry of activity in the kinetic theory of gases. E.G.D. Cohen and co-workers extended the kinetic theory by including repeated collisions that form a ring. This difficult calculation was successful in explaining the long-time tail.

By the early 1970s, three different explanations for the long-time tail in the velocity TCF were available, and all three played important roles in the future development of time-dependent statistical mechanics. As already mentioned, the kinetic theory approach with multiple collisions of the tagged particle with a group of neighbors such that the collisions can be described as a ring could explain the long-time tail.

A paper by de Gennes opened the door for extending the kinetic theory of gases to the liquid state. It was observed by van Vleck and co-workers that the Rayleigh peak in the Rayleigh-Brillouin spectrum (discussed in the last chapter) undergoes a pronounced narrowing at large wavenumbers. The maximum narrowing was found at the value of wavenumber close to $2\pi/\sigma$, where σ is the molecular diameter. In a paper of far-reaching importance, de Gennes pointed out that at such large wavenumbers or small length scales, momentum and energy relaxations pose no constrain to the relaxation dynamics and only the number conservation is relevant. Thus, one can write a diffusion equation in the force field of the neighbors, which is essentially a Smoluchowski equation, also known as the Smoluchowski-Vlasov equation. Such a description could explain the sharp departure from the hydrodynamic prediction.

The importance of de Gennes's work stems from the fact that it suggested a way to extend Navier-Stokes hydrodynamics to large wavenumbers or small length scales. This was indeed achieved many years later by de Schepper and Cohen but by using the kinetic theory of gas approach. A part of de Schepper-Cohen's work was anticipated in the theory of molecular hydrodynamics, as described in the well-known monograph by Boon and Yip.

Because of the difficulty faced in extending Boltzmann's direct approach to the kinetic theory of gases, several semi-empirical approaches were initiated. One of them replaced the Collison kernel (discussed below) by a Fokker-Planck operator. In another approach, the collision operator was approximated by a relaxation kernel. This last approach is known as the Bhatnagar-Gross-Krook or BGK equation.

In this chapter, we will first discuss the kinetic theory results for dilute gases and subsequently discuss Boltzmann's transport equation (also known as Boltzmann's kinetic equation).

6.2   POSTULATES OF THE KINETICS THEORY OF GASES

The scope of the kinetic theory of gases is better appreciated if we start with the postulates that form the basis. Also, these postulates serve the purpose of systematizing the concepts involved. There are five postulates, which are stated differently in different texts but essentially contain the same statements. We state them in order of merits.

(i)   The system consists of atoms and molecules that are in constant motion.
(ii) The density of the gas is low, and the size of constituent particles is point-like such that the volume fraction of these particles is small compared to the total volume of the system. The average distance between any two particles is much larger than the size of each particle.
(iii) The particles do not interact with each other. However, they do suffer elastic collisions with each other and with the surface of the wall. Thus, the collisions are instantaneous.
(iv) Each atom/molecule possesses kinetic energy because of its state of motion but no potential energy (as they do not interact). The kinetic energy is dependent on the temperature of the system.
(v)   The average distance traversed between two consecutive collisions is much larger than the size of the atoms/molecules involved.

These postulates are stated differently, and in a different order, in different books. The reader should note the contradiction that although the particles are stated not to interact, they are allowed to have collisions. These collisions are elastic, they change the direction of velocities of the participants but not the magnitude.

Often not realized, the kinetic theory of gases provides us with some of the most important relations and concepts of natural science, such as the universal gas constant (R), Avogadro number, Boltzmann constant k_B, the equipartition theorem, to name a few. We first discuss these important subjects, and then more complex topics like Boltzmann's kinetic equation.

6.3   IDEAL GAS LAW, AVOGADRO NUMBER, AND TEMPERATURE

Experiments of Boyle, Gay-Lussac, and Avogadro showed that all gases in the limit of very low density follow the perfect gas law,

$$pV = nRT \tag{6.1}$$

where V is the volume, p the pressure and $n = \dfrac{N}{N_A}$ is the number of moles of gas, $R = N_A k_B$ is the universal gas constant, and T is the absolute temperature. At the time of writing Eq. (6.1) the universality of R was established through experimental observation only. It was the kinetic theory of gases that gave a microscopic justification of Eq. (6.1) and $R.N$ is the number of particles present in the gas, N_A is Avogadro's number, and k_B is Boltzmann's constant.

By using elementary considerations, that the pressure exerted on a wall is the force per unit area and that the force is given by the rate of change of momentum, we can obtain an expression or the pressure that a gas exerts on its container wall. This expression in turn can be expressed in terms of the *average translational* kinetic energy of gas particles (atoms or molecules) by,

$$pV = \frac{1}{3}Nm\bar{c}^2 = \frac{2}{3}N\left(\frac{1}{2}m\bar{c}^2\right) \tag{6.2}$$

where N is the number of gas particles, m is the mass of a gas particle, and $\bar{c}^2$ is the mean square velocity of gas particles. The quantity in parenthesis represents the average translational kinetic energy of a gas particle.

On equating with the perfect gas law, we obtain the kinetic definition of temperature,

$$T = \frac{2}{3N} \frac{N}{nR} \left(\frac{1}{2} m\bar{c}^2 \right) = \frac{2}{3k_B} \left(\frac{1}{2} m\bar{c}^2 \right) \tag{6.3}$$

Therefore, the average translational kinetic energy of a particle is,

$$\frac{1}{2} m\bar{c}^2 = \frac{3}{2} k_B T \tag{6.4}$$

The above equation provides a quantitative definition of temperature that is general so long as classical mechanics remains valid. *That is, even though we have used the ideal gas law on the way to derive Eq.(6.4), its validity goes far beyond ideal gas law. This is an example of how a general law can be derived by using a limited approach!*

If the gas consists of N particles then the average translational kinetic energy of the system is

$$E_{trans} = \frac{3}{2} Nk_B T \tag{6.5}$$

For perfect gases (consisting of phantom atoms or molecules), as the total internal energy (E) is equal to the average translation kinetic energy of the gas particles, we can write,

$$E = E_{trans} = \frac{3}{2} Nk_B T \tag{6.6}$$

$$T = \frac{2E}{3Nk_B}. \tag{6.7}$$

Note that the internal energy of gases consisting of an ensemble of monatomic molecules (noble gases) is entirely expressed in the form of their translational energy, provided the temperature of the system is not too high (i.e., temperatures below a few thousand degrees) in order for the internal motions of molecules (i.e., motions of electrons in the case of monatomic molecules) to be neglected. In effect, the above equation relating temperature with the internal energy is generally valid for all monatomic gases under normal pressure and temperature.

6.4 MAXWELL VELOCITY DISTRIBUTION AND EQUIPARTITION THEOREM

As the motions of the gas particles are random, they should not have a preferred direction of movement (i.e., isotropy of the velocity space). This implies that in Cartesian coordinates the motions in all three (x, y, and z) directions are equivalent and the components of the mean squared velocity of the gas particles are equal and can be written as,

$$\langle v_x^2 \rangle = \langle v_y^2 \rangle = \langle v_z^2 \rangle = \frac{1}{3} \bar{c}^2 \tag{6.8}$$

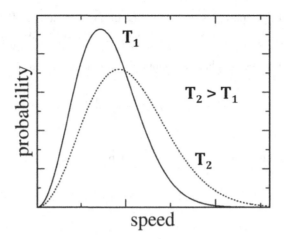

FIGURE 6.1   Plot for the Maxwell velocity distribution against speed that shows the distribution is shifted to higher speeds and is broadened at higher temperatures.

Thus, the average kinetic energy of translational motion of each particle is partitioned equally into three components corresponding to motions in three directions, i.e.,

$$\frac{1}{2}m\left\langle v_x^2 \right\rangle = \frac{1}{2}m\left\langle v_y^2 \right\rangle = \frac{1}{2}m\left\langle v_z^2 \right\rangle = \frac{1}{2}k_B T \qquad (6.9)$$

This is one of the most important results of the kinetic theory.

We already mentioned that the velocities of gas particles are random. This randomness exists both in direction and speed. In one of the most important equations of natural sciences, Maxwell formulated the distribution that goes by his name and is given by,

$$P(v)\ dv\ =\ A\ \exp\left(-mv^2 / 2k_B T\right)\ dv \qquad (6.10)$$

which gives the probability of finding a particle between velocity $\mathbf{v}$ and $\mathbf{v} + d\mathbf{v}$. Just like the equipartition theorem, Maxwell's velocity distribution is universally valid for a classical system. We depict the temperature dependence of the Maxwell distribution in Figure 6.1.

6.4.1   AN ALTERNATIVE DERIVATION OF MAXWELL'S VELOCITY DISTRIBUTION

An alternative derivation of the Maxwell velocity distribution has recently been presented by using the central limit theorem of the probability theory. We have discussed the central limit theorem in Chapter 2. In the present scenario, velocity of a particle at any time t can be regarded as sum of the velocity change of the same in the earlier times. The change occurs due to collisions and is random. Since velocity is a vector, we need to decompose the sum into individual components. Therefore, the central limit theorem gives a Gaussian distribution for the velocity. It can also be shown by using Boltzmann's formula relating entropy (S) to the total number of microstates of the system, Ω, that the standard deviation $\langle v^2 \rangle = k_B T/m$, where the terms have their usual meaning. This completes the proof of the Maxwell distribution.

6.5   TRANSPORT PROPERTIES

6.5.1   MEAN FREE PATH

Mean free path λ is defined as the average distance a molecule travels between two collisions. Suppose a molecule collides with frequency z. Therefore, the distance traveled between two collisions is

$$\lambda = \frac{\bar{c}}{z} \tag{6.11}$$

Suppose σ is the diameter of the molecule, then collision frequency can be written as

$$z = \sqrt{2}\,\pi\sigma^2 \bar{c}\,\bar{n} \tag{6.12}$$

where $\bar{n}$ is the number of molecules per unit volume.

Thus, from Eqs. (6.11) and (6.12), the expression of the mean free path can be written as

$$\lambda = \frac{1}{\sqrt{2}\pi\sigma^2\,\bar{n}} \tag{6.13}$$

6.5.2   Viscosity

Viscosity is related to the flux of the momentum, and from the kinetic theory of gas one can obtain the following expression:

$$\eta = \frac{1}{3}\bar{n}\,m\bar{c}\,\lambda \tag{6.14}$$

where λ is the mean free path.

6.5.3   DIFFUSION

The diffusion coefficient of gas can be written in terms of average speed and mean free path as

$$D = \frac{1}{3}\bar{c}\,\lambda \tag{6.15}$$

We substitute the expressions of $\bar{c}$ and λ in Eq. (6.15) to obtain

$$D = \frac{1}{3}\sqrt{\frac{8k_B T}{\pi m}}\,\frac{1}{\sqrt{2}\pi\sigma^2\bar{n}} \tag{6.16}$$

On the other hand, for hard sphere potential, the Maxwell friction is given by

$$\zeta_M = \left(\frac{8}{3m}\right)\sqrt{2\pi m k_B T}\rho\sigma^2. \tag{6.17}$$

Now, Eq. (6.16) can be rewritten in terms of friction as follows:

$$D = \frac{16}{9\pi} \frac{k_B T}{\zeta_M}. \tag{6.18}$$

The above expression shows that the celebrated Einstein relation between diffusion and friction gets modified in the Maxwellian limit. Later we discuss a more accurate friction due to Enskog.

6.5.4 THERMAL CONDUCTIVITY

Thermal conductivity (κ) is related to heat flux, and it determines the ability to conduct heat. According to the kinetic theory of gas, the expression of thermal conductivity can be written as follows:

$$\kappa = \frac{1}{3} \frac{N}{V N_A} \overline{c}_v \overline{c} \lambda = \frac{1}{3} \rho \overline{c}_v \overline{c} \lambda \tag{6.19}$$

Here ρ is concentration in moles per unit volume, $\overline{c}_v$ is the heat capacity per molecule, and λ is the mean free path.

6.6 BOLTZMANN'S KINETIC EQUATION (BKE)

As already mentioned, Boltzmann's kinetic equation (BKE), also known as the Boltzmann transport equation, is considered as the starting point of the dynamical theory of *interacting systems*. The ideal gas laws fail as density is increased from very small values. This limit can be best understood by introducing a dimensionless quantity $\rho^* = \rho \sigma^3$, where ρ is the number density and σ is the molecular diameter. For the ideal gas limit to be valid, the reduced density ρ^* needs to be much less than unity, maybe even less than 0.05, so that the average distance between two molecules is much greater than the molecular diameter, such that effects of intermolecular interactions are negligible. But, real gas and liquids are found at much larger densities.

Boltzmann set out to extend Maxwell's theory by including intermolecular interactions by letting molecules follow Newton's equations of motion, where the force on each molecule arises from intermolecular interactions with surrounding molecules. He introduced a series of probabilistic assumptions that were revolutionary at that time when the mechanical view was believed to be approached to be adopted. In the following, we briefly describe the steps that led to the kinetic equation derived by Boltzmann. But before we proceed, we would like to state the BKE in a simple symbolical form that should be understandable to students because BKE is a complicated equation.

The first point is that BKE is an equation of motion for single-particle phase space density, $f^{(1)}(\mathbf{r}, \mathbf{p}, t)$, where $(\mathbf{r}, \mathbf{p})$ is a point in phase space and f gives the probability of occupation per unit volume at that point, at time t. As a density, it provides the number of molecules in a small volume element centered around $\mathbf{r}$ and $\mathbf{p}$. Second, BKE at heart is essentially a continuity equation of the type we already discussed in the last chapter, except that we are dealing with phase space density. Thus, we write the BKE as

$$\frac{\partial f^{(1)}}{\partial t} + \frac{p}{m} . \nabla_r f^{(1)} + \frac{1}{m} \mathbf{F}(\mathbf{r}) . \nabla_u f^{(1)} = I[f^{(1)}] \tag{6.20}$$

where the term on the right-hand side describes the change in the singlet phase space density $f(\mathbf{r}, \mathbf{p}, t)$ due to collisions. This density function changes both by the inertial motion of the molecules [as the molecules move in and out of the phase space volume element around $(\mathbf{r}, \mathbf{p})$, a position in the phase space] and also by the forces acting on it.

The important point about the collision integral is that it must involve two particles. Thus, a two-particle phase space density is required. BKE involves careful decomposition of the two-particle distribution into a product of two single-particle distributions.

Let a force F acts on the particles. If we consider an infinitesimal time interval δt from time t, the same molecules can be found in a volume element centered around $\mathbf{r}'$ and $\mathbf{p}'$. Thus, we can write

$$f^{(1)}(\mathbf{r}, \mathbf{p}, t)\, d\mathbf{r}\, d\mathbf{p} = f^{(1)}(r + v\delta t,\ p + F\delta t,\ t + \delta t)\, dr'\, dp' \tag{6.21}$$

where the velocity $v = \dfrac{p}{m}$, m being the mass of the particles. F is the force acting on the particle.

One can easily show that $drdp = dr'dp'$. (To prove this, consider a simple two-dimensional square around position r, the X and Y arms get shifted with time by different amounts to create a parallelogram whose area can be shown to be the same as $drdp$.)

Thus, we can write,

$$f^{(1)}(r, p, t) = f^{(1)}(r + v\delta t,\ p + F\delta t,\ t + \delta t) \tag{6.22}$$

But in the presence of collisions, equality is not maintained, as the number density changes. So, we can write

$$f^{(1)}(\mathbf{r} + v\delta t, \mathbf{p} + F\delta t, t + \delta t) - f^{(1)}(\mathbf{r}, \mathbf{p}, t) = \left(\frac{\partial f^{(1)}}{\partial t}\right)_{coll} \delta t \tag{6.23}$$

We expand the first term on the left to first order in the Taylor series to obtain

$$\boxed{\left[\frac{\partial}{\partial t} + \frac{\mathbf{p}}{m}.\nabla_r + \mathbf{F}.\nabla_p\right] f^{(1)}(\mathbf{r}, \mathbf{p}, t) = \left(\frac{\partial f^{(1)}}{\partial t}\right)_{coll}} \tag{6.24}$$

where ∇_r and ∇_p are the gradient operator with respect to $\mathbf{r}$ and $\underset{\sim}{\nabla}_p$.

Sometimes the above equation is called the Boltzmann transport equation, and this serves as the starting point of many approximate treatments. Precise determination of $\left(\dfrac{\partial f}{\partial t}\right)_{coll}$ is difficult. The determination of this collision term was a major accomplishment of Boltzmann, and in the process, he introduced many important concepts and ideas that guided us for almost over a century. In the following, we briefly outline the main steps.

We now calculate the change in the probability density $f(\mathbf{r}, \mathbf{p}, t)$ due to collisions. While some collisions scatter particles to outside the box $d\mathbf{r}'d\mathbf{p}'$, some particles outside $d\mathbf{r}d\mathbf{p}$ reach inside the box due to collisions that occur outside (say, in the volume element $d\mathbf{r}'d\mathbf{p}'$). Thus one has the following master equation to account for the change due to collision:

$$\left(\frac{\partial f^{(1)}}{\partial t}\right)_{coll} \delta t = (\bar{R} - R)\delta t \tag{6.25}$$

where $\bar{R}$ is the addition (the source term) due to a collision outside the original box in phase space, while R is a loss term where one particle gets kicked out of the box – both due to binary collisions in the time interval between t and $t + \delta t$.

Next, we need to compute $\bar{R} - R$. This is an involved process that requires consideration of the details of the collision. We make the following approximations.

   i.   We consider the gas so dilute that only binary (that is, two particle) collisions occur during the time interval δt. That is, three or higher particle collisions are rare.

   ii.   Collisions are elastic, both energy and momentum are conserved

$$\mathbf{p}_1 + \mathbf{p}_2 = \mathbf{p}_1' + \mathbf{p}_2' \tag{6.26}$$

$$E = E_1 + E_2 = E_1' + E_2' \tag{6.27}$$

   iii.   Energy is only kinetic. This follows from the assumption of low density.

We note that the collision causes changes of the momenta and hence the kinetic energy, but both the sums need to be conserved. Let us assume that during time δt, the initial number of the colliding spheres within our volume element is δN_{12}. Then the number of transitions is $\delta N_{12} \delta P_{12 \to 1'2'} \delta t$. We write δN_{12} as,

$$\delta N_{12} = f_{12}^{(2)}(\mathbf{r}, \mathbf{p}_1, \mathbf{p}_2, t) d^3 r d^3 \mathbf{p}_1 d^3 \mathbf{p}_2 \tag{6.28}$$

where $f_{12}^{(2)}(\mathbf{r}, \mathbf{p}_1, \mathbf{p}_2, t)$ is a joint probability distribution of the two particles 1, and 2 with a small volume element centered around $\underset{\sim}{r}$ with momentum $\mathbf{p}_1$ and $\mathbf{p}_2$. Let $\delta P_{12 \to 1'2'}$ be the transition probability during the collision process. So, outward flow can be written as,

$$R \delta t d r d \mathbf{p}_1 = \delta t d r d \mathbf{p}_1 \int d \mathbf{p}_2 \delta P_{12 \to 1'2'} f_{12}^{(2)}(\mathbf{r}, \mathbf{p}_1, \mathbf{p}_2, t) \tag{6.29}$$

Similarly,

$$\bar{R} \delta t d r d \mathbf{p}_1' = \delta t d \mathbf{r}' d \mathbf{p}_1' \int d \mathbf{p}_2 d \mathbf{p}_1' d \mathbf{p}_2' \delta P_{1'2' \to 12} f_{1'2'}^{(2)}(\mathbf{r}, \mathbf{p}_1', \mathbf{p}_2', t) \tag{6.30}$$

Note that both the events are taking place at the position $\underset{\sim}{r}$,

We now have two unknowns, transition probability δP, and the two-particle distribution function. Thus, we have landed in a situation where to obtain an equation of motion for singlet density function $f_1^{(1)}(\mathbf{r}, \mathbf{p}_1, t)$. We need two-particle density function $f_{12}^{(2)}(\mathbf{r}, \mathbf{p}_1, \mathbf{p}_2, t)$. This is the general character of interacting systems because the dynamics of one tagged particle is determined by interactions with the rest. This was also one of the earliest examples of the hierarchy of the distribution functions that is a common bottle-neck in theoretical science. The resolution proposed by Boltzmann has shaped our understanding of time-dependent statistical mechanics. In the following, we shall outline the steps without describing them in great detail.

We next make the observations that $\delta P_{12 \to 1'2'}$ and $\delta P_{1'2' \to 12}$ depend on the relative velocity. The following considerations are needed to obtain expressions for these transition probabilities. (i) The transition must obey momentum and energy conservation; (ii) the initial to the final state by scattering can be calculated easily by using the Fermi golden rule. The latter was not available to Boltzmann who had to use detailed scattering geometry with impact parameter, the radius of particle, etc. The transition probabilities can be written as,

$$\delta P_{12 \rightarrow 1'2'} = d\mathbf{p}'_1 \, d\mathbf{p}'_2 \, \delta(P - P') \, \delta(E - E') | T_{fi} |^2 \tag{6.31}$$

where $| T_{fi} |^2$ is the transition probability in the same form as the Fermi golden rule. The other two delta function terms here are just the constrains of conservation of momentum and energy during the collision process. And for the reverse process

$$\delta P_{1'2' \rightarrow 12} = d\mathbf{p}'_1 \, d\mathbf{p}'_2 \, \delta(P - P') \, \delta(E - E') | T_{if} |^2 \tag{6.32}$$

The use of these last two expressions in Eqs. (6.31) and (6.32) lead to the following equation of motion:

$$\left(\frac{\partial f}{\partial t} \right)_{coll} = \iiint d\mathbf{p}_2 d\mathbf{p}'_1 d\mathbf{p}'_2 \delta(\mathbf{P} - \mathbf{P}') \delta(E - E') | T_{fi} |^2 \left(f^{(2)}_{1'2'} - f^{(2)}_{12} \right) \tag{6.33}$$

Here we have introduced the notation $f^{(2)}_{12}$ for a two-particle distribution function, with the subscript containing the indices of the momenta of the two particles.

The above elegant expression can be understood in a rather simple fashion. The first thing to note is that the left hand contains the singlet distribution function $f(r, p_1, t)$. That is, the number density is evaluated at position r. The second thing to note is that the scattering or transition is occurring into $\mathbf{p}'_1$ and $\mathbf{p}'_2$. We also need to average over the momentum of the incoming second particle that is colliding with our tagged molecule. The collision process conserves energy and momentum. Thus, this equation is easy to understand.

This is however not the expression that was derived by Boltzmann when quantum mechanics was not available. His expression was expressed in terms of the collision integral and was given as

$$\boxed{ \left(\frac{\partial f^{(1)}}{\partial t} \right)_{coll} = \iint d\mathbf{p}_2 \, d\Omega \left| \mathbf{v}_1 - \mathbf{v}_2 \right| \left(\frac{d\sigma}{d\Omega} \right) \left(f^{(2)}_{1'2'} - f^{(2)}_{12} \right) } \tag{6.34}$$

This equation is not closed because the singlet distribution requires the solution of the two-particle distribution function.

6.6.1 Assumption of Molecular Chaos and Boltzmann Kinetic Equation

We find above a situation (common in theoretical treatments of many-body physics) where a quantitative description of a single-particle distribution function requires an input of a two-particle distribution function. We shall face this in the next chapter when we study the BBGKY hierarchy. As at the time of Boltzmann, we do not have an accurate way to advance even today. In order to make further progress, and to obtain a closed equation of motion for $f(\mathbf{r}, \mathbf{p}_1, t)$, Boltzmann assumed a decomposition of the following form:

$$f^{(2)}_{12}(\mathbf{r}, \mathbf{p}_1, \mathbf{p}_2, t) = f^{(1)}_1(\mathbf{r}, \mathbf{p}_1, t) \, f^{(1)}_2(\mathbf{r}, \mathbf{p}_2, t) \tag{6.35}$$

This means that in a small volume element centered on point r, the momenta of two particles 1 and 2 are uncorrelated. In the presence of strong interaction between the two particles, the validity of this assumption is doubtful.

This is a famous assumption in the theory of transport processes. It is called the "assumption of molecular chaos" and "*Stosszahlansatz*" in many textbooks. It was controversial from the beginning and Boltzmann was criticized for this assumption.

With the assumption of molecular chaos, we can now write down a closed-form equation for $f_1^{(1)}(\mathbf{r}, \mathbf{p}_1, t)$ as,

$$\left(\frac{\partial}{\partial t} + \frac{\mathbf{p}_1}{m} \cdot \nabla_r + F \cdot \nabla_p\right) f_1^{(1)}(\mathbf{r}, \mathbf{p}_1, t) = \iiint d\mathbf{p}_2 d\mathbf{p}_1' d\mathbf{p}_2' \delta(P - P') \delta(E - E') |T_{fi}|^2$$
$$\left(f_{1'}^{(1)} f_{2'}^{(1)} - f_1^{(1)} f_2^{(1)}\right)$$

(6.36)

Although the Boltzmann kinetic equation has the character of a master equation, it is *a nonlinear* equation with microscopic transition probability between the states. It is a difficult equation to solve. Considerable progress has been made in the last century that we discuss below.

6.7 THE *H*-THEOREM

This is one of the most well-known theorems of statistical mechanics, even though it is fairly restrictive and applicable only to an ideal gas. The theorem considers the following function and H is defined by the following integration:

$$H(t) = -\int d\mathbf{p}\, f(\mathbf{p}, t) \log f(\mathbf{p}, t)$$

(6.37)

where $f(\mathbf{p}, t)$ is the time-dependent single-particle momentum distribution, the same as the one used above. The famous Boltzmann H-theorem states that if the distribution $f(\mathbf{p}, t)$ satisfies Boltzmann's transport equation, then

$$\frac{dH}{dt} \geq 0$$

(6.38)

The equality sign is satisfied only at equilibrium when the distribution attains the Maxwell form. That is, when $f(\mathbf{p}, t)$ is not an equilibrium distribution, the function H will continuously increase till the equilibrium distribution is reached.

We must point out that the H-function involves the time-dependent distribution of velocity or momentum, and does not contain the position. Thus, the distribution function involved is not the phase space density.

This unidirectional nature of the H function led to calling it an arrow of time, which is also applied to entropy. However, these are mostly qualitative descriptions.

There have been several different definitions of the H-theorem in an attempt to answer the repeated criticisms of H-theorem and also several attempts to put the H-theorem on firmer ground. However, one needs to remember that Boltzmann's H-theorem is strictly valid only for ideal gas. In the following, we present the proof, which is routinely used.

The equilibrium distribution function $f_0(\mathbf{p})$ is the solution to the equation

$$\partial f^{(1)}(\mathbf{p}, t) / \partial t = 0.$$

(6.39)

Therefore, $f_0(\mathbf{p})$ must satisfy the following integral equation:

$$0 = \int d^3\mathbf{p}_2\, d^3\mathbf{p}_1'\, d^3\mathbf{p}_2'\, \delta^4\left(P_f - P_i\right)\left|T_{fi}\right|^2 \left(f_{1'2'}^{(2)} - f_{12}^{(2)}\right). \tag{6.40}$$

We note that the binary collision obeys the following probability conservation condition:

$$f_0^{(1)}\left(\mathbf{p}_2'\right)f_0^{(1)}\left(\mathbf{p}_1'\right) - f_0^{(1)}\left(\mathbf{p}_2\right)f_0^{(1)}\left(\mathbf{p}_1\right) = 0 \tag{6.41}$$

Therefore, the Boltzmann transport equation has to be solved with the above condition. In order to prove Eq. (6.40) with the help of Eq. (6.41), we introduce the H-function defined in Eq. (6.37), where $f(\mathbf{p},t)$ is the distribution function at time t that satisfies

$$\partial f_0^{(1)}(\mathbf{p},t)/\partial t = \int d^3 p_2\, d^3 p_1' d^3 p_2' \delta^4\left(P_f - P_i\right)\left|T_{fi}\right|^2 \left(f_{1'2'}^{(2)} - f_{12}^{(2)}\right) \tag{6.42}$$

Differentiation of H-function is given by

$$\frac{dH(t)}{dt} = -\int d^3\mathbf{p}\, \frac{\partial f^{(1)}(\mathbf{p},t)}{\partial t}\left[1 + \log f^{(1)}(\mathbf{p},t)\right] \tag{6.43}$$

Since, at equilibrium, $\partial f / \partial t = 0$ implies $\dfrac{dH}{dt} = 0$.

If f satisfies the Boltzmann transport equation, then we show below that the H-function satisfies Eq. (6.38). Substituting Eq. (6.42) into the integrand of Eq. (6.43), we have,

$$\frac{dH(t)}{dt} = -\int d^3\mathbf{p}_2\, d^3\mathbf{p}_1'\, d^3\mathbf{p}_2'\, \delta^4\left(P_f - P_i\right)\left|T_{fi}\right|^2 \left(f_{1'2'}^{(2)} - f_{12}^{(2)}\right)\left(1 + \log f_1\right). \tag{6.44}$$

One can write,

$$\frac{dH(t)}{dt} = -\frac{1}{2}\int d^3\mathbf{p}_2\, d^3\mathbf{p}_1'\, d^3\mathbf{p}_2'\, \delta^4\left(P_f - P_i\right)\left|T_{fi}\right|^2 \times \left(f_{1'2'}^{(2)} - f_{12}^{(2)}\right)\left(2 + \log\left(f_1^{(1)}f_2^{(1)}\right)\right). \tag{6.45}$$

This integral is invariant under the interchange of $\{\mathbf{p}_1,\mathbf{p}_2\}$ and $\{\mathbf{p}_1',\mathbf{p}_2'\}$ because for every collision, there is an inverse collision with the same T matrix.

Hence,

$$\frac{dH(t)}{dt} = \frac{1}{2}\int d^3\mathbf{p}_2\, d^3\mathbf{p}_1' d^3\mathbf{p}_2'\, \delta^4\left(P_f - P_i\right)\left|T_{fi}\right|^2 \left(f_{1'2'}^{(2)} - f_{12}^{(2)}\right)\times\left[2 + \log\left(f_1^{(1)}f_2^{(1)}\right)\right] \tag{6.46}$$

where P_f and P_i are the final and initial probability, respectively. We take the half of the sum of Eq. (6.45) and Eq. (6.46), we obtain

$$\frac{dH}{dt} = -\frac{1}{4}\int d^3\mathbf{p}_2\, d^3\mathbf{p}_1' d^3\mathbf{p}_2'\delta^4\left(P_f - P_i\right)\left|T_{fi}\right|^2 \left(f_{2'}^{(1)}f_{1'}^{(1)} - f_2^{(1)}f_1^{(1)}\right)\times$$
$$\left[\log\left(f_2^{(1)}f_1^{(1)}\right) - \log\left(f_{2'}^{(1)}f_{1'}^{(1)}\right)\right]. \tag{6.47}$$

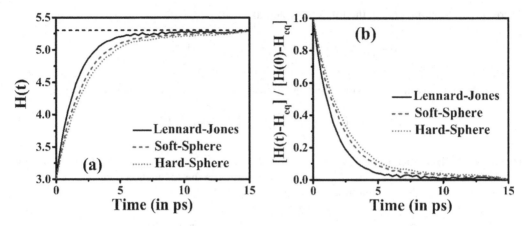

FIGURE 6.2 (a) The computed time evolution of $H(t)$ obtained via computer simulations (in NVE ensemble) for 3D systems of dilute gases (at reduced density, $\rho^* = 0.10$, and average reduced temperature, $T^* = 2.0$) interacting with Lennard-Jones, soft sphere, and hard sphere potentials. In all the cases, the H-function increases monotonically and then attains equilibrium at a longer time, which is the equilibrium value at the final temperature, shown by black dashed line. (b) Time evolution of the normalized $H(t)$ for the systems shown in (a). The time scales have been converted from reduced unit to real unit (in ps) using the argon parameters. Note the different time scales in each case.

Source: Taken with permission from [17].

The integrand of the integral in Eq. (6.47) is never positive. This completes the proof of Boltzmann's H-theorem.

Figure 6.2(a) demonstrates the validity of the H-theorem where we plot the time-dependent H-function for three-dimensional systems of dilute gases (at reduced density, $\rho^* = 0.10$ and average reduced temperature, $T^* = 2.0$, which is obtained by a procedure described below) interacting with Lennard-Jones, soft sphere, and hard sphere potentials. The initial nonequilibrium state is created by taking the amplitude of the velocities of all the particles exactly the same; the magnitude is in accordance with the equipartition theorem and temperature. This allows us to carry out the simulation in the microcanonical (NVE) ensemble. We see that in all the cases, the H-function sharply increases in a short time and subsequently monotonically approaches the equilibrium value (shown by a black dotted line) at a longer time, which is the equilibrium value at the chosen temperature. Note that the time scales have been converted from reduced unit to real unit (in ps) using the argon parameters. In Figure 6.2(b), we show the time evolution of normalized $H(t)$ defined as $\left(\left(H(t) - H_{eq}\right)/\left(H(0) - H_{eq}\right)\right)$. The relaxation times of the systems can be obtained by exponential fit of the normalized $H(t)$. It is observed that the relaxation of $H(t)$ is faster for the systems interacting with the Lennard-Jones potential than those interacting via the hard sphere potential.

6.7.1 Relationship between H-Function and Entropy

H-function and entropy have been associated with each other from the beginning because both define a direction of time. Clausius' statement asserts that the entropy of an isolated system can only increase with time. In fact, for the one-component ideal gas one can derive a simple relation between the two, which establishes that while H decreases with time, S increases with the same.

For ideal gas at equilibrium, the velocity distribution is Maxwellian and given by Eq. (6.10), where $p = mv$ to be used to get $f(p)$). Since $f(p)$ is known, one can evaluate the H-function at equilibrium trivially to obtain

$$H_{eq} = -\ln\left(\frac{1}{2\pi m k_B T}\right)^{3/2} + \frac{3}{2}$$ (6.48)

We can obtain the entropy per particle of an ideal gas from the Sackur-Tetrode equation

$$S^{id} = k_B \ln\left(2\pi m k_B T\right)^{3/2} + \frac{5}{2}$$ (6.49)

Thus, we obtain the following relation:

$$S^{id} = k_B H_{eq} + \text{constant}$$ (6.50)

We emphasize that this remarkable relation holds only for an ideal gas at equilibrium. There is an ln V term absorbed in the "constant" factor of Eq. (6.50).

The assertion of a relationship between H and S functions on the other hand extends to the evolution of the two functions from a given nonequilibrium state to an equilibrium state. There has been no convincing proof. While the H-function can be regarded as a mechanical quantity, and we could define for a distribution $f(p, t)$, the same is not true for entropy. While a strictly valid definition for time-dependent entropy may not be possible, one can attempt to define evolving entropy of a subsystem that is in contact with a bath that is governed by faster dynamics. For example, we can change the temperature of the system in a controlled manner, slowly, such that one can define entropy in the intermediate states. However, that remains problematic.

6.8　BHATNAGAR-GROSS-KROOK EQUATION

In 1954, Bhatnagar, Gross, and Krook (BGK) proposed a kinetic model that is based on the Boltzmann kinetic equation. This model approximates the effects of molecular collisions by replacing the collision integral by a simpler, relaxation term. This paves the way for a much simpler derivation of the transport equations of macroscopic variables.

The BGK equation is a popular model to solve the complex integro-differential equation presented by Boltzmann. It is widely used in computer simulations.

The Boltzmann kinetic equation has the following general form:

$$\frac{\partial f}{\partial t} + \mathbf{v}.\nabla_r f + \frac{1}{m}\mathbf{F}(\mathbf{r}).\nabla_v f = \left(\frac{\partial f}{\partial t}\right)_{collision}$$ (6.51)

where $\mathbf{v}$ is the velocity of a molecule. Here and below, we use, for convenience, $f(\mathbf{r}, \mathbf{v}, t) = f^{(1)}(\mathbf{r}, \mathbf{v}, t) = f$ is the same number density in the phase space used above. The external force acting on the molecule is given as $\mathbf{F}(\mathbf{r})$ and m is the molecular mass. In the simplest kinetic model, the collision integral, which is on the right-hand side of Eq. (6.51) is approximated as

$$\left(\frac{\partial f}{\partial t}\right)_{collision} = \frac{\left(f_{eq} - f\right)}{\tau}$$ (6.52)

So the gas-kinetic Bhatnagar-Gross-Krook (BGK) model equation is of the form:

$$\frac{\partial f}{\partial t} + \mathbf{v}.\nabla_r f = \frac{f_{eq} - f}{\tau}$$ (6.53)

where f is the distribution function, which is a function of time t of the spatial location $\mathbf{x}$, the particle velocity $\mathbf{v}$, and the translational and rotational degrees of freedom. The change in the distribution function due to the free flowing of molecules in space is represented by the second term in the left-hand side of Eq. (6.53) and the simplified collision term of the Boltzmann equation is denoted in the right-hand side. In BGK model, the collision operator term contains τ, which is the single relaxation time. This is the time taken for a nonequilibrium state to evolve to an equilibrium state, represented as f_{eq}. In the special case, where there is no spatial gradient in the distribution function, say in a homogeneous system, the approach to equilibrium occurs by the collision term and a single relaxation mechanism. Note that the relaxation time is not known *a priori*.

The BGK equation finds many applications in phenomenological statistical mechanics. Hynes *et al.* have used it to study solvent friction effects on the rate of barrier crossing rate in chemical kinetics.

6.9 SOLUTION OF BOLTZMANN'S EQUATION: HILBERT AND CHAPMAN-ENSKOG THEORY

The first reliable solution of the nonlinear Boltzmann equation was presented by Hilbert, and subsequently improved up on by Enskog and Chapman. The procedure is rather involved, and we refer the interested students to the book by Resibois and Deleener, and by Chapman and Cowling. These two books provide detailed derivations of the solution to the Boltzmann equation.

These solutions employ two limiting conditions that apparently look contradictory but are expected to hold over a range of density in dilute gas with short range interactions. A reduced density parameter $\varepsilon = n\sigma^3$ is introduced where n is the number density and σ is the range of interaction. ε acts as the smallness parameter. Now one assumes that we can write the Boltzmann equation as

$$\frac{\partial f}{\partial t} + \mathbf{v}.\nabla_r f + \frac{\mathbf{F}}{m}.\nabla_v f = \frac{1}{\varepsilon} I[f] \tag{6.54}$$

where $I[f]$ is the Boltzmann collision operator. Small ε implies that the streaming term on the left-hand side contributes little to the evolution of the phase space (single particle) density. This limit assures convergence to Maxwell's velocity distribution. We then assume the following expansion of the distribution function in the smallness parameter ε:

$$f = f^{(0)} + \varepsilon f^{(1)} + \varepsilon^2 f^{(2)} + ... \tag{6.55}$$

We hope that the reader has realized the contradictory assumptions: the density is low yet collision integral dominates. However, this can hold true when the external force is zero or negligible and *the gradient term is small. That is, the system is nearly homogeneous.* Under such combined conditions, the zeroth order term obeys the following expression:

$$f^{(0)}(\mathbf{r}, \mathbf{v}, t) = n(\mathbf{r}, t) \left[\frac{m}{2\pi k_B T(\mathbf{r}, t)} \right]^{3/2} \exp\left(-\frac{m(\mathbf{v} - \mathbf{v}(\mathbf{r}, t))^2}{2\pi k_B T(\mathbf{r}, t)} \right) \tag{6.56}$$

which is a local equilibrium Maxwell-Boltzmann distribution that accounts for a possible local flow term $v(\mathbf{r},t)$ and a position-dependent temperature, $T(\mathbf{r},t)$, which is defined through a local equipartition theorem relation between energy density and temperature. $n(\mathbf{r},t)$ is the number density. We already encountered the local equilibrium assumption in Chapter 5, on hydrodynamics.

An important goal achieved by Hilbert was the derivation of Navier-Stokes equations from the Boltzmann equation. This is achieved through the above equation and by introducing *a set of collision invariants, which are nothing but number density, three momentum densities, and the energy density. These quantities are conserved during a collision.* These are also the conserved quantities in hydrodynamics. We refer to the text by Chapman and Cowling referred to in the Bibliography.

Chapman and Enskog introduced a systemic way to implement improvement over the local equilibrium approximation, by using the above series expansion in ε. In the standard procedure, we can substitute the expansion in the equations above to obtain the following two equations:

$$\frac{\partial f}{\partial t} = \frac{\partial f^{(0)}}{\partial t} + \varepsilon \frac{\partial f^{(1)}}{\partial t} + \varepsilon^2 \frac{\partial f^{(2)}}{\partial t} + \dots\dots, \tag{6.57}$$

$$I_S\left[f^{(0)} + \varepsilon f^{(1)} + \varepsilon^2 f^{(2)} + \dots\right] = \varepsilon I_C\left[f^{(0)} + \varepsilon f^{(1)} + \varepsilon^2 f^{(2)} + \dots\right] \tag{6.58}$$

In the above I_S is the streaming term and I_C is the collision term. When we equate the powers in ε we obtain equations for $f^{(1)}$ in terms of $f^{(0)}$, equation for $f^{(2)}$ in terms of $f^{(0)}$ and $f^{(1)}$ and so on. We already know $f^{(0)}$.

This procedure leaves us with an equation that involves two operators. We next expand the nth order distribution in terms of collision invariants so that the right-hand side vanishes, leaving us with an equation, which can be solved. The details are variable in the text by Chapman and Cowling.

We discussed in the previous chapter the important role played by separation of time scales, especially in gases and liquids. The shortest time scale is the time during a collision τ_{dc}. The next shortest time is the free time that a particle experiences between two successive collisions. We shall call it τ_{fc}. In a dilute gas, $\tau_{fc} \gg \tau_{dc}$. Note that this strong inequality breaks down in liquids.

Consider a low-density gas (what we call a dilute gas) with $n^* \ll 1$. Let us consider an instantaneous state of the system whose evolution is described by the Boltzmann kinetic equation,

$$\frac{\partial f}{\partial t} + \mathrm{v}.\nabla_r f + \frac{\mathbf{F}}{m}.\nabla_v f = \frac{1}{n^*} I[f], \tag{6.59}$$

where $I[f]$ is the collision operator.

Let us analyze the left-hand side of the equation. In the absence of an external force, the third term of the left-hand side is zero. The second term is expected to be small when the distribution is mildly inhomogeneous when n^* is small, the time evolution of $f(\mathbf{r}, \mathbf{p}, t)$ is given by the right-hand side.

6.9.1 VERIFICATION OF BOLTZMANN'S *H*-THEOREM

Boltzmann's *H*-theorem has been studied numerically, by different models, although an extensive systematic study seems to be missing. We discuss below certain generalization that have been pursued. Whatever studies available, the theorem has been found to be valid. The approach to the equilibrium seems to have an initial rapid approach, followed by a slow tail, as shown in Figure 6.2.

6.9.2 DERIVATION OF THE NAVIER-STOKES EQUATION FROM THE BOLTZMANN EQUATION

As mentioned above, a major result of the series expansion solution of the Boltzmann kinetic equation of the distribution function $f(\mathbf{r}, \mathbf{v}, t)$ is that it also leads to a derivation of the Navier-Stokes equation. This is a highly instructive exercise, which an interested student should certainly carry out. We direct the student to the textbook by Chapman and Cowling for this exercise, as it is out of the scope of this book.

6.9.3 ENSKOG APPROXIMATION FOR FRICTION

For the hard sphere potential, the celebrated Enskog expression is given by,

$$\zeta_{E,\,HS} = \frac{8\pi}{3}\rho\sigma^2 g(\sigma)\sqrt{\frac{mk_B T}{\pi}}.$$

(6.60)

This expression finds wide use. The above expression differs from the Maxwell-Boltzmann by having the factor $g(\sigma)$, which is the radial distribution function at contact. At liquid densities $\left(\rho\sigma^3 \approx 0.8\right)$, $g(\sigma)$ is large, about 4.80, it makes a big impact on the friction. If we increase the density further, $g(\sigma)$ sharply rises as it is related to the pressure by,

$$\frac{P}{\rho k_B T} = 1 + \frac{2\pi}{3}g(\sigma).$$

(6.61)

The above relation for $g(\sigma)$ is exact for hard spheres.

For arbitrary potential, there is no closed form expression. In the low-density limit, one finds the following expression:

$$\zeta_E = \frac{16\rho\sqrt{\pi}}{3\left(mk_B T\right)^{3/2}}\int_0^\infty dp \int_0^\infty db\, b\; p^5 . e^{-p^2/mk_B T}\left[1 - Cos\chi(p,b)\right].$$

(6.62)

$$\chi(p,b) = \pi - 2b\int_{r_{min}}^\infty dr\,\frac{1}{r^2\sqrt{1 - \dfrac{m\phi(r)}{p^2} - \dfrac{b^2}{r^2}}}.$$

(6.63)

where r_{min} is the turning point in the collision process (distance of closest approach).

In Figure 6.3, we compare simulation results with Enskog prediction for argon gas. The agreement is excellent in low density $\rho\sigma^3 \approx 0.8$, $T^* = 1.5$.

Enskog friction, however, fails to describe diffusion via the Einstein relation $D = \dfrac{k_B T}{\zeta}$, except at small densities $\left(\rho^* \leq 0.2\right)$. At intermediate densities, contribution from the flow of currents adds to the diffusion, which can be described by the mode coupling theory as described later.

6.10   AN ALTERNATIVE REPRESENTATION OF BOLTZMANN'S H-FUNCTION

There has been various generalization of Boltzmann's H-function in the attempt to capture the approach of a nonequilibrium state to the final equilibrium one. Not all of them seem to be based on a kinetic equation of the type pioneered by Boltzmann. However, this approach is quite popular. For example, the following form has been used,

$$H_{GB}(t) = -\int d\mathbf{x}P(\mathbf{x},t)\ln\left(\frac{P(\mathbf{x},t)}{P_{eq}(\mathbf{x})]}\right)$$

(6.64)

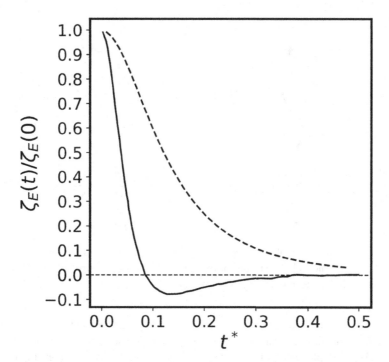

FIGURE 6.3 The calculated time dependent Enskog friction plotted at a low reduced density of $\rho^* = 0.1$, at the reduced temperature $T^* = 1.5$, for Lennard-Jones spheres. We also show the exponential approximation by the dashed line. See Reference [18] for further details.

This generalized form of Boltzmann serves similar purpose as the original Boltzmann's *H*-function. Here $P_{eq}(\mathbf{x})$ is the equilibrium distribution of a given variable **x**. The variable "**x**" has often been assumed to be a position variable or a combination thereof.

6.11 THREE-BODY EFFECTS

An interesting recent extension of Enskog theory was presented recently that included three-body effects on friction. This was found to be particularly important at intermediate densities, for particles with attractive interactions, like Lennard-Jones potential. This study could explain the observed difference in diffusion between hard sphere and LJ fluids. In a nutshell, this effect is again like in a Virial series where the third virial coefficient can be important at intermediate densities, and vastly different between hard spheres and LJ particles. These effects have been discussed in the paper by Miyazaki *et al.* [16].

6.12 INCLUSION OF RING COLLISIONS

Boltzmann's kinetic equation includes the change of probability distribution, $f(\mathbf{r}, \mathbf{p}, t)$ through a binary collisions, and through subsequent use of molecular chaos that allows us to decompose a two-particle phase space distribution into the product of two single-particle distribution. In the Enskog treatment, the decay of the velocity time correlation function (discussed later in Chapter 10 of this book) is exponential, for hard spheres.

There was a long gap before any progress on this front was made, partly because of the difficulty of solving the nonlinear Boltzmann kinetic equation. Subsequent to the discovery of the long-time tail in the velocity time correlation function of hard disks and hard spheres, it became apparent that neither the

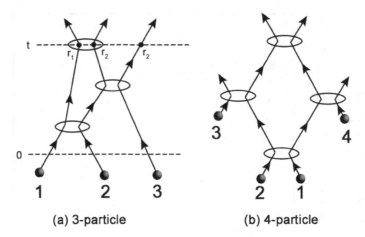

(a) 3-particle   (b) 4-particle

FIGURE 6.4 We depict two examples of possible sequences of binary collisions among the neighbouring particles. In this figure, we show ring collisions of two types (i) involving 3 particles and (ii) 4 particles. The rings depict repeated collisions among the same group of neighbours.

Boltzmann kinetic equation nor the generalization due to Enskog can explain these new features. Cohen and Dorfman in the 1970s showed that one could take into consideration a sequence of binary collisions that forms a ring. For example, consider the following series of binary collisional events $1 + 2$:

$$1 + 2 \rightarrow 1' + 2'$$
$$2' + 3 \rightarrow 2'' + 3'$$
$$3' + 4 \rightarrow 3'' + 4'$$
$$4' + 1' \rightarrow$$

Examples of ring collisions have been depicted in Figure 6.4. Such collisions introduce intermolecular correlations not included in Boltzmann's equation. Fortunately, when the repeated collisions form only rings (and chains that are already included in Boltzmann's equation), it can be solved to obtain such quantities as the velocity TCF. We refer to the relevant works cited in the Bibliography.

6.13 CAGE EFFECT

A useful physical picture to visualize a liquid is to imagine it as a collection of spheres at high density, as was studied by Bernal many years ago. In this model of liquid, a tagged molecule is surrounded by a large number of neighbors who are essentially slightly away from the contact and the minimum distance of approach. Thus, these neighboring molecules form a cage around our tagged central molecule and slow down the outward movement of this molecule. As a result repeated collisions take place, including the ring collisions discussed above. Figure 6.5 depicts formation of such a cage by repeated collisions.

The cage is manifested in a sharp peak in the radial distribution function. Interestingly, some of the effects of the cage can be included through an effective force derived from the density-functional theory that we described in our equilibrium statistical mechanics book, references are given below.

In the high-density limit one can successfully adopt a simple description initiated by de Gennes, where a Smoluchowski equation of motion for the time-dependent density can be used to describe semi-quantitatively the density relaxation. We have discussed this topic later in Chapter 23.

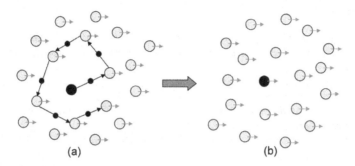

FIGURE 6.5 Here the tagged particle (filled black circle) gets repeatedly kicked around by the other particles (filled gray circles) moving in some positive direction. It is intuitively clear that because of these collisions, the tagged particle loses memory of its initial velocity at some positive rate. The tagged particle couples ("feels") the collective hydrodynamic modes, like current and density, through these repeated collisions.

The cage gives rise to repeated collisions, as described above. Unfortunately, the collisions are more complex than can be described by simple rings. In the high-density liquid, one needs to generalize further.

6.14 KINETIC THEORY APPROACH TO MODE COUPLING THEORY

We have discussed this topic in Chapter 23, which is a review of mode coupling theory. As pointed out earlier in this chapter and elaborated upon in Chapter 23, the Boltzmann kinetic equation breaks down in dense gases and in the liquid state because of repeated collisions between the same pair of molecules. This cage needs to relax in order for the tagged particle to move away. This is described by using density relaxation. Thus, the coupling between a particle's velocity and the density mode of the liquid are coupled, and such couplings are addressed by the mode coupling theory.

6.15 SUMMARY

In this (difficult) chapter, we briefly discussed the basic concepts of a highly developed area of time-dependent statistical mechanics – the kinetic theory of gases. The area started with the seminal works of Maxwell (and several others) who introduced the language of velocity distribution function in a gas, and could obtain, for the first-time microscopic expressions for thermodynamic quantities like pressure, specific heat, and transport coefficients like diffusion, thermal conductivity, and similar quantities. Boltzmann realized the great scope of Maxwell's work and attempted to include intermolecular interactions in terms of collision interactions between any two atoms. The efforts of Boltzmann led to the birth of statistical mechanics. However, Boltzmann's kinetic-equation-based approach turned out to be very difficult and kept theoreticians busy for the next hundred and fifty years. The task is still not finished.

Perhaps the most recent important work on kinetic theory of gases was carried out by Cohen, Dorfman, and Ernst (referenced below). They together included the effects of ring collisions. This was a remarkable feat. This leads to the description of the long-time power-law tail.

Subsequently, an extended kinetic theory was developed by de Schepper and Cohen who introduced extended kinetic theory that can be considered an extension of the Enkog theory. We have discussed this work in the chapter on mode coupling theory (Chapter 23).

BIBLIOGRAPHY

1. Chapman, S., and T. G. Cowling. 1990. *The Mathematical Theory of Non-Uniform Gases: An Account of the Kinetic Theory of Viscosity, Thermal Conduction and Diffusion in Gases*. Cambridge: Cambridge University Press.
2. Résibois, P. M., M . De Leener, and M. F. De Leener. 1977. *Classical Kinetic Theory of Fluids*. New York: John Wiley & Sons.
3. Cercignani, C. 1969. *Mathematical Methods in Kinetic Theory*. New York: Springer.
4. Huang, K. 1991. *Statistical Mechanics*. Hoboken: Wiley
5. Ernst, M. H., E. H. Hauge, and J. M. J. van Leeuwen 1970. Asymptotic time behavior of correlation functions. *Physical Review Letters 25,* 1254.
6. Dorfman, J. R., and E. G. D. Cohen. 1970. Velocity correlation functions in two and three dimensions. *Physical Review Letters 25,* 1257.
7. Ernst, M. H. 1991. Mode-coupling theory and tails in CA fluids. *Physica D 47,* 198.
8. Dorfman, J. R., and H. van Beijeren. 1977. *The Kinetic Theory of Gases, in Statistical Mechanics, Time-Dependent Processes,* ed. B. J. Berne, ch 3. New York: Plenum Press.
9. Stewart, H. 1971. *An Introduction to the Theory of the Boltzmann Equation*. New York: Dover Books. p. 221.
10. Hynes, J. T., R. Kapral, and M. Weinberg. 1979. Molecular theory of translational diffusion: microscopic generalization of the normal velocity boundary condition. *Journal of Chemical Physics 70,* 1456.
11. Kauzmann, W. 2012. *Kinetic Theory of Gases*. New York: Dover Publications.
12. Cartwright, M. 2017. *Kinetic-Molecular Theory*. Visionlearning. Vol. CHE-4 (2).
13. Tong, D. *Kinetic Theory*. University of Cambridge Graduate Course (www.damtp.cam.ac.uk/user/tong/kintheory/kintheory.pdf)
14. Alder, B. J. and T. E. Wainwright. 1970. Decay of the velocity autocorrelation function. *Physical Review A 1,* 18.
15. Bhattacharyya, S. and B. Bagchi. 1997. Anomalous diffusion of small particles in dense liquids. *Journal of Chemical Physics 106,* 1757.
16. Miyazaki, K., G. Srinivas, and B. Bagchi. 2001. The Enskog theory for self-diffusion coefficients of simple fluids with continuous potentials. *Condensed Matter Physics, 4,* 315.
17. Kumar, S., S. Acharya, and B. Bagchi. 2023. Examination of Boltzmann's *H*-function: dimensionality and interaction sensitivity dependence, and a comment on his *H*-theorem. *Physical Review E 107,* 024138.
18. Bagchi, B., G. Srinivas, and K. Miyazaki. 2001. The Enskog theory for classical vibrational energy relaxation in fluids with continuous potentials. *Journal of Chemical Physics, 115,* 4195.

Part II

Fundamentals

7 Liouville's Theorem, Liouville's Equation, and BBGKY's Hierarchy

Joseph Liouville [1809–1882]

OVERVIEW

Liouville's equation, together with Liouville's theorem, forms the backbone of time-dependent statistical mechanics. Liouville's equation (LE) is an equation of motion of the time-dependent N-particle, $6N$-dimensional (when particles are atoms) phase space probability density. It builds on the description that already existed in classical mechanics, like Hamilton's equations. LE serves as the starting point for derivation of useful equations like reduced distribution functions, which are more useful than the full LE equation itself. In addition, many exact formal relations are also derived by using Liouville's equation. Examples include the linear response theory, projection operator technique. Liouville's equation plays an essential role in developing the time correlation function formalism of transport properties like viscosity and diffusion coefficient. In this chapter, we introduce and discuss Liouville's equation and Liouville's theorem. In the subsequent chapters we use LE in many applications of TDSM.

7.1 INTRODUCTION

In the year 1838, the equation that bears his name was derived by Liouville [1]. It was J.W. Gibbs who first pointed out the importance of the Liouville equation as the fundamental equation of statistical mechanics [2]. It not only forms the conceptual backbone of all subsequent developments, but it also serves as the starting point of many other derivations and approximate treatments. Students are suggested to go through this chapter carefully and develop a thorough understanding of the concepts.

Let us start with a simple thought-experiment. Let us consider opening a small bottle of dense iodine gas at the corner of a room. The reason we want it dense is that we can see the gas because of its unique purple color. Let us consider the bottle to have a large opening so that the gas comes out fast, and we can watch the gas to become disbursed and the color disappear with time. We understand that the iodine atoms are now distributed uniformly in the air of the room. Now imagine that we could follow individual atoms, and also the oxygen and nitrogen molecules of the air. What would we see? We shall see that the iodine atoms are undergoing collisions with the surrounding oxygen and nitrogen atoms that are abundant. These atoms move in the three-dimensional space with random velocities and the collisions appear to be random. However, we also know that these

DOI: 10.1201/9781003157601-9

atoms and molecules move in space following Newton's equations of motion. But it is hard to visualize the motions of so many molecules. Nevertheless, the motion of an iodine atom is coupled to the surrounding atoms and molecules of the air. The motion of an individual iodine atom can be described by using a six-dimensional space spanned by the coordinates and momenta of the tagged iodine atom. The motion of the N molecules however requires $6N$-dimensional space for positions and momenta of the N molecules. We can think of a $6N+6$-dimensional space for a complete description of the time-dependent evolution of the system one iodine plus N air molecules). Such a detailed every-particle-based description however is too complex and might not be needed. This $6N$-dimensional space is called phase space of the N-particle system.

The main advantage of introducing the phase space is that it allows us to build distribution functions and their evolution with time. A phase space distribution $\rho(\Gamma)$ gives the probability of observing the system in a given region of the space.

As already mentioned, we deal with two types of phase space. One is a single particle 6-dimensional space to follow the individual atom's motion, like that of an iodine molecule. This 6-dimensional space is spanned by three momenta and three coordinates. This single-particle phase space is called μ-space. The instantaneous state of the particle is denoted by a point in this μ-space. The $6N$-dimensional phase space is called Γ-space. In our given example, the motion in the μ-space of a single iodine atom is coupled to the motion of the $6N$-dimensional Γ-space of air molecules.

To repeat, a point in N-particle phase space at time t denotes the instantaneous positions and momenta of the N particles at that point of time. As time evolves, the point moves depicting the movement of the molecules of the system. The curve that joins the points is called a trajectory.

While the evolution of positions and momenta are obtained by solving either Newton's equations of motion or Hamilton's equations, they provide too detailed information that is not required (most of the time). Even if we could solve for the trajectory, we would not know what to do with so much data. A trajectory is not an observable except the special case of Brownian motion of a large colloidal particle, which is visible under a microscope. In the latter example, we may be able to follow the motion of the colloid in its μ-space while the motion of the solvent in its Γ-phase space, which is not visible, as it has been averaged out.

A trajectory in phase space contains information about the microscopic motion of the system. To build a quantitatively reliable theory, we need to obtain a reduced description, but yet need to start from the phase space description. This starting point is given by Liouville's equation.

To gain some insight into the phase space, let us consider the Hamiltonian of a simple harmonic oscillator

$$H = KE + PE = \frac{p^2}{2m} + \frac{1}{2}kq^2 \tag{7.1}$$

We next plot the trajectory of the harmonic oscillator in phase space. For constant energy, the trajectory is an ellipse. *Such closed trajectories are signatures of bound states.* On one elliptical curve, the energy of the particle is constant. Figure 7.1 shows the phase space trajectories of the harmonic oscillator for two constant energy values, E_1 and E_2. Three points (A, B, and C) on the outer ellipsoid represent three points in the phase space indicating widely separated positions of the system in its phase space, even at constant energy E_2.

Next, we may consider the trajectory of tagged particles (atom or molecule) in a liquid. Our tagged particle interacts with the rest of the particles in the system and as a result, its positions and momenta change with time. Figure 7.2 shows trajectories of three tagged particles in liquids. These trajectories show the difference in the trajectory of a free particle from that in a bound state. When the trajectory is dispersed and covers a large domain, we consider the system to be ergodic.

Another interesting example of a trajectory is provided by the case where the system escapes from one bound state to another. We show the trajectory of such a trajectory in Figure 7.3.

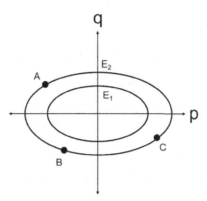

FIGURE 7.1   Phase space trajectories of the harmonic oscillator for two constant energy values, E_1 and E_2. It is to be noted that three points (A, B, and C) on the ellipsoid denote three points in phase space indicating widely separated positions of the system in its phase space, even at constant energy E_2.

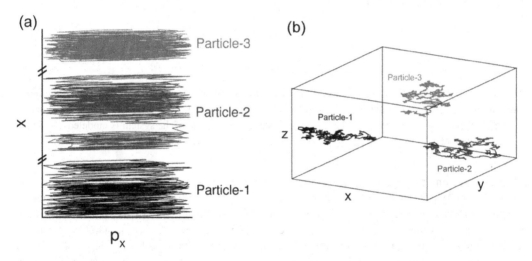

FIGURE 7.2   (a) Schematic representation of the trajectory in μ phase space. (b) The trajectory of three tagged spherical particles from molecular dynamics simulations. The continuous movement of the position of the particles in the configuration space is a consequence of interactions with surrounding liquid molecules.

7.2   ENSEMBLE

We now move towards developing the concept of phase space density, ρ (Γ,t), which in general is time-dependent. We accomplish this by employing the concept of ensembles. We have discussed the concept and the utility of ensembles in our textbook on *Equilibrium Statistical Mechanics*. We shall briefly go through the same here. The concept of an ensemble is a brilliant mental construction of J.W. Gibbs. This is based on the realization that a system at equilibrium must have a very large number of microscopic states. And the natural motion of the system at nonzero temperature takes the system through a finite fraction of these states in a time that is comparable to the time of measurement of the macroscopic properties. If we now consider that the system remains at *fixed constant energy all these times*, then the trajectory moves on a constant energy surface. Thus, as the particles of the system move in accordance with the laws of mechanics (classical or quantum), the microscopic state of the system changes. This is the same trajectory we discussed above. Macroscopic

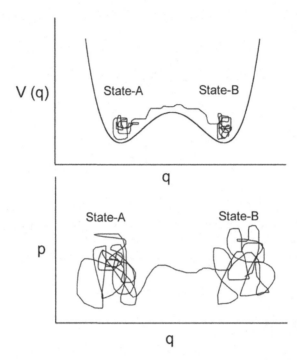

FIGURE 7.3 Trajectory depicting the transition from one bound state to another bound state. In the upper panel, the potential energy is shown against the configurational coordinate q while the lower panel shows the space trajectory with momentum p being the ordinate. Here the state of the system is modeled as a solute particle in a potential minimum where the system executes a transition from one minimum to another.

properties are time averages over this trajectory. We now construct billions and billions of mental replicas of the system such that each of our replicas has the same macroscopic control parameters, like the number of particles and the volume. Importantly, each mental replica is expected to reside at a given time in a different microscopic state. Because the number of microscopic states is enormous, can be made larger than the number of mental replicas. This is an important point. This collection of metal replicas is called an ensemble. Thus, an ensemble is a collection of a large number of mental replicas or copies of the original system, each is expected to reside at a different microscopic state at a given time.

We can represent the microscopic state of each of these replicas by a point in the phase space. Thus, the time average can be replaced by an ensemble average, which means an average of the property in question (say, X) over the members of the ensemble. This is achieved by constructing a probability distribution in phase space that gives the probability of finding the system in that region of phase space. One can understand the merit of this construction by noting that the probability distribution is different for a gas and a liquid state.

Suddenly, the problem becomes tractable. While we cannot calculate the trajectory and the time average, we can think of constructing the probability distribution of the system being in a microscopic state and we can calculate that property. Sometimes such a distribution can be quite simple (like the Boltzmann distribution). The main point is that we do not need to calculate trajectories for time-independent or equilibrium properties.

To summarize, an ensemble represents a collection of a mental replica of the original system that inhabits different microscopic states of the same. For example, let us consider an ensemble of systems with constant number N at constant volume V and energy E. Such a system is called an

(*NVE*) system (also known as a *microcanonical ensemble*). In such a system, the trajectory of the molecule moves on a constant energy surface. Therefore, all the points on this trajectory are equally probable. An average of any property of the system is equal to the time average.

7.3 FROM ENSEMBLES TO DISTRIBUTION: DEFINITION OF PHASE SPACE DENSITY

Once the concept of the ensemble is understood, one can define the distribution in phase space. Since each member of the ensemble has the same number of particles (in the microcanonical or the *NVE* ensemble), each is represented by a point in the phase space. Each system is in a different microscopic state, which means they are at different locations in phase space. *Thus, we can define a volume element in the phase space and count the number of points in the volume element and thus define a phase space density.* The density can be normalized.

Clearly, at equilibrium, this density is time-independent. Each point is still moving following Newton's or Hamilton's equation of motion, but at equilibrium, the density is time invariant. However, as we disturb this distribution, for example, by applying an external field, the distorted distribution becomes time-dependent as it evolves back to the equilibrium distribution as we remove the perturbation. Thus, we do not need to follow the motion of each individual molecule of the system and obtain the time average. We can now talk of the distribution and fluctuation of thermodynamic quantities in terms of phase space density.

The phase space density provides the probability of finding the system at a given phase point, $(\mathbf{q}^{(N)}, \mathbf{p}^{(N)})$ at time t, with an initial starting point $(\mathbf{q}^{(N)}(0), \mathbf{p}^{(N)}(0))$ at $t = 0$. We denote the phase space density function by $\rho(\mathbf{q}^{(N)}, \mathbf{p}^{(N)}, t)$, where we have used the vector notation for position and momentum; $\mathbf{q}^N$ and $\mathbf{p}^N$ both denote $3N$-dimensional column vectors containing values of the positions and momenta of all the N particles of the system. As already mentioned, this gives the *probability of finding the system within a small volume element centered around,* $\mathbf{q}^N$ *and* $\mathbf{p}^N$. We now formally define the phase space density of the system by product of Dirac delta functions.

$$\rho(\mathbf{q}^{(N)}, \mathbf{p}^{(N)}, t) = \delta(\mathbf{q}^{(N)} - \mathbf{q}^{(N)}(t)) \delta(\mathbf{p}^{(N)} - \mathbf{p}^{(N)}(t)) \tag{7.2}$$

where, $\mathbf{q}^{(N)}(t)$ and $\mathbf{p}^{(N)}(t)$ represent the time-dependent positions and momenta of the system along the phase space trajectory of the system. The phase space density function is normalized as the system has to be somewhere in phase space.

$$\iint \rho(\mathbf{q}^{(N)}, \mathbf{p}^{(N)}, t) d\mathbf{q}^{(N)} d\mathbf{p}^{(N)} = 1 \tag{7.3}$$

The above formal definition of phase space density is exact and includes all the information about a system that we could possibly need. Liouville's equation (LE) is the equation of motion of this quantity.

To end this discussion, we remind the reader that *in this 6N-dimensional phase space a point denotes the instantaneous state of the system and* is called a *representative point*. As time passes, the atoms and molecules of the system execute their natural motion, which can be represented by a line joining the phase space points that the system travels through.

We need to learn some pre-requisites before we can start on the derivations of the Liouville theorem and the Liouville equation. In the flow chart (as shown in Figure 7.4), we show a concise representation of the origin and the applicability of the Liouville equation. This lays out the basic framework for the ensuing discussion in the subsequent sections.

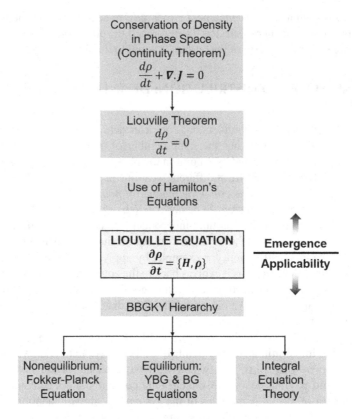

FIGURE 7.4 Flow-chart of derivation and applicability of the Liouville equation.

7.4 HAMILTON'S EQUATIONS OF MOTION

In order to follow the time evolution, that is, the motion of the system in its phase space, we require an equation of motion for the positions and the momenta of the particles of the system. This motion can be described either by Newton's equations or Hamilton's equations – the choice is dictated by the nature of constraints in the system. Both of these equations describe the temporal evolution of the system directly in terms of $6N$ momenta and position coordinates. While Newton's equations are convenient in computer simulations, they do not allow for formal manipulations about probability distribution functions and therefore are somewhat useless for theoretical and formal treatments where Hamilton's equations of motion (expressed in terms of the Hamiltonian of the system and thus energy conservation can be easily included, unlike in Newton's equations of motion) provide a straightforward approach as described below. Students may consult Goldstein's *Classical Mechanics* for more details [4].

In a conservative system, the energy is conserved as a function of time [4]. Hamilton's equations of motion are given by a set of coupled first-order partial differential equations,

$$
\begin{aligned}
\frac{\partial q_i}{\partial t} &= \frac{\partial H}{\partial p_i} \\
\frac{\partial p_i}{\partial t} &= -\frac{\partial H}{\partial q_i}
\end{aligned}
\tag{7.4}
$$

Thus, one can express the temporal evolution of $6N$ coordinates in terms of the derivatives of the system Hamiltonian with respect to the conjugate coordinates. These elegant equations are fairly universal [4].

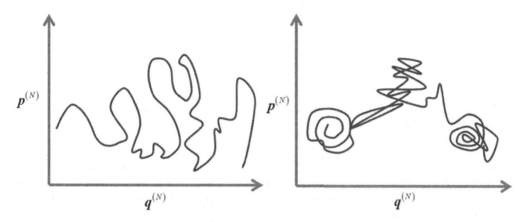

FIGURE 7.5 (a) Free particle phase space trajectory. (b) Quasi-bound state phase space trajectory.

In Figure 7.5, we depict the phase space trajectory of (a) a free particle, and (b) a quasi-bound particle. These are single particle trajectories. In a many-body system, the particles interact with each other and their motion draws a trajectory in the phase space. We discuss these concepts below.

7.5 PHASE SPACE DENSITY AND LIOUVILLE'S THEOREM

Liouville's theorem is a statement of the conservation of the number of particles of the N-particle system along its trajectory in the phase space. *This is the same as the continuity theorem of hydrodynamics*, discussed in detail in Chapter 5 on hydrodynamics. This is based on the observation that in a conservative system, particles can neither be created nor destroyed. Thus, there is no source or sink term in the phase space in a closed system. In other words, *the phase space density is a constant of motion*. The mathematical form of the Liouville theorem is given by

$$\boxed{\frac{d\rho}{dt} = 0} \qquad (7.5)$$

If we consider a volume element $dq^{(N)}dp^{(N)}$ in phase space around the point $(q^{(N)}, p^{(N)})$ then $\rho(q^{(N)}, p^{(N)}, t)dq^{(N)}dp^{(N)}$ gives the number of representative points (weight in the phase space) in the volume element. Now, the Liouville theorem states that *along the trajectory of the system in its phase space, the density is conserved*. We now show that this theorem leads to the Liouville equation (LE).

An alternative statement of the Liouville theorem is that *for a collection of N particles, the phase space volume occupied by them at any given time is constant*. That is, the volume is conserved along the motion of the system. An elegant discussion of this theorem is given in the book of Tuckerman [5].

7.6 LIOUVILLE'S EQUATION (LE)

Liouville's equation is an equation of motion for the phase space density (or, probability) distribution function, $\rho(q^{(N)}, p^{(N)}, t)$. At the level of information content, it is equivalent to Newton's or Hamilton's equations of motion, but the translation to the probabilistic description opens the door for better understanding and useful approximations. It provides a complete description of the system both at equilibrium and also away from equilibrium. Figure 7.6 represents a schematic description of the continuity equation in fluid mechanics where the density is conserved in a laminar flow.

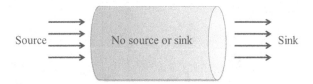

FIGURE 7.6 Schematic description of the continuity equation in fluid mechanics where the density is conserved in a laminar flow.

This equation represents the evolution of phase space distribution function for the conservative Hamiltonian systems. In short, it is a continuity equation for flux, as discussed below. The main idea behind the continuity equation is depicted in Figure 7.6. The dynamics in the phase space occurs in the absence of a source or sink.

Let $\rho\left(\mathbf{q}^{(N)}, \mathbf{p}^{(N)}, t\right)$ represent the phase space density. The derivation of the Liouville equation employs the condition that the total derivative of the phase space density at a given phase point is a sum of partial derivatives, one provides change while the position is kept fixed while the other provides the change of density due to infinitesimal change in position. They together provide the total change in phase space density as given below (in the following, we shall suppress the superscript N)

$$\frac{d\rho}{dt} = \frac{\partial \rho}{\partial t} + \sum_{i=1}^{N} \nabla_q \left(\rho \dot{q}_i\right) + \sum_{i=1}^{N} \nabla_p \left(\rho \dot{p}_i\right) \tag{7.6}$$

The left-hand side of Eq. (7.6) is identically zero, due to Liouville's theorem [see Eq. (7.5)]. Therefore, the equation of motion for the evolution of phase space density is given as

$$\frac{\partial \rho}{\partial t} = -\sum_{i=1}^{N} \nabla_q \left(\rho \dot{q}_i\right) - \sum_{i=1}^{N} \nabla_p \left(\rho \dot{p}_i\right).$$

Or, writing the derivatives explicitly

$$\frac{\partial \rho}{\partial t} = -\sum_{i=1}^{N} \left(\frac{\partial q_i}{\partial t} \frac{\partial \rho}{\partial q_i} + \frac{\partial p_i}{\partial t} \frac{\partial \rho}{\partial p_i} \right) - \rho \left[\sum_{i=1}^{N} \left(\frac{\partial \dot{q}_i}{\partial q_i} + \frac{\partial \dot{p}_i}{\partial p_i} \right) \right], \tag{7.7}$$

where we have regrouped terms for convenience (see below). We next use Hamilton's equation of motion to obtain

$$\frac{\partial \dot{q}_i}{\partial q_i} = \frac{\partial^2 H}{\partial q_i \partial p_i}; \quad \frac{\partial \dot{p}_i}{\partial p_i} = -\frac{\partial^2 H}{\partial p_i \partial q_i}. \tag{7.8}$$

When we add them

$$\rho \left[\sum_{i=1}^{N} \left(\frac{\partial \dot{q}_i}{\partial q_i} + \frac{\partial \dot{p}_i}{\partial p_i} \right) \right] = \sum_{i=1}^{N} \left(\frac{\partial^2 H}{\partial q_i \partial p_i} - \frac{\partial^2 H}{\partial p_i \partial q_i} \right) = 0. \tag{7.9}$$

Thus only the first term survives, which can now be written as

$$\frac{\partial \rho}{\partial t} = -\sum_{i=1}^{N} \left(\frac{\partial \rho}{\partial q_i} \frac{\partial H}{\partial p_i} - \frac{\partial \rho}{\partial p_i} \frac{\partial H}{\partial q_i} \right) \qquad (7.10)$$

This is the classical Liouville equation.

It is a highly nontrivial equation, where the momenta and the coordinates of all the N particles of the system are, in principle, coupled with each other. We can write the classical Liouville equation symbolically as

$$\boxed{\frac{\partial \rho}{\partial t} = -\{\rho, H\} = \{H, \rho\}} \qquad (7.11)$$

where $\{H, \rho\}$ is Poisson's bracket. For two arbitrary dynamical variables A and B of coordinates and momenta, the Poisson bracket is given by

$$\{A, B\} = \sum_{i=1}^{N} \left(\frac{\partial A}{\partial q_i} \frac{\partial B}{\partial p_i} - \frac{\partial A}{\partial p_i} \frac{\partial B}{\partial q_i} \right). \qquad (7.12)$$

Now, we define the Liouville operator, L, by

$$L = i\{H, \quad\} \qquad (7.13)$$

$i\{H, \quad\}$ is called Liouville operator denoted by, L, and defined as

$$iL = -\sum_{i=1}^{N} \left(\frac{\partial H}{\partial q_i} \frac{\partial}{\partial p_i} - \frac{\partial H}{\partial p_i} \frac{\partial}{\partial q_i} \right). \qquad (7.14)$$

This leads to the following symbolic equation:

$$\frac{\partial \rho}{\partial t} = \{H, \rho\} = -iL\rho, \qquad (7.15)$$

which has a formal solution,

$$\rho(t) = e^{-iLt}\rho(0). \qquad (7.16)$$

The solution (7.16) is highly useful in the formulation of time-dependent statistical mechanics. Students are advised to go through Refs. [6–9] for discussions and applications. The book by Feynman [6] is mainly suggested for students.

7.7 LIOUVILLE'S EQUATION FOR DYNAMICAL VARIABLES

We have already derived the equation of motion of phase space density function

$$\rho = \rho^{[N]}\left(\mathbf{q}^{(N)}, \mathbf{p}^{(N)}; t\right) \qquad (7.17)$$

Alternatively, one can derive an equation of motion for a dynamical variable A of phase space defined as, $A = A\left(\mathbf{q}^{(N)}(t), \mathbf{p}^{(N)}(t)\right)$. Note that we do not have the continuity theorem because the dynamical variable need not be a conservable quantity (such as magnetic or electric polarization). Hence, we proceed to write the equation of motion directly. The rate of change of the instantaneous value of A at a given p, q is given by

$$\left.\frac{\partial A}{\partial t}\right|_{p^N, q^N} = \sum_{i=1}^{N} \left(\frac{\partial A}{\partial q_i} \cdot \frac{\partial q_i}{\partial t} + \frac{\partial A}{\partial p_i} \cdot \frac{\partial p_i}{\partial t} \right) \tag{7.18}$$

Using Hamilton's equation of motion again we get

$$\frac{\partial A}{\partial t} = \sum_{i=1}^{N} \left(\frac{\partial A}{\partial q_i} \cdot \frac{\partial H}{\partial p_i} - \frac{\partial A}{\partial p_i} \cdot \frac{\partial H}{\partial q_i} \right) = iLA \tag{7.19}$$

Note that this equation has an expression similar to that of the Liouville equation but there is a difference in sign.

Note also that for a quantum system A is an operator and ρ is a density matrix. Then our representation of the dynamical variable advocates implicit time dependence of the phase space coordinates. This is known as Heisenberg representation. In the subsequent section, we shall discuss the quantum Liouville equation.

7.8   QUANTUM LIOUVILLE EQUATION (QLE)

The physical state k of an N-particle system is fully characterized by the wave function Ψ^k. The entire time evolution of Ψ^k is determined by the time-dependent Schrödinger equation,

$$\widehat{H}\left|\Psi^k(t)\right\rangle = i\hbar \frac{\partial}{\partial t}\left|\Psi^k(t)\right\rangle \tag{7.20}$$

As in time-dependent perturbation theory, the wave function of the system can be expanded in terms of a set of ortho-normal wave functions $\{\left|\phi_n\right\rangle\}$ (that forms the basis set) as

$$\left|\Psi^k(t)\right\rangle = \sum_n a_n^k(t)\left|\phi_n\right\rangle \tag{7.21}$$

where.

$$a_n^k(t) = \left\langle \phi_n \mid \Psi^k \right\rangle \tag{7.22}$$

By taking derivative on both sides and subsequently multiplying both sides by $i\hbar$ we obtain

$$\begin{aligned}
i\hbar \dot{a}_n^k(t) &= i\hbar \left\langle \phi_n \mid \dot{\Psi}^k(t) \right\rangle \\
&= \left\langle \phi_n \mid \widehat{H} \mid \Psi^k(t) \right\rangle \\
&= \left\langle \phi_n \mid \widehat{H} \mid \sum_m a_m^k(t)\phi_m \right\rangle \\
&= \sum_m H_{nm} a_m^k(t)
\end{aligned} \tag{7.23}$$

where the matrix element H_{nm} is expressed as

$$H_{nm} = \left\langle \phi_n \mid \widehat{H} \mid \phi_m \right\rangle.$$

We know that total probability amplitude is conserved. Therefore,

$$\sum_n |a_n^k(t)|^2 = 1 \tag{7.24}$$

The matrix elements (mn) of the density operator ρ, ρ_{mn}, is now given by

$$\rho_{mn}(t) = \frac{1}{N} \sum_k a_m^k(t) a_n^k *(t). \tag{7.25}$$

From Eqs. (7.24) and (7.25), it is clear that the conservation of probability density (that the system can be in any state n) leads to the following relation (or, constrain):

$$\sum_n \rho_{nn} = 1. \tag{7.26}$$

Now taking the derivative on both sides of Eq. (7.25) we get

$$i\hbar \dot{\rho}_{mn} = \frac{1}{N} \sum_{k=1}^N \left[i\hbar \left\{ \dot{a}_m^k(t) a_n^k *(t) + a_m^k(t) \dot{a}_n^k *(t) \right\} \right].$$

Use of the time-dependent Schrodinger equation gives the following equality:

$$i\hbar \dot{\rho}_{mn} = \frac{1}{N} \sum_{k=1}^N \left[\left\{ \sum_l H_{ml} a_l^k(t) \right\} a_n^k *(t) - a_m^k(t) \left\{ \sum_l H *_{nl} a_l^k *(t) \right\} \right]$$

$$= \sum_l \left\{ H_{ml} \rho_{ln}(t) - \rho_{ml}(t) H_{ln} \right\}$$

$$= (\widehat{H}\hat{\rho} - \hat{\rho}\widehat{H}). \tag{7.27}$$

Here we used the Hermitian property of operator $\widehat{H}$. That is, $H *_{nl} = H *_{ln}$

Using the commutator of the form [Eq. (7.27)] the quantum Liouville equation (QLE) becomes

$$i\hbar \dot{\hat{\rho}} = [\widehat{H}, \hat{\rho}]. \tag{7.28}$$

If we replace $\hat{\rho}$ with an observable A, then we obtain the quantum Liouville equation in the Heisenberg representation. The difference between the two representations is that in the Heisenberg picture there is a negative sign before the commutator.

If you are familiar with quantum mechanics, then you might have already recognized that the quantum Liouville equation [Eq. (7.28)] can be straightforwardly obtained from the classical Liouville equation [Eq. (7.15)] by using Bohr's correspondence principle, where a classical equation can be transformed into a quantum equation by replacing functions by their respective operator forms and Poisson bracket by the commutator.

Nevertheless, Eq. (7.28) is a useful form for practical use. Students are again referred to Feynman's elegant book for further discussions. We also suggest the two articles by Zwanzig – references are given at the end of the chapter.

Note that $\dot{\rho}_{mn} = 0$ indicates stationary states where (1) $\hat{\rho}(\widehat{H})$ and (2) $\dot{\widehat{H}} = 0$

In this case, the Liouville operator is denoted as

$$\hat{L} = i\left[\widehat{H}, \quad\right] \tag{7.29}$$

7.9   A FEW USEFUL COMMENTS ABOUT LIOUVILLE'S EQUATION

Note that the Liouville equation is the $6N$-dimensional analog of the equation of continuity of an incompressible fluid. It describes the fact that phase points of the ensemble are neither created nor destroyed. In the simple case of a single nonrelativistic particle moving in Euclidean space under a force field F with coordinates r and p, Liouville's theorem can be written as

$$\frac{\partial \rho}{\partial t} + \frac{\mathbf{p}}{m} \cdot \nabla_r \rho + \mathbf{F} \cdot \nabla_p \rho = 0. \tag{7.30}$$

The 1D version of this equation is quite useful for the purpose of understanding.

$$\frac{\partial \rho}{\partial t} + \frac{p}{m} \frac{\partial}{\partial x} \rho + F \frac{\partial}{\partial p} \rho = 0. \tag{7.31}$$

This simple equation finds wide use, particularly in the three-dimensional form. In astrophysics this is called the Vlasov equation, or sometimes the collisionless Boltzmann equation. This is used to describe the evolution of phase space density of a large number of collisionless particles moving in a gravitational potential. In problems of interest in chemistry, materials science and biology, not only the number of particles N is very large (typically of the order of Avogadro's number, for a laboratory-scale system), but also they are strongly interacting. So, one may think that Eq. (7.30) is not too useful. However, in dense liquids, one sometimes can use force F as a time-dependent force arising from nearest-neighbor molecules. Such an approach has been used in many applications.

When we set $\partial \rho / \partial t = 0$, we obtain an equation for the stationary states of the system and can be used to find the density of microstates accessible in a given statistical ensemble. The stationary state equation is satisfied by ρ, which is equal to any function of the Hamiltonian H. In particular, it is satisfied by the Maxwell-Boltzmann distribution $\rho = e^{-H/k_B T}$, where T is the temperature and k_B is the Boltzmann constant.

7.10   APPLICABILITY OF LIOUVILLE EQUATION

Liouville's equation finds widespread applications. Below we list a few of these applications.

(i)   Kubo used the Liouville equation to derive an elegant expression for the dynamical response of a system to a time-dependent perturbation of the system. This leads to the foundation of the fluctuation-dissipation theorem, which is one of the most profound results of theoretical science. The interested student is referred to read Kubo's paper [7].

(ii)   The Liouville equation is valid for both equilibrium and nonequilibrium systems. It is also the starting point of the derivation of Green-Kubo relations for linear transport coefficients, such as shear viscosity, thermal conductivity, or electrical conductivity [1].

(iii) The eigenfunctions of the Liouville operator form a Hilbert space as discussed by von Neumann. This property was used by Zwanzig to derive the relaxation equation by using the projection operator technique [9, 10].

(iv) After a sufficiently long time, a dynamical system will return to a state very close to its initial state. It is known as the **Poincaré recurrence** theorem, which is one of the most important applications of the Liouville theorem. The time that is elapsed for the recurrence is known as the recurrence time. The theorem was first discussed by Henri Poincaré in 1890.

7.11 FROM LIOUVILLE'S EQUATION TO BBGKY'S HIERARCHY

The Liouville equation gives a formal approach in terms of the phase space density to the time-dependent statistical mechanics. However, this is an exact formal equation that is not greatly useful in applications as it is not clear how systematic and controlled approximation can be made to treat N interacting atoms and molecules. The Liouville equation is converted to an exact and equivalent system of equations for reduced distribution functions, which are known as the BBGKY hierarchy. This hierarchy serves as an extremely useful approach to both equilibrium and time-dependent statistical mechanics. While we shall derive the equation later, below we present the equation so as to allow us to compare it with the Liouville equation.

$$\frac{\partial f^{(n)}}{\partial t} + \sum_{i=1}^{n} \frac{\mathbf{p}_i}{m} \cdot \frac{\partial f^{(n)}}{\partial \mathbf{q}_i} + \sum_{\substack{i,j=1 \\ i \neq j}}^{n} \nabla v_{ij} \cdot \frac{\partial f^{(n)}}{\partial \mathbf{p}_i} = \sum_{i=1}^{n} \int .. \int \nabla v_{i,n+1} \cdot \frac{\partial f^{(n+1)}}{\partial \mathbf{p}_i} d\mathbf{q}_{n+1} d\mathbf{p}_{n+1} \qquad (7.32)$$

where $f^{(n)}(t)$ is an n-particle distribution function in the N-particle phase space where $n \leq N$, and the rest of the terms have their usual meaning.

Also, here we have changed the notation. We shall keep $\rho(\Gamma)$ as a compact notation for the full N-particle phase space distribution. This is also in keeping with the notations introduced by Liouville and Boltzmann.

Eq. (7.32) was first derived by Born in 1935 and is known as the Bogoliubov, Born, Green, Kirkwood, and Yvon (BBGKY) hierarchy. An important aspect of this equation is that it expresses one-particle distribution in terms of two-particle distribution, and two-particle distribution in terms of three-particle distribution and so on. To use BBGKY, we need to close the hierarchy. However, the closure of this hierarchy often introduces a serious approximation.

As will be discussed later in more detail, the transformation of the Liouville equation to the BBGKY hierarchy brings out the essence of the difficulty we face in treating interacting N body problems. A position-dependent single-particle density depends on its nearest neighbors and therefore introduces the conditional probability for the existence of another particle at a distance, r, which is the two-particle distribution function. Similarly, the probability of observing two particles separated by distance, r, depends on other particles surrounding it.

An enormous amount of effort has been directed in understanding the few-body distribution functions (singlet two article and three-particle) in the interacting many-body system because most of our experimental observables can be expressed in terms of the few-body distribution function.

For a closed system of N identical particles let us consider the phase space probability density function $f^{(N)}(\mathbf{q}^N, \mathbf{p}^N, t)$. The temporal evolution of this function is given by Liouville's equation as

$$\frac{\partial f^{(N)}}{\partial t} + \left\{ f^{(N)}, H \right\} = 0 \qquad (7.33)$$

where H is the Hamiltonian of the system given by

$$H = \sum_i \frac{\mathbf{p}_i^2}{2m} + \sum_{i<j} u_{ij}(\mathbf{q}_i, \mathbf{q}_j) \qquad (7.34)$$

and $\{f^{(N)},H\}$ denotes the Poisson bracket,

$$\{f^{(N)},H\} = \sum_{i=1}^{N}\left(\frac{\partial H}{\partial \mathbf{p}_i}\frac{\partial f^{(N)}}{\partial \mathbf{q}_i} - \frac{\partial H}{\partial \mathbf{q}_i}\frac{\partial f^{(N)}}{\partial \mathbf{p}_i}\right)$$
$$= \sum_{i=1}^{N}\left(\frac{\mathbf{p}_i}{m}\frac{\partial f^{(N)}}{\partial \mathbf{q}_i} - \sum_{i<j}\nabla_q u_{ij}(\mathbf{q}_i,\mathbf{q}_j)\cdot\frac{\partial f^{(N)}}{\partial \mathbf{p}_i}\right) \quad (7.35)$$

$u_{ij}(\mathbf{q}_i,\mathbf{q}_j)$ is the potential for the interaction between particles at q_i and q_j. For an n-particle subspace in the above N-particle system, one can define the reduced probability density function, $f^{(n)}(\mathbf{q}_1...\mathbf{q}_n,\mathbf{p}_1...\mathbf{p}_n)$, as

$$f^{(n)}(\mathbf{q}_1...\mathbf{q}_n,\mathbf{p}_1...\mathbf{p}_n) = \frac{N!}{(N-n)!}\int\cdots\int f^{(N)}d\mathbf{q}_{n+1}...d\mathbf{q}_N d\mathbf{p}_{n+1}...d\mathbf{p}_N. \quad (7.36)$$

Differentiating this with respect to time, we get

$$\frac{\partial f^{(n)}(\mathbf{q}_1...\mathbf{q}_n,\mathbf{p}_1...\mathbf{p}_n;t)}{\partial t} = \frac{N!}{(N-n)!}\int..\int\frac{\partial f^{(N)}(\mathbf{q}_1...\mathbf{q}_n,\mathbf{p}_1...\mathbf{p}_n;t)}{\partial t}d\mathbf{q}_{n+1}...d\mathbf{q}_N d\mathbf{q}_{n+1}...d\mathbf{p}_N. \quad (7.37)$$

Hence, from Eqs. (7.33), (7.35), and (7.37), we get

$$\frac{\partial f^{(n)}(\mathbf{q}_1...\mathbf{q}_n,\mathbf{p}_1...\mathbf{p}_n;t)}{\partial t} = -\frac{N!}{(N-1)!}\left[\int...\int d\mathbf{q}_{n+1}...d\mathbf{q}_N d\mathbf{p}_{n+1}...d\mathbf{p}_N\sum_{i=1}^{N}\left(\frac{\mathbf{p}_i}{m}\cdot\frac{\partial f^{(N)}}{\partial \mathbf{q}_i} - \sum_{j>i}\nabla_q u_{ij}\cdot\frac{\partial f^{(N)}}{\partial \mathbf{p}_i}\right)\right]. \quad (7.38)$$

The first term in Eq. (7.38), by definition, can be written as,

$$-\frac{N!}{(N-n)!}\sum_{i=1}^{N}\int...\int d\mathbf{q}_{n+1}...d\mathbf{q}_N d\mathbf{p}_{n+1}...d\mathbf{p}_N\frac{\mathbf{p}_i}{m}\cdot\frac{\partial f^{(N)}}{\partial \mathbf{q}_i} = -\sum_{i=1}^{n}\frac{\mathbf{p}_i}{m}\cdot\frac{\partial f^{(n)}}{\partial \mathbf{q}_i}. \quad (7.39)$$

The second term in Eq. (7.38) can be transformed in the following way.

$$\frac{N!}{(N-1)!}\int...\int d\mathbf{q}_{n+1}...d\mathbf{q}_N d\mathbf{p}_{n+1}...d\mathbf{p}_N\sum_{i=1}^{N}\sum_{j>i}^{N}\nabla_q u_{ij}\cdot\frac{\partial f^{(N)}}{\partial \mathbf{p}_i}$$
$$= \frac{N!}{(N-1)!}\left(\sum_{i=1}^{N}\sum_{\substack{j=1\\j>i}}^{n}\nabla_q u_{ij}\cdot\int...\int d\mathbf{q}_{n+1}...d\mathbf{q}_N d\mathbf{p}_{n+1}...d\mathbf{p}_N\frac{\partial f^{(N)}}{\partial \mathbf{p}_i}\right).$$
$$+ \frac{N!}{(N-1)!}\left(\sum_{i=}^{N}\sum_{\substack{j=n+1\\i<j}}^{N}\int...\int d\mathbf{q}_{n+1}...d\mathbf{q}_N d\mathbf{p}_{n+1}...d\mathbf{p}_N\nabla_q u_{ij}\cdot\frac{\partial f^{(N)}}{\partial \mathbf{p}_i}\right) \quad (7.40)$$
$$= \sum_{\substack{j=1\\j>i}}^{n}\nabla_q u_{ij}\cdot\frac{\partial f^{(N)}}{\partial \mathbf{p}_i} + \int...\int d\mathbf{q}_{n+1}d\mathbf{q}_{n+1}\nabla_q u_{i,n+1}\cdot\frac{\partial f^{(n+1)}}{\partial \mathbf{p}_i}$$

Hence, the final equation of motion for the reduced n-particle distribution function becomes

$$\frac{\partial f^{(n)}}{\partial t} + \sum_{i=1}^{n} \frac{\mathbf{p}_i}{m} \cdot \frac{\partial f^{(n)}}{\partial \mathbf{q}_i} - \sum_{\substack{i,j=1 \\ j>i}}^{n} \nabla_q u_{ij} \cdot \frac{\partial f^{(n)}}{\partial \mathbf{p}_i} = \sum_{i=1}^{n} \int .. \int d\mathbf{q}_{n+1} d\mathbf{p}_{n+1} \nabla u_{i,n+1} \cdot \frac{\partial f^{(n+1)}}{\partial \mathbf{p}_i}. \qquad (7.41)$$

Note that Eq. (7.41) is an exact relation between the n-particle reduced probability density function and the $n+1$-particle reduced probability density function. Approximations need to be considered in order to render this a closed form.

For the important single-particle distribution ($n = 1$), we have the following equation:

$$\frac{\partial f^{(1)}}{\partial t} + \frac{\mathbf{p}}{m} \cdot \frac{\partial f^{(1)}}{\partial \mathbf{q}} + \mathbf{F} \cdot \frac{\partial f^{(1)}}{\partial \mathbf{p}} = \int .. \int d\mathbf{q}_2 d\mathbf{p}_2 \nabla u_{1,2} \cdot \frac{\partial f^{(2)}}{\partial \mathbf{p}_i}. \qquad (7.42)$$

This the same equation was faced by Boltzmann, and we discussed it in Chapter 6, with the following differences. The streaming term here can contain a force term (F) from either an external source or from intermolecular interactions. We added this by hand here to bring out the correspondence with Boltzmann, which, otherwise is absent in $f^{(1)}$ – the interaction term has been absorbed in the right-hand side.

Eq. (7.42) has a proud history. Not only Boltzmann used this, it is also the starting point of Vlasov's equation, and we can derive Smoluchowski equation from this form when we replace the right-hand side by a stochastic operator. It also forms the basis of the Bhatnagar-Gross-Crook (BGK) equation as discussed in Chapter 6 already.

7.12 SUMMARY

Liouville's equation describes time evolution of phase space density in an N-body interacting system. It has its roots in Hamilton's equations. The significance of the Liouville equation lies not only in its inherent fundamental value and in its ability to explain dynamics on a constant energy surface, but the multitude of systematic approximations that it allows. We have already discussed in our text entitled *Equilibrium Statistical Mechanics* how its sibling, BBGKY's hierarchy, provided the first systematic approach towards the evaluation of equilibrium distribution functions that are so essential to our statistical mechanical description of many-body systems.

In the next chapters we discuss many aspects of time-dependent statistical mechanics, starting with time correlation function formalism, and then the projection operator technique.

PROBLEMS

1. Starting $\rho = e^{-\beta H}$ from where the free particle Hamiltonian is $H = \dfrac{P^2}{2m}$ show that $\dfrac{\partial \rho}{\partial t} = 0$.

   (Hint: Write down the Liouville equation $\rightarrow$ Substitute the Hamiltonian $\rightarrow$ Take derivative w.r.t. momentum.)

2. For one-dimensional classical harmonic oscillator, solve the equation of motion to get the time evolution of co-ordinate and momentum and also solve the Liouville equation at equilibrium. (Hint: Solve the Hamiltonian equation of motion to obtain position and momentum co-ordinate evolution $\rightarrow$ Write down the Liouville equation $\rightarrow$ take derivative w.r.t. momentum and position $\rightarrow$ Use equilibrium condition, i.e., $\dfrac{\partial \rho}{\partial t} = 0$)

3. Consider a toy model where Hamiltonian has the following form, $H = \frac{1}{2}(p^2 + q^2)^2$ where q is the generalized position and p is the conjugate generalized momentum. This is actually a nonlinear harmonic oscillator model.
   (a) Solve the time evolution of co-ordinate and momentum.
   (b) Solve Liouville's equation (LE) for the system at equilibrium.
   (Hint: Similar as previous problem).

4. Suppose a spin system is placed in a constant magnetic field of magnitude B_0, direction is defined along the z direction. There local perturbing electromagnetic fields, which change randomly with time, act on the spins. Hamiltonian has the following form $H = \frac{1}{2}\omega(p^2 + q^2) + pF_y - pF_x$, where ω is the contribution of the z component of the local perturbing field B and q is the generalized position co-ordinate and p is the conjugate momentum coordinate. F_x and F_y are components of the local fields, where $F_x = \gamma M_z B$, $F_y = \gamma M_z B$ and $q = M_y$, $p = M_x$. M is the magnetization.
   (a) Write down the time evolution of position and momentum
   (b) Write down the Liouvillian for the Hamiltonian.

5. Consider for one particle system, the equation of motion in one spatial direction is given as $\dot{x} = -ax$
   Here $x(t)$ is the position of the particle at time t and a is the constant. Since compressibility of the system is nonzero the ensemble distribution function $g(x,t)$ satisfies a Liouville equation of the form,

$$\frac{dg}{dt} - \alpha x \frac{dg}{dx} = \alpha g.$$

At $t = 0$ the initial distribution has the following Gaussian form:

$$g(x,0) = \frac{1}{\sqrt{2\pi\sigma^2}} \exp\left(-x^2/2\sigma^2\right)$$

   (a) Find out a solution of the Liouville equation that satisfies the initial distribution. (Hint: Make a suitable transformation of the differential equation such that the r.h.s. will be zero. Multiply x^2 in the initial distribution with an arbitrary function, which is equal to 1 at $t = 0$ and substitute it into the new equation to solve the differential equation for arbitrary function.)
   (b) Check out the normalization of your solution.
       (Hint: $\int dx g(x,t) = 1$)

6. Calculate the density matrices for free particle and simple harmonic oscillator. Free particle Hamiltonian and the Hamiltonian of simple harmonic oscillator is given as $H = \frac{P^2}{2m}$ and $H = \frac{P^2}{2m} + \frac{1}{2}m\omega^2 x^2$ respectively. (Hint: Consult Ref. [5])

7. Derive BBGKY's hierarchy from Liouville's equation.
   (Hint: Write down the Liouville equation using Poisson bracket → Time derivative of reduced probability density → Combine these two steps).

BIBLIOGRAPHY

1. Liouville, J. 1838. Note on the theory of the variation of arbitrary constants. *Journal de Mathematiques Pures et Appliquees 3*, 342–49
2. Gibbs, J. W. 1884. On the fundamental formula of statistical mechanics, with applications to astronomy and thermodynamics. *Proceedings of American Association of Advanced Science, 33*, 57–58.
3. Bagchi, B. 2018. *Statistical Mechanics for Chemistry and Materials Science*. Boca Raton: CRC Press.
4. Goldstein, H. 2011. *Classical Mechanics* (3rd edition). Indian Subcontinent: Pearson.
5. Tuckerman, M. E. 2010. *Statistical Mechanics: Theory and Molecular Simulation*. Oxford: Oxford University Press.
6. Feynman, R. P. 1972. *Statistical Mechanics: A Set of Lectures*. Massachusetts: The Benjamin/Cummings Publishing Company.
7. Kubo, R. 1957. Statistical-mechanical theory of irreversible processes. I. General theory and simple applications to magnetic and conduction problems. *Journal of the Physical Society of Japan, 12*, 570–86.
8. Hansen, J. P., and I. R. McDonald. 1976. *Theory of Simple Liquids*. New York: Academic Press.
9. Zwanzig, R. W. 2001. *Non-Equilibrium Statistical Mechanics*. Oxford: Oxford University Press.
10. Zwanzig, R. W. 1961. *Lectures in Theoretical Physics*. Vol. III. New York: Wiley-Interscience.
11. Pathria, R. K. 1996. *Statistical Mechanics*. Oxford: Butterworth-Heinemann.
12. Arnold, W. I. 1989. *Mathematical Methods of Classical Mechanics*. New York: Springer-Verlag.

8 Time Correlation Function Formalism

OVERVIEW

Time correlation functions (TCFs) constitute essential ingredients of nonequilibrium statistical mechanics and irreversible processes. *TCFs provide the microscopic link between the experimental observables and the molecular details of the system under study.* For example, viscosity is given by time integration over the stress-stress time correlation function, which is expressed in terms of the intermolecular interaction potential. The diffusion coefficient is given by integration over the velocity TCF. As discussed in Chapter 3 of Part I, the results of light and neutron scattering experiments are expressed and interpreted in terms of appropriate TCFs. The theoretical studies of TCFs naturally constitute important parts of nonequilibrium statistical mechanics. The study is divided into several parts. In one part, we study the properties of TCFs. In another part, we study the calculations of TCFs. The latter was discussed in the chapter on hydrodynamics, where we presented expressions of density-density TCF. More accurate calculations reliable at molecular length scales are routinely carried out not only using more sophisticated approaches but increasingly using molecular dynamics simulations. However, a study of the TCFs requires an understanding of several broad, general properties that prove immensely helpful in assuring accuracy. In this chapter, we study these general properties, such as symmetry properties, time-reversal symmetry, and several others. These properties are derived starting from the Liouville equation, as discussed below. In addition, we include a mathematical algorithm that will help students in the computation of the time correlation function of a given variable using computer programs. We also direct students to the classic review of TCF by Zwanzig.

8.1 INTRODUCTION

As emphasized in Chapter 3, time correlation functions (TCFs) connect experimentally observed variables to the microscopic world. At the same time, they provide an exact, physical, and intuitive way to represent the dynamics of a system from a molecular level description. It is one of the most successful and essential tools that we employ in the study of time-dependent statistical mechanics. Although we now have computational schemes to calculate the TCFs from intermolecular interactions using molecular dynamics simulations, TCFs are often obtained from a phenomenological description. These phenomenological descriptions use a random or stochastic approach and are highly instructive. Examples are provided by hydrodynamics (as discussed in Chapter 5) and also by the Langevin equation approach, to be described later. They are of great use and relevance in the understanding of spectroscopy, as described in Chapter 3. Moreover, TCFs are of great interest in their own merit.

DOI: 10.1201/9781003157601-10

We can draw a helpful analogy between the time correlation functions of nonequilibrium statistical mechanics with the partition function we study in equilibrium statistical mechanics. All the thermodynamic properties can be calculated if we know the partition function (PF). For example, free energy is given by $A = -k_B T \ln Q(N,V,T)$, where $Q(N,V,T)$ is the canonical partition function (I direct the students to our textbook on "*Equilibrium Statistical Mechanics*", referenced below [9]). It is particularly helpful if we can analytically derive an expression of the PF of our system. However, such calculations are exceedingly difficult for real systems. Instead, we often try to obtain reduced distribution functions, which are directly related to the experimental observable. These distribution functions can be obtained through computer simulations.

Nevertheless, PF provides us with a clear understanding of where to start. Time correlation functions play quite a similar role in the theory of the transport process. Once we know the TCF, we can calculate the transport properties that determine the relaxation phenomena. And the starting point is the Liouville equation.

To develop an important and useful perspective, it is worthwhile to mention a few important relations between TCFs and the transport coefficients. Below, we enlist the exact expressions of some physical observables in terms of the integration over time correlation function. We have described the expressions of transport coefficients in terms of the appropriate TCF. *Often it is rather easy to guess the definition because it is always the TCF of the conjugate dynamical property.* For diffusion, which is the displacement of a tagged particle in the coordinate (say, x) space, the relevant TCF is the velocity-velocity autocorrelation function (ACF). For viscosity, it is the stress TCF, which is also the transverse current TCF. For the dielectric constant, it is the TCF of the total dipole moment TCF.

(i)   The diffusion coefficient in terms of velocity (**v**) autocorrelation function (ACF),

$$D = \frac{1}{3}\int_0^\infty dt \left\langle \mathbf{v}(0).\mathbf{v}(t)\right\rangle. \tag{8.1}$$

(ii)   The coefficient of viscosity (η) in terms of transverse current (σ_{ab}, $a,b = x,y,z$) ACF,

$$\eta = \frac{\beta}{5V}\sum_a \sum_b \int_0^\infty dt < \sigma_{ab}(0)\,\sigma_{ab}(t) >. \tag{8.2}$$

Here V denotes the volume of the system and $\beta = \dfrac{1}{k_B T}$.

(iii)   The infra-red absorption coefficient $\alpha(\omega)$ at a frequency ω can be written in terms of total dipole moment (**M**) ACF,

$$\alpha(\omega) = \frac{2\pi\omega^2\beta}{3nc}\int_{-\infty}^\infty dt e^{-i\omega t}\left\langle \mathbf{M}(0).\mathbf{M}(t)\right\rangle_{eq}. \tag{8.3}$$

Here c is the velocity of light in vacuum, β is the inverse temperature, and n denotes the index of refraction.

(iv)   The electrical conductivity κ in terms of the current $\mathbf{J}(t)$ ACF is defined as,

$$\kappa = \frac{\beta}{3V} \int_0^\tau dt \, \langle \mathbf{J}(0).\mathbf{J}(t) \rangle. \tag{8.4}$$

(v)   The friction coefficient ζ in terms of force $(\mathbf{F})$ ACF via Kirkwood's formula,

$$\zeta = \frac{\beta}{3} \int_0^\tau dt \, \langle \mathbf{F}(0).\mathbf{F}(t) \rangle. \tag{8.5}$$

One can state that many of the dynamical observables are described in terms of relevant TCFs. In Chapter 3, we discussed the important role of rotational and vibrational time correlation functions in explaining many spectroscopic experimental results. We shall now discuss definitions of TCFs.

8.2   DEFINITION OF TIME CORRELATION FUNCTION

The definition of the time correlation function is simple. It involves time averaging over a long time trajectory of the dynamical quantity whose TCF is being calculated. In the case of autocorrelation, it is particularly simple and is given by, for a dynamical quantity $A(t)$ by

$$\boxed{\langle A(0)A(t) \rangle = \lim_{T \to \infty} \frac{1}{T} \int_0^T ds \, A(s) \, A(t+s)}. \tag{8.6}$$

The following is the operational definition of the above equation. Let us first note that the system is in equilibrium and executing its natural motion in phase space through the motions of atoms and molecules that constitute the system. As the system moves along its trajectory through its phase space through the motions of its atoms and molecules, the variable A changes with time. We generally describe any dynamical state of the system by specifying phase space coordinates $\mathbf{q}_1, \mathbf{q}_2, \ldots, \mathbf{q}_N$ and conjugate momenta $\mathbf{p}_1, \mathbf{p}_2, \ldots, \mathbf{p}_N$ of all degrees of freedom of the N-particle system at any particular time. Our dynamical variable A is a function of coordinates $\mathbf{p}_1, \mathbf{p}_2, \ldots, \mathbf{p}_N$ and conjugate momenta $\mathbf{p}_1, \mathbf{p}_2, \ldots, \mathbf{p}_N$.

The time evolution of any mechanical property is given by the Liouville equation, $A(t) = e^{iLt} A(t=0)$. Thus, we have a quantitative measure of the variation of A with time. However, the system inhabits adjacent regions in a short time, so the values of A are correlated at short time separations. It shall go $\langle A \rangle^2$ at long-time separations when the variables become uncorrelated.

In Figure 8.1(a), we plot the energy of a system of atoms interacting via Lennard-Jones' interaction potential. The system is at equilibrium. Thus, the energy fluctuates around an average value,

$$\delta E(t) = E(t) - \langle E \rangle. \tag{8.7}$$

And the relevant TCF is $C_E(t) = \langle \delta E(0) \, \delta E(t) \rangle$ In Figure 8.1(b), we plot the normalized time correlation function of energy fluctuation, defined as a time average over a long trajectory. The decay of energy TCF shows behavior typical to many dynamical variables. The decay usually has three stages. It is slow and Gaussian in the beginning (not clear from the above figure, though), then becomes faster and exponential, and often exhibits a nonexponential tail as it approaches zero, within only a few ps in the example shown. This TCF plays an important role in spectroscopy and

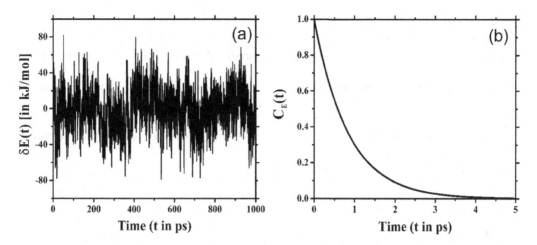

FIGURE 8.1 (a) Temporal evolution of energy fluctuation and (b) the normalized time correlation function of energy plotted against time. This is obtained from molecular dynamics simulations of liquid argon at $T = 120$ K. Note that the decay is complete within ~4 ps.

also in determining the frequency-dependent specific heat. The latter is measured by specific heat spectroscopy.

One often needs TCF between two different dynamical variables, which could be defined as,

$$\langle A(0)B(t)\rangle = \lim_{T\to\infty}\frac{1}{T}\int_0^T ds\, A(s)\,B(t+s).$$ (8.8)

Such a correlation often allows an understanding of the anticorrelation between two quantities, like potential energy and kinetic energy, in a system where energy is not globally conserved.

Sometimes it helps in the numerical calculation of a TCF if we include an averaging over multiple initial configurations sampled from an equilibrium distribution of the initial starting configuration. Let $\rho_0(\mathbf{p},\mathbf{q})$ be canonical probability density. We then define the TCF in the following fashion:

$$\langle A(0)A(t)\rangle = \int d\Gamma \rho_0(\Gamma)\lim_{T\to\infty}\frac{1}{T}\int_0^T ds\, A(s;\Gamma)\,A(t+s;\Gamma)$$ (8.9)

where $\rho_0(\Gamma) = \dfrac{e^{-\beta H(\Gamma)}}{Q}$ and Q denotes the canonical partition function. In Eq. (8.9), $A(t;\Gamma)$ is

defined as $A(t;\Gamma) \equiv A(\mathbf{q}^N(t),\mathbf{p}^N(t))$.

By notation we use, $A(t=0) = A$. That is, the variable $t=0$ for the initial time is not carried forward.

8.3 IMPORTANT AND USEFUL PROPERTIES OF TCF

As we have indicated earlier, TCFs provide microscopic connections to many different measurable dynamical properties of equilibrium systems. However, TCFs themselves have a number of interesting properties that additionally find wide use in time-dependent statistical mechanics.

8.3.1 Self-Adjoint Property

We shall first derive several exact mathematical relations that *all* TCFs satisfy by virtue of the following properties of the Liouville operator (L), reproduced here for the reader's convenience.

(i) Note that Liouville operator L as defined, is an imaginary operator, with an "i" in front.

(ii) Note that $A(t) = e^{iLt}A(0) \equiv e^{iLt}A$. And, $A^*(t) = \left(e^{iLt}A\right)^* = e^{iLt}A^*$ because iL is real.

This is important as it finds frequent use in theoretical manipulations and formal proof. We proceed to prove the property now.

Because iL is real, it immediately follows that

$$(iL)^* = iL \text{ and } \left(e^{\pm iLt}\right)^* = e^{\pm iLt}.$$

By definition,

$$\langle A\,|\,LB \rangle = \int d\Gamma \rho(\Gamma)\left(A(\Gamma)\right)^* LB(\Gamma). \tag{8.10}$$

Since $iL = \dot{\Gamma}.\partial/\partial\Gamma$, it follows that $L = -i\dot{\Gamma}.\partial/\partial\Gamma$, and so

$$\langle A\,|\,LB \rangle = -i\int d\Gamma \rho(\Gamma)A^*(\Gamma)\dot{\Gamma}.\frac{\partial}{\partial\Gamma}B(\Gamma).$$

If we integrate this equation by parts to obtain,

$$\begin{aligned}\langle A\,|\,LB \rangle &= -i\left[\dot{\Gamma}\rho(\Gamma)A^*(\Gamma)B(\Gamma)\right]^s + i\int d\Gamma B(\Gamma)\frac{\partial}{\partial\Gamma}.\dot{\Gamma}\rho(\Gamma)A^*(\Gamma)\\ &= -\int d\Gamma B(\Gamma)\rho(\Gamma)LA^*(\Gamma)\\ &= \langle LA\,|\,B\rangle\end{aligned}$$

We have discarded the surface term [i.e., $\dot{\Gamma}\rho(\Gamma)A^*(\Gamma)B(\Gamma)$] in the above integration. In the limits $\{q^N\} \to \infty, \{p^N\} \to \infty$, we can safely assume that there will be no microstates with infinite momenta; hence $\rho(\Gamma)$ in this limit vanishes. *This derivation establishes the Hermiticity property* of the Liouville operator. At the same time, because $d\Gamma$ and $\rho(\Gamma)$ both are real, we can write the above equation as

$$\begin{aligned}\langle A\,|\,LB \rangle &= \left(\int d\Gamma \rho(\Gamma)B^*(\Gamma)LA(\Gamma)\right)^*.\\ &= \langle B\,|\,LA\rangle^*\end{aligned}$$

Therefore,

$$\langle A\,|\,LB \rangle = \langle B\,|\,LA \rangle^*. \tag{8.11}$$

This property is useful when the correlation function is complex.

8.3.2 Stationarity Property

This valuable property holds at equilibrium. Let us consider a general time correlation function

$$C_{AB}(t_1, t_2) = \int d\Gamma \rho(\Gamma)\left(e^{it_1 L}A(\Gamma)\right)^* e^{it_2 L}B(\Gamma). \tag{8.12}$$

From the adjoint property of e^{iLt}, i.e., $\left(A^{*}e^{iLt}B\right)=\left(e^{-iLt}A^{*}B\right)$ we have

$$\begin{aligned}
C_{AB}(t_1,t_2) &= \int d\Gamma \rho(\Gamma)\left(e^{-iLt_1}e^{iLt_2}B(\Gamma)\right)A^{*}(\Gamma) \\
&= \int d\Gamma \rho(\Gamma)A^{*}(\Gamma)\left(e^{-iLt_1}e^{iLt_2}B(\Gamma)\right). \\
&= \left\langle A \mid B(t_2-t_1)\right\rangle \equiv C_{AB}(t_2-t_1)
\end{aligned}$$

(8.13)

In equilibrium, the TCF for a property defined by values of the dynamical variables A and B at two different times depends only on the *difference* between the two times, and not on their absolute values. So TCFs are stationary.

This important property often finds use in TDSM is dependent on the adjoint property of the operator e^{iLt}.

8.3.3 TIME REVERSAL SYMMETRY OR PARITY

Let us consider the time correlation function

$$\left\langle A \mid B(t)\right\rangle = \int d\Gamma \rho(\Gamma)\left(A(\Gamma)\right)^{*} e^{iLt}B(\Gamma).$$

(8.14)

We remind the reader that we drop $t=0$ and write $A(t=0)$ as A. Expanding the exponential function in series form, we get

$$\begin{aligned}
\left\langle A \mid B(t)\right\rangle &= \sum_{n=0}^{\infty} \frac{(it)^{n}}{n!} \int d\Gamma \rho(\Gamma)\left(A(\Gamma)\right)^{*} L^{n}B(\Gamma) \\
&= \sum_{n=0}^{\infty} \frac{(it)^{n}}{n!} \int d\Gamma \rho(\Gamma)\left(A(\Gamma)\right)^{*} LL^{n-1}B(\Gamma) \\
&= \sum_{n=0}^{\infty} \frac{(it)^{n}}{n!}\left\langle A \mid LL^{n-1}B\right\rangle
\end{aligned}$$

(8.15)

Using the Hermitian property of L (i.e., $\left\langle A \mid LB\right\rangle = \left\langle LA \mid B\right\rangle$), we can write as,

$$\left\langle A \mid B(t)\right\rangle = \sum_{n=0}^{\infty} \frac{(it)^{n}}{n!}\left\langle LA \mid L^{n-1}B\right\rangle.$$

(8.16)

Iterating in this way, we finally reach the following equation:

$$\left\langle A \mid B(t)\right\rangle = \sum_{n=0}^{\infty} \frac{(it)^{n}}{n!}\left\langle L^{n}A \mid B\right\rangle.$$

Recalling that L is imaginary, i.e., $\left(L^{n}\right)^{*} = (-1)^{n} L^{n}$ we obtain

$$\begin{aligned}
\left\langle A \mid B(t)\right\rangle &= \int d\Gamma \rho(\Gamma)\left(e^{-iLt}A^{*}(\Gamma)\right)B(\Gamma) \\
&= \int d\Gamma \rho(\Gamma)\left(e^{-iLt}A(\Gamma)\right)^{*} B(\Gamma). \\
\left\langle A \mid B(t)\right\rangle &= \left\langle e^{-itL}A \mid B\right\rangle \equiv \left\langle A(-t) \mid B\right\rangle
\end{aligned}$$

(8.17)

This relation, $\langle A | B(t) \rangle = \langle A(-t) | B \rangle$ is called the time-reversal symmetry. Note that this is trivially true for dynamical variables in deterministic systems, like Newton's equations of motion. It is useful to envisage two arrows emanating from the $t = 0$ point in two directions. Thus, it is essentially a statement of time translation property derived above.

8.3.4   TIME DERIVATIVE OF TCF

Consider the time correlation function $C_{AB}(t)$, defined as

$$C_{AB}(t) = \int d\Gamma \rho(\Gamma) A^*(\Gamma) e^{itL} B(\Gamma). \tag{8.18}$$

We take the time derivative on both sides of the above equation to obtain

$$\frac{\partial C_{AB}(t)}{\partial t} = \int d\Gamma \rho(\Gamma) A^*(\Gamma) iL e^{iLt} B(\Gamma),$$

on the other hand, $B(t) = e^{iLt} B(\Gamma)$, hence $\dfrac{dB(t)}{dt} = iL e^{iLt} B(\Gamma).$

$$\dot{C}_{AB}(t) = \int d\Gamma \rho(\Gamma) A^*(\Gamma) \dot{B}(t) \tag{8.19}$$

In the case of the autocorrelation function, i.e., $A=B$ it can be shown using the time symmetry property of the time correlation

$$C_{AA}(t) = C_{AA}(-t) \text{ and } \dot{C}_{AA}(t) = -\dot{C}_{AA}(-t). \tag{8.20}$$

This means the autocorrelation functions are an even function of time, whereas the time derivatives of the autocorrelation functions are an odd function of time.

8.3.5   CONSERVATION OF AMPLITUDE

The above identity on time-reversal symmetry has interesting implications. We define the norm of a dynamical variable A as $\langle |A(t)|^2 \rangle$. What we mean by this definition is that

$$\langle |A(t)|^2 \rangle = \int d\Gamma \rho(\Gamma) A^*(t) A(t)$$
$$= \int d\Gamma \rho(\Gamma) \left(e^{iLt} A(\Gamma) \right)^* e^{iLt} A(\Gamma).$$

From the adjoint condition, we can rewrite the above equation as

$$\langle |A(t)|^2 \rangle = \langle e^{-itL} e^{itL} A | A \rangle = \int d\Gamma \rho(\Gamma) A^*(\Gamma) A(\Gamma).$$
$$= \langle |A|^2 \rangle \tag{8.21}$$

The above relation reveals that the norm at time t, i.e., the ensemble average of the magnitude of the square of $A(t)$, remains unchanged from its initial norm ($t = 0$). Therefore, it can be stated that e^{iLt} is a unitary operator or norm preserving operator. It basically rotates the dynamical variable leaving its magnitude unchanged, which is illustrated in Figure 8.2.

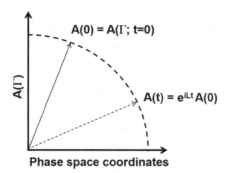

Phase space coordinates

FIGURE 8.2 e^{iLt} is a unitary operator $\left(\left|e^{iLt}\right|=1\right)$. Hence it preserves the norm of any vector. So, e^{iLt} behaves like a rotating operator and changes the orientation of the initial vector keeping its magnitude constant.

8.4 PHYSICAL OBSERVABLES FROM TIME CORRELATION FUNCTIONS

As discussed in Chapter 3, many different and fairly common experiments directly probe different time correlation functions. For example, Rayleigh-Brillouin spectroscopy and neutron scattering techniques probe density-density TCF, Raman scattering, and infra-red spectroscopy probe rotational and vibrational TCFs through line shape measurements, dielectric relaxation measurements study total dipole moment TCF. In recent times, time domain experiments have been developed, like solvation dynamics (time-dependent fluorescence Stokes shift), fluorescence depolarization, 2D infra-red spectroscopy, etc. These techniques allow the determination of TCFs directly in the time domain. This section discusses a few such examples.

8.4.1 DIFFUSION

The most common application of the time correlation function involves the velocity time correlation function whose time integration gives us the diffusion coefficient D. The convenient way to establish this connection is to start with a one-dimensional diffusion equation for space (x) and time (t) dependence of the concentration of a tagged particle $n(x,t)$

$$\frac{\partial n(x,t)}{\partial t} = D\frac{\partial^2}{\partial x^2}n(x,t). \tag{8.22}$$

The expectation value of x^2 can be determined by

$$\left\langle x^2 \right\rangle = \int dx\, x^2 n(x,t). \tag{8.23}$$

Multiplying both sides of Eq. (8.22) by x^2 and integrating over x we find

$$
\begin{aligned}
\frac{\partial}{\partial t}\left\langle x^2 \right\rangle &= \int dx x^2 \frac{\partial}{\partial t}n(x,t)\\
&= D\int dx\, x^2 \frac{\partial^2}{\partial x^2}n(x,t)\\
&= 2D\int dx\, n(x,t) + D\left[x^2 \frac{\partial n(x,t)}{\partial x}\Bigg|_{-\infty}^{\infty}\right] - 2D\left[xn(x,t)\big|_{-\infty}^{\infty}\right]\\
&= 2D \qquad \text{since } \int dx\, n(x,t) = 1.
\end{aligned}
\tag{8.24}
$$

In the above derivation, we have discarded the last two terms since the probability distribution function $n(x, t)$ tends to zero as $x \to \pm\infty$. Now, the net displacement of the position of a particle can be related to the velocity correlation function as follows. We start with the definition to write

$$
\begin{aligned}
\langle x^2 \rangle &= \left\langle \int_0^t ds_1 v(s_1) \int_0^t ds_2 v(s_2) \right\rangle \\
&= \int_0^t ds_1 \int_0^t ds_2 \langle v(s_1)v(s_2) \rangle.
\end{aligned}
\tag{8.25}
$$

We take time derivative and combine the two equivalent terms on the right-hand side we get

$$
\frac{\partial}{\partial t} \langle x^2 \rangle = 2 \int_0^t ds \langle v(t)v(s) \rangle.
\tag{8.26}
$$

We now use time stationarity property of TCF to introduce a variable $u = t - s$

$$
\frac{\partial}{\partial t} \langle x^2 \rangle = 2 \int_0^t ds \langle v(t-s)v(0) \rangle = 2 \int_0^t du \langle v(u)v(0) \rangle.
\tag{8.27}
$$

Next, we remember that diffusion is a long-time property when the mean displacement becomes independent of time. This limit is easily achieved by noting that the above integral becomes independent of time in the long-time limit. We compare Eq. (8.27) with Eq. (8.24) to give us the desired relation

$$
D = \int_0^\infty dt \langle v(t)v(0) \rangle.
\tag{8.28}
$$

The diffusion coefficient in three dimensions can be defined in general

$$
D = \frac{1}{3} \int_0^\infty dt \langle \mathbf{v}(t) . \mathbf{v}(0) \rangle
\tag{8.29}
$$

where $\mathbf{v}$ is the velocity vector.

An alternative derivation starts directly with Einstein definition of diffusion coefficient

$$
D = \lim_{t \to \infty} \frac{[r(t) - r(0)]^2}{6t} = \lim_{t \to \infty} \frac{[\Delta r(t)]^2}{6t}.
$$

The subsequent steps in derivation are the same. We write the displacement as an integral over time-dependent velocity, and the desired relation follows.

8.4.2 VISCOSITY

It was Green who first developed microscopic expressions for the coefficient of shear viscosity η and bulk viscosity η_v by means of Fokker-Planck theory. The expressions for shear viscosity and bulk viscosity are given in terms of the time correlation function as

$$\eta = \frac{\beta}{5V} \sum_a \sum_b \int_0^\infty dt \langle \sigma_{ab}(t)\, \sigma_{ab}(0) \rangle$$
$$\eta_v = \frac{\beta}{9V} \sum_a \sum_b \int_0^\infty dt \langle \sigma_{aa}(t)\, \sigma_{bb}(0) \rangle$$

(8.30)

where V denotes the volume of the system, $\beta = \dfrac{1}{k_B T}$ and the tensor σ_{ab} ($a, b = x, y, z$) is given by

$$\sigma_{ab} = \sum_j \frac{p_{ja} p_{jb}}{m_j} + \sum_j R_{ja} F_{jb} - \langle \sigma_{ab} \rangle.$$

(8.31)

Here, R_{ja} denotes the ath Cartesian coordinate of the position of j molecule of mass m_j, p_{ja} corresponds to the corresponding momentum, F_{ja} is the ath component of total force acting on j molecule and the subtracted part $\langle \sigma_{ab} \rangle$ is a function of internal energy E, pressure $\bar{P}$, and particle number N needed only for bulk viscosity. The subtracted part is needed only for bulk viscosity to take into account the ensemble used for the averages.

In the expression for shear viscosity (η), the stress tensor σ_{ab} has the following five independent components: σ_{xy}, σ_{yz}, σ_{zx}, $(\sigma_{xx} - \sigma_{yy})/2$, and $(\sigma_{yy} - \sigma_{zz})/2$. For isotropic systems, the off-diagonal elements of the stress tensor σ_{xy}, σ_{yz}, and σ_{zx} are equivalent. Due to rotational invariance, the terms $(\sigma_{xx} - \sigma_{yy})/2$ and $(\sigma_{yy} - \sigma_{zz})/2$ are also equivalent. However, for the confined system, the degeneracy is lifted. The bulk viscosity, expressed by Eq. (8.30), has nine components.

8.4.3 Dielectric Susceptibility

In accordance with Kubo's method, for an isolated and isotropic spherical specimen, the scalar dielectric susceptibility $\varepsilon(\omega)$ is related to the net polarizability $\alpha_s(\omega)$ as

$$\frac{\varepsilon(\omega) - 1}{\varepsilon(\omega) + 2} = \frac{4\pi}{3} \alpha_s(\omega).$$

(8.32)

Eq. (8.32) is the well-known Clausius-Mossotti equation, which is an exact expression for spherical systems suspended in vacuum. Here $\alpha_s(\omega)$ is defined as

$$\alpha_s(\omega) = \frac{\beta}{3V} \left[-\frac{d}{dt} \int_0^\infty dt e^{-i\omega t} \langle M(0).M(t) \rangle \right].$$

(8.33)

The derivation of this expression requires the linear response theory, which is detailed later in Chapter 11. Here M corresponds to the total electric moment of the sphere of volume V. It is important to note that the expression of dielectric constant is strongly dependent on the geometry of the system, and exact expressions for different geometries are not always available. Hence one often seeks the help of certain approximations.

For theoretical studies, we often use a simpler expression, which is exact for an infinite system where the dipolar liquid is contained in a rectangular box/container. This expression is given by

$$\varepsilon(\omega) - 1 = \frac{4\pi}{3V} \mathcal{L}\left(-\frac{d\varphi_{MM}(t)}{dt} \right)$$

(8.34)

where $\varphi_{MM}(t)$ is the moment-moment correlation function $\langle M(0)M(t)\rangle$. $\mathcal{L}$ here denotes the Laplace transformation, with $i\omega$ as the Laplace variable conjugate to time t.

8.4.4 ELECTRICAL CONDUCTIVITY

In terms of current correlation function, electrical conductivity is defined as

$$\kappa = \frac{1}{3k_B TV} \int_0^\tau dt \langle \mathbf{J}(0).\mathbf{J}(t)\rangle \tag{8.35}$$

where k_B is the Boltzmann constant, V denotes the volume of the system at temperature T and $\mathbf{J}(t)$ denotes the charge current at time t. $\mathbf{J}(t)$ is given by $\mathbf{J}(t) = \sum_i q_i \mathbf{v}_i$, where $\mathbf{v}_i$ is the velocity of the ion i with charge q_i. This formula was introduced by Kubo in 1953. If the charge carrier is an electron, and if we assume that motions of different electrons is uncorrelated, then Eq. (8.35) leads trivially to the Drude model of electrical conductivity of metals, when we assume that the single-particle velocity TCF decays exponentially with a time constant τ. Thus, we certainly get a cleaner derivation if we start from Kubo's expression.

8.4.5 FRICTION COEFFICIENT FROM KIRKWOOD'S FORMULA

Here we discuss a somewhat approximate but often useful expression of friction coefficient, known as Kirkwood's formula. This expression gives friction in terms of a force-force time correlation function. We know from Stokes law of hydrodynamics that the applied constant external force $\mathbf{F}$ on a large particle suspended in liquid generates a velocity $\mathbf{v}$, which is related to the force by $\mathbf{F} = \zeta \mathbf{v}$. Thus, we expect friction to be given by force-force TCF. However, this relationship turns out to be a bit tricky. Read below.

For Brownian motion of a tagged solute particle, the frictional force is a retarding force $-\zeta \mathbf{v}$ by the surrounding solvent molecules as there is no constant external force. Its motion is governed by the Langevin equation,

$$m\dot{v} = -\zeta v + R(t) \tag{8.36}$$

where $R(t)$ denotes the random force, $\mathbf{v}$ indicates the velocity of the particle, and ζ is the friction constant. The random force obeys the fluctuation-dissipation theorem that gives an expression for the friction in terms of the integral over the TCF $\langle R(0)R(t)\rangle$. We shall discuss these aspects in Part III of the book.

The first statistical mechanical derivation for friction coefficient was introduced in terms of force-force correlation function by Kirkwood and is given by

$$\zeta = \frac{1}{3k_B T} \int_0^\tau dt \langle \mathbf{F}(0).\mathbf{F}(t)\rangle. \tag{8.37}$$

In the Kirkwood's expression $\mathbf{F}(t)$ denotes the total force exerted by all solvent molecules on the tagged solute molecule. The upper limit of integration τ is supposed to be macroscopically short but long compared with characteristic molecular times, for a reason discussed below.

The formulation of Kirkwood theory is not exact. In simulations, one often find that in the long time, the force-force TCF decays to a constant value, sometimes referred to as the "Kirkwood plateau", which is an artifact because the average of the force $\langle \mathbf{F}(t)\rangle$ is nonzero because of inadequate averaging, or due to the presence of a slowly decaying component. Also, the force that enters Kirkwood's formula needs to be made orthogonal to the position and the velocity, as required by the projection operator technique described in the later chapters. This problem can be partly circumvented by defining a fluctuating force by subtracting the average force from $\mathbf{F}(t)$.

A precise definition of friction starting from Zwanzig's projection operator technique can be developed and has been developed by Sjogren and Sjolander (SS) that shall be described in Chapter 10. One of the main results of the SS analysis is the derivation of an equation for friction, which is of great interest

$$\frac{1}{\zeta} = \frac{1}{\zeta_{bin} + \zeta_{\rho\rho}} + \frac{1}{\zeta_{tt}}. \tag{8.38}$$

Here the inverse of total friction ζ is expressed as the sum of inverse of frictions from microscopic, local interactions and from transverse current term. The first denominator on the right-hand side of the above equation is composed of binary friction (like the Enskog term) and a density term that takes care of correlated collisions. The second term on the right-hand side, the transverse current term, reduces to the Stokes friction in the size of the large size of the tagged solute whose friction is being calculated. We shall defer further discussion of these terms until Chapter 10.

A working definition of friction widely used in computer simulations is to assign the random force on a tagged solute as the force obtained by keeping the solute's motion fixed. This is an approximate but simple way to exclude mixing of the solute's dynamics with solvent dynamics. The estimates so obtained are fairly accurate. In fact, one can derive exactly the density part of the MCT expression by using this procedure where the force is obtained from the time-dependent density functional theory as discussed in the latter chapters.

8.5   SHORT TIME EXPANSION OF TIME CORRELATION FUNCTIONS

It is practically impossible to obtain exact expression of a TCF that is valid over the entire time range. Derivation of useful analytical expressions often involves approximations. While it is hard to predict the full-time evolution of a TCF, *we can often obtain the exact expression of its behavior in the short time by using a short-time expansion.* The short-time expansion is a simple Taylor expansion around time $t = 0$, and although meant to be valid for short times, can often be extended to intermediate times, and accounting for more than 50% of the decay. *Importantly, the coefficients of the short-time expansion are given by equilibrium correlation functions and are often known accurately.*

If $C_{AA}(t)$ is the autocorrelation function of a dynamical variable A,

$$C_{AA}(t) = \langle A(0)A(t) \rangle, \tag{8.39}$$

a short-time expansion by using the Taylor expansion for the quantity $A(t)$

$$\begin{aligned}
C_{AA}(t) &= \left\langle \left(A(0) + t \left. \frac{dA(t)}{dt} \right|_{t=0} + \frac{t^2}{2!} \left. \frac{d^2 A(t)}{dt^2} \right|_{t=0} + \text{Higher order terms} \right) A(0) \right\rangle \\
&= C_{AA}(0) + \frac{t^2}{2!} \left\langle \left. \frac{d^2 A(t)}{dt^2} A(0) \right\rangle \right|_{t=0} + \frac{t^4}{4!} \left\langle \left. \frac{d^4 A(t)}{dt^4} A(0) \right\rangle \right|_{t=0} + \cdots \\
&= \sum_{n=0}^{\infty} \frac{t^{2n}}{(2n)!} C_{AA}^{(2n)}(0) \\
&= \sum_{n=0}^{\infty} (-1)^n \frac{t^{2n}}{(2n)!} \langle A^{(2n)}(0)A(0) \rangle,
\end{aligned} \tag{8.40}$$

where $(2n)$ represents the $2n$ fold derivatives with respect to time. All the odd-valued derivatives of $A(t)$ become zero when evaluated at $t = 0$, i.e., $\langle A^{(2n+1)}A(0)\rangle = 0$ because of time-reversal symmetry around $t = 0$. As a result, we can drop out all the odd terms from the above TCF expression. The expression of TCF given by Eq. (8.39) can be further simplified by using the stationary property of TCF as

$$\frac{d^{2n}}{dt^{2n}}\langle A(0)A(0)\rangle = \langle A^{(2n)}(0)A(0)\rangle = \langle A^{(n)}(0)A^{(n)}(0)\rangle. \qquad (8.41)$$

Finally, TCF at the short time is given by

$$C_{AA}(t) = \sum_{n=0}^{\infty} \frac{t^{2n}}{(2n)!}\langle A^{(n)}A^{(n)}\rangle. \qquad (8.42)$$

Any explicit time expression with the quantity $A^{(n)}(t)$ on right-hand side of Eq. (8.41) is left out because it concerns the value of $A(t)$ at $t = 0$

To evaluate the moments of the short-time expansion, we need the power spectrum or Fourier transform of the TCF $C_{AA}(t)$, defined as

$$C_{AA}(t) = \frac{1}{2\pi}\int_{-\infty}^{\infty} \tilde{C}_{AA}(\omega)e^{-i\omega t}d\omega. \qquad (8.43)$$

Rewriting Eq. (8.42) again as

$$C_{AA}(t) = \sum_{n-0}^{\infty} \frac{t^{(2n)}}{(2n)!}C_{AA}^{(2n)}(0). \qquad (8.44)$$

We now use the time derivative property of TCF discussed above,

$$C_{AA}^{(2n)}(t) = (-1)^n \left(\frac{\partial}{\partial t}\right)^{(2n)}\langle A(0)A(t)\rangle$$

$$= (-1)^n \left(\frac{\partial}{\partial t}\right)^{(2n)}\frac{1}{2\pi}\int_{-\infty}^{\infty} \tilde{C}_{AA}(\omega)e^{-i\omega t}d\omega. \qquad (8.45)$$

$$C_{AA}^{(2n)}(0) = \frac{1}{2\pi}\int_{-\infty}^{\infty} \tilde{C}_{AA}(\omega)\omega^{2n}d\omega = \langle \omega^{2n}\rangle$$

The zeroth, second, and fourth moments are given by 1, $\langle \omega^2\rangle$ and $\langle \omega^4\rangle$, respectively. In the next two chapters, we shall derive explicit expressions for these moments for important TCFs.

8.6   LONG-TIME TAIL OF TIME CORRELATION FUNCTIONS

Alder and Wainwright observed that the long-time tail of velocity time correlation function (VTCF) of hard disk or hard sphere decays as $(\eta t)^{-d/2}$, where η is the viscosity and d is the dimensionality of the system. They compared the VTCF from molecular dynamics with the hydrodynamic predictions and found that these two start to converge after a certain time. The average linear fit of these two after the convergence shows a time (t) exponent of $-3/2$ for the hard sphere system.

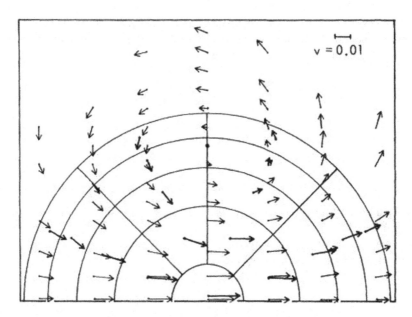

FIGURE 8.3 Model for the vortex flow pattern around a hard disk. The arrows indicate the velocity field in that region. Half of the system is shown because of the inherent symmetry.

Source: Figure is taken with permission from [10].

This long-time behavior of the VTCF is qualitatively explained in terms of a vortex flow pattern (Figure 8.3) on a microscopic scale around the moving particle. The hydrodynamics model of Alder and Wainwright explains the persistence of this circulatory motion in terms of the separation of compression and rarefaction in the stationary fluid through which the particle is moving. At late times, the velocity of the particle decays solely due to the shear viscosity of the incompressible fluid. We have discussed this aspect in Chapter 10.

8.7 AN EXPLICIT CALCULATION OF TIME CORRELATION FUNCTIONS

Based on the theories that we have developed we now discuss a problem that is although somewhat artificial is illustrative. The problem is the calculation of the position autocorrelation function of a single harmonic oscillator connected with a thermal heat reservoir at temperature T. The simplicity of this model lies in the fact that we shall treat the oscillator as an isolated system despite the system being allowed to interact with the heat bath. Thus, our system is a single one-dimensional harmonic oscillator of mass m with Hamiltonian of the form

$$H = \frac{p^2}{2m} + \frac{1}{2}kx^2. \tag{8.46}$$

Here, k represents the spring constant of the oscillator. Our goal is to calculate the TCF $\langle x \mid x(t) \rangle$, where $x(t)$ is the position of the oscillator at a later time t starting with initial position $x(0)$, which we denote here by x. We start with the definition of the TCF

$$\langle x \mid x(t) \rangle = \int_{-\infty}^{\infty} dx \int_{-\infty}^{\infty} dp \rho(x, p) x x(t). \tag{8.47}$$

In the above expression $\rho(x, p)$ denotes the phase space density, which we need to evaluate along with the expression of $x(t)$. To do this, we recall Hamilton's equations of motion for the time evolution of the trajectory as the system is isolated, i.e.,

$$\dot{x} = \frac{\partial H}{\partial p} = \frac{p}{m} \text{ and } \dot{p} = -\frac{\partial H}{\partial x} = -kx. \tag{8.48}$$

The above two equations can be combined to obtain the following equation:

$$\ddot{x} = -\frac{k}{m} x. \tag{8.49}$$

This is a second-order differential equation and we can solve easily with the aid of two initial conditions; $x(0) = x$ and $p(0) = p$.

An alternative, equally illustrative way to solve this problem is to start with Liouville's equation. From the definition of Liouville's operator we know

$$iL = \dot{x} \frac{\partial}{\partial x} + \dot{p} \frac{\partial}{\partial p} = \frac{p}{m} \frac{\partial}{\partial x} - kx \frac{\partial}{\partial p}. \tag{8.50}$$

Formally, $x(t)$ is now given by the expression $x(t) = e^{iLt} x(0)$, which using the series representation of the exponential, leads to

$$
\begin{aligned}
x(t) &= \left[1 + t \left(\frac{p}{m} \frac{\partial}{\partial x} - kx \frac{\partial}{\partial p} \right) + \frac{t^2}{2} \left(\frac{p}{m} \frac{\partial}{\partial x} - kx \frac{\partial}{\partial p} \right) \left(\frac{p}{m} \frac{\partial}{\partial x} - kx \frac{\partial}{\partial p} \right) + \dots \right] x \\
&= x + t \frac{p}{m} - \frac{t^2}{2} \frac{kx}{m} - \frac{t^3}{3!} \frac{pk}{m^2} + \frac{t^4}{4!} \frac{k^2 x}{m^2} + \dots \\
&= x \left(1 - \frac{\left(t \sqrt{k/m} \right)^2}{2!} + \frac{\left(t \sqrt{k/m} \right)^4}{4!} - \dots \right) + \frac{p}{m} \sqrt{\frac{m}{k}} \left(\left(t \sqrt{k/m} \right) - \frac{(t \sqrt{k/m})^3}{3!} + \dots \right).
\end{aligned}
\tag{8.51}
$$

We introduce the frequency factor $\omega = \sqrt{\frac{k}{m}}$ and resuming the series we find the above equation reduces to

$$x(t) = x \cos(\omega t) + \frac{p}{m\omega} \sin(\omega t). \tag{8.52}$$

Using the above expression of $x(t)$ and the expression of equilibrium probability of the system, which is given by $\rho(x, p) = \frac{\exp(-\beta H)}{\int dx dp \exp(-\beta H)}$, we obtain

$$
\begin{aligned}
\langle x | x(t) \rangle &= \frac{1}{Q} \int_{-\infty}^{\infty} dx \int_{-\infty}^{\infty} dpx \left(x \cos(\omega t) + \frac{p}{m\omega} \sin(\omega t) \right) \exp \left(-\frac{\beta k x^2}{2} - \frac{\beta p^2}{2m} \right) \\
&= \frac{k_B T}{k} \cos(\omega t).
\end{aligned}
$$

Therefore, finally, we have

$$C_{xx}(t) = \langle x(0)x(t) \rangle = \frac{k_B T}{k} \cos(\omega t). \tag{8.53}$$

The average of the product does not decay to zero even in the very long time as this is a deterministic model. A Langevin-equation-based analysis indeed shows that in a stochastic environment the product $\langle x(0)x(t) \rangle$ indeed goes to zero. This has been discussed in this book in Chapters 5 and 10.

8.8 ALGORITHM FOR COMPUTATION OF TIME CORRELATION FUNCTIONS

Computation of time correlation functions is an integral part of research involving time-dependent statistical mechanics. In this section we give a concise account of the computer algorithm to compute TCF of any two time-dependent variables, say $A(t)$ and $B(t)$. For the case where $A = B$, the TCF is known as autocorrelation function (ACF).

Let us consider the time progression of these two variables, $A(t)$ and $B(t)$. The time line is shown schematically in Figure 8.4. Here δt is a single timestep and N is the total number of time steps. We explain the algorithm for computation of the TCF $\langle A(0)B(t) \rangle$ with the help of a sample FORTRAN 90 program given below.

```
1  do j = 1, tcor
2  s = j - 1
3  cnt = 0
4  tcf = 0.0
5  do k = 1, N-s, gap
6  tcf = tcf + A(k) * B(k+s)
7  cnt = cnt + 1
8  enddo
9  tcf = tcf / float(cnt)
10 enddo
```

In the above algorithm, we show the nested loops (do loop in FORTRAN 90) responsible for the computation of TCF. Here, tcor is the timestep till which the correlation is to be calculated. Typically it is taken as a third or fourth of the total simulation time (here, the total time T is given by N.δt), but it of course depends on the type of TCF that is being calculated. The larger the value of tcor with respect to N, the poorer the statistics becomes for the long-time correlation. Here, the

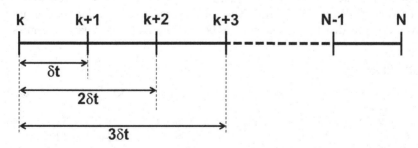

FIGURE 8.4 Schematic time line for the time-dependent variables $A(t)$ and $B(t)$ showing the individual time steps and the shift in time origin.

running index **s** represents the shift in time origin. For clarity of understanding this, let us revisit the definition of TCF given by Eq. (8.6).

$$\langle A(0)A(t) \rangle = \lim_{T \to \infty} \frac{1}{T} \sum_{s} A(s)\, A(t+s).$$

Here, we see that the change in s incorporates the time averaging with respect to the shifted time origin. The term `gap` is included in the loop for the following reason. If two time steps are very close to each other, say in the order of 1 fs, then the values of the variables $A(t)$ and $B(t)$ will show almost no change. Hence, it is advisable to skip a few steps in between to make the code computationally less expensive. `cnt` is a parameter that counts the number of steps for which the TCF is computed. The total `tcf` is divided by `cnt` for the proper time averaging outside the loop. The TCF thus obtained is plotted against **s** to obtain Figure 8.1(b).

8.9   ROTATIONAL TIME CORRELATION FUNCTIONS

As has been emphasized in Chapter 3, and also discussed in detail in Chapter 22 (rotational diffusion chapter), rotational TCFs of different kinds are of great importance in correlating experimental results to the underlying molecular dynamics. For example, the study of dielectric relaxation to a great extent is a study of rotational TCF. Anisotropic depolarization and NMR also studies the orientational dynamics. More recently, studies of solvation dynamics have provided information about coupled rotational-translational motions. Ionic conductivity of electrolytic solutions is also strongly coupled to the rotational motions of surrounding water molecules. A preliminary introduction to the above relations has already been discussed in Chapter 3.

8.10   SUMMARY

In this important chapter, we have introduced the definition of time correlation function that is central to time-dependent statistical mechanics. We have derived many of its important properties. We have already discussed the intimate relation between experimental observables and TCF in Chapter 3. We have described expressions of transport coefficients in terms of the appropriate TCF. Often it is rather easy to guess the definition because it is always the conjugate dynamical property. For diffusion, which is the displacement in the coordinate (say, x) space, the relevant TCF is the velocity-velocity TCF. For viscosity, it is the transverse current TCF. For dielectric relaxation, it is the total dipole moment-dipole moment TCF.

Time dependence of TCF also reveals important information about the dynamics of a system, and is sensitive to intermolecular correlations, present in the system. Often the time dependence is nonexponential, as in glassy liquids.

Many experiments measure the time correlation function in the frequency domain, like Raman line width or Rayleigh-Brillouin spectrum. Time domain and frequency domain measurements can be sensitive to different dynamic range, and thus both are useful to understand dynamics.

BIBLIOGRAPHY

1. Zwanzig, R. 1965. Time-correlation functions and transport coefficients in statistical mechanics. *Annual Review of Physical Chemistry, 16*(no. 1), 67–102.
2. Zwanzig, R. 2001. *Nonequilibrium Statistical Mechanics*. Oxford: Oxford University Press.
3. Hansen, J., and I. R. McDonald. 1990. *Theory of Simple Liquids*. Amsterdam: Elsevier.
4. Berne, B., and G. D. Harp. 1970. On the calculation of time correlation functions. *Advances in Chemical Physics, 17*, 63–227.

5. Luban, M., and H. L. James. 1999. Equilibrium time correlation functions and the dynamics of fluctuations. *American Journal of Physics, 67*(no. 12), 1161–1169.

6. McQuarrie, D. A. 1976. *Statistical Mechanics*. New York: Harper & Row.

7. Allen, M. P., and D. J. Tildesley. 1987. *Computer Simulation of Liquids*. Oxford: Clarendon Press.

8. Böttcher, C. J. F., and P. Bordewijk. 1978. *Theory of Electric Polarization*. Amsterdam: Elsevier Science Limited.

9. Bagchi, B. 2018. *Statistical Mechanics for Chemistry and Materials Science*. Boca Raton: CRC Press.

10. Alder, B. J., and T. E. Wainwright. 1970. Decay of the velocity autocorrelation function. *Physical Review A, 1*(1), 18.

9 Density-Density and Current-Current Time Correlation Functions

OVERVIEW

In this chapter, we discuss two important time correlation functions that find extensive use in time-dependent statistical mechanics. These are the density-density time correlation function (DD-TCF) and the current-current time correlation function (CC-TCF), which deserve special attention. The frequency and wave-number-dependent DD-TCF is also called dynamic structure factor. We encounter two types of dynamic structure factor, self, denoted by $F_s(k,t)$ and coherent $F(k,t)$. The CC-TCF is also decomposed into two types, longitudinal, $C_l(k,t)$ and transverse, $C_t(k,t)$. We have already encountered them in the hydrodynamics chapter (Chapter 5), especially in the calculation of the Rayleigh-Brillouin spectrum obtained from light scattering. In this chapter we are concerned mostly with intermediate-to-large wave number region where hydrodynamic description breaks down and where molecular level descriptions are necessary. This region is accessible to neutron scattering experiments.

9.1 INTRODUCTION

Density-density and current-current time correlation functions find wide use in the study of relaxation phenomena. In this chapter we provide certain basic results with appropriate explanations. Extensive work has been carried out in the calculation of these two correlation functions after computer simulations become available. Many results have been discovered that were not anticipated from hydrodynamic theory.

In Chapter 5, we discussed these time correlation functions. The density-density time correlation functions describe the Rayleigh-Brillouin spectroscopy in one of the early applications of TCF, although the description in terms of TCF had to wait to be formulated by Green and Kubo. The transverse current-current TCF is connected with viscosity.

The intermediate correlation function $F(k,t)$ is defined as the autocorrelation function between the Fourier components of density, with k as the wave number

$$F(k,t) = \frac{1}{N}\langle \rho(-\mathbf{k})\rho(\mathbf{k},t)\rangle. \tag{9.1}$$

Time and position-dependent density $\rho(\mathbf{r},\ t)$ can be decomposed into a sum of a constant average density ρ_0 and a fluctuating density $\delta\rho(\mathbf{r},\ t)$. A pictorial representation of the fluctuation in density is shown in Figure 9.1. Here we are measure the number of particles that enter and leave the sphere of diameter L in the fluid. Crudely wave vector $\mathbf{k}$ in $\delta\rho(\mathbf{k},t)$ is related to the length scale as $k = 2\pi / L$.

DOI: 10.1201/9781003157601-11

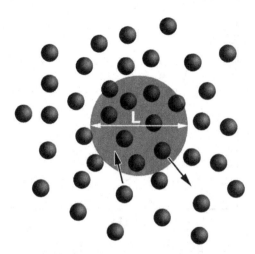

FIGURE 9.1 Pictorial representation of number density fluctuations over a length scale L. The fluctuations in the number density of fluid are measured from the average density and given by the instantaneous number of particles in a given volume. These are caused by the entrance in and the exit out of the sphere of a certain diameter. The length scale L at is related to wavenumber by $k \simeq 2\pi / L$. In the hydrodynamic regime (over r a very large length scale L), the fluctuation in the density becomes slow as many molecules need to move in or out to cause a fluctuation.

At large k, i.e., at a very small length scale L, fluctuations can decay fast because movement of one molecule can cause large fluctuation. Here number is conserved but not momentum and energy. In the other extreme, the $k \to 0$ (hydrodynamic limit) the fluctuation in density becomes small as the exit of a few molecules causes small perturbation. In this limit, both energy and momentum, in addition to the number, need to be conserved. Similarly, we can use this procedure to discuss fluctuations in different microscopic quantities, which can be measured (at least in computer simulations) at different lengths and time scales.

The density-density correlation function itself is decomposed into two correlation functions: the self-dynamic structure factor and the coherent dynamic structure factor. The terminology changes depending on the use of time or frequency domain functions. This is partly because experimentally we measure these functions in the frequency domain. The time domain density-density correlation function, $F(k,t)$, is called the intermediate scattering function while the frequency domain part is called the coherent dynamic structure factor. Typical behavior of $F(k,t)$ with time is shown in Figure 9.2 and Figure 9.3 for fluid and glassy states, respectively. Similarly, the frequency domain function $F_s(k,z)$ is called the self-dynamic structure factor.

The current-current correlation function is also decomposed into two functions: longitudinal and transverse, usually denoted by $C_l(k,t)$ and $C_t(k,t)$. This decomposition arises because the current $\mathbf{J}$ itself is a vector, so the correlation function $\langle \mathbf{J}^\alpha(-\mathbf{k},0)\mathbf{J}^\beta(\mathbf{k},t)\rangle$ is a tensor that is decomposed into components parallel and perpendicular to $\mathbf{k}$.

In the following sections, we discuss these four correlation functions.

9.2   SELF-INTERMEDIATE SCATTERING FUNCTION $F_s(k,t)$

The self-intermediate scattering function $F_s(\mathbf{k},t)$ describes how the dynamics or the motion of a single particle evolves with time in a many-body system. It finds use in theoretical descriptions to understand the dynamics of the liquid, such as the slowing down of relaxation near a glass transition.

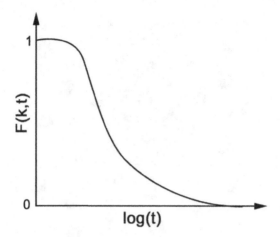

FIGURE 9.2 Time dependence of a typical intermediate scattering function for a fluid state. The figure shows how the correlation between fluctuations of number density for a normal liquid decay with time at a fixed value of the wavenumber k that is representative of a length L in space, as illustrated in Figure 9.1. At long time, relaxation is governed by an exponential decay.

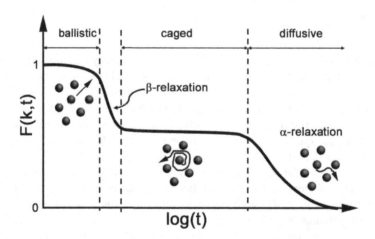

FIGURE 9.3 Schematic representation of intermediate scattering function $F(k,t)$ at a fixed temperature for a glassy state, characterized by β and α relaxations, described below. In contrast to Figure 9.2 (normal liquid), $F(k,t)$ for a glassy liquid *shows a multi-step relaxation pattern*. At short times, particles undergo ballistic motion. At the intermediate time (relaxation is called the β-relaxation regime) a plateau develops due to the cage effect (particle finds itself in a transient freezing state because of the cage formation by the neighboring particles), during this regime $F(k,t)$ remains constant. This cage persists for longer and longer time with decreasing temperature but the height of the plateau remains the same, i.e., the size of the cage does not change drastically with temperature. At a sufficiently long time fluctuations become decorrelated and the correlation function decays to zero. This final relaxation process is called the α-relaxation and this relaxation is not a simple exponential as in normal fluid. It is a more slow decay process called the stretched exponential behavior of the form $\exp(-t / \tau)^{\beta}$, with $0 < \beta < 1$ (this β is not same as β-relaxation).

The importance of $F_s(\mathbf{k},t)$ is that it is one of the quantities that can be measured in experiments. The time-dependent density of a tagged particle (*say*, α) at position r_α is given by

$$n(\mathbf{r},t) = \delta(\mathbf{r} - \mathbf{r}_\alpha(t)).$$
(9.2)

The Fourier transform of Eq. (9.2) is given as

$$n(\mathbf{k},t) = e^{i\mathbf{k}\cdot\mathbf{r}_\alpha(t)}.$$
(9.3)

It is important to note the following two points. First, the assumption of identical particles is for simplicity and can be relaxed. Second, the tagged or single-particle density is denoted by $n(\mathbf{r}, t)$ while that of number density by $\rho(\mathbf{r}, t)$. It is important to remember the distinction between the two. The average single-particle density is $\langle n(\mathbf{r},t) \rangle = 1/V$, while the average collective density, $\langle \rho(\mathbf{r}, t) \rangle = N/V$.

Tagged particle correlation function or self-intermediate scattering function is defined as,

$$F_s(\mathbf{k},t) = \langle n(-\mathbf{k},0)n(\mathbf{k},t) \rangle = \left\langle e^{i\mathbf{k}\cdot(\mathbf{r}_\alpha(t)-\mathbf{r}_\alpha(0))} \right\rangle.$$
(9.4)

Generally, we focus on the dynamics of a system in two limits: first is the hydrodynamic limit where we explore how a system behaves at the low wavenumber $k \to 0$ and also the low frequency $\omega \to 0$ (long time). The second is the opposite limit, i.e., we study the dynamical response or behavior of a system at large wavenumber and short times-to-intermediate times. This is the range studied by computer simulations. The large wavenumber usually means that the wavenumber k is comparable to or larger than $2\pi/\sigma$, where σ is the molecular diameter. We are also interested at long times, much longer than the collision times in liquids, not accessible to simulations but often studied easily by experiments.

To give a qualitative idea of how $F(k,t)$ and $F_s(k,t)$ evolve with time, we take a one-component hard sphere potential system, which is the simplest model of fluid. The pair potential $V(r)$ at separation r is given by

$$\begin{aligned} V(r) &= \infty, r \leq \sigma \\ &= 0, r > \sigma \end{aligned}$$
(9.5)

where σ is the diameter of the hard sphere particle. In order to study the behavior of correlation functions of the fluctuations of dynamical variables, first, we need to discuss these functions at different special points but at the same point (like a snapshot of the system). The correlation of the fluctuation (e.g., number density) at a fixed time represents the static or structural properties of the liquid. For a high-density liquid, the radial distribution function $g(r)$ is the most basic quality characteristic of its structural properties. The results shown here are in the *k-space*. The static properties of the system are studied in terms of the quantity $S(k)$ [the Fourier transform of $g(r)$], which is termed as the static structure factor for the isotropic liquid.

Figure 9.4 represents the static structure factor for a dense liquid system at a fixed packing fraction. To study the dynamical correlation functions, we often study dynamics at the value of the wave vector corresponding to the first peak of $S(k)$, shown by a solid circle and the dashed line on $k\sigma$-axis is the corresponding value. Figure 9.5 depicts the emergence of slow dynamics in the self-dynamic structure factor, measured at the first peak of static structure factor.

For a detailed discussion, first we focus on the short time, i.e., in large k and ω limit. The short time expansion for $F_s(\bar{k},t)$ around $t = 0$ is obtained as

$$\begin{aligned} F_s(\mathbf{k},t) &= 1 - \frac{t^2}{2!}\langle \dot{n}(-\mathbf{k},0)\dot{n}(\mathbf{k},0) \rangle + \frac{t^4}{4!}\langle \ddot{n}(-\mathbf{k},0)\ddot{n}(\mathbf{k},0) \rangle - \cdots \\ &= 1 - \frac{t^2}{2!}\langle \omega^2 \rangle + \frac{t^4}{4!}\langle \omega^4 \rangle - \cdots \end{aligned}$$
(9.6)

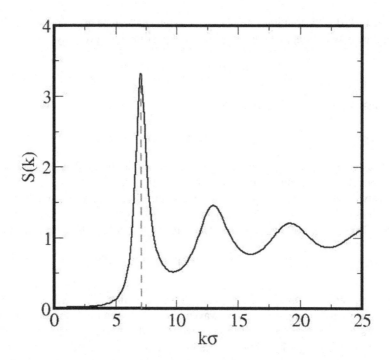

FIGURE 9.4 Static structure factor $S(k)$ at packing fraction 0.520 for a one-component liquid obtained from Percus-Yevick solution with Verlet-Weiss correction. The dashed line gives the value of $S(k)$ corresponding to the position of the first peak at $k_m = 7.15$. Unit of length is taken in terms of the diameter of the hard sphere σ.

where $\langle \omega^{2n} \rangle$ with $n = 0, 1, 2 \cdots$ are the even moments of the expansion (9.6). For convenience, we shall consider a system with continuous potential, like Lennard-Jones or soft sphere potential so that the odd moments vanish because of time parity property of TCF discussed in the previous chapter.

The calculation of these moments is given below. The second moment is straight-forward to calculate

$$\langle \omega^2 \rangle = \langle \dot{n}(-\mathbf{k}, 0)\dot{n}(\mathbf{k}, 0) \rangle = \langle (\mathbf{k}\dot{\mathbf{r}}(0))^2 e^{i\mathbf{k} \cdot (\mathbf{r}_\alpha(0) - \mathbf{r}_\alpha(0))} \rangle = \omega_0^2. \tag{9.7}$$

Here we have defined the frequency $\omega_0^2 = k_B T k^2 / m$, which gives us the second moment of self-correlation function $F_s(\mathbf{k}, t)$ as $\langle \omega^2 \rangle = \omega_0^2$.

The fourth moment of this $F_s(\mathbf{k}, t)$ is obtained in terms of the second derivatives of $n(\mathbf{k}, t)$ at $t = 0$.

$$\langle \omega^4 \rangle = \langle \ddot{n}(\mathbf{k}, 0)\ddot{n}(-\mathbf{k}, 0) \rangle = \left\langle \frac{d}{dt}\left(-i\mathbf{k}\dot{\mathbf{r}}(0)e^{-i\mathbf{k} \cdot \mathbf{r}_\alpha(0)}\right) \frac{d}{dt}\left(i\mathbf{k}\dot{\mathbf{r}}(0)e^{i\mathbf{k} \cdot \mathbf{r}_\alpha(0)}\right) \right\rangle. \tag{9.8}$$

These derivatives can be easily evaluated to obtain,

$$\langle \omega^4 \rangle = 3\omega_0^4 + \langle (\mathbf{k}\dot{\mathbf{v}})^2 \rangle. \tag{9.9}$$

The $\langle (\mathbf{k}\dot{\mathbf{v}})^2 \rangle$ term involves the time derivative of velocity and can be evaluated by using Newton's equation $F = ma$, where a is the acceleration. The force obtained in terms of the gradient of the potential as $F = -\nabla U$, $\dot{v} = F/m = -\nabla U/m$.

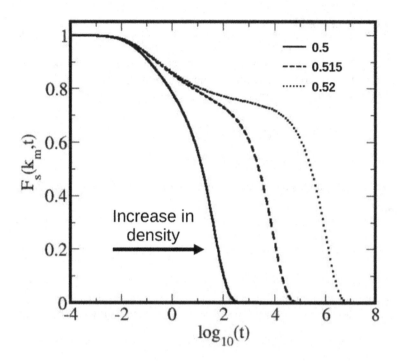

FIGURE 9.5 Self-dynamic structure factor $F(k_m,t)$ for a one-component hard spheres at packing fractions 0.500 (solid line), 0.515 (dashed line), and 0.520 (dotted line), and at wavenumber k_m where the static structure factor is peaked (shown by the dashed line in Figure 9.4). $F(k_m,t)$ also called the tagged or single-particle correlation function and can be obtained directly in the frequency plane from the measurement of $S_s(k,\omega)$ in neutron scattering experiments. The nature of the relaxation is discussed in Figure 9.2 and Figure.

$$\langle (\mathbf{k}\dot{\mathbf{v}})^2 \rangle = \frac{k^2}{m^2}\langle \nabla U.\nabla U \rangle = k^2 \frac{k_B T}{m}\left\langle \frac{\nabla^2 U}{m} \right\rangle = \omega_0^2 \Omega_0^2. \tag{9.10}$$

Here Ω_0^2 is usually termed as an Einstein frequency and defined as

$$\Omega_0^2 = \frac{1}{3m}\langle \nabla^2 U \rangle. \tag{9.11}$$

The fourth moment, thus, has the following expression:

$$\langle \omega^4 \rangle = 3\omega_0^4 + \omega_0^2 \Omega_0^2. \tag{9.12}$$

Therefore, the short-time expansion for $F_s(\mathbf{k},t)$ is given by the following expansion:

$$F_s(\mathbf{k},t) = 1 - \frac{t^2}{2!}\omega_0^2 + \frac{t^4}{4!}\left[3\omega_0^4 + \omega_0^2 \Omega_0^2\right] - \cdots \tag{9.13}$$

This is an exact short-time expansion of the self-dynamic structure factor.

9.3   FREE PARTICLE CASE

An interesting limit of Eq. (9.13) is the case of the free particles, i.e., the particles behave as an ideal gas. For the ideal gas, there will be no attraction or repulsion between the particles, i.e., $\nabla U = 0$. We have only the kinetic energy. Einstein frequency Ω_0^2 from equation (9.11) is zero in free particle limit. Then $F_s(\mathbf{k},t)$ has the given form,

$$F_s(\mathbf{k},t) = 1 - \frac{t^2}{2!}\omega_0^2 + \frac{t^4}{4!}3\omega_0^4 - \cdots \qquad (9.14)$$

Interestingly, this series is an exponential function of $-\frac{1}{2}\omega_0^2 t^2$, i.e., $F_s(\mathbf{k},t)$ have the given following Gaussian form:

$$\boxed{F_s(\mathbf{k},t) = e^{-\frac{1}{2}\omega_0^2 t^2}}. \qquad (9.15)$$

Eq. (9.15) is rigorously true for the ideal case, but $F_s(\mathbf{k},t)$ in general, can have a Gaussian form at short times and the third term of Eq. (9.13) follows a certain inequality $\Omega_0^2 \ll 3\omega_0^2$ i.e.,

$$k \gg \frac{\Omega_0}{\sqrt{\dfrac{3k_B T}{m}}}. \qquad (9.16)$$

The quantity on the right-hand side of Eq. (9.16) has the dimension of inverse length l (denominator on right-hand side is the average thermal velocity and the numerator is collision frequency). We can treat this length as the mean free path of the particles or distance traveled by particles with thermal velocity before experiencing a collision. Now, we have $k \gg 1/l$ or $\lambda \ll l$. This inequality relation implies that a system can be treated in a free particle regime when the distance involved is so short that the particles move almost independent of each other.

9.4   DENSITY-DENSITY CORRELATION FUNCTION $F(\mathbf{k},t)$

We now return to the case of collective (that is, space-dependent total) density TCF, $F(\mathbf{k}, t)$. This TCF allows us to study the *collective properties* (as opposed to the single-particle dynamics discussed above) of a system. The collective density ρ is simply defined as the sum of self-density function $n(\mathbf{r},t)$

$$\rho(\mathbf{r},t) = \sum_\alpha \delta(\mathbf{r} - \mathbf{r}_\alpha(t)). \qquad (9.17)$$

Fluctuation in density ρ is defined as

$$\delta\rho(\mathbf{r},t) = \rho(\mathbf{r},t) - \langle\rho\rangle \equiv \rho(\mathbf{r},t) - \rho_0 \qquad (9.18)$$

Students should note that $\langle\rho\rangle$ is usually denoted by ρ_0 or ρ_ℓ.

Just as in the case of single-particle density, it is often more convenient to describe the density-density TCF in terms of its Fourier transform. So, we define intermediate scattering function $F(\mathbf{k},t)$ as a correlation function of $\rho(\mathbf{k},t)$ at two different times

$$F(\mathbf{k},t) = \frac{1}{N}\langle\rho(-\mathbf{k})\rho(\mathbf{k},t)\rangle. \tag{9.19}$$

Once again, we can write the Taylor expansion for the density $\rho(\mathbf{k},t)$ and find the short-time behavior of $F(\mathbf{k},t)$. We anticipate the expansion and write it in terms of the moments as follows:

$$F(\mathbf{k},t) = F(\mathbf{k},0) - \frac{t^2}{2!}\langle\omega^2\rangle + \frac{t^4}{4!}\langle\omega^4\rangle - \cdots \tag{9.20}$$

We can find the zeroth moment of $F(k,t)$ by solving at $t = 0$

$$F(\mathbf{k},0) = \frac{1}{N}\langle\rho(-\mathbf{k},0)\rho(\mathbf{k},0)\rangle = \frac{1}{N}\left\langle\sum_j e^{i\mathbf{k}\mathbf{r}_j(0)}\sum_i e^{-i\mathbf{k}\mathbf{r}_i(0)}\right\rangle. \tag{9.21}$$

This is nothing but the static structure factor. So, $F(\mathbf{k},0) = S(\mathbf{k})$.

In this case, the higher moments are different from $F_s(\mathbf{k}, t)$. In order to calculate the second moment $\langle\omega^2\rangle$ and the fourth moment $\langle\omega^4\rangle$, we consider the interaction between the particles a sum of two separate parts. The first part is the self-part involving interactions of each particle with itself, as in this case $i = j$. The second part involves interactions of each particle with other particles. The latter is called the distinct part as interactions involve particles for which $i \neq j$. To evaluate the effects for self-part we use the results we calculated in the self-intermediate scattering function $F_s(\mathbf{k}, t)$ section.

Fortunately, the entire second moment $\langle\omega^2\rangle$ for density-density $F(\mathbf{k},t)$ is given in terms of only self-part because the $i \neq j$ terms involve velocity correlation among different particles and those which are zero and do not contribute, so we have the following nice result:

$$\langle\omega^2\rangle = \frac{1}{N}\langle\dot{\rho}(-\mathbf{k},0)\dot{\rho}(\mathbf{k},0)\rangle = \omega_0^2. \tag{9.22}$$

But this is not the case with the fourth moment $\langle\omega^4\rangle$. The contribution for $\langle\omega^4\rangle$ comes from both self- and distinct parts, i.e.,

$$\begin{aligned}\langle\omega^4\rangle &= \frac{1}{N}\langle\ddot{\rho}(-\mathbf{k},0)\ddot{\rho}(\mathbf{k},0)\rangle\\ &= \frac{1}{N}\left\langle\sum_i\left[-i(\mathbf{k}\dot{\mathbf{v}}_i) - (\mathbf{k}v_i)^2\right]e^{-i\mathbf{k}\mathbf{r}_i(0)}\sum_j\left[i(\mathbf{k}\dot{\mathbf{v}}_j) - (\mathbf{k}v_j)^2\right]e^{i\mathbf{k}\mathbf{r}_j(0)}\right\rangle.\end{aligned} \tag{9.23}$$

When $i = j$ we can calculate the contribution from self-part, which is given by Eq. (9.12). For distinct, i.e., when $i \neq j$ (after doing some algebra), we have

$$\langle\omega^4\rangle = 3\omega_0^4 + \omega_0^2\Omega_0^2 - \omega_0^2\Omega_L^2, \tag{9.24}$$

where Ω_0^2, Ω_L^2 has the following form in terms of the pairwise potential $u(r)$ and the pair correlation function $g(r)$,

$$\Omega_0^2 = \frac{\rho}{3m}\int d\mathbf{r}\nabla^2 u(g(\mathbf{r})) \tag{9.25}$$

$$\Omega_L^2 = \frac{\rho}{m}\int dr e^{-ik\cdot r}\nabla^2 u(g(\mathbf{r})). \tag{9.26}$$

Therefore, the short-time expansion for intermediate scattering function $F(\mathbf{k},t)$ has the following form:

$$F(\mathbf{k},t) = S(\mathbf{k}) - \frac{t^2}{2!}\omega_0^2 + \frac{t^4}{4!}\left[3\omega_0^4 + \omega_0^2\Omega_0^2 - \omega_0^2\Omega_L^2\right] - \cdots \tag{9.27}$$

In contrast to the Gaussian time dependence of $F_s(\mathbf{k},t)$, $F(\mathbf{k},t)$ is not just non-Gaussian but shows departure in the slightly longer time. The two functions remain surprisingly close to each other even in glassy liquids to the extent that in theoretical studies we focus on the single-particle dynamic structure factor, which allows tremendous simplification.

9.5   LONG-TIME LIMITS OF $F(k, t)$ AND $F_s(k, t)$

In the previous sections, we explored the short-time behavior of the time correlation functions $F(\mathbf{k},t)$ and $F_s(\mathbf{k},t)$, through the expansion around $t = 0$. This short-time domain is referred to as the kinetic regime. *One of the most important conclusions drawn from the short-time expansion is that the evolution of these functions can be predicted from static properties of the system like structure factor and radial distribution function $g(r)$.* This is however not the case for describing the long-time decay behavior of these correlation functions. In the long-time, transport properties like diffusion coefficient D, viscosity η play an important role in characterizing the time evolution of $F(\mathbf{k},t)$ and $F_s(\mathbf{k},t)$.

An approximate description of the long-time behavior of the correlation function $F_s(\mathbf{k},t)$ can be predicted through the diffusion of a tagged particle. The microscopic single particle density $n(\mathbf{r},t)$ and corresponding current $J_s(\mathbf{r},t)$ satisfies a continuity equation of the following form:

$$\frac{\partial n(\mathbf{r},t)}{\partial t} + \nabla.J_s(\mathbf{r},t) = 0 \tag{9.28}$$

and corresponding Fick's rule of diffusion is

$$J_s(\mathbf{r},t) = -D\nabla n(\mathbf{r},t) \tag{9.29}$$

where D is the self-diffusion coefficient. Combination of Eqs. (9.28) and (9.29) gives the diffusion equation

$$\frac{\partial n(\mathbf{r},t)}{\partial t} - D\nabla^2 n(\mathbf{r},t) = 0. \tag{9.30}$$

Taking Fourier transform over real space

$$\frac{\partial n(\mathbf{k},t)}{\partial t} = -Dk^2 n(\mathbf{k},t). \tag{9.31}$$

The equation for $F_s(\mathbf{k},t)$ can be obtained by multiplying the above equation by $n(-\mathbf{k},0)$ and taking the average

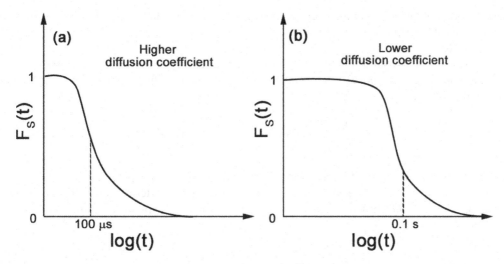

FIGURE 9.6 Self-intermediate scattering function $F_s(t)$ with faster and slower dynamics in left and right panel, respectively. Eq. (9.33) shows that the diffusion coefficient D can be obtained from the decay rate of $F_s(t)$ at a fixed wave vector k. $F_s(t)$ in (a) decays to $1/e$ times of its initial value at the time of the order of $100\mu s$ while in (b) it decays 0.1 s, which is three orders of magnitude greater than that of (a). It means the diffusion coefficient D of the system shows in ((b) is a thousand times higher than that of (a). Note that the scales of the time axis are different in the two cases.

$$\frac{\partial F_s(\mathbf{k},t)}{\partial t} = -Dk^2 F_s(\mathbf{k},t). \tag{9.32}$$

The solution of this equation is exponential,

$$F_s(k,t) = e^{-Dk^2 t}. \tag{9.33}$$

The self-dynamic structure can be obtained by taking Laplace transform of Eq. (9.33) and using the relation

$$S_s(\mathbf{k},\omega) = \frac{1}{\pi}\mathrm{Re}[F_s(z,t)]_{z=i\omega}, \tag{9.34}$$

$$S_s(\mathbf{k},\omega) = \frac{1}{\pi}\frac{Dk^2}{\omega^2 + (Dk)^2}, \tag{9.35}$$

This equation represents a single, Lorentzian curve with half-width height $2Dk^2$ and cantered at $\omega = 0$. This relation shows that we can also relate the diffusion coefficient to incoherent neutron scattering. In the hydrodynamic limit $k \to 0$ and $\omega \to 0$, the coefficient D is related to $S_s(\mathbf{k},\omega)$ in the following way:

$$\boxed{D = \lim_{\omega\to 0}\lim_{k\to 0}\frac{\omega^2}{k^2}\pi S_s(k,\omega)} \tag{9.36}$$

Here the order of taking the limit is very crucial, i.e., first take the limit over the wave vector then over ω. Qualitative estimate of diffusion of a single-particle dynamic structure factor $F_s(t)$ is shown in Figure 9.6.

For the coherent dynamic structure $F(k, t)$, we adopt a similar procedure. Now we need to deal with the number density $\rho(\mathbf{r}, t)$. The equation of motion is given by the Smoluchowski-Vlasov mean-field equation

$$\frac{\partial \rho(\mathbf{r}, t)}{\partial t} - \bar{D}\nabla \cdot [\nabla - F(\mathbf{r}, t)]\rho(\mathbf{r}, t) = 0 \tag{9.37}$$

where $F(\mathbf{r}, t)$ is the mean-field force given by the gradient of effective potential energy $V_{eff}(\mathbf{r}, t)$, i.e., $F(\mathbf{r}, t) = -\nabla V_{eff}(\mathbf{r}, t)$. Here the effective potential energy is determined from mean spherical approximation and has the following form in terms of a two-particle direct correlation function $c(\mathbf{r})$

$$\beta V_{eff}(\mathbf{r}, t) = -\int d\mathbf{r}' c(|\mathbf{r} - \mathbf{r}'|)\rho(\mathbf{r}', t). \tag{9.38}$$

Substituting Eq. (9.38) into Eq. (9.37) and taking the Fourier transform in the inverse k we get the solution for $F(k, t)$ as

$$F(k, t) = S(k)e^{-\frac{\bar{D}}{S(k)}k^2 t} \tag{9.39}$$

where $S(k)$ is the static structure factor. This expression is not valid at small k, which is the hydrodynamic limit, but is found to be semi-quantitively accurate at intermediate to large wavenumbers. *The above expression was first derived by de Gennes who used this to explain the drastic narrowing of $S(k, \omega)$ observed in 1950s for the first time by neutron scattering.*

Eq. (9.39) has a nice physical interpretation. At intermediate to large wavenumbers, we probe local, short wave length dynamics where conservation of neither the momentum nor the energy serves to constrain the dynamics, but the density conservation remains always a constrain. In the language of the field (that has its origin in the Rayleigh-Brillouin spectrum), it is commonly stated that at *large wavenumbers, the heat mode degenerates into a self-diffusion mode.*

9.6 CONTINUED FRACTION REPRESENTATION OF DYNAMIC STRUCTURE FACTORS

We most often need analytical expressions of dynamic structure factors that are valid not just at short and long times, but over the entire time evolution. As we know, the short-time expression is Gaussian or nearly Gaussian, while the long-time part is exponential or stretched exponential. Therefore, there is a need to extend the short-time expansions into longer times This is achieved through continued fraction representation.

Short-time expansions of $F(\mathbf{k}, t)$ and $F_s(\mathbf{k}, t)$ given by Eqs. (9.6) and (9.20). In principle, one needs complete knowledge of all the moments for fully evaluating the series at short times. In the present form of the series, the correlation functions get diverge for large t.

To avoid this unphysical situation one can use the well-known Mori continued fraction expansion in the following form [to keep the notation simple we are suppressing wavenumber (k) dependence],

$$F(z) = \cfrac{1}{z + \cfrac{\langle [\dot{\rho}(0)]^2 \rangle / \langle [\rho(0)]^2 \rangle}{z + \cfrac{\langle [\ddot{\rho}(0)]^2 \rangle / \langle [\dot{\rho}(0)]^2 \rangle - \langle [\dot{\rho}(0)]^2 \rangle / \langle [\rho(0)]^2 \rangle}{z + \cdots}}} \qquad (9.40)$$

The continued fraction in Eq. (9.40) is valid only for the autocorrelation function, which are even in time and all the odd frequency moments vanish. In the present section, we shall show how to evaluate $F(k,t)$ by using the continued fraction truncating at second order. From Eq. (9.40) for $F(k,t)$ we have a zeroth moment $\langle \rho(-k,0)\rho(k,0) \rangle = S(k)$, second moment $\langle \dot{\rho}(-k,0)\dot{\rho}(k,0) \rangle = \omega_0^2$ and the fourth moment is $\langle \ddot{\rho}(-k,0)\ddot{\rho}(k,0) \rangle = \omega_0^2(3\omega_0^2 + \Omega_0^2 - \Omega_L^2)$. Substituting, all these moments we get,

$$\boxed{\phi(k,z) = \cfrac{1}{z + \cfrac{\omega_0^2 / S(k)}{z + \cfrac{\Delta_k}{z + 2\sqrt{\Delta_k / \pi}}}}} \qquad (9.41)$$

where $\phi(k,z) = F(k,z) / S(k)$ and $\Delta_k = 2\omega_0^2 + \Omega_0^2 - \Omega_L^2$. Eq. (9.41) is an important expression. Note that $\phi(k,z)$ is the normalized dynamic structure factor. Figure 9.7 below is obtained by solving Eq. (9.41).

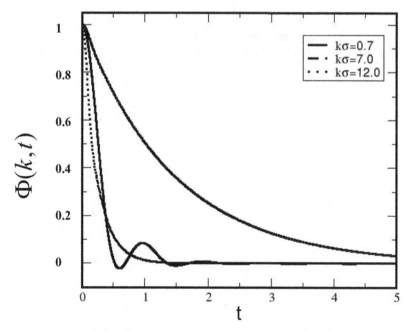

FIGURE 9.7 Numerical solution of Eq. (9.41) for the Lennard-Jones potential at different values of wavenumber $k\sigma$, where σ is the diameter of the particle.

9.7   CURRENT-CURRENT TIME CORRELATION FUNCTION

The particle current associated with microscopic density $\rho(\mathbf{r},t)$ is defined as

$$J(\mathbf{r},t) = \sum_{i=1}^{N} \mathbf{u}_i(t)\delta(\mathbf{r}-\mathbf{r}_i(t)) \tag{9.42}$$

where $\mathbf{u}_i$ is the velocity of the ith particle. In the Fourier transformed space

$$J(\mathbf{k},t) = \sum_{i=1}^{N} \mathbf{u_i}(t)\exp(\mathbf{k}\cdot\mathbf{r}_i(t)). \tag{9.43}$$

The current is a vector. Thus, the current-current correlation function is a tensor whose components are given in terms of three components of a velocity. For an isotropic system, the Fourier component of $J(\mathbf{k},t)$ may be separated into two parts: longitudinal and transverse part. The longitudinal and transverse modes are in which the velocity vector points parallel and perpendicular to the wave vector $\mathbf{k}$, respectively. For example, suppose the wave vector $\mathbf{k}$ points in the z-direction then $u_z(t)$ is the longitudinal current and $u_x(t)$ and $u_y(t)$ are the transverse currents. The geometry is illustrated in Figure 9.8.

This decomposition of current is helpful in the calculation because only the longitudinal component is related to the microscopic density.

Therefore, the longitudinal autocurrent correlation function is defined by

$$C_l(k,t) = \frac{k^2}{N}\left\langle J^z(-k,0)J^z(k,t)\right\rangle. \tag{9.44}$$

The pre-factor has been placed to serve two purposes. First, the current carries a dimension of length by time. So, k^2 removes the length dependence. The denominator is placed to make it an intensive property. Similarly, the transverse autocurrent correlation function is defined by

$$C_t(k,t) = \frac{k^2}{N}\left\langle J^x(-k,0)J^x(k,t)\right\rangle. \tag{9.45}$$

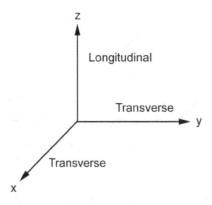

FIGURE 9.8   Velocity components in k-space. Here the direction of propagation is along the z-direction, so the component of the current along the wave vector k is the longitudinal mode. Components along the x- and y-axes are in xy-plane that is perpendicular to the direction of propagation and these modes are termed transverse modes.

Using continuity equation for microscopic density $\rho(\mathbf{r},t)$ and stationary property of TCF we get

$$C_l(k,t) = \frac{1}{N}\langle\dot{\rho}(-k,0)\dot{\rho}(k,t)\rangle = -\frac{d^2 F(k,t)}{dt^2}. \tag{9.46}$$

We may get the relationship between longitudinal current $C_l(k,\omega)$ and dynamic structure factor $S(k,\omega)$ by taking Laplace transform of Eq. (9.46) and using Eq. (9.34)

$$C_l(k,\omega) = \omega^2 S(k,\omega). \tag{9.47}$$

This result gives us information on how fluctuations in density are related to fluctuations in longitudinal current and have no effect on the transverse current. It has been found that in classical statistical mechanics both longitudinal and transverse currents have the same zero-time values as

$$C_l(k,0) = C_t(k,0) = \omega_0^2, \tag{9.48}$$

where $\omega_0^2 = (k_B T / m)k^2$. It should be noticed that the zeroth-order moment of current autocorrelation is the same as the second moment of $F_s(\mathbf{k},t)$ and $F(\mathbf{k},t)$ represented by Eqs. (9.7) and (9.22). Using Eq. (9.6), we can write a short-time expansion of these longitudinal and transverse current autocorrelation function $C_l(k,0)$ and $C_t(k,0)$

$$C_l(k,t) = \omega_0^2\left(1 - \omega_L^2 \frac{t^2}{2!} + \cdots\right) \tag{9.49}$$

$$C_t(k,t) = \omega_0^2\left(1 - \omega_T^2 \frac{t^2}{2!} + \cdots\right) \tag{9.50}$$

The zeroth moment ω_0^2 of current is a consequence of continuity equation, i.e, the conservation of the particles, so it is independent of the interaction's continuity between them. But, the higher moments like ω_L^2 or ω_T^2 indeed depend on the potential between the particles.

In the small k limit, the longitudinal and transverse frequencies ω_L^2 and ω_T^2 are related to the speed of sound c_L^2 and c_T^2 by the relation

$$\begin{aligned} c_L^2 &= k^2\omega_L^2 \\ c_T^2 &= k^2\omega_T^2 \end{aligned} \tag{9.51}$$

where the speed of sounds in both cases is given by,

$$c_L^2 = 3\omega_0^2 + \frac{\rho_0}{2m}\int d\mathbf{r}g(r)(1 - \cos kz)\frac{\partial^2 u(\mathbf{r})}{\partial z^2} \tag{9.52}$$

and

$$c_T^2 = \omega_0^2 + \frac{\rho_0}{m}\int d\mathbf{r}g(r)(1 - \cos kz)\frac{\partial^2 u(\mathbf{r})}{\partial x^2}. \tag{9.53}$$

Therefore, in the small k limit $\omega_L^2 \sim 3\omega_T^2$. Equations (9.49) and (9.50) are valid for continuous potential. It is important to discuss some crucial relations between longitudinal current and density correlation function. Here higher moment of longitudinal current ω_L^2 contribution in terms of the density correlation function is given by

$$\omega_0^2 \omega_L^2 = -\frac{d^2 C_l(k,t)}{dt^2}\bigg|_{t=0} = \frac{d^4 F(k,t)}{dt^4}\bigg|_{t=0} = \frac{1}{N}\langle \ddot{\rho}(-k,0)\ddot{\rho}(k,0)\rangle. \tag{9.54}$$

Hence, one can write,

$$\omega_L^2 = \frac{\langle \omega^4 \rangle_{\rho\rho}}{\langle \omega^2 \rangle_{\rho\rho}}. \tag{9.55}$$

For the transverse current correlation function

$$\omega_0^2 \omega_T^2 = -\frac{d^2 C_t(k,t)}{dt^2}\bigg|_{t=0}. \tag{9.56}$$

9.8 TRANSVERSE CURRENT CORRELATION FUNCTION AND VISCOSITY

In the hydrodynamic limit, i.e., in the small k and long-time limit, the decay behavior of the transverse current fluctuations is governed by the equation

$$\frac{\partial}{\partial t}J^x(k,t) + vk^2 J^x(k,t) = 0. \tag{9.57}$$

Note that this equation has the same form as the diffusion equation given by Eq. (9.31). While the quantity v has the same dimension as the diffusion coefficient D, it is two orders of magnitude larger than D. v is given by $\eta/\rho m$ where η is the shear viscosity, m is the mass, and ρ is the number density.

Transverse current correlation $C_t(k_m,t)$ can be obtained by multiplying Eq. (9.57) by $J^x(-k,0)$ and take the thermal average

$$\frac{\partial}{\partial t}C_t(k,t) + vk^2 C_t(k,t) = 0. \tag{9.58}$$

The wavenumber and time-dependent transverse correlation function has been studied extensively by computer simulations. In dense liquids, the TCF exhibits an oscillatory decay. Exponential decay sets in only in the long time.

9.8.1 USEFUL RELATIONS INVOLVING $S_s(k, \omega)$ AND $S(k, \omega)$

Experiments often provide wavenumber (k) and frequency (ω) dependent time correlation functions. Here derive several useful relations.

Let us define the Fourier components of the microscopic current associated with the tagged particle i that have velocity u_i as follows:

$$\mathbf{j}_i(\mathbf{k},t) = \mathbf{u}_i e^{-i\mathbf{k}\cdot\mathbf{r}}. \tag{9.59}$$

The self-current autocorrelation function is given by

$$C_s(k,t) = \langle \mathbf{k.j}_i(-k,0)\mathbf{k.j}_i(k,t)\rangle. \tag{9.60}$$

The velocity autocorrelation function of the tagged particle is related to self-current correlation function and self-dynamic structure factor by the following two interesting relations:

$$\begin{aligned}
Z(t) &= \langle u_i(0)u_i(t)\rangle \\
&= \lim_{k\to 0}\frac{1}{k^2}C_s(k,t) \\
&= -\lim_{k\to 0}\frac{1}{k^2}\frac{d^2}{dt^2}F_s(k,t).
\end{aligned} \tag{9.61}$$

The limit $k \to 0$ procedure removes $\exp(i\,\mathbf{k.r})$. In the Laplace plane, the relation between $C_s(k,t)$ and $F_s(k,t)$ is given by

$$\tilde{C}_s(k,z) = z^2\tilde{F}_s(k,z) - iz. \tag{9.62}$$

From Eq. (9.61), the power spectra for $Z(t)$ for a single particle is given by

$$\begin{aligned}
Z(\omega) &= \frac{\omega^2}{2\pi}\lim_{k\to 0}\frac{1}{k^2}\int_{-\infty}^{\infty}dt F_s(k,t)e^{i\omega t} \\
&= \omega^2\lim_{k\to 0}\frac{S_s(k,\omega)}{k^2}.
\end{aligned} \tag{9.63}$$

In general, the generalized Langevin equation of the velocity autocorrelation function has the following form:

$$\dot{Z}(t) = -\int_0^t \zeta(t-\tau)Z(\tau)d\tau, \tag{9.64}$$

where $\xi(t)$ is called the memory function for the velocity autocorrelation $Z(t)$. The Laplace transform of Eq. (9.64) has the following form:

$$Z(z) = \frac{k_B T}{-iz + \zeta(z)}. \tag{9.65}$$

For consistency with the hydrodynamic result, we require

$$\zeta(0) = \frac{k_B T}{mD}, \tag{9.66}$$

where D is the self-diffusion coefficients, and $\zeta(0)$ is the friction which we denote just by ζ. Eq. (9.66) is the celebrated Einstein relation.

Next we introduce a function, $M_s(k,0) = k^2 Z(0)$. A Markovian approximation is to replace $\zeta(t)$ by a quantity independent of time t. This approximation is equivalent to assuming an exponential form for $M_s(k,t)$ as

$$M_s(k,t) = \omega_0^2 e^{-|t|/\tau_s(k)} \tag{9.67}$$

where $\omega_0 = k_B T k / m$ and $\tau_s(k)$ is a wavenumber-dependent relaxation time. The constraint then requires to satisfy

$$\tau_s(0) = \frac{mD}{k_B T}.$$

(9.68)

So far the discussion has been done for the tagged particle correlation function $F_s(k,t)$ in terms of the self-current autocorrelation function $C_s(k,t)$. Similar relation can be obtained for density correlation function $F(k,t)$ in terms of the longitudinal correlation function $C_l(k,t)$ as

$$C_l(k,t) = -\frac{d^2}{dt^2} F(k,t)$$

(9.69)

and the Laplace transform is

$$\tilde{C}_l(k,z) = z^2 \tilde{F}(k,z) - iz S(k).$$

(9.70)

9.9 SUMMARY

The density-density TCF is measured by Rayleigh-Brillouin spectroscopy at small wavenumbers and by coherent and incoherent neutron scattering at large wavenumbers. As already emphasized several times, these two experimental techniques provide information about relaxation at two different length and time scales. The density-density TCF, commonly known as dynamic structure factor, plays a pivotal role in the microscopic description of time-dependent phenomena of liquids and amorphous systems. The current-current TCF on the other hand, is intimately connected to transport properties. The transverse current-current TCF connects to viscosity. The longitudinal TCF is related to dynamic structure factor as discussed here. We discuss these two TCFs separately for their importance in nonequilibrium phenomena.

BIBLIOGRAPHY

1. Zwanzig, R. 1965. Time-correlation functions and transport coefficients in statistical mechanics. *Annual Review of Physical Chemistry, 16*(no. 1), 67–102.
2. Hansen, J. P., and I. R. McDonald, 1976. *Theory of Simple Liquids*. New York: Academic Press.
3. Boon, J., and S. Yip, 1991. *Molecular Hydrodynamics*. New York: Dover.
4. Kob, W., and H. C. Andersen. 1994. Scaling behavior in the β-relaxation regime of a supercooled Lennard-Jones mixture. *Physical Review Letters, 73*, 1376.
5. Konijendijk, H. H. U., 1977. *Short Time Behaviour of Density Correlation Functions. Doctral Thesis*.
6. Rahman, A. 1972. *Statistical Mechanics: New Concepts, New Problems, New Applications*. eds S. A. Rice, K. F. Freed and J. C. Light. Chicago: University of Chicago Press.
7. Percus, J. K., and G. J. Yevick. 1958. Analysis of classical statistical mechanics by means of collective coordinates. *Physical Review, 110*, 1.

10 Velocity Time Correlation Function

OVERVIEW

Although not directly observable in experiments, velocity time correlation function $C_v(t)$ has a special place in the dynamical theory of liquids because it reveals a wealth of information about the microscopic dynamical events that a tagged particle experiences. The time dependence of the velocity correlation function often exhibits a rich structure. The short-time decay is dominated by binary interactions with its neighbors, the long-time decay, at normal liquid density, is dominated by coupling to hydrodynamic modes, especially the transverse current and the density modes. At very high density, like close to the glass transition, the current modes decay fast, and the density modes determine the decay of velocity TCF. *Not only do we get the value of the self-diffusion coefficient through the integration of $C_v(t)$, but it also provides information about the density of states through the power spectrum.* It forms an essential part of the mode coupling theory and in theories of chemical kinetics and various relaxation phenomena.

10.1 INTRODUCTION

Because neither the translational nor the rotational velocity TCF is accessible experimentally, computer simulations have been used extensively to access these TCFs. The first determination of the velocity TCF was made in 1964 for liquid argon by Rahman in a landmark work [1]. Later the velocity TCF of hard spheres was computed by Alder and Wainright in the late 1960s [2]. These "measurements" created a lot of excitement as it was found that in dense liquids, the velocity TCF behaved markedly different from the prediction of the ordinary Langevin equation. The latter predicts an exponential decay. On the other hand, the computed velocity TCF showed two unique features. First, the TCF was negative in the intermediate time range (at times about an order of magnitude larger than the collision time), and there was a power-law tail in a long time. It can be stated now, after more than 50 years, that these discoveries provided the impetus for the subsequent rapid development of time-dependent statistical mechanics.

Among the most notable theoretical developments that followed the computer-aided discoveries on velocity TCF, the studies by Dorfman and Cohen using kinetic theory of gases [4], the study by Pomeau and Resibois using hydrodynamic mode coupling theory [5], and by Zwanzig and Bixon using the generalized hydrodynamics [6] were the most prominent ones. All these approaches witnessed further developments and applications.

Despite the flurry of activity, a fully accurate theoretical description of the velocity TCF had to wait for the development of the microscopic mode coupling theory, which happened in the 1980s. The difficulty faced by all three approaches mentioned above is that an accurate description of velocity TCF requires the inclusion of the dynamics of surrounding liquid molecules

DOI: 10.1201/9781003157601-12

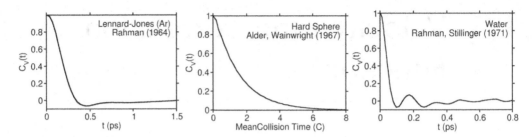

FIGURE 10.1 Velocity autocorrelation functions of some systems, taken from the classical papers by the pioneers of molecular dynamics simulation.

Source: The graphs have been replotted from the data taken from these papers. Left to right: argon (Lennard-Jones) [1], hard spheres [2], and water [3].

that form a cage with a lifetime much larger than the collision time. In a dense liquid, the collision time is very short. In fact, the lifetime of the cage becomes longer as the collision time becomes shorter.

The microscopic definition of any time correlation function has been detailed in Chapter 8. Accordingly, the velocity time correlation function or V-TCF ($C_v(t)$) is defined by Eq. (10.1)

$$C_v(t) = \langle \mathbf{v}(0).\mathbf{v}(t) \rangle = \lim_{T\to\infty} \frac{1}{T} \int_0^T ds\, \mathbf{v}(s).\mathbf{v}(t+s) \qquad (10.1)$$

where T is the time duration of the trajectory. V-TCF exhibits distinct features in the short, intermediate, and long times. In Figure 10.1, we reproduce the famous velocity TCFs from Rahman and Alder-Wainright, and also for molecular liquid water.

Theory of V-TCF has a long history and here we discuss only a small part of the developments. From the properties of the time correlation function discussed in the previous two chapters, the short-time behavior can be obtained by a simple Taylor expansion around $t = 0$.

$$C_v(t) = \langle \mathbf{v}(0)\bullet\mathbf{v}(0) \rangle - \frac{t^2}{2!}\langle \dot{\mathbf{v}}(0)\bullet\dot{\mathbf{v}}(0) \rangle + \frac{t^4}{4!}\langle \ddot{\mathbf{v}}(0)\bullet\ddot{\mathbf{v}}(0) \rangle - \cdots \qquad (10.2)$$

The first term is simply the mean square velocity of the particle. Thus,

$$\langle \mathbf{v}(0) \cdot \mathbf{v}(0) \rangle = v_0^2 = \frac{3k_B T}{m} \qquad (10.3)$$

by the equipartition theorem $\frac{1}{2}m < v^2 > = \frac{3}{2}k_B T$. The second term can be evaluated by using Newton's equation of motion $\mathbf{F} = m\mathbf{a}$, where $\mathbf{a}$ is the acceleration on the particle and $a = \dot{v}$. $\mathbf{F}$ is a force defined as the gradient of the potential energy U experienced by the particle, i.e., $\mathbf{F} = -\nabla U$ and $\mathbf{a} = \dot{\mathbf{v}} = \frac{\mathbf{F}}{m} = -\frac{\nabla U}{m}$. Therefore, the contribution of the second moment in terms of potential energy U is given by

$$\langle \dot{\mathbf{v}}(0).\dot{\mathbf{v}}(0) \rangle = \left\langle \frac{\nabla U.\nabla U}{m^2} \right\rangle. \qquad (10.4)$$

In order to evaluate this mean square average, we follow a procedure that finds repeated use in statistical mechanics, which is to evaluate both equilibrium and time-dependent averages. Here we first note that the derivatives of Hamiltonian H and U with respect to the position coordinate are equal because the kinetic part of the Hamiltonian is position-independent. We proceed as follows, with f_{eq} as the equilibrium phase space Boltzmann distribution ($f_{eq} = e^{-\beta H}/Q_N$, where Q_N is the N-particle canonical partition function).

$$
\begin{aligned}
\langle \nabla U_\alpha . \nabla U_\alpha \rangle &= \frac{1}{Z_N} \int \partial r_\alpha \left[\nabla_\alpha U . \frac{\partial U}{\partial r_\alpha} \right] \exp(-\beta U) \\
&= -\frac{1}{\beta Z_N} \int \partial r_\alpha \left[\nabla_\alpha U . \frac{\partial}{\partial r_\alpha} \exp(-\beta U) \right] \\
&= \frac{k_B T}{Z_N} \int \partial r_\alpha \left[\frac{\partial}{\partial r_\alpha} . \nabla_\alpha U \right] \exp(-\beta U) \\
&= \frac{k_B T}{Z_N} \int \partial r_\alpha \nabla_\alpha^2 U \exp(-\beta U) \\
&= k_B T \langle \nabla_\alpha^2 U \rangle.
\end{aligned}
\tag{10.5}
$$

To go from the second to the third step, we have performed integration by parts and set the first of the two terms to zero as the velocity can be set to zero at the boundary. Note the consequent change in sign. In the above equation, Z_N denotes the configurational integral. Eq. (10.5) expresses the result that the second moment is proportional to the second derivative of U with respect to the position of the αth particle, i.e., the curvature of the potential energy. If the particles in the fluid are interacting through pairwise-additive continuous interacting potential, the total potential energy U can be expressed as pairwise sum of pairwise interactions $u(r_1, r_2) \equiv u(12)$ between particles 1 and 2. From Eq. (10.5) one can further write,

$$
\begin{aligned}
\langle \nabla_\alpha U . \nabla_\alpha U \rangle &= k_B T \int dr_1 \int dr_2 \nabla_1^2 u(12) \left\{ (N-1) \int dr_3 \cdots dr_N \frac{e^{-\beta U}}{Z_N} \right\} \\
&= \frac{1}{N} k_B T \int dr_1 \int dr_2 \nabla_1^2 u(12) \frac{N(N-1) \int dr_3 \dots dr_N e^{-\beta U}}{Z_N} \\
&= \frac{1}{N} k_B T \int dr_1 \int dr_2 \nabla_1^2 u(12) \rho_0^2 g(12) \\
&= \rho_0 k_B T \int dr \left(\nabla^2 u(r) \right) g(r),
\end{aligned}
\tag{10.6}
$$

where ρ_0 is the average number density and $g(r)$ is the pair correlation function. When we substitute the above expression in the second term of Eq. (10.2), we get

$$
\begin{aligned}
\langle \dot{v}(0) . \dot{v}(0) \rangle &= \frac{1}{m^2} \rho_0 k_B T \int dr \left(\nabla^2 u(r) \right) g(r) \\
&= \frac{k_B T}{m} \Omega_0^2 = v_0^2 \Omega_0^2,
\end{aligned}
\tag{10.7}
$$

where Ω_0^2 is termed as the Einstein frequency and defined in terms of the pair correlation function $g(r)$ in the following way:

$$
\Omega_0^2 = \frac{\rho_0}{m} \int dr \left(\nabla^2 u(r) \right) g(r).
\tag{10.8}
$$

For short time, it represents the average collision frequency of the tagged particle in the system. Finally, we can write the short-time velocity autocorrelation function

$$C_v(t) = v_0^2 \left(1 - \frac{t^2}{2}\Omega_0^2 + \cdots\right). \tag{10.9}$$

If we compare Eq. (10.9) with the short-time expansion of velocity time correlation function (V-TCF) in a harmonic potential energy surface, we recognize that the second term, which is equal to $\frac{k_B T}{m}\Omega_0^2$ suggests that Ω_0 is a frequency. As already mentioned, this is the Einstein frequency, and it reflects the cage surrounding the tagged particle as seen in the V-TCF.

Although a large degree of our attention and effort goes into understanding the long-time decay behavior of V-TCF, the short-time dynamics play a very important role in a large number of processes. For example, in activated barrier crossing dynamics, it is the high-frequency response that plays an important role, as we shall see in Chapter 20.

10.2   LONG-TIME TAIL OF VELOCITY TIME CORRELATION FUNCTION

Unlike short-time decay, which can be calculated accurately from the structural properties of a liquid, the long-time decay poses both conceptual and technical difficulties. We have an exact starting point in the case of short-time decay in the form of a Taylor expansion. In contrast, we do not have an *a priori* idea about the functional form in the long-time decay. Nevertheless, the long decay is crucial as it determines not just the amplitude of the transport coefficients but also dictates our understanding of phenomena such as glass transition.

While the short-time decay of TCF is determined by a few particle collisions and was captured by Boltzmann through the binary collision interaction, the long-time decay involves many-particle interactions. One important event that determines the long-time behavior of TCFs in liquids is the correlated collisions where several particles undergo repeated collisions. Such events are needed to describe the slow structural relaxation of the cage formed by the nearest neighbors.

The long-time decay of TCFs exhibits several startling behaviors. For example, the long-time decay of velocity TCF decays as $t^{-d/2}$, here d is the dimension of the system. The sign of this dependence also depends on several factors. This $t^{-d/2}$ dependence leads to a logarithmic divergence of the mean square displacement in two-dimensional systems. Current-current TCF also exhibits interesting nontrivial decay behavior in the long-time limit. The density-density TCF of a liquid exhibits a highly nonexponential, stretched exponential-type decay in supercooled liquids, near its glass transition temperature.

Theoretical approaches to understanding long-time behavior can be grouped into two categories.

(i)   The first group of theories is based on the kinetic theory of gases that essentially generalizes the historically important treatment of Boltzmann and Enskog. These theories were developed in the late 1960s and early 1970s and were pioneered by Cohen, Dorfman, and others.

(ii) The second group of theories is based on the generalization of hydrodynamics and was developed by Zwanzig, Pomeau, and Resibois and others. A big role was played by molecular dynamics simulations that, in fact, were the first to observe the power-law long-time decay of the TCFs (Alder-Wainright, Rahman).

To put it in a nutshell, the power-law decay in a long time arises from the coupling of the position of the tagged particle to the flows or the current modes in the system. Neither the extended kinetic theory of gases nor the generalized or molecular hydrodynamic theories can predict the long-time decay correctly. The most accurate theory at this point is the mode coupling theory (MCT), which takes into account both the structure of the liquid and its correlated dynamics in a consistent fashion. We shall discuss MCT in detail in Chapter 23.

These TCFs exhibit interesting behavior also at the intermediate times. For example, the velocity TCF contains a pronounced negative region [Figure 10.1(a)] that serves to reduce the value of the time integral of the velocity TCF and thus, reduces the value of the self-diffusion coefficient of a molecule in a dense liquid state. The negative region itself arises from the rebound of a molecule with the nearest-neighbor molecules that form the cage.

10.3   VELOCITY AUTOCORRELATION FROM GENERALIZED HYDRODYNAMICS

Hydrodynamics provides a somewhat simpler approach to study the velocity TCF. This is achieved through an elegant expression for the frequency-dependent friction that provides V- TCF through the generalized Einstein relation, which is discussed below. This hydrodynamic expression finds use in many applications, like the calculation of the rate of barrier crossing, velocity-time correlation function, and several others. The most succinct derivation was provided by Zwanzig and Bixon, whose expression for friction involves frequency-dependent shear and bulk viscosity.

$$\hat{\zeta}(z) = \left(\frac{4\pi}{3}\right)\eta_s(z)RX^2\left[2(X+1)P+(1+Y)Q\right] \tag{10.10}$$

where

$$X = \left(z\rho_0/\eta_s\right)^{1/2}R \tag{10.11}$$

$$Y = z\left[c^2 + \frac{z\eta_l}{\rho_0}\right]^{-1/2}R \tag{10.12}$$

$$P = \frac{3}{\Delta}\left(3+3Y+Y^2\right) \tag{10.13}$$

$$Q = \frac{3}{\Delta}\left(3+3X+X^2+\frac{X^2(1+X)}{2+\beta/\eta_s}\right) \tag{10.14}$$

$$\Delta = 2X^2\left[3+3Y+Y^2\right]+Y^2\left[3+3X+X^2\right]+\frac{3X^2(1+X)(2+2Y+Y^2)}{2+\beta/\eta_s} \tag{10.15}$$

Here, η_s is the frequency-dependent shear viscosity, R is the radius of sphere, ρ_0 is the solvent density, c is the sound velocity, and β is the slip parameter. The value of this parameter is zero for slip and infinity for stick boundary conditions. $\eta_l(z)$ is the longitudinal viscosity, which is expressed in terms of shear ($\eta_s(z)$) and bulk ($\eta_v(z)$) viscosities [Eq. (10.16)].

$$\eta_l(z) = \frac{4}{3}\eta_s(z)+\eta_v(z). \tag{10.16}$$

The expressions for shear and bulk viscosities are given by Maxwell forms as

$$\eta_s(z) = \frac{\eta_s^0}{1 + z\tau_s} \tag{10.17}$$

$$\eta_v(z) = \frac{\eta_v^0}{1 + z\tau_v} \tag{10.18}$$

where η_s^0 and η_v^0 are the zero-frequency shear and bulk viscosities and the time constants τ_s and τ_v are the relaxation times for shear and bulk modes of the solvent.

This approach by Zwanzig and Bixon generalized the standard Navier-Stokes treatment by introducing viscoelasticity into the theory by making the viscosity time-dependent. A detailed discussion of the frequency-dependent friction is given later. Once we get the frequency-dependent friction, we can compute the frequency-dependent diffusion coefficient using the generalized Einstein equation [Eq. (5.61)], which in turn gives the frequency-dependent velocity autocorrelation [Eq. (5.62)] that encapsulates the essence of generalized hydrodynamics.

$$D(z) = \frac{k_B T}{m[z + \zeta(z)]} \tag{10.19}$$

$$\boxed{C_V(z) = \frac{k_B T}{m[z + D(z)]}} . \tag{10.20}$$

Figure 10.2 shows the ability of the generalized hydrodynamics to capture the initial and the intermediate time decay behavior of the velocity TCF of liquid argon as was obtained by Rahman. While the ability to describe the negative region of V-TCF is quite impressive, the hydrodynamic approach fails to capture the long-time tail. One needs the mode coupling theory to understand the long-time decay as it arises due to the coupling of the tagged particle's motion to the collective motion (described by the hydrodynamic modes) of the liquid.

10.4 COMPARISON BETWEEN TRANSLATIONAL AND ROTATIONAL V-TCF

While much of the studies of velocity TCF centers around atomic liquids, like argon or model systems of hard spheres, in the real world, we deal with molecules. Also, many molecules like carbon monoxide, nitrogen, ethanol, methanol, and other industrially relevant liquids and solvents are all molecular in nature. There appears to be a deep-seated tendency to avoid these liquids as they are too complex to treat in theory. However, advances in molecular dynamics simulations have removed this difficulty, and interesting results are becoming available.

We showed above in Figure 10.1, the velocity TCF of water molecules. It is interesting to enquire if we can learn something interesting by comparing the normalized linear velocity and angular velocity TCFs. Obviously, the initial values are given in both cases by the equipartition theorem.

Any binary interaction that determines the short-time behavior relaxes both linear and angular velocity TCF, but could be by different degrees. Thus, a comparison between the two could be enlightening. While the initial decay parts reflect the binary interaction, the intermediate negative regions could be quite different. Especially different could be the long-time tails, if any, for the two TCFs.

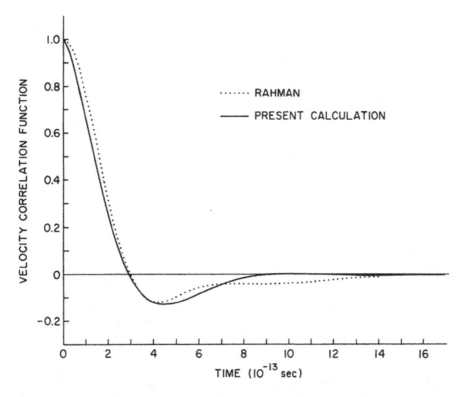

FIGURE 10.2   Comparison of V-TCF of argon computed from the generalized hydrodynamics approach of Zwanzig and Bixon [7] with the MD simulation results of Rahman [1].

In Figure 10.3, we compare the linear and angular velocity correlation functions of a small linear molecule, nitric oxide.

Thus, both the two functions exhibit a noticeable minimum, which is, in the case of translational velocity, due to back-scattering. However, for rotation it is a mix of both inertial motion and back-scattering.

10.5   POWER SPECTRUM

The Fourier transform of velocity TCF is usually called power spectrum in statistical mechanics. Note that power spectrum is also routinely used to denote the Fourier transform of fluctuating quantities.

$$S(\omega) = \frac{1}{2\pi} \int_{-\infty}^{\infty} dt [e^{i\omega t} C_v(t)] = \frac{1}{2\pi} \int_{-\infty}^{\infty} dt [C_v(t)\cos(\omega t)] \qquad (10.21)$$

The importance of the power spectrum is that it gives the density of states. For atomic systems, this density of states corresponds to the intermolecular vibrational density of states. For molecular liquids, the DOS is a mixture of the rotational and translational density of states. In Figure 10.4, we show the velocity time correlation function of water and also the power spectrum. The latter clearly reveals the contributions of the translational and rotational contributions.

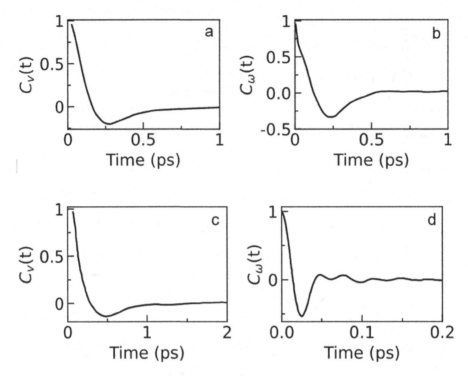

FIGURE 10.3 Top: (a) linear velocity correlation function and (b) angular velocity correlation function of nitric oxide at 300 K. Bottom: (c) linear velocity time correlation function of water at 298 K (taken with permission from Andriy and Nahtigal [10]) and (d) angular velocity time correlation function of water at 300 K.

Source: Figure taken with permission from [11].

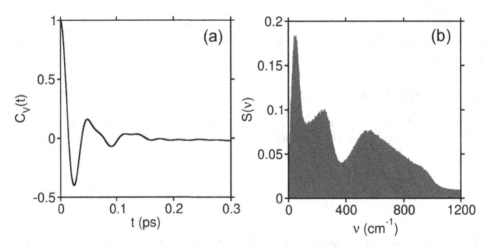

FIGURE 10.4 (a) Velocity time correlation function (V-TCF) and (b) power spectrum of water at 300 K and ambient pressure. These are computed from molecular dynamics simulations. Power spectrum is obtained by Fourier transform of VACF.

10.6 ADVANCED THEORETICAL METHODS

In this chapter, we have refrained from going deep into any advanced method. Instead, we emphasized the general properties of the velocity TCF. We have discussed the advanced methods in Chapter 23, which is the MCT chapter. However, certain basic steps can be discussed. We essentially adopt two approaches. In the first approach, one evaluates the V-TCF directly by projecting the velocity of the tagged particle on the current mode. This approach was initiated by Pomeau and Resibois and also by Gaskell and Miller. The second approach, which we attribute to Sjogren and Sjolander, calculates the frequency-dependent friction and uses that in the generalized Einstein relation to obtain the V-TCF. This latter method is in the spirit of the hydrodynamic approach of Zwanzig and Bixon. This approach has a simple interpretation. The generalized diffusion coefficient is decomposed into the contributions from the three hydrodynamic modes: the density and the two current modes. Thus, we can write $D(z) = D_{mic}(z) + D_{curr}(z)$. The longitudinal current mode is found to make a small contribution in dense liquids.

10.7 SUMMARY

The velocity-time correlation function of atomic and molecular liquids is not accessible experimentally. However, they play a vitally important role in our theoretical and conceptual understanding of dynamical processes in liquids. Because of their pivotal role in the study of nonequilibrium phenomena, we have devoted a separate chapter to this TCF.

Because this TCF is not accessible experimentally, computer simulations have been used to obtain them, and the results, as discussed here, have thrown new light on microscopic aspects of dynamics of molecular motion. Since the diffusion constant is a time integral of velocity TCF, understanding of the TCF also leads to the understanding of the diffusion process.

In this chapter, we have discussed not only the basic aspects of V-TCF. But also certain advanced topics like short-time expansion, and long-time power-law decay.

Velocity TCF mirrors the dynamics of the liquid. At high density, in particular, it couples to the hydrodynamic modes, both current and density modes. Thus, we find that single-particle dynamics behave like collective property. The velocity TCF, thus, provides an easy way to study dynamics of the system, especially by computer simulations.

We shall discuss in the mode coupling theory chapter (Chapter 23) how to understand the coupling between the single particle and the collective dynamics.

BIBLIOGRAPHY AND SUGGESTED READING

1. Rahman, A. 1964. Correlations in the motion of atoms in liquid argon. *Physical Review*, *136*(2A), A405.
2. Alder, B. J., and Wainwright, T. E. 1967. Velocity autocorrelations for hard spheres. *Physical Review Letters*, *18*(23), 988.
3. Rahman, A., and Stillinger, F. H. 1971. Molecular dynamics study of liquid water. *Journal of Chemical Physics*, *55*(7), 3336–3359.
4. Dorfman, J. R., and Cohen, E. G. D. 1970. Velocity correlation functions in two and three dimensions. *Physical Review Letters*, *25*(18), 1257.
5. Pomeau, Y., and Resibois, P. 1975. Time dependent correlation functions and mode-mode coupling theories. *Physics Reports*, *19*(2), 63–139.
6. Gaskell, T., and Miller, S. 1978. *Journal of Physics C, 11*, 3749.
7. Zwanzig, R., and Bixon, M. 1970. Hydrodynamic theory of the velocity correlation function. *Physical Review A*, *2*(5), 2005.

8. Zwanzig, R. 2001. *Nonequilibrium Statistical Mechanics*. Oxford: Oxford University Press.
9. Hansen, J. P., and McDonald, I. R. 1990. *Theory of Simple Liquids*. Amsterdam: Elsevier.
10. Andriy, P. and Nahtigal, I. 2006. Spatial hydration structures and dynamics of phenol in sub- and super-critical water. *Journal of Chemical Physics*, *124*, 024507
11. Rog, T., and Murzyn, K. 2003. *Journal of Compututational Chemistry, 24*(5), 657–67.

11 Linear Response Theory and Fluctuation-Dissipation Theorems

OVERVIEW

The linear response theory (LRT) provides us a quantitative method to calculate the response of a system to an external perturbation, in terms of the properties of the system at equilibrium. The fluctuation-dissipation theorems (FDT) establish quantitative relations for transport properties in terms of time correlation functions. These are of great practical value. In fact, the linear response theory (LRT) can be considered as the direct generalization of the statistical mechanical expressions of the equilibrium response functions like specific heat, compressibility, susceptibility, etc., to the time domain. Both LRT and FDT are invariably connected with the names of Einstein, Onsager, and Kubo who contributed to the development of these theories from fundamental principles and put them on a firm basis. In fact, Kubo built on the regression hypothesis of Onsager, which has been discussed in Chapter 4 where we also discussed linear response theory briefly, albeit qualitatively. In this chapter, we develop the basic formalism starting from the Liouville equation for the density matrix. The development parallels the perturbation theories routinely used in quantum mechanics. Linear response theory is also a first-order perturbation. The same formalism can be used to derive a relation between fluctuation and dissipation, which is called the fluctuation-dissipation theorem. We encounter FDT in the context of the Langevin equation where the random force term (which is a fluctuation) gives rise to dissipation that is described by the friction term. FDT and LRT together form the cornerstone and provide the very basis of time-dependent statistical mechanics.

11.1 INTRODUCTION

The linear response theory (LRT) can be considered as the direct generalization of the equilibrium response functions like specific heat, compressibility, susceptibility, etc. to the time domain. In equilibrium statistical mechanics, these response functions are given exactly by the mean square fluctuations of the relevant properties. For example, specific heat is the mean square fluctuation in energy, the dielectric constant is the mean square fluctuation of total dipole moment, and isothermal compressibility is the mean square fluctuation in density. In the time domain, they translate into relations between dynamic response functions that are given by the relevant TCFs. To elaborate further, when we supply heat to a system, the increase in temperature is given by the specific heat,

$$\Delta T = \frac{1}{C_V} Q,$$ where Q is the heat supplied and C_V is the specific heat. Thus, when heat is supplied

from outside, the response, which is the increase in temperature, is given by specific heat, which is an equilibrium property. The latter is possible because the response is linear in the perturbation. In time-dependent response to a changing perturbation, the linear response is again given by the

DOI: 10.1201/9781003157601-13

properties of the system at equilibrium, and by a time correlation function. Clearly, establishment of this relationship is of profound importance.

Historically speaking, the early indication of LRT appeared in the Stokes relation ($F = \zeta v$) between the force F and the velocity v. This relation constitute a marked departure from Newton's equations, because here a constant force acting on a tagged large particle in liquid generates a constant velocity, that is, the acceleration is absent. This is because the friction makes the energy dissipate.

As explained by Einstein, the incessant motion of a large tagged particle, much larger in size than that of the solvent molecules, is due to random collisions with the surrounding molecules. These collisions give random forces on the tagged particle. *In a remarkable work, Einstein established that diffusion of the tagged particle is determined by the frictional forces that again originate from the same random forces.* Since friction is a measure of dissipation, it is natural that the friction is quantitatively related to these random collisional forces. This relationship is a part of a more general and robust relationship known as the fluctuation-dissipation theorem (FDT). Einstein's relation between mobility (or, diffusion) and friction is considered the first fluctuation-dissipation theorem. This is further quantified in the ordinary Langevin equation where the force-force time correlation function gives the friction. *We discuss these aspects in Chapter 16 in more detail.*

This relationship between random forces or fluctuations and dissipation is made quantitative through linear response theory. In a certain sense, the linear response theory established by Kubo is the most important basic result of nonequilibrium statistical mechanics, although it was already being used in several earlier works.

Kubo described linear response theory as the response being determined by the "*natural motion*" of the system. As discussed in more detail later, the statistical mechanical linear response theory bears a close similarity to the first-order perturbation theory of quantum mechanics. And the derivation of the fluctuation-dissipation theorem is similar to the time-dependent quantum mechanics.

It is important to realize that both the linear response theory and the fluctuation-dissipation theorem originate from the regression hypothesis proposed by Onsager on his way to establishing the reciprocal relations, discussed in Chapter 4. While we discuss the inter-relationship later, let us briefly mention here that the linear coefficients considered by Onsager are the linear response functions. All these concepts are valid in the linear regime near equilibrium.

11.2   DERIVATION OF THE LINEAR RESPONSE THEORY

The main outcome of the linear response theory is the demonstration that the response of a physical property (or, the change suffered due) to an external perturbation *is given by the properties of the unperturbed system, and most importantly, the latter can be expressed as a time correlation function. Linear response means that the response is proportional to the external perturbation.*

We consider two properties A and B, like the additional energy (due to perturbation) of the system due to an external electric field (A) and resultant polarization (B). These are also dynamic variables that are both functions of positions and momenta of the constituent particles. The idea is to couple variable A to an external field and to measure the response in variable B, which need not be the same as A.

Let there be an applied time-dependent field $F(t)$ that is weakly coupled to the dynamical variable A. The Hamiltonian of the system can be written as

$$H = H_0 + H'(t) \tag{11.1}$$

where H_0 is the time-independent Hamiltonian of the unperturbed system and $H'(t)$ is the Hamiltonian due to the external perturbation which is given as

$$H'(t) = -\int d\mathbf{r}\, A(\mathbf{r})F(\mathbf{r},t).$$

We next assume that the field is a combination of monochromatic plane waves. It is thus sufficient to treat the external perturbation as a single plain wave

$$F(r,t) = \frac{1}{V} F_k \exp\left[i(k.r - \omega t) \right] \tag{11.2}$$

With this functional form, the external Hamiltonian reduces to

$$H'(t) = -A_{-k} F_k \exp(-i\omega t),$$

where A_k is defined as

$$A_k = \frac{1}{V} \int dr\, e^{ik.r} A(r) \tag{11.3}$$

The time evolution of phase space probability $f^{[N]}(t) \equiv f^{[N]}(r^N, p^N, t)$ is determined by the following Liouville equation in the presence of field:

$$\frac{\partial f^{[N]}(t)}{\partial t} = -iL f^{[N]}(t) = \left\{ H_0 + H', f^{[N]}(t) \right\}$$
$$= -iL_0 f^{[N]}(t) - \left\{ A, f^{[N]}(t) \right\} F(t) \tag{11.4}$$

where the curly brackets, $\{....\}$ stand for Poisson brackets and L_0 denotes the Liouville operator corresponding to the unperturbed Hamiltonian. We next decompose the probability density as

$$f^{[N]}(t) = f_0^{[N]} + \Delta f^{[N]}(t) \tag{11.5}$$

We next substitute Eq. (11.3) into Eq. (11.4) to obtain

$$\frac{\partial}{\partial t} f_0^{[N]} + \frac{\partial}{\partial t} \Delta f^{[N]}(t)$$
$$= -iL_0 f_0^{[N]} - iL_0 \Delta f^{[N]}(t) - \left\{ A, f_0^{[N]}(t) \right\} F(t) - \left\{ A, \Delta f^{[N]}(t) \right\} F(t) \tag{11.6}$$

We now observe that the first term on the left and the first term on the right sides are equal by the Liouville equation for the unperturbed system. Next, we keep terms only up to the linear order in the external field to obtain the following equation:

$$\frac{\partial \Delta f^{[N]}(t)}{\partial t} = -iL_0 \Delta f^{[N]}(t) - \left\{ A, f_0^{[N]} \right\} F(t) \tag{11.7}$$

We can easily obtain the formal solution of the above equation in the following form:

$$\boxed{\Delta f^{[N]}(t) = -\int_{-\infty}^{t} ds \exp\left[-i(t-s)L_0 \right] \left\{ A, f_0^{[N]} \right\} F(s)} \tag{11.8}$$

That Eq. (11.6) is indeed the solution of Eq. (11.5) can be verified by direct substitution. Eq. (11.8) provides the expression for the distortion of the phase space density distribution due to the perturbation, to the first order. The name linear response theory arises from the preceding steps where we retained terms only up to the first order in perturbation.

We next use the definition of the Poisson bracket (vide Hamilton's equations) to express the Poisson bracket on the right-hand side of Eq. (11.6) and carry out a few straightforward steps,

$$
\begin{aligned}
\left\{ A, f_0^{[N]} \right\} &= \sum_{i=1}^{N} \left(\frac{\partial A}{\partial r_i} \cdot \frac{\partial f_0^{[N]}}{\partial p_i} - \frac{\partial A}{\partial p_i} \cdot \frac{\partial f_0^{[N]}}{\partial r_i} \right) \\
&= -\beta \sum_{i=1}^{N} \left(\frac{\partial A}{\partial r_i} \cdot \frac{\partial H_0}{\partial p_i} - \frac{\partial A}{\partial p_i} \cdot \frac{\partial H_0}{\partial r_i} \right) f_0^{[N]} \\
&= -\beta \left(iL_0 A \right) f_0^{[N]} = -\beta \dot{A} f_0^{[N]}
\end{aligned}
\tag{11.9}
$$

where we have used the definition of the N-particle phase space distribution $e^{-\beta H_0}$.

Next, we observe that the mean change in the dynamic variable B can be described in terms of the change in the distribution function. Therefore, we can write the following important expression for the response B:

$$
\begin{aligned}
\left\langle \Delta B(t) \right\rangle &= \iint dr^N dp^N B(r^N, p^N) \Delta f^{[N]}(t) \\
&= \beta \int_{-\infty}^{t} ds\, F(s) \iint dr^N dp^N f_0^{[N]} \, \mathrm{B} \exp\left[-i(t-s)L_0 \right] \dot{\mathrm{A}} \\
&= \beta \int_{-\infty}^{t} ds\, F(s) \iint dr^N dp^N f_0^{[N]} \, \dot{\mathrm{A}} \exp\left[-i(t-s)L_0 \right] \mathrm{B} \\
&= \int_{-\infty}^{t} ds\, \phi_{BA}(t-s) F(s)
\end{aligned}
\tag{11.10}
$$

In the above, we have introduced the after-effect function, which is defined as

$$
\phi_{BA}(t) = \beta \left\langle B(t)\dot{A} \right\rangle = -\beta \left\langle \dot{B}(t)A \right\rangle.
\tag{11.11}
$$

Note that $\phi_{BA}(t)$ gives the change in the property in response to the effect of the external field on property A. *The right-hand side of Eq. (11.11) is the time correlation function.* In deriving this equation, we used the phase space average representation of TCF. In many applications, we need $B = A$, as discussed below.

Eqs. (11.10) and (11.11) show that within the first-order perturbation theory, the response is linear in the perturbing field. *But the important point is that the proportionality is given by a time correlation function evaluated in the absence of the perturbation. This is the essence of the linear response theory.*

Note that the thermal averages that are involved in the above equations need to be performed over the unperturbed system because in the linear approximation the variable evolves in time under the influence of an unperturbed system propagator $e^{iL_0 t}$. *These two equations together constitute the central expression of Kubo's linear response theory (LRT).*

The above equation can be generalized to the situation when the external field is both space and time-dependent. Now the response function can be expressed in terms of after-effect function through the relation,

$$
\left\langle \Delta B(r,t) \right\rangle = \int_{-\infty}^{t} \iint ds\, dr'\, \phi_{BA}\left(r - r', t - s \right) F\left(r', s \right).
$$

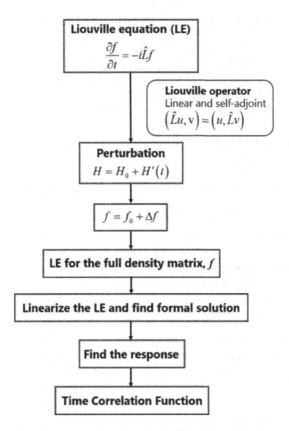

FIGURE 11.1 Flowchart showing the steps involved in the derivation of the linear response theory.

This expression is derived assuming the linear response. The above expression is now Fourier transformed to obtain the following well-known form of the response in property B,

$$\left\langle \Delta B_k(t) \right\rangle = \int_{-\infty}^{t} ds\ \phi_{BA}\left(\mathbf{k}, t-s\right) F_k\left(s\right)$$

$$\text{where } \phi_{BA}(\mathbf{k},t) = -\frac{\beta}{V}\left\langle \dot{B}_k(t) A_{-k} \right\rangle. \tag{11.12}$$

We are now in a position to derive an important relation between the changes affected by the external field F on the property A,

$$\left\langle \Delta A_k \right\rangle = \frac{\beta}{V}\left\langle A_k A_{-k} \right\rangle F_k. \tag{11.13}$$

The above equation is one of the forms of fluctuation-dissipation theorem. Here $\left\langle \Delta A_k \right\rangle$ is the change affected by the field. An appropriate example is provided by the relation between the dielectric constant of a liquid and the mean square fluctuation of the total dipole moment [6].

In Figure 11.1, we have provided a flowchart as a guide to shows the steps we need to take from Liouville equation to the time correlation function formalism.

11.3 FLUCTUATION-DISSIPATION THEOREM (FDT)

The most obvious statement of the fluctuation-dissipation theorem is the Langevin equation that provides a relation between acceleration of a particle in terms of a systematic force and a random force. We shall discuss in Chapter 15. In the absence of any external force, the systematic force describes retardation due to dissipation of energy and given by the friction. The random force is the fluctuation. These two are related by the FDT, as described below and also in Chapter 15.

Fluctuation-dissipation theorem is inevitably connected with the name of Kubo who systematized many of the already known results and provided certain exact theorems discussed below.

In fact, Kubo proposed two fluctuation-dissipation theorems.

(i)   The first FD theorem is the relation between mobility and velocity TCF. This has its origin in the hydrodynamic relation (Stokes relation) that gives velocity $v = \dfrac{F}{\zeta}$, or, $F = \zeta v$ Note that $D = \mu k_B T$, where μ is the mobility. So, we can regard mobility and diffusion as almost the same. However, usually the mobility is defined as the terminal steady velocity achieved, divided by the external force exerted. Now that the relation between mobility and diffusion is established, *we recall that D is given in terms of velocity TCF, discussed at length in Chapter 10.* Note further that $\zeta = \dfrac{k_B T}{D} = \dfrac{1}{\mu}$

(ii) The second FDT is a relation between friction and the integral over the random force TCF. This holds both in ordinary Langevin equation (LE) (discussed above) and the generalized LE.

To summarize, while the first FDT is a relation between diffusion and velocity TCF, the second FDT is a relation between friction and the random force TCF. This usage is made clear in Chapter 15 where we discuss Brownian motion and Langevin equation.

More generally, the fluctuation-dissipation theorem (FDT) is a statement of relations between natural, spontaneous fluctuations in a system at equilibrium, and various dissipation functions, like friction, viscosity, energy loss during time-dependent perturbation. In fact, FDT follows directly and easily from the linear response theory described above, and further illustrated below.

We start with Eq. (11.13). In the case of isotropic fluid, we can assume that the external field has a monochromatic form. Following the procedure outlined above, we express the response function in frequency dependence given below,

$$\left\langle \Delta B_k(t) \right\rangle = \int_{-\infty}^{t} ds\, \phi_{BA}(k, t - s) F_k \exp\left[-i(\omega + i\varepsilon)s\right]$$

$$= F_k \exp\left[-i(\omega + i\varepsilon)t\right] \int_{-\infty}^{t} ds\, \phi_{BA}(k, t - s) \exp\left[-i(\omega + i\varepsilon)(s - t)\right] \qquad (11.14)$$

$$= F_k \exp\left[-i(\omega + i\varepsilon)t\right] \int_{0}^{\infty} dt\, \phi_{BA}(k, t) \exp\left[i(\omega + i\varepsilon)t\right]$$

We need a positive nonzero epsilon to ensure convergence of the integral. We ultimately take the limit, $\varepsilon \to 0$ when the above result reduces to

$$\left\langle \Delta B_k(t) \right\rangle = \chi_{BA}(k, \omega) F_k \exp(-i\omega t). \qquad (11.15)$$

Here, $\chi_{BA}(\omega)$ is the complex susceptibility or dynamic response function.

Integrating the above equation by parts and using the functional form of $\phi_{BA}(k, t)$ gives,

$$\boxed{\chi_{BA}(k, \omega) = \frac{\beta}{V}\left[C_{BA}(k, t = 0) + i\omega \tilde{C}_{BA}(k, \omega)\right].}$$

$$(11.16)$$

As the dynamic response function is a complex function, we can write

$$\chi_{BA}(k,\omega) = \chi'_{BA}(k,\omega) + i\chi''_{BA}(k,\omega).$$
(11.17)

When A and B are the same, we obtain from Eq. (11.16)

$$\boxed{\chi''_{AA}(k,\omega) = \frac{\beta\omega}{V} C_{AA}(k,\omega)}.$$
(11.18)

The zero-frequency limit of Eq. (11.14) can be used to obtain the static susceptibility

$$\chi_{AA}(k) \equiv \chi_{AA}(k,\omega=0) = \frac{\beta}{V} C_{AA}(k,t=0)$$
(11.19)

Eq. (11.18) derived above can be regarded as a generalized, and quantitative, statement of fluctuation-dissipation theorem (FDT), sometimes also referred to as fluctuation-dissipation relation (FDR). The reason for such a name as FDT, to re-emphasize, is that the right-hand side, the decay of the TCF, due to fluctuation. The left-hand side, the imaginary part of χ gives the dissipation or energy loss. Remember that the system, assumed to be a classical oscillator, is driven at a frequency ω. While the fluctuation part is easily describable in terms of TCF, the relation of χ'' with dissipation requires further comment.

As a specific example, we can recall that the imaginary part of the dielectric function, $\varepsilon''(\omega)$, is called the dielectric loss. The real part gives the absorption while the imaginary part gives the dissipation. The ordinary Langevin equation explicitly connects fluctuation in velocity with friction and perhaps is the best known example of the relationship between fluctuation and dissipation. The following relation between the force-force TCF and friction ζ:

$$\langle F(0).F(t)\rangle = d\, k_B T\, \zeta\, \delta(t)$$
(11.20)

where d is the dimensionality. We discuss later in Chapter 15 that the Langevin equation, with the help of the above relation between friction and the force-force TCF indeed leads to the Einstein relation. The non-Markovian extension of the above relation with the generalized Langevin equation is related to Eq. (11.15) in the sense that we discuss frequency dependence of dissipation and transport properties. The generalized Langevin equation, discussed in Chapter 15, leads to a generalized diffusion constant, $D(z)$. These kinds of relations, along with Eq. (11.16) are referred to as the second fluctuation-dissipation theorem.

However, these nomenclatures essentially refer to historical aspects and have no other deeper meaning.

What is important to remember is that as in the Langevin equation, that fluctuation and dissipation are correlated – fluctuations do not just give rise to relaxation, they are also responsible for dissipation.

11.4   RELATIONSHIP WITH ONSAGER'S REGRESSION HYPOTHESIS

Onsager's regression hypothesis states that relaxation of the fluctuation in a macroscopic quantity, like fluctuation in polarization, follows the same dynamical law as the average of the macro-variable itself. The average of the macro-variable is often the observed quantity, like dielectric relaxation for polarization. In Kubo's language, the relaxation is determined by the *natural motion of the system* that determines both the fluctuation, a microscopic quantity, and the average, a macroscopic quantity, given by the equation

$$\frac{\langle \delta M(0)\, \delta M(t)\rangle}{\langle \delta M(0)^2\rangle} = \frac{\langle \Delta M(t)\rangle}{\langle \Delta M(0)\rangle}. \tag{11.21}$$

where $\Delta M(t)$ is the macroscopic displacement from the average caused by an external perturbation while $\delta M(t)$ is the natural microscopic fluctuation at equilibrium state of the system. Clearly, $\delta M(t)$ is determined by the natural motion of the unperturbed system.

Kubo's linear response theory puts the regression hypothesis on a firm basis by starting from the Liouville equation. As discussed above, LRT shows that if the disturbance created by an external perturbation is treated within the first-order perturbation theory, then indeed the dynamics is given by that of the unperturbed liquid. That is the effects of the perturbed liquid are reflected only in second and higher-order theory. As already mentioned, this is equivalent to the first-order perturbation theory of quantum mechanics. Note that the transition rate in the Fermi golden rule uses only the unperturbed wave function.

Fermi golden rule is known to be valid only for weak perturbation. In the case of linear response theory or the regression hypothesis, the validity is found to be fairly robust. This is attributed to the fact that in a macroscopic system, the displacement by perturbation is often small as it is not easy to perturb a macroscopic system far from equilibrium without causing major change.

11.5 APPLICATIONS

Both LRT and FDT are widely used in the study of relaxation phenomena and in the study of transport phenomena in general. As emphasized repeatedly, these are our major arsenals in the study of time-dependent statistical mechanics. We shall just present two representative applications below. These are well-known, but let us briefly dwell on them, to gain further understanding.

11.5.1 Derivation of Hydrodynamic Stokes Law using Linear Response Theory

According to the linear response theory described above (and repeated below), the expectation value of any observable (B) in the presence of an external time-dependent force $F(t)$ can be written as

$$\langle \Delta B(t)\rangle = \int_0^t ds\, \varphi_{BA}(t-s)F(s) \tag{11.22}$$

Here the after-effect function $\varphi_{BA}(t)$ is given by

$$\varphi_{BA}(t) = \frac{1}{k_B T}\langle B(t)\dot{A}\rangle \tag{11.23}$$

Here A denotes an observable of the system coupled with the external perturbation. Let us now assume that a tagged particle is pulled with a constant force F_{ex} along the X-direction from time $t=0$. Then the total Hamiltonian of the system at time t becomes

$$H(t) = H_0 + H'(t) \tag{11.24}$$

where H_0 denotes the unperturbed Hamiltonian and H' is given by $H' = -F_{ex}x(t)$. In the presence of this pulling, the conjugate observable drift velocity can be determined as

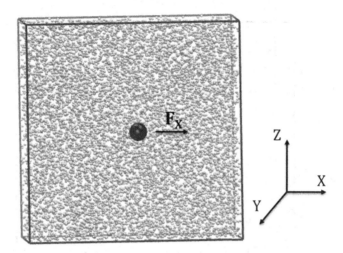

FIGURE 11.2 Schematic diagram of the system consisting of soft sphere particles. Here a tagged sphere is randomly picked, tagged and pulled along the X-direction with a constant force F_X. The steady state drift velocity is then measured.

Source: Taken from [12].

$$\left\langle v_x(t) \right\rangle = \frac{F_{ex}}{k_B T} \int_0^t ds \left\langle v_x(s)\dot{x} \right\rangle. \tag{11.25}$$

As is required by LRT, we have neglected the higher-order terms. In the long-time limit, Eq. (11.25) reduces as

$$\begin{aligned}
\left\langle v_x(t) \right\rangle &= \frac{F_{ex}}{k_B T} \lim_{t \to \infty} \int_0^t ds \left\langle v_x(s)v_x(0) \right\rangle \\
&= \frac{F_{ex}}{k_B T} D.
\end{aligned} \tag{11.26}$$

According to LRT, the averaging needs to be taken over the initial conditions, i.e., in the absence of the external field. Here D is the self-diffusion coefficient of the system. Now, we invoke Einstein's relation between diffusion D and friction ζ, i.e., $D = \dfrac{k_B T}{\zeta}$ is now used to obtain

$$\boxed{\left\langle v_x(t) \right\rangle = \frac{1}{\zeta} F_{ex}} \tag{11.27}$$

Eq. (11.27) is the celebrated Stokes' relation well known in hydrodynamics. This equation also provides a mechanical definition of friction. *In fact, the relationship between Stokes' friction and Einstein's friction by diffusion should be regarded the very first application of LRT.* In the following we illustrate the validity of the LRT by using the pulling experiment used above. In our example, we pull, in computer, a tagged atom in the X-direction in a solvent of a large number of atoms of the same size as the tagged and pulled atom. The geometry of the computer experiment is shown below in Figure 11.2. The evidences of the surprising validity of Stokes law and linear response theory are given in Figure 11.3 and Table 11.1.

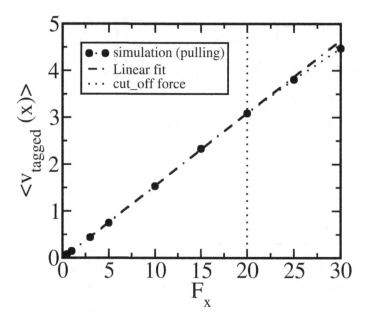

FIGURE 11.3 Plot of the X-component of the drift velocity of the tagged particle against the pulling force F_X along the X-direction. Here all the atoms interact with each other using the soft sphere, $\left(\dfrac{\sigma}{r}\right)^{12}$ potential. The black dots are the values obtained from pulling simulations. Here the effective friction is obtained by a linear fit using the Stokes relation. One can observe the emergence of slight deviation from the linearity when the perturbative force is large, above 20. Both the velocity and the force are in reduced units, as is usual in MD simulations. Here the unit is that of an argon liquid.

Source: Taken from [12].

TABLE 11.1
Agreement Between Stokes Friction and Einstein Friction Obtained by Computer Experiments in Two Model Systems. This is a Langevin Equation Simulation with a Small Bare Friction to Ensure Stability of the System

$T^*=0.8$	LJ system		Soft sphere	
Bare friction	Einstein friction	Stokes friction	Einstein friction	Stokes friction
$\zeta_{bare}=1.0$	$\zeta_{Einstein}=8.0$	$\zeta_{Stokes}=8.0$	$\zeta_{Einstein}=6.4$	$\zeta_{Stokes}=6.4$

11.5.2 SOLVATION DYNAMICS

A topical subject where LRT is routinely used is the phenomenon of nonequilibrium solvation dynamics (SD) of a newly created ion or a dipole in dipolar solvent, like water or acetonitrile. Solvation dynamics measures the time-dependent change in the polarization of the solvent. Here the fluorescence Stokes shift of an excited probe gives a dynamic measure of the progression of solvation of a newly created charge distribution. In application of the LRT, this rate of nonequilibrium solvation is calculated by using the property of the liquid at equilibrium. The nonequilibrium time-dependent change is expressed in terms of equilibrium time correlation function of the longitudinal polarization fluctuation. The latter is determined by the frequency-dependent dielectric function of the unperturbed liquid, and usually, a good agreement is obtained. Theories of solvation dynamics have been discussed in our previous book "*Molecular Relaxation in Liquids*" (Oxford).

LRT is also used widely in defining the rate of a chemical reaction, discussed in Chapter 20. Most often, in these applications, the transport properties and relaxation behavior of the unperturbed liquid are used to understand relaxation in the presence of the perturbation. LRT thus can be used to measure the properties of the unperturbed liquid. In fact, such applications have been going on for a long time, and LRT provides a microscopic understanding of diverse applications. Below we discuss a reason for the validity of LRT.

11.6 REASONS FOR THE WIDE VALIDITY OF THE LRT

Eqs. (11.5)–(11.7) articulate the main assumptions of linear response theory. The linearization employed limits the validity of the relations derived above. Nevertheless, linear response theory has found wide use and success in a variety of situations, ranging from such phenomena as dielectric relaxation to chemical kinetics.

Let us first try to understand the reason for the wide validity of LRT. In many applications, we are involved with the macroscopic response, such as in the measurement of dielectric relaxation by switching on and off a time-dependent electric field. Or, measure density relaxation by light scattering experiments like Rayleigh-Brillouin spectroscopy. In many of these applications, the liquid or the solvent system seldom gets perturbed too far away from equilibrium *because of a large force constant of fluctuation that resists large change and tries to keep the system near its equilibrium.* For example, in a system with large specific heat, one cannot increase the temperature significantly by supplying a small heat. A similar condition applies to systems with large dielectric constant. *These large values of macroscopic fluctuations indeed ensures the validity of LRT.* However many limitations of linear response theory and FDT remain. One major difficulty is imposed by the selection of the reference state. When the reference state becomes locally inhomogeneous or anisotropic, then the validity of linear response relations like the Stokes-Einstein relation becomes limited.

LRT becomes questionable near a phase transition, such as critical temperature. Here the force constant of fluctuation mentioned above becomes very small, allowing large distortion to a small perturbation. Thus, the central assumption of LRT breaks down.

Another limitation is posed by the nonequilibrium systems, such as a liquid under steady shear flow, or a system under temperature gradient. Here again, the assumptions made in Eq. (11.5) are not valid. There have been considerable discussions on this point in the literature. In most of the discussions, the linear response is assumed to be a response from an equilibrium isotropic liquid. When the system is in a steady state far from equilibrium, we do not have the analog of Boltzmann distribution. Thus, while one can adopt a similar approach, practical applicability is unclear.

We now turn to a large body of work that has attempted to extend Kubo's linear response theory. We have already discussed Prigogine's celebrated work that discusses the stability of a nonequilibrium steady state in terms of the rate of entropy production. We have discussed this study in Chapter 4. There have also been studies that attempted to extend LRT by following Kubo's approach, but with mixed success. More interestingly, a set of fluctuation relations (FR) have been derived. These FR relations are expected to be valid at far from equilibrium. We refer to the work of Giuliani *et al.* for detailed discussions.

11.7 SUMMARY

Linear response theory and fluctuation-dissipation theorem are the two pillars of time-dependent statistical mechanics. Thus, the importance of this chapter cannot be over-emphasized. These two ideas form the basis of our time correlation function approach to the study of observables, like transport properties and line shape functions. These concepts and principles are used even in defining the rate of a chemical reaction as in the Yamamoto-Chandler theory.

In addition to their importance, these two theories/theorems are two of the most elegant parts of TDSM. Students are encouraged to go through Kubo's original articles cited below, in this chapter. Although we have closely followed Kubo's articles, a study of the original articles can prove to be immensely beneficial.

Onsager also introduced the concept of microscopic reversibility, which is the microscopic analog of the principle of detailed balance. Such a micro-macro relation proves useful in the first-principles calculations starting from the Liouville equation type description. Microscopic reversibility holds only in equilibrium. Thus, the rates that enter in the description are the rates to be evaluated at equilibrium. Hence, in certain sense, concepts like linear response have been in vogue in chemical kinetics for quite some time.

BIBLIOGRAPHY

1. Kubo, R. 1957. Statistical-mechanical theory of irreversible processes. I. General theory and simple applications to magnetic and conduction problems. *Journal of the Physical Society of Japan, 12,* 570–586.
2. Kubo, R. 1966. The fluctuation-dissipation theorem. *Reports on Progress in Physics, 29,* 255–284.
3. Onsager, L. 1931. Reciprocal relations in irreversible processes. I. *Physical Review, 37*(4), 405.; Onsager, L. Reciprocal relations in irreversible processes. II. *Physical Review, 38*(12), 2265.
4. Zwanzig, R. W. 2001. *Non-Equilibrium Statistical Mechanics.* Oxford: Oxford University Press.
5. Zwanzig, R. W. 1965. Time-correlation functions and transport coefficients in statistical mechanics. *Annual Review of Physical Chemistry, 16,* 67–102.
6. Forster, D. 1975. *Hydrodynamic Fluctuations, Broken Symmetry, and Correlation Functions.* Massachusetts: The Benjamin/Cummings Publishing Company, Inc. (Advanced Book Program Reading).
7. Pathria, R. K. 1996. *Statistical Mechanics.* Oxford: Butterworth-Heinemann.
8. Hansen, J. P., and I. R. McDonald. 1976. *Theory of Simple Liquids.* New York: Academic Press.
9. Prigogine, I.1967. *Introduction to Thermodynamics of Irreversible Processes.* 3rd edition. New York: John Wiley & Sons.
10. Bagchi, B. 2012. *Molecular Relaxation in Liquids.* New York: Oxford University Press.
11. Giuliani, A., F. Zamponi, and G. Gallavotti. 2005. Fluctuation relation beyond linear response theory. *Journal of Statistical Physics, 119,* 909–944.
12. Acharya, S., and B. Bagchi, unpublished

12 Projection Operator Technique

OVERVIEW

Projection operator technique is simple in purpose. It allows us to separate the dynamical evolution of a quantity into two parts: known and unknown, the latter to be evaluated. However, this technique serves multiple purposes in TDSM. It is used to derive a reduced equation of motion for the relaxation of a dynamical variable. It is also used to derive expression for an experimentally observed TCF in terms of theoretically known TCFs. This technique was developed by Zwanzig, who was inspired by Kubo's approach of "reduction in degrees of freedom" in an interacting many-body system. In the projection operator method, we divide the *time evolution of a dynamical variable A(t)* into two parts: (i) one through a relevant subspace that is closely related to *A* and varies on a similar time scale, and (ii) an orthogonal subspace, which could be fast and therefore can be averaged out to obtain a time-dependent friction kernel. The idea then is that the first contribution is known while the second may be described accurately by projecting it onto the relevant subspace [that is, (i)]. In the derivation of the reduced equation of motion, one starts with the Liouville equation and projects out irrelevant degrees of freedom to obtain, after specific algebraic manipulations, an equation of motion of the dynamical variable. The final relaxation equation has the form of the Langevin equation with microscopic expressions for the terms involved (frequency term, the friction, and the random force term). This final equation of motion is often termed relaxation equation or memory function equation. It finds wide use in the study of time-dependent statistical mechanics. In the present chapter we derive the memory function equation and discuss a few applications.

12.1 INTRODUCTION

The motivation behind the projection operator technique is simple and straightforward. It allows us to separate the evolution dynamics of a dynamical variable into two parts: a part that we know and a part that remains to be evaluated. It is a general technique, but in TDSM it works with the Liouville equation. Prior to statistical mechanics, it was routinely used in quantum mechanics. However, the use of the technique is more elaborate as it allows us to derive a generalized relaxation equation for a TCF where the known part, usually the short-time decay part obtained by the usual Taylor expansion (as discussed in Chapters 7–10) and the long-time, unknown part, by a memory function, which remains to be evaluated and forms the crux of the problem.

DOI: 10.1201/9781003157601-14

However, in TDSM, the PO technique serves more than one purpose. It allows us to connect to the generalized Langevin equation, which we discuss later in Chapter 15 in detail, but here we introduce for the sake of clarity.

Langevin's equation is synonymous with the Brownian motion. Langevin's equation is phenomenological and often written simply as a pair of equations in position x and velocity v of a tagged particle, and can be written as follows:

$$\dot{x}=v$$
$$\dot{v}=F(x)-\zeta v+f \qquad (12.1)$$
$$<f>=0, <f(0)f(t)>=2k_B T\zeta\delta(t).$$

In the above equation, the total force on the particle is divided into two parts: a systematic force, which is the combination $F(x)$-ζv and a random force term $f(t)$. Here $F(x)$ is the external force, like from an electric field on a charged particle. The random force is supposed to take into account collisions and other fast interactions that we have no way to include in any detailed fashion most of the time. The systematic force term assured correct equilibrium distribution and also accounts for dissipation. We shall discuss this equation several times during this text.

Eq. (12.1) is an example of a reduced description. The question naturally arises: can we derive such an equation from first principles such that we have a better understanding of the random force term and the friction?

The answer, fortunately is yes, and this is partly the goal of the projection operator technique. This is a particularly valid point because any first principles study of time-dependent statistical mechanics needs to start with the Liouville equation (except for those numerical studies that employ computer simulations to solve the trajectory by solving or propagating Newton's equations of motion) for analytical studies, which are plagued by the difficulty posed by the Liouville equation for the system of many interacting particles. The projection operator technique provides a way to tackle the difficulty.

Liouville equation for a dynamical variable A is $dA(t)/dt = iLA(t)$. This is an exact equation for an interacting system of N atoms or molecules, but the equation is 6N-dimensional (for atoms) because when the atoms interact, their motions in the phase space get coupled. However, as already discussed in the first chapter, one often does not need a detailed solution of the entire many-body phase space probability distribution. Most often, we require an equation of motion for a collective quantity, like the density and/or the momentum density, or a two-body joint distribution function, or, as in the case of chemical reactions, the reaction coordinate, the goal *is to derive a reduced equation of motion of a dynamical quantity or, of a probability distribution.* The question then is, how to achieve this goal?

Fortunately, we already have guidance from successful theories like those derived in hydrodynamics, where conservation laws are combined with constitutive relations. In hydrodynamics, we already have carried out the average over "other degrees of freedom", and left with equations of motion of conserved variables. Following this cue, we thus need to retain the relevant degrees of freedom, and remove irrelevant degrees by integrating over the latter degrees of freedom. While we know the relevant or desired degrees of freedom or the dynamical variable, how does one define the irrelevant degrees of freedom?

Much insight and help come from the Langevin equation where the relevant variables are the position x and the velocity v of the tagged particle, and the irrelevant variables are the positions and the momenta of the solvent molecules. When we start from the Liouville equation, we first need to eliminate the irrelevant degrees of freedom. Actually, this procedure can be done accurately when the motions of (x, v) are slower than those of the solvent molecules. We have discussed such applications later in this chapter and Chapter 13.

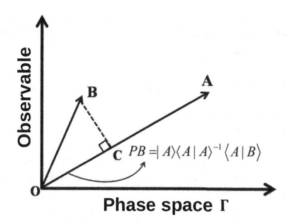

FIGURE 12.1 *A* and *B* are two vectors in the Liouville space. OC is the projection of *B* onto *A* engineered by projection operator, *P*. $PB = |A\rangle\langle A|A\rangle^{-1}\langle A|B\rangle$ denotes the projection of *B* onto *A*. Here *X*-axis formally denotes the eigenvectors in the Liouville space.

12.2 THE SEPARATION INTO FAST AND THE SLOW DEGREES OF FREEDOM

These are some of the nomenclatures, which can be confusing but relatively simple. The fast variables are the ones whose dynamics we know, and of the slow variables we know partly.

In liquid, the shortest time scale is the duration between two collisions suffered by a tagged molecule. This time scale is of the order of a few fs. Next in the hierarchy is the time scale of momentum relaxation, which is also in the sub-ps range. However, density and energy relaxation are relatively slow processes. The conserved quantities, like the number density and the current densities, vary on a much slower time scale.

We often need to understand time dependence of experimentally observed variables like diffusion of a tagged particle, solvation energy of a newly created charge, or width of a vibrational line shape (Raman effect) or NMR line shape. We call the dynamical variable in quest $B(t)$. As stated above, and we stress once more, the motion of $B(t)$ is influenced by the interactions and dynamics of the surrounding medium. But let us assume that we know about the dynamics of another set of variables $A(t)$. Examples of $A(t)$ could include conserved variables like density and momentum. These are also slow variables. The reduced description proceeds in the following fashion. First, project $B(t)$ on these slow variables, which define our "relevant" subspace. The rest of the variables are considered "fast" variables. These fast variables can be products of the slow variables. This orthogonal subspace is included through a "friction" also called a memory function.

This procedure is explained in Figure 12.1. If this process can be carried out starting from the Liouville equation, one could keep certain details about microscopic correlations that are missing in the hydrodynamic description.

12.3 RELEVANT VERSUS IRRELEVANT DEGREES OF FREEDOM

In a language that is more general than the fast and slow variables (but essentially not much different in approach), one divides the set of dynamical variables into two parts: a relevant subspace (the one that we are interested in) and the rest that can be termed irrelevant. *Now, the strategy would be to exploit this separation and derive an equation of motion for the variables in the relevant subspace only. The orthogonal subspace gives the memory function and the random force term.*

Note that the Langevin equation (LE) provides a good, although extreme, example of this separation. In LE only the position and the velocity of the tagged particle form the relevant subspace, while the positions and the velocities of the other particles form the irrelevant subspace. The acceleration on the tagged particle is also clubbed with the irrelevant subspace. Thus, both the time scale and the necessity dictate the separation.

12.4   REDUCTION OF DEGREES OF FREEDOM

In order to remove the explicit presence of the orthogonal degrees of freedom from the equation of motion, and yet keep their important contributions (such as maintaining the equilibrium distribution), we need to devise a method where the orthogonal degrees of freedom are grouped and propagated separately in Liouville's equation. The latter provides an average effect, thus affording a reduction in the degree of freedom. Note that we remove the irrelevant degrees of freedom, but their influence is retained through a friction term (in our applications). This is a standard technique in theoretical science. It is with this goal, that *one uses a projection operator to separate the irrelevant degrees of freedom from the relevant degrees of freedom.* In other words, the irrelevant degrees of freedom are made orthogonal to the system variables, in a manner similar to the Gram-Schmidt orthogonalization procedure often employed in quantum mechanics. We need to remember that *the goal is to find the time evolution of B(t), like v(t) of the tagged particle* in the Langevin equation. This evolution is influenced by any known force on our particle and by the position and velocities of all other particles in the system. The latter are treated an irrelevant and absorbed in the noise term [Eq. (12.1)]. The solution of the equation of motion of course requires the influence of the noise whose impact is included through the friction term. Also, insights can be gathered from the Navier-Stokes hydrodynamics equation. Say we are interested in the time evolution of the position and time-dependent density. Our relevant subspace consists of the five conserved variables. These are the slow ones. The rest of the variables is orthogonal to this set. It is assumed that there is a clear separation of time scales. The irrelevant degrees of freedom are absorbed into viscosity and noise terms – these two are related by the fluctuation-dissipation theorem. The idea is that we know the dynamics of relevant variables accurately and we can make approximations for the irrelevant variables. This is the reduction of variable articulated by Kubo. The projection operator (PO) technique discussed next was pioneered by Zwanzig.

12.5   THE GENERAL RELAXATION EQUATION

In this section, we present a microscopic derivation of the generalized relaxation equation (GRE). This relaxation equation is exact and provides a powerful formalism that can be used to compute the time correlation functions under circumstances where the phenomenological equations (like the Langevin equation) do not apply. *While LE is phenomenological, all the terms in GRE are expressed in terms of the Liouville operator, namely, the relaxation frequency, the memory kernel, and the random force term.*

Figure 12.1 shows a schematic representation of the two vectors in the Liouville space. A projection operator P is defined such that it projects an arbitrary vector $\mathbf{B}(q)$ on $\mathbf{A}(q)$. P can be written as

$$\boxed{PB \equiv \mid \mathbf{A} \rangle \langle \mathbf{A} \mid \mathbf{A} \rangle^{-1} \langle \mathbf{A} \mid B \rangle}. \tag{12.2}$$

The brackets denote the classical scalar product. Another projection operator Q is defined in such a way that it projects an arbitrary vector on the subspace, which is orthogonal to $\mathbf{A}(q)$. Thus Q can be written as

$$Q \equiv 1 - P. \tag{12.3}$$

It can be shown that operators P and Q are Hermitian and Idempotent (i.e., $P^2 = P$ and $Q^2 = Q$).

Below we provide a general treatment following Zwanzig. Let us consider a vector $\mathbf{A}(t)$ with multiple dynamical components (like density, momentum etc.) that describe dynamical evolution of the system. This subset $\mathbf{A}$ contains the dynamical variable $B(t)$ and other slow variables (like number density, momentum density, etc.).

Since the time evolution of the system is given by the Liouville operator, the time evolution of the vector can be written as $\mathbf{A}(t) = e^{iLt}\mathbf{A}$, where $\mathbf{A}$ is the initial value of the vector and L corresponds to Liouville's operator. Not mentioned explicitly but A depends on the phase space, as in Liouville's equation.

A crucially important point here is that the Liouville operator, and as a result, the propagator e^{iLt} operates in the full phase space and involves all the N particles of the system. We overcome this formidable difficulty in two steps. First, as already discussed, we describe orthogonal space dynamics by a set of relevant variables, including in $\mathbf{A}(t)$. Second, we project these variables back on A to get the effects of propagation in the orthogonal space. This step mixes the two sets of variables and allows us to describe the time dependence of $A(t)$ in terms of A variables and the overlap functions (called vertex functions, discussed below).

The basic philosophy is simple. We separate the dynamical evolution in two parts by using the projection operator, as shown below. The time derivative of $\mathbf{A}(q,t)$ can be written as,

$$\frac{\partial \mathbf{A}(t)}{\partial t} = e^{iLt}iL\mathbf{A}. \tag{12.4}$$

The identity, $P + Q = 1$ when inserted in the above equation, can be rewritten as

$$\begin{aligned}\frac{\partial \mathbf{A}(t)}{\partial t} &= e^{iLt}(P+Q)iL\mathbf{A} \\ &= e^{iLt}PiL\mathbf{A} + e^{iLt}QiL\mathbf{A}.\end{aligned} \tag{12.5}$$

We want this equation to look like the Langevin equation of motion for a dynamical variable $\mathbf{A}(t)$. Let us first consider the first term on the right-hand side of Eq. (12.5).

$$\begin{aligned}e^{iLt}PiL\mathbf{A} &= e^{iLt}|\mathbf{A}\rangle\langle\mathbf{A}|\mathbf{A}\rangle^{-1}\langle\mathbf{A}|iL\mathbf{A}\rangle \\ &= e^{iLt}|\mathbf{A}\rangle\left[\langle\mathbf{A}|\mathbf{A}\rangle^{-1}\langle\mathbf{A}|\dot{\mathbf{A}}\rangle\right] \\ &\equiv ie^{iLt}|\mathbf{A}\rangle\mathbf{\Omega} \\ &= i\mathbf{\Omega}\,|\mathbf{A}(t)\rangle.\end{aligned} \tag{12.6}$$

Here the frequency term $\mathbf{\Omega}$ is a tensor and is defined as

$$\mathbf{\Omega} \equiv \langle\mathbf{A}|\mathbf{A}\rangle^{-1}\langle\mathbf{A}|L\mathbf{A}\rangle. \tag{12.7}$$

Note that $\mathbf{A}$ is a vector, for example, (x, v). If $\mathbf{A}$ contains a single variable, then the frequency term is zero. Thus, the frequency matrix contains contributions of other components of the vector A on a given component.

Now, we want to evaluate the second term on the right-hand side of Eq. (12.5). We use the following exact identity:

$$e^{-Mt} = e^{-(M+N)t} + \int_0^t dt_1 e^{-M(t-t_1)} N e^{-(M+N)t_1}. \tag{12.8}$$

This interesting identity can be verified by differentiation. One needs to use Leibnitz's rule of differentiation under the integral sign. Probably a better way to understand this partitioning is through the following Laplace transformed representation (with z as the Laplace variable conjugate to time t)

$$\frac{1}{z+M} = \frac{1}{z+M+N} + \frac{1}{z+M}N\frac{1}{z+M+N}.$$

This can also be verified easily.

If we substitute M by $-iL$, and N by iPL, and use $1-P=Q$, we get the following identity:

$$e^{iLt} = e^{iQLt} + \int_0^t d\tau e^{iL(t-\tau)}iPLe^{iQL\tau}. \tag{12.9}$$

Using this identity, Eq. (12.5) can be rewritten in the following form:

$$\frac{\partial \mathbf{A}(t)}{\partial t} = e^{iLt}Pi L\mathbf{A} + e^{iQLt}Qi L\mathbf{A} + \int_0^t d\tau e^{iL(t-\tau)}iPLe^{iQL\tau}Qi L\mathbf{A}. \tag{12.10}$$

Let us now define the following quantities. We define a force term $F(q,\tau)$ as

$$\mathbf{F}(\tau) \equiv e^{iQL\tau}Qi L\mathbf{A} = Qe^{iQL\tau}Qi L\mathbf{A} = Q\mathbf{F}(\tau). \tag{12.11}$$

We see that this force remains orthogonal to $\mathbf{A}(t)$ at all times. The last integral in Eq. (12.10) involves the term $iPL\mathbf{F}(\tau)$, which can be rewritten as

$$iPL\mathbf{F}(\tau) = iPLQ\mathbf{F}(\tau) = (1-Q)iLQ\mathbf{F}(\tau). \tag{12.12}$$

Since Q and L are both Hermitian, $iPLF(\tau)$ can be written as

$$\begin{aligned}
PiL\,|\,\mathbf{F}(\tau)\rangle &= iLQ\,|\,\mathbf{F}(\tau)\rangle - (1-|\,\mathbf{A}\rangle\langle\mathbf{A}\,|\,\mathbf{A}\rangle^{-1}\langle\mathbf{A}\,|)iLQ\,|\,\mathbf{F}(\tau)\rangle \\
&= -|\,\mathbf{A}\rangle\langle\mathbf{A}\,|\,\mathbf{A}\rangle^{-1}\langle iLQ\mathbf{A}\,|\,\mathbf{F}(\tau)\rangle \\
&= -|\,\mathbf{A}\rangle\langle\mathbf{A}\,|\,\mathbf{A}\rangle^{-1}\langle\mathbf{F}(0)\,|\,\mathbf{F}(\tau)\rangle.
\end{aligned} \tag{12.13}$$

To get this expression for $PiLF(\tau)$, we used the following three important properties:
(i) $\langle\mathbf{A}\,|\,LF(\tau)\rangle = \langle LA\,|\,\mathbf{F}(\tau)\rangle$; (ii) P acting on L gives us zero; and (iii) $\langle F(0)\,| = -\langle QiLA\,|$.

Eq. (12.13) defines the memory function $\Gamma(\tau)$ as

$$\Gamma(\tau) \equiv \langle\mathbf{A}\,|\,\mathbf{A}\rangle^{-1}\langle\mathbf{F}\,|\,\mathbf{F}(\tau)\rangle. \tag{12.14}$$

That is, the memory function is proportional to the random force autocorrelation function. This is the well-known second fluctuation-dissipation theorem.

With the definition of the frequency and memory function, Eq. (12.10) can be rewritten as

$$\boxed{\frac{\partial \mathbf{A}(t)}{\partial t} = i\Omega.A(t) - \int_0^t d\tau\Gamma(\tau)\mathbf{A}(t-\tau) + \mathbf{F}(t)}. \tag{12.15}$$

This is a fundamental and exact equation. It is a nonlinear Langevin equation of motion for the dynamical variables $\mathbf{A}(t)$ *in terms of the free streaming term* Ω, *which is static and is a matrix. The memory function* $\Gamma(t)$ *is obtained in terms of the autocorrelation function. The technique described* above provides a general method, which can be used to extract an equation of motion for any arbitrary dynamical variable. If we multiply both sides of Eq. (12.15) by $\mathbf{A}(q)$ and take the ensemble average with the condition that $(A \mid F^*(t)) = 0$, we get the equation for the time correlation function as follows:

$$\frac{\partial \mathbf{C}(t)}{\partial t} = i\mathbf{\Omega}\,\mathbf{C}(t) - \int_0^t d\tau\,\mathbf{\Gamma}(\tau)\,\mathbf{C}(t - \tau) \tag{12.16}$$

where $\mathbf{C}(t) = \big(\mathbf{A}\big|\mathbf{A}(t)\big)$ is the correlation matrix. Eq. (12.16) is the generalized relaxation equation (GRE) and is also known as the memory function equation.

In the above derivations, $\mathbf{A}$ can be a single variable but then the frequency term is zero. When $\mathbf{A}$ is a vector or a column matrix, the time evolution of any component of the time correlation matrix is governed by the following equation:

$$\frac{dC_{\mu v}(t)}{dt} = i\sum_\lambda \mathbf{\Omega}_{\mu\lambda} C_{\lambda v}(t) - \sum_\lambda \int_0^t d\tau\, \mathbf{\Gamma}_{\mu\lambda}(t - \tau)C_{\lambda v}(\tau). \tag{12.17}$$

The generalized relaxation equation in terms of its components and in the frequency plane can be written as

$$zC_{\mu v}(z) - i\Omega_{\mu\lambda}C_{\lambda v}(z) + \Gamma_{\mu\lambda}(z)C_{\lambda v}(z) = \tilde{C}_{\mu v}. \tag{12.18}$$

In the above expression, we used the notation like $\tilde{C}_{\mu v} = \tilde{C}_{\mu v}(z = 0)$. The matrix elements of the frequency and memory function are given by

$$\Omega_{v\lambda} \equiv \sum_k \big(LA_v\big|A_k^*\big)\big(A_k\big|A_\lambda^*\big)^{-1} \tag{12.19}$$

and

$$\Gamma_{v\lambda}(z) = \sum_k \big(F_v(z)\big|F_k^*\big)\big(A_k\big|A_\lambda^*\big)^{-1}. \tag{12.20}$$

Eq. (12.18) is commonly written in the following form:

$$\mathbf{C}(z) = \frac{\tilde{\mathbf{C}}}{z\mathbf{1} - i\mathbf{\Omega} + \mathbf{\Gamma}(z)}. \tag{12.21}$$

This important equation shall be used in Chapter 13 to develop a continued fraction representation for the dynamic structure factor, $S(k, \omega)$. We have already discussed some aspects in Chapter 9.

12.6   LOCAL RELAXATION OF A SINGLE CONSERVED VARIABLE

We shall now present an illustrative application of the projection operator technique where we derive the expression for the memory function term for a single *conserved* variable, A. Since it is conserved, it does not have a time dependence globally, but it can show time dependence as a local position-dependent variable $A(r, t)$, such as local density, defined as a Fourier transform, by

$$A(q,t) = \sum_i a_k^i \delta\left(r - r_i\right) e^{i\mathbf{q}\cdot\mathbf{r}_j} = \sum_{j=1,N} a_j(t) e^{i\mathbf{q}\cdot\mathbf{r}_j} = \sum_{j=1,N} a_j(t) e^{iqz_j} \tag{12.22}$$

where a_j indicates some property (such as the dipole moment) associated with the jth molecule at position r, and the sum is over all the molecules of the system. Here wave vector q is taken along the Z-axis of the Cartesian coordinate system. $A(q,t)$ is the spatial Fourier transform of specific density of interest, $A(r,t) = \sum_i a_k^i \delta\left(r - r_i\right)$, that is, furthermore, we assume that we have an isotropic system. So, the variables depend on the magnitude q of the wave vector.

We now proceed to determine the memory function, which is defined, as outlined above, by $\Gamma(q, \tau) \equiv \langle A \mid A \rangle^{-1} \langle F(-\mathbf{q}) \mid F(\mathbf{q}, \tau) \rangle$. It can be shown (and discussed below in later section) that $\Gamma(q, \tau)$ is an even function of q, i.e., $\Gamma(q) = \Gamma(-q)$. We can then perform Taylor expansion of $\Gamma(q)$ around $q = 0$,

$$\Gamma(q,t) = \Gamma^0(t) + q^2 \Gamma^2(t) + \ldots\ldots \tag{12.23}$$

As we are dealing with a conserved variable, the memory function is zero in the q going to the zero limit. This is simply because $q = 0$ means that we are summing over all the molecules of the system, so there is no decay. Therefore, the first term of the above expansion is zero.

We next define a flux term given by

$$J(q) = \frac{1}{iq} iLA(q) \tag{12.24}$$

where $J(q)$ is the flux of the variable $A(q)$. Therefore, the following identities follow in a straightforward fashion with the aid of Eq. (12.25),

$$\dot{A} = iLA(q) = iqJ(q). \tag{12.25}$$

Eq. (12.26) can be rewritten as

$$J(q) = \frac{1}{iq} \dot{A}(q) \tag{12.26}$$

$$= \frac{1}{iq} \sum_j \left(\dot{a}_j + iqa_j \dot{z}_j\right) e^{iqz_j} \tag{12.27}$$

In the above expression, we used the definition of $A(q)$, i.e., Eq. (12.22). Now, the exponential function can be expanded as

$$e^{iqz_j} = 1 + iqz_j + \frac{\left(iqz_j\right)^2}{2!} + \ldots$$

Therefore, Eq. (12.27) can be rewritten as

$$
\begin{aligned}
J(q) &= \frac{1}{iq} \sum_j \left(\dot{a}_j + iq a_j z_j \right) \left(1 + iq z_j + \ldots \right) \\
&= \sum_j \dot{a}_j z_j + a_j \dot{z}_j + o(q^2) \\
&= J^0(q) + o(q^2)
\end{aligned}
\tag{12.28}
$$

where $J^0(q)$ we have retained up to the zeroth order in flux that is independent of q. Our next goal is to express the generalized force $F(\tau)$ in terms of flux. We can write $F(\tau)$ in the following way:

$$
\begin{aligned}
F(\tau) &= e^{iQLt} i (1 - P) LA \\
&= e^{iQLt} (1 - P) \frac{iq}{iq} iLA \\
&= iq e^{iQLt} J(q) \quad .
\end{aligned}
\tag{12.29}
$$

We employed the condition $PiLA = P\dot{A} = 0$ in order to obtain the simplified expression for $F(\tau)$. With the aid of the above equation, we obtain the following expression for the force-force correlation function:

$$
\langle F \mid F(\tau) \rangle = q^2 \langle J(q) e^{iQLt} J(q) \rangle.
\tag{12.30}
$$

To simplify Eq. (12.30) further we recall the following identity relation: $e^{iQLt} = e^{iLt} - \int_0^t d\tau e^{iL(t-\tau)} iPL e^{iQL\tau}$

and we also know $iPL = i \mid A \rangle \langle A \mid A \rangle^{-1} \langle A \mid L$. Therefore, we can write $iLA(q) = iqJ(q)$. It is important to note that in the $q \to 0$ limit, the identity relation reduces to $e^{iQLt} \to e^{iLt}$. As a result, the force term becomes $F(\tau) = iq e^{iLt} J^0(q) = iq J^0(q,t)$. Under this condition, Eq. (12.30) can be rewritten as follows.

$$
\langle F \mid F(\tau) \rangle = q^2 \langle J^0(0) J^0(\tau) \rangle
$$

Therefore, memory function becomes,

$$
\boxed{\Gamma(q, \tau) = q^2 \langle A(q) \mid A(q) \rangle^{-1} \langle J^0(0) J^0(\tau) \rangle.}
\tag{12.31}
$$

Again, from Eq. (12.23) we know $\Gamma(q) = q^2 \Gamma^{(2)}$. One should note that the quadratic dependence on q of the memory function is a characteristic of the conserved variable.

Thus, the generalized relaxation equation of property $A(q)$ can be written as

$$
\frac{\partial A(q,t)}{\partial t} = -q^2 \Gamma^{(2)} A(q,t) + F(q,t).
\tag{12.32}
$$

12.7   GASKELL-MILLER EVALUATION OF VELOCITY AUTOCORRELATION FUNCTION

In the following, we discuss an important application of the projection operator technique, presented by Gaskell and Miller, which is illustrative and simple. *This also constitutes one of the early and simplest examples of the use of the projection operator technique to mode coupling theory.* The Gaskell-Miller treatment provides us with one of the clearest expositions of the PO technique.

The velocity-time correlation function is the ensemble average for a tagged particle i characterized by initial position r_0 and velocity v_0. The time correlation of the velocities of a tagged particle at two different times is defined as

$$C_V(t) = \left(v_{ix}, \exp(iLt)v_{ix}\right).$$ (12.33)

Since it is practically impossible to use the full evolution operator, we proceed by using its projection on the subspace of product variables $\rho_{ki}\, j_{-k}^{\alpha}$ we obtain

$$\exp(iLt) \approx P \exp(iLt) P.$$ (12.34)

Eq. (12.34) is an approximation. *We next need to define the subset on which we can project the velocity.* We can easily see that linear hydrodynamic modes do not couple to the tagged particle velocity. We then search for appropriate binary products of two hydrodynamic modes. One appropriate choice is $\rho_{ki}\, j_{-k}^{\alpha}$, although other choices are also possible, but let us proceed with this one. The projection is defined by

$$PB = \sum_k \sum_\alpha (\rho_{ki}\, j_{-k}^{\alpha}, B)(\rho_{ki}\, j_{-k}^{\alpha}, \rho_{ki}\, j_{-k}^{\alpha})^{-1} \rho_{ki}\, j_{-k}^{\alpha}.$$ (12.35)

We need the expression for $\left(v_{ix}, P e^{iLt} P v_{ix}\right)$
First, we take,

$$e^{iLt} P v_{ix} = \sum_{k'} \sum_\beta \left(\rho_{k'i}\, j_{-k'}^{\beta}, v_{ix}\right)\left(\rho_{k'i}\, j_{-k'}^{\beta}, \rho_{k'i}\, j_{-k'}^{\beta}\right)^{-1} e^{iLt} \rho_{k'i}\, j_{-k'}^{\beta}$$

$$P e^{iLt} P v_{ix} = \sum_{k'} \sum_\beta \left(\rho_{k'i}\, j_{-k'}^{\beta}, v_{ix}\right)\left(\rho_{k'i}\, j_{-k'}^{\beta}, \rho_{k'i}\, j_{-k'}^{\beta}\right)^{-1} \cdot$$
$$\sum_k \sum_\alpha \left(\rho_{ki}\, j_{-k}^{\alpha}, e^{iLt} \rho_{k'i}\, j_{-k'}^{\beta}\right)\left(\rho_{ki}\, j_{-k}^{\alpha}, \rho_{k'i}\, j_{-k'}^{\alpha}\right)^{-1} \rho_{ki}\, j_{-k}^{\alpha}.$$ (12.36)

Equation of $C_V(t)$ becomes

$$\begin{aligned}
C_V(t) &= \left(v_{ix}, P \exp(iLt) P v_{ix}\right) \\
&= \sum_{kk'} \sum_\alpha \sum_\beta (\rho_{k'i}\, j_{-k'}^{\beta}, v_{ix})(\rho_{k'i}\, j_{-k'}^{\beta}, \rho_{k'i}\, j_{-k'}^{\beta})^{-1} \\
&\quad \times (\rho_{ki}\, j_{-k}^{\alpha}, \exp(iLt)\rho_{k'i}\, j_{-k'}^{\beta}) \times (\rho_{ki}\, j_{-k}^{\alpha}, \rho_{ki}\, j_{-k}^{\alpha})^{-1}(v_{ix}, \rho_{ki}\, j_{-k}^{\alpha}).
\end{aligned}$$ (12.37)

Since

$$\left\langle \rho_{ki}\, j_{-k}^{\alpha} \rho_{-ki}\, j_k^{\alpha} \right\rangle = N(k_B T / m) \text{ and } \left\langle v_{ix}, \rho_{ki}\, j_{-k}^{\alpha} \right\rangle = \left(k_B T / m\right)\delta_{\alpha x},$$ (12.38)

which follows

$$\left\langle \rho_{ki}\, j_{-k}^{\alpha} \rho_{-ki}\, j_k^{\alpha} \right\rangle \left\langle v_{ix}, \rho_{ki}\, j_{-k}^{\alpha} \right\rangle = \frac{1}{N}\delta_{\alpha x}.$$ (12.39)

Four-variable correlation functions are factorized into products of two-variable correlation functions

$$\rho_{ki}\, j^{\alpha}_{-k}, \exp(iLt)\rho_{k'i}\, j^{\beta}_{-k'} \approx \left(\rho_{ki},\ \exp(iLt)\rho_{k'i}\right)\left(j^{\alpha}_{-k}, \exp(iLt)j^{\beta}_{-k}\right)\delta_{k,k'}$$
$$\equiv \left\langle \rho_{ki}(t)\rho_{-ki}\right\rangle\left\langle j^{\beta}_{-k}(t)j^{\alpha}_{k}\right\rangle. \tag{12.40}$$

Therefore,

$$C_V(t) = \frac{1}{N^2}\sum_k \left\langle \rho_{ki}(t)\rho_{-ki}\right\rangle\left\langle j^{\beta}_{-k}(t)j^{\alpha}_{k}\right\rangle. \tag{12.41}$$

The first factor is the self-intermediate scattering function $F_s(k,t)$ and the second is a current correlation function. We have discussed these functions in the preceding Chapters (8–10).

Switching from a sum to an integral and replacing the current correlation function by its average over a sphere

$$\boxed{C_V(t) = \frac{1}{3\rho}(2\pi)^{-3}\int dk\, F_s(k,t)\frac{1}{k^2}\left[C_l(k,t)+2C_t(k,t)\right].} \tag{12.42}$$

However, the above expression has a serious problem! As $t \to 0$, $C_V(t)$ diverges, since $F_s(k,t=0)=1$ and $C_l(k,t=0)=C_t(k,t=0)=k^2\left(\dfrac{k_BT}{m}\right)$. This is rectified by introducing a wavenumber-dependent "form factor" $f(q)$, which ensures a cut-off at large wavenumbers to avoid difficulty. This was discussed by Gaskell and Miller. Subsequent studies do not need to introduce this form factor in such an *ad hoc* manner.

The expression for $C_V(t)$ becomes

$$C_V(t) = \frac{1}{3\rho}(2\pi)^{-3}\int dk\, \hat{f}(k)F_s(k,t)\frac{1}{k^2}\left[C_l(k,t)+2C_t(k,t)\right]. \tag{12.43}$$

This interesting expression was derived by Gaskell and Miller. This has been discussed in other places too, even repeated as this treatment serves to explain the procedure in several different ways.

12.8 IMPORTANT SYMMETRY PROPERTIES OF TIME CORRELATION FUNCTIONS IN PROJECTION OPERATOR TECHNIQUE

12.8.1 IF A_i AND A_j HAVE THE SAME SIGNATURES UNDER TIME-REVERSAL SYMMETRY, THEN THE FREQUENCY MATRIX Ω ALWAYS VANISHES

Proof. As discussed earlier, a dynamical variable A_i transforms under time reversal as $\gamma_i A_i$ where γ_i is the signature of the dynamical variable A_i. Under time-reversal symmetry, L transforms as $L \to -L$. By definition, the frequency matrix $\Omega_{ij} = \sum_k \langle A_i \mid A_k\rangle^{-1}\langle A_k \mid LA_j\rangle$. So, after time reversal, the frequency matrix transforms as

$$\Omega_{ij} \to \gamma_i\gamma_k\gamma_k\gamma_j \sum_k \langle A_i \mid A_k\rangle^{-1}\langle A_k \mid LA_j\rangle = -\gamma_i\gamma_j\Omega_{ij}.$$

Therefore, γ_i and γ_j must have different signatures to obtain a nonvanishing frequency matrix.

12.8.2   If the Variable A_i Transforms Under Time Reversal as $\gamma_i A_i$, where, $\gamma_i = \pm 1$, then the Random Force Matrix $\langle F \mid F(t) \rangle$ Transforms as $\langle F \mid F(t) \rangle \rightarrow \gamma_i \gamma_j \langle F_i \mid F_j(-t) \rangle$

Proof. Time reversal performs the operation $\{\mathbf{q},\mathbf{p}\} \rightarrow \{\mathbf{q},-\mathbf{p}\}$. Therefore, the projection operator becomes $P \rightarrow \gamma_i \gamma_i \gamma_j \gamma_j P = P$. Under the same condition Q transforms as Q. By definition, the random force-time correlation function $\langle \mathbf{F} \mid \mathbf{F}(t) \rangle$ is given as

$$\langle F_i \mid F_j(t) \rangle = \int d\Gamma f_0(\Gamma) \left(QiLA_i \right)^* e^{QiLt} QiLA_j$$

After time reversal force-force correlation function becomes,

$$\langle F_i \mid F_j(t) \rangle \rightarrow \int d\Gamma f_0(\Gamma) \left(Qi(-L)\gamma_i A_i \right)^* e^{Qi(-L)t} Qi(-L)\gamma_j A_j .$$
$$= \gamma_i \gamma_j \int d\Gamma f_0(\Gamma) \left(QiLA_i \right)^* e^{QiL(-t)} QiLA_j$$
$$= \gamma_i \gamma_j \langle F_i \mid F_j(-t) \rangle$$

12.8.3   Under the Inversion of Parity Operation, the Memory Function Vanishes when A_i and A_j Have the Same Signatures

Proof. The inversion of the parity operator changes the sign of both positions and momentum, i.e., $\{\mathbf{q},\mathbf{p}\} \rightarrow \{-\mathbf{q},-\mathbf{p}\}$. Therefore, the elements of memory function will transform under the inversion of parity operation as $\langle F_i \mid F_j \rangle \rightarrow \gamma_i \gamma_j \langle F_i \mid F_j \rangle$. It survives unless A_i and A_j have different signatures.

By the same way, we can prove the following.

Under reflection if A_i transforms as $\alpha_i A_i$ (where $\alpha_i = \pm 1$), then the symmetry matrix vanishes (Ω_{ij}) if A_i and A_j have different signatures. This completes the proof.

12.9   SUMMARY

In this chapter, we have presented the projection operator technique and its application to the derivation of the relaxation equation. The projection operator formalism is one of the central tools of nonequilibrium statistical mechanics. It allows us to build up systematic approximations starting from the Liouville equation, which is fully microscopic. The most crucial point for a student to remember is that the technique of the projection operator allows us to calculate the time correlation function (TCF) of a given dynamical quantity in terms of other known TCFs. Thus, we can find the TCF of the velocity of a tagged solute molecule in terms of the hydrodynamic TCFs. We have presented quite a few applications in this chapter. In the forthcoming chapter, we shall discuss more applications of the projection operator technique that will be of great use for the students.

BIBLIOGRAPHY

1. Berne B. and R. Pecora. 1976. *Dynamic Light Scattering*. Hoboken: Wiley.
2. Zwanzig, R. W. 2001. *Non-Equilibrium Statistical Mechanics*. Oxford: Oxford University Press
3. Kubo, R. 1966. The fluctuation-dissipation theorem, *Reports of Progress in Physics 29*, 255.
4. Zwanzig, R. 1961. *in Boluder Lectures in Theoretical Physics*, Vol. III, p.106, New York.
5. Zwanzig, R. 1965. Time-correlation functions and transport coefficients in statistical mechanics. *Annual Review of Physical Chemistry*, *16*(1), 67–102.
6. Mori, H. 1965. Transport, collective motion, and Brownian motion, *Progress of Theoretical Physics 33*, 423.
7. Gaskell, T., and Miller, S. 1978. Longitudinal modes, transverse modes and velocity correlations in liquids. I. *Journal of Physics C: Solid State Physics*, *11*(18), 3749.

13 Mori Continued Fraction and Related Applications

OVERVIEW

The projection operator technique introduced in the previous chapter, remains a valuable tool since it provides a systematic and frequently used theoretical approach to the derivation of relaxation equations. This technique serves as the much-needed bridge between the phenomenology and the microscopic world of TDSM. We discuss here how this job is accomplished by the Mori continued fraction representation of a time correlation function. The continued fraction representation allows evaluation of the correlation function by systematically introducing higher moments, and avoiding the drawbacks of a direct short-time expansion. It is important to note here that a continued fraction is an analytically well-behaved function. Furthermore, we describe three additional applications of projection operator technique in this chapter that can be used by students to acquire proficiency in the projection operator technique. It is vital to note that this method permits time correlation functions to be calculated from the microscopic structure.

13.1 INTRODUCTION

In the previous chapter, we present the projection operator technique and its application to the derivation of the fundamental relaxation equation of time-dependent statistical mechanics. The relaxation equation serves as the starting point of many applications of TDSM. *The most difficult and sought-after quantity in TDSM is the memory function.* As the projection operator technique demonstrates, this memory function is determined by dynamics in the orthogonal subspace – orthogonal to the vector of dynamical variables whose time dependence is known to us. Ultimately of course this orthogonal subspace is mapped into our known hydrodynamic set of slow variables through binary product formation and through the use of certain self-consistency conditions, as we discuss in Chapter 23 on mode coupling theory. As shown in Figure 12.1 in the last chapter, when the dynamical variable of interest, $B(q, t)$ is projected on the subset $A(q, t)$, we expect to know this projected part. It is the orthogonal subspace that remains unknown, and requires special attention.

The construction of the orthogonal subspaces is not easy in practice. It has been discussed at length by Keyes and Oppenheim. *The usual technique is to replace the orthogonal subspace by certain binary products of the hydrodynamic modes.* For example, if the wavenumber-dependent density, ρ_k, is the dynamical variable, we can consider the following two binary products, $\rho_q \rho_{k-q}$ and $\rho_q j_{k-q,}$ where j_k is the wavenumber-dependent current density. While the procedure appears to be ill-defined at first, it becomes a routine procedure after a certain amount of practice.

After constructing the orthogonal subspace, the evaluation of the memory function requires certain manipulation, again becomes clear with practice. Basically, we have two ways to go. One approach has been to construct a short-time expansion directly of the memory function. This leads

DOI: 10.1201/9781003157601-15

to a continued fraction representation, as detailed below. The other approach is based on mode coupling theory that will be discussed later in Chapter 23.

13.2   CONSTRUCTION OF THE CONTINUED FRACTION REPRESENTATION OF TIME CORRELATION FUNCTION

We first note that a continued fraction is an analytically well-behaved function, and easy to use in numerical evaluation. In the present case, it allows us to build on the exactly known short-time expansion of the relevant TCF, which are in turn given by the moments of the TCF.

Let us assume that A is the dynamical variable of interest, and $C(t)$ is the time correlation function, $\langle A(0)A(t)\rangle$. As described in the last chapter, according to Mori-Zwanzig formalism time correlation function $C(t)$ satisfies the following equation:

$$\frac{\partial C(q,t)}{\partial t} = i\Omega(q)C(q,t) - \int_0^t d\tau \Gamma(q,\tau)C(q,t-\tau) \tag{13.1}$$

where $C(q,t) = \langle\langle A(-q,0)A(q,t)\rangle\rangle$.

After rearranging the Laplace transform of both sides of the above Eq. (13.1), $\hat{C}(q,z)$ takes the following form:

$$\hat{C}(q,z) = \frac{C(q,0)}{z1 - i\Omega(q) + \hat{\Gamma}(q,z)} \tag{13.2}$$

where, as derived in the last chapter,

$$\hat{\Gamma}(q,z) = \langle A \mid A \rangle^{-1}\langle F \mid \frac{1}{z - iLQ} \mid F \rangle \text{ and } \mid F \rangle = iLQ \mid A \rangle$$

and $C(q,0)$ represents the value of $\hat{C}(q,z)$ at $t = 0$. $\hat{\Gamma}(q,z)$ represents the memory function.

The solution of the relaxation equation is nontrivial because we do not have easy access to the memory function. The calculation of memory function requires the use of mode coupling theory, described later in Chapter 23. It requires specification of the orthogonal subspace, and evaluation of time correlation functions, which are determined self-consistently.

Mode coupling theory however requires as input an accurate short-time description of the time correlation function, and of the memory function. Mori developed a continued fraction representation that performs precisely this task. In fact, we have described an alternative approach of achieving this goal of short-time expansion and transforming the TCF to a continued fraction in Chapter 9. What follows is a systematic approach towards the same goal.

First, we define a new projection operator

$$P_1 = \mid F \rangle\langle F \mid F \rangle^{-1}\langle F \mid \text{ and } Q_1 = 1 - P_1.$$

We use these two projection operators P_1 and Q_1 to transform $\hat{\Gamma}(q,z)$ in the following fashion:

$$\begin{aligned}
\hat{\Gamma}(q,z) &= \langle A \mid A \rangle^{-1}\langle F \mid \frac{1}{z - QiLQ} \mid F \rangle, where \ Q \mid F \rangle = \mid F \rangle \\
&= \langle A \mid A \rangle^{-1}\langle F \mid \frac{1}{z - QiLQ(P_1 + Q_1)} \mid F \rangle.
\end{aligned} \tag{13.3}$$

Let us recall the following exact operator identity:

$$\frac{1}{(M+N)} = \frac{1}{M} - \frac{1}{M}\frac{N}{(M+N)}. \tag{13.4}$$

We now set $M = z - QiLQQ_1$ and $N = QiLQP_1$. We can rewrite Eq. (13.3) with the aid of identity relation, i.e., Eq. (13.4) as follows:

$$\begin{aligned}\hat{\Gamma}(z) &= \langle A \mid A \rangle^{-1}\langle F \mid \frac{1}{z - QiLQQ_1} + \frac{1}{z - QiLQQ_1}\frac{QiLQP_1}{z - QiLQ}\mid F\rangle \\ &= \hat{C}_1(z) + \hat{C}_2(z).\end{aligned} \tag{13.5}$$

Using power series expansion, $\hat{C}_1(z)$ can be defined as,

$$\hat{C}_1(z) = \langle A \mid A \rangle^{-1}\langle F \mid \frac{1}{z - QiLQQ_1}\mid F\rangle = \langle A \mid A \rangle^{-1}\frac{1}{z}\langle F \mid \left(1 + \frac{QiLQQ_1}{z} + ...\right)\mid F\rangle. \tag{13.6}$$

Now, using the definition $Q_1 \mid F \rangle = 0$ (by construction), we can simplify Eq. (13.6) as follows:

$$\hat{C}_1(z) = \langle A \mid A \rangle^{-1}\frac{1}{z}\langle F \mid F\rangle. \tag{13.7}$$

Let us simplify another term involved in Eq. (13.5), which is given by,

$$\begin{aligned}\hat{C}_2(z) &= \langle A \mid A \rangle^{-1}\langle F \mid \frac{1}{z - QiLQQ_1}\frac{QiLQP_1}{z - QiLQ}\mid F\rangle \\ &= \langle F \mid F\rangle^{-1}\langle F \mid \frac{QiLQ}{z - QiLQQ_1}\mid F\rangle\hat{\Gamma}(z) \\ &= \langle F \mid F\rangle^{-1}\hat{C}_3(z)\hat{\Gamma}(z)\end{aligned} \tag{13.8}$$

where

$$\hat{C}_3(z) = \langle F \mid \frac{QiLQ}{z - QiLQQ_1}\mid F\rangle. \tag{13.9}$$

We perform series expansion to rewrite Eq. (13.9) as

$$\begin{aligned}\hat{C}_3(z) &= \frac{1}{z}\langle F \mid QiLQF\rangle + \frac{1}{z}\langle F \mid \frac{1}{z}QiLQQ_1\left(1 + \frac{1}{z}QiLQQ_1 + ..\right)QiLQ\mid F\rangle \\ &= \frac{i}{z}\langle F \mid F\rangle\Omega_1 + \frac{1}{z}\langle F \mid QiLQQ_1\frac{Q_1QiLQ}{z - Q_1QiLQ}\mid F\rangle\end{aligned} \tag{13.10}$$

where Ω_1 is defined as

$$\Omega_1 = \langle F \mid F\rangle^{-1}\langle F \mid LF\rangle.$$

Let us define a new generalized random force as

$$|f\rangle = Q_1 Q i L Q |F\rangle. \tag{13.11}$$

We can rewrite Eq. (13.10) with the definition of new random force as follows:

$$\hat{C}_3(z) = \frac{i}{z}\langle F|F\rangle\Omega_1 - \frac{1}{z}\langle f| \frac{1}{z - Q_1 Q i L Q}|f\rangle$$

$$= \frac{i}{z}\langle F|F\rangle\Omega_1 - \frac{1}{z}\langle F|F\rangle\hat{\Gamma}_1(z) \tag{13.12}$$

where, $\hat{\Gamma}_1(z) = \langle F|F\rangle^{-1}\langle f|(z - Q_1 Q i L Q)^{-1}|f\rangle.$

We put all pieces together to obtain the following expression for the memory function:

$$\hat{\Gamma}(z) = \frac{\Gamma(0)/C(0)}{z - i\Omega_1 + \hat{\Gamma}_1(z)}. \tag{13.13}$$

Using Eq. (13.13), the equation of $\hat{C}(z)$ can be rewritten as

$$\hat{C}(z) = \frac{C(0)}{z - i\Omega + \dfrac{\Gamma(0)/C(0)}{z - i\Omega_1 + \hat{\Gamma}_1(z)}}. \tag{13.14}$$

The above sequence of steps can be repeated with each new memory function that is generated by the procedure. The result is a continued fraction representation of $\hat{C}(z)$.

The continued fraction, initially developed by Mori, is known as Zwanzig-Mori formulation and has been widely used.

13.3 FURTHER APPLICATIONS OF PROJECTION OPERATOR TECHNIQUE

13.3.1 MOTION OF A PARTICLE WITH LARGE MASS IN A FLUID OF SMALL BATH PARTICLES

Consider we have a system that consists of one large, heavy particle of mass M immersed in a bath of N number of small particles of mass m (where $m \ll M$).

We are interested in knowing the motion of tagged large particle irrespective of the bath particles. That is, we want to achieve a reduced description where we eliminate the bath degrees of freedom. We furthermore assume that all the interactions between bath particles and tagged particles are pairwise additive in nature. Hence, the total Hamiltonian of the whole system can be given by

$$H = \frac{p_0^2}{2M} + \sum_{i=1}^{N}\frac{p_i^2}{2m} + \sum_{i=1}^{N}V(|r_0 - r_i|) + \sum_{i=1}^{N-1}\sum_{j=i+1}^{N}V(|r_i - r_j|) \tag{13.15}$$

where p_o and r_0 are the momentum and position of the tagged large particle, respectively, and p_i denotes the momentum of the ith bath particle. We can partition the total Hamiltonian into two parts as $H = H_0 + H_B$ where

$$H_0 = \frac{p_0^2}{2M} + \sum_{i=1}^{N}V(|r_0 - r_i|)$$

$$H_B = \sum_{i=1}^{N}\frac{p_i^2}{2m} + \sum_{i=1}^{N-1}\sum_{j=i+1}^{N}V(|r_i - r_j|).$$

Now, the corresponding Liouvillian can be expressed as

$$iL = iL_o + iL_B$$

where,

$$iL_0 = \frac{p_0}{M}.\nabla_{r_0} + F_0.\nabla_{p_0}$$ (13.16)

$$iL_B = \sum_{i=1}^{N} \frac{p_i}{m}.\nabla_{r_i} + F_i.\nabla_{p_i}$$

where F_0 and F_i are the total forces acting on the tagged massive particle and the ith bath particle of the medium. We next introduce a projection operator P, defined by,

$$P = | p_{0x} \rangle \langle p_{0x} | p_{0x} \rangle^{-1} \langle p_{0x} |.$$ (13.17)

P projects an arbitrary dynamical variable onto the x-component momentum of the tagged particle. Now from the generalized relaxation equation, we can write,

$$\frac{d | p_{0x} \rangle}{dt} = i\Omega | p_{0x} \rangle - \int_0^t dt' \Gamma(t-t') | p_{0x}(t') \rangle + | f_x(t) \rangle$$ (13.18)

where

$$\Omega_0 = \langle p_{0x} | p_{0x} \rangle^{-1} \langle p_{0x} | Lp_{0x} \rangle,$$
$$| f_x(t) \rangle = e^{QiLt} QiL | p_{0x} \rangle$$

and the memory function is given by

$$\Gamma(t) = \langle p_{0x} | p_{0x} \rangle^{-1} \langle f_x | f_x(t) \rangle$$

Since, $iL | p_{0x} \rangle = | \dot{p}_{0x} \rangle$, hence by symmetry $\Omega_0 = 0$. In the same way, it can be shown

$$| f_x(t) \rangle = e^{QiLt} Q | \dot{p}_{0x} \rangle$$
$$= e^{QiLt} | \dot{p}_{0x} \rangle$$ (13.19)
$$= e^{QiLt} | F_{0x} \rangle$$

where F_{0x} is the total (actual) force acting on the tagged particle along the x-axis, to be distinguished from $f_x(t)$, which is the projected force (the memory function is determined by the projected force – an important distinction to be remembered). Let us now introduce a scaled momentum of the tagged particle $P_0 = \sqrt{\frac{m}{M}} p_0 = \varepsilon p_0$, where $\varepsilon = \sqrt{\frac{m}{M}}$. In terms of scaled variables, Liouvillian of the tagged particle Eq. (13.16) becomes

$$iL_0 = \sqrt{\frac{m}{M}} \left(\frac{P_0}{m}.\nabla_{r_0} + F_0.\nabla_{P_0} \right)$$ (13.20)

Now, we recall the identity relation described in the last chapter $e^{QiLt} = e^{-iLt} - \int_0^t dt' e^{QiL(t-t')} PiLe^{iLt'}$. Using this identity, Eq. (13.19) can be rewritten as,

$$e^{QiLt} \mid F_{0x} \rangle = e^{iLt} \mid F_{0x} \rangle - \int_0^t dt' e^{QiL(t-t')} PiLe^{iLt'} \mid F_{0x} \rangle. \qquad (13.21)$$

We can simplify the second term of the right-hand side of Eq. (13.21) further as

$$\int_0^t dt' e^{QiL(t-t')} PiLe^{iLt'} \mid F_{0x} \rangle$$

$$= -\int_0^t dt' e^{QiL(t-t')} \mid p_{0x} \rangle \langle p_{0x} \mid p_{0x} \rangle^{-1} \langle iLp_{0x} \mid e^{iLt'} F_{0x} \rangle.$$

Since, $iL = iL_0 + iL_B$, and iL_B acts only on phase space variables of the bath particles. So the terms $\mid iLp_{0x} \rangle$ in the integrand reduces to $\mid iL_0 p_{0x} \rangle$. As we have shown in Eq.(13.20) this is a term of order ε can therefore be neglected in comparison to leading terms in Eq. (13.21), which now becomes $e^{QiLt} \mid F_{ox} \rangle \approx e^{iLt} \mid F_{ox} \rangle$.

Using the same trick again, we finally obtain,

$$\mid f_x(t) \rangle = e^{QiLt} \mid F_{0x} \rangle \simeq e^{iL_B t} \mid F_{0x} \rangle. \qquad (13.22)$$

Therefore, the random force may be interpreted as the intermolecular force exerted on the tagged molecule by all the solvent molecules when the tagged particle is kept stationary. According to this interpretation, $\mid f_x(t) \rangle$ can be considered as a "fast" variable that loses its correlation with its initial value very quickly. In other words, the memory function $\Gamma(t)$ in the GRE for the momentum of the tagged particle can be assumed to decay to 0 extremely fast. That being the case, the GRE itself can be rewritten as,

$$\frac{\partial \mid p_{0x}(t) \rangle}{\partial t} = -\int_0^t dx \Gamma(x) \mid p_{0x}(t-x) \rangle + \mid f_x(t)$$

$$\simeq -\int_0^\infty dx \Gamma(x) \mid p_{0x}(t) \rangle + \mid f_x(t) \rangle. \qquad (13.23)$$

If we introduce a parameter ζ as $\zeta = \int_0^\infty \Gamma(t) dt$, we see that the evaluation equation of $\mid p_{0x}(t) \rangle$ takes the form

$$\frac{\partial p_{0x}(t)}{\partial t} = -\zeta p_{0x}(t) + f_x(t). \qquad (13.24)$$

Eq. (13.24) is the well-known Langevin equation and ζ is referred to as the friction coefficient of the system.

13.3.2   Relation Between Inverse Relaxation Time and Memory Function

The discussion in this section follows closely that presented initially by Yamamoto and later developed by Chandler, reference given below.

Suppose we consider a reversible chemical reaction like $A \rightleftarrows B$. k_1 and k_{-1} are the forward and the backward rate constants of this reaction, respectively. At low a concentration of solute, the macroscopic theory of chemical kinetics provides the rate equation as

$$\frac{d\langle N_A \rangle_t}{dt} = -k_1 \langle N_A \rangle_t + k_{-1} \langle N_B \rangle_t \tag{13.25}$$

where $\langle N_A \rangle_t$ represents the concentration of A at time t. The solution of this differential equation for a closed system (i.e., $\langle N_A \rangle_t + \langle N_B \rangle_t$ = constant at any time) takes the following form:

$$\frac{\langle N_A \rangle_t}{\langle N_A \rangle_0} = \frac{k_{-1} + k_1 \exp\left(-\dfrac{t}{\tau}\right)}{k_{-1} + k_1} \tag{13.26}$$

where inverse relaxation time is defined as $\tau^{-1} = k_{-1} + k_1$. In a reversible reaction, inverse relaxation time τ^{-1} plays a role that is similar to that of transport coefficient in hydrodynamics. With this analogy, we seek an expression for inverse relaxation time involving flux autocorrelation function. To derive such an expression we study hydrodynamics and memory function. Consider an equilibrium time correlation function

$$C(t) = \langle \delta N_A(0) \delta N_A(t) \rangle \text{ where,} \delta N_A(t) = N_A(t) - N_A(0). \tag{13.27}$$

Now, let us define two projection operators P and Q as $P = | \delta N_A \rangle \left(\delta N_A^2 \right)^{-1} \langle \delta N_A |$ and $Q = 1 - P$. Before proceeding further, we need to make the following two assumptions:

(i) $N_A(t)$ is assumed to be a slow variable, which means it relaxes with a longer time scale than the other dynamical variables of the system, like motion of the solvent molecules that do not need to cross any barrier. (ii) This dynamical slow variable cannot couple with another slow variable but it can couple with other faster variables of the system.

Time dependence in $N_A(t)$ arises from the fact of how reactants cross the potential barrier to reach into the product side. $N_A(t)$ can become a slow variable when the barrier separating the reactant and product is much larger than $k_B T$. By using Heaviside functions, the number of molecules at time t in state A is defined as

$$N_A(t) = \sum_{i=1}^{N} H_A\left[q_i(t)\right], H_A(r) = \begin{cases} 1, & r < r_c \\ 0, & r > r_c \end{cases} \tag{13.28}$$

$q_i(t)$ terms represent internal coordinate of the ith solute molecule at time t. Here r_c corresponds to the location of transition point at the barrier top (see Chapter 20 for an illustration). Using the Mori-Zwanzig formalism we can write an expression for $C(t)$ in the Laplace plane as

$$\hat{C}(z) = \frac{\langle (\delta N_A)^2 \rangle}{\left(z + \hat{\Gamma}(z) \right)} \text{ where } \hat{\Gamma}(z) = \langle (\delta N_A)^2 \rangle^{-1} \langle \dot{N}_A e^{QLt} \dot{N}_A \rangle. \tag{13.29}$$

In the above expression, the frequency matrix is taken as zero since

$$\langle \delta N_A L \delta N_A \rangle = \langle \delta N_A \dot{N}_A \rangle = 0.$$

For a closed system, one can write

$$N_A + N_B = \text{constant}$$
$$C(0) = \left\langle \left(\delta N_A\right)^2 \right\rangle^{-1} = \frac{\langle N_A \rangle \langle N_B \rangle}{N}.$$

Based on the transition state theory approximation, inverse relaxation time is defined by,

$$\tau_{TS}^{-1} = \left(2x_A x_B\right)^{-1} \left\langle |\dot{q}_1| \right\rangle_{r_c} \left\langle \delta(r_c - q_1) \right\rangle,$$

where x_A and x_B are the mole fraction of the A and B molecule, respectively. $\left\langle \delta(r_c - q_1) \right\rangle$ gives a probability distribution for finding a molecule at position r_c with internal coordinate q_1. $\left\langle |\dot{q}_1| \right\rangle_{r_c}$ denotes equilibrium ensemble average of $\dot{q}_1$ when q_1 is fixed at r_c. From the microscopic point of view, we can relate relaxation time with time correlation function $C(t)$ by the following equation:

$$\tau = \int_0^\infty dt \, \frac{\left\langle \delta N_A \delta N_A(t) \right\rangle}{\left\langle \left(\delta N_A\right)^2 \right\rangle}. \tag{13.30}$$

The first derivative of time correlation function using Eq. (13.27) at low concentration of solute is given by,

$$\frac{dC(t)}{dt} = \left[\delta N_A \dot{N}_A(t) \right]$$
$$= -\left\langle N H_A(q_1) \dot{q}_1(t) \delta\left[r_c - q_1(t) \right] \right\rangle. \tag{13.31}$$

The above equation is valid only at a low concentration of solute where the solute molecules are not correlated. We are interested to look at the behavior of $C(t)$ at time $t = 0$. Let us now calculate $\dfrac{dC(t)}{dt}$ at $t = 0$ and it is found to be $\dot{C}(0) = \dfrac{1}{2}\left\langle N\dot{q}_1 \delta\left[r_c - q_i(t) \right] \right\rangle = 0$. Here the second equality comes from the trivial fact that the classical equilibrium distribution is an even function of velocities.

From Eq. (13.31) it is clear that $\dot{C}(t)$ is not continuous near $t = 0$. We find that $\dot{C}(t)$ is precisely zero at $t = 0$ but exhibits nonzero behavior near $t = 0$. This implies that the second derivative of $C(t)$ contains the delta function. One can relate the second derivative of $C(t)$ and τ_{TS}^{-1} by the following equation:

$$\lim_{\delta \to 0+} \int_0^\delta dt \left[\ddot{C}(t) / C(0) \right] = -\tau_{TS}^{-1}. \tag{13.32}$$

The behavior of $C(t)$ at time $t \to 0$ can be derived by studying large z behavior of the memory function.

$$\frac{\hat{C}(z)}{C(0)} \sim \left(z + \tau^{-1}\right)^{-1}. \tag{13.33}$$

On the other hand, according to the standard Mori-Zwanzig formalism, we can write $\hat{C}(z) = \left[z + \hat{\Gamma}(z)\right]^{-1} \langle(\delta N_A)^2\rangle$ with $\Gamma(t) = \langle(\delta N_A)^2\rangle^{-1}\langle\dot{N}_A e^{QLt}\dot{N}_A\rangle$. The singular behavior of $C(t)$ discussed earlier implies that $\Gamma(t)$ contains a delta function at $t = 0$. In order to verify this, we perform the time integral of $\Gamma(t)$ to find

$$\lim_{\varepsilon\to 0^+}\int_0^\varepsilon dt\,\Gamma(t) = \langle(\delta N_A)^2\rangle^{-1}\lim_{\varepsilon\to 0+}\langle\dot{N}_A[N_A(\varepsilon) - N_A(0)]\rangle$$
$$= \tau_{TS}^{-1} \tag{13.34}$$

where the first equality comes from the fact that $\exp(QL\varepsilon)\dot{N}_A = \dot{N}_A(\varepsilon) + O(\varepsilon)$ and the second equality arises after some algebraic manipulation. It is important to note that the singularity at $t = 0$ does not hinder the applicability of Mori-Zwanzig formalism as $\Gamma(t)$ is integratable.

Comparing Eq. (13.30) and Eq. (13.34) we can write

$$\tau^{-1} = \hat{\Gamma}(0) = \int_0^\infty dt\,\Gamma(t). \tag{13.35}$$

From Eq. (13.35) one can conclude that the delta function part of memory function contributes τ_{TS}^{-1} to $\hat{\Gamma}(0)$. Then we have

$$\tau^{-1} = \tau_{TS}^{-1} + \int_0^\infty dt\,\Delta\Gamma(t) \tag{13.36}$$

where $\Delta\Gamma(t)$ presents the nonsingular part of the memory function.

These two equations [Eqs. (13.35) and (13.36)] are equivalent to each other. If the integral over $\Delta\Gamma(t)$ is small compared to τ_{TS}^{-1}, then the transition state theory approximation is valid. Note that the memory function evolves with e^{QiLt}. In the standard hydrodynamics application of the Mori-Zwanzig formalism, projected dynamics and real dynamics are related in the long wavelength limit.

13.3.3 Kinetic Equation for Orientational Relaxation

The discussion in this subsection closely follows that given in the book by Berne and Pecora.

In the case of dilute gases or dilute solution of macromolecules, individual molecule scatters light independently. But in the case of dense liquids, a strong spatial and orientational correlation between nearest neighbors are present. For such a system spectrum of scattered light consists of widths and relative intensities of various bands. According to the molecular theory of light scattering intensity of scattered light from the monatomic liquid is given as follows, $I_{if}^\alpha(q, \omega) = (n_i.n_f)^2\,\alpha^2 S(q, \omega)$ where, n_i and n_f corresponds to initial and final direction of polarization, α and $S(q, \omega)$ represent polarizability and spectrum of density fluctuation of wave vector q, respectively. In light scattering spectroscopy, I_{VH} is commonly known as the depolarized component. By definition, it is expected I_{VH} to be zero for a spherical molecule. But, experiments predict the nonzero value of I_{VH} even for inert gas like Ar, Ne, etc. The experimental depolarized spectrum of pure liquid appears to be split.

In this section, we are going to present a theoretical interpretation for such splitting present in the spectrum using the Mori-Zwanzig projection operator technique. Moreover, it is found that xy and yz components of polarizability fluctuation tensor $\delta\alpha_{\alpha\beta}$ are only responsible for polarization fluctuation that gives rise to VH scattering.

$$\delta\alpha_{VH}(q,t) = \delta\alpha_{yx}(q,t)\sin\frac{\theta}{2} - \delta\alpha_{yz}(q,t)\cos\frac{\theta}{2} \tag{13.37}$$

$\alpha\beta$ component of polarizability fluctuation tensor is given by,

$$\delta\alpha_{\alpha\beta}(q,t) = \alpha\sum_{j=1}^{N}\exp iqz_j + \beta\sum_{j=1}^{N}\left[u_\alpha^j u_\beta^j - \frac{1}{3}\delta_{\alpha\beta}\right]\exp iqz_j. \tag{13.38}$$

In the above expression, q is taken along the z-direction, θ indicates the angle between the initial and the final polarization direction and u_α^j specifies the orientation of α axis of jth molecule.

In order to apply the projection operator formalism, we assume that $\delta\alpha_{\alpha\beta}(q,t)$ is our slow (i.e., it has a long relaxation time) variable and primary (i.e., it cannot be expressed in terms of other slow variables) property of interest. Thus, our main goal is to find separate expressions for $\delta\alpha_{yx}$ and $\delta\alpha_{yz}$. Now, we introduce $v_y(q)$, which represents Fourier transform of y-component of the velocity and is given by

$$v_y(q) = \sum_{j=1}^{N}v_y^j\exp(iqz_j). \tag{13.39}$$

Both $v_y(q)$ and $\delta\alpha_{yz}$ have the same signature under reflections in xz and yz planes and translation along z-axis.

Now according to the Mori-Zwanzig formalism for the two-component system we define,

$$\begin{aligned}A_1(q) &= v_y(q)\\A_2(q) &= \delta\alpha_{yz}(q).\end{aligned} \tag{13.40}$$

Let us consider the frequency matrix $\Omega(q)$. The diagonal elements of the matrix would be zero because $A_1(q)$ and $A_2(q)$ have opposite time-reversal symmetry. The off-diagonal terms of the frequency matrix would be determined by the term $\left(Lv_y(q), \delta\alpha_{yz}(q)\right)$. Application of Liouville operator changes the $v_y(q)$ operator as follows:

$$iLv_y(q) = \sum_{j=1}^{N}\left[\frac{1}{m}F_{yj} + iqv_z^j v_y^j\right]\exp iqz_j \tag{13.41}$$

where F_{yj} represents y-component of the force acting on particle j of the system. After integrating by parts of the off-diagonal term it takes the form $i\beta^{-1}\left(\sum_j\exp iqz_j, \frac{\partial}{\partial y_j}\alpha_{yz}(q)\right)$. This term vanishes as $\alpha_{yz}(q)$ is independent of q (in the absence of intermolecular correlations). So, finally, the frequency matrix turns out to be a null matrix, i.e.,

$$\Omega(q) = \begin{pmatrix}0 & 0\\0 & 0\end{pmatrix}.$$

After the frequency matrix, the next important task in projection operator formalism is the calculation of the components of the random force matrix.

Let us define,

$$F_1 = QiLA_1, F_2 = QiLA_2.$$

Before proceeding further, following the Liouville theorem we can write

$$
\begin{aligned}
iLA_1 &= \frac{\partial A_1}{\partial t} \\
&= \frac{\partial}{\partial t} \sum_{j=1}^{N} v_y^j \exp(iqz_j) \\
&= \sum_{j} \left[a_y^j + iqv_y^j v_z^j \right] \exp iqz_j
\end{aligned}
\tag{13.42}
$$

where a_y^j represents y-component of the acceleration of particle j. From this expression, it is clear that iLA_1 exhibits linear dependence on q in the limit $q \to 0$. It can be shown that the fourth term of the random force matrix does not exhibit any q dependency in a small q limit. On the other hand, second and third terms are odd functions of q. So, when there is short-ranged intermolecular forces between molecules of the fluid, we can assume that the second and the third terms are at least linear in q. Below we present a complete random force matrix showing explicit dependence on q of each term. Here, prime terms are independent of q.

$$
\Gamma(q) = \begin{pmatrix} q^2 \Gamma'_{11}(q) & q\Gamma'_{12}(q) \\ q\Gamma'_{21}(q) & \Gamma_{22}(q) \end{pmatrix}.
$$

Substituting final expression of $\Gamma(q)$ and $\Omega(q)$ in the Mori-Zwanzig formalism we get,

$$
\begin{aligned}
\frac{dA_1}{dt} &= -q^2 \Gamma''_{11}(q) A_1 - q\Gamma'_{12}(q) A_2 + F_1 \\
\frac{dA_2}{dt} &= -q^2 \Gamma_{22}(q) A_2 - q\Gamma'_{21}(q) A_2 + F_2.
\end{aligned}
\tag{13.43}
$$

After solving the above two equations by the method of Laplace transformation we get the following expression as the inner product of A_2 as follows:

$$
\langle A_2^*(q,0)\widehat{A}_2(q,z)\rangle = \langle A_2 \mid A_2^+ \rangle \left[\frac{(z + q^2 \Gamma'_{11}(q))}{z^2 + \Gamma_z(q)z + w_s^2(q)} \right]
\tag{13.44}
$$

where

$$
\Gamma_z(q) = \Gamma_{22}(q) + q^2 \Gamma'_{11}(q)
$$
$$
w_s^2(q) = q^2 \left[\Gamma'_{11}(q)\Gamma_{22}(q) - \Gamma'_{12}(q)\Gamma'_{21}(q) \right]
$$

yz component of spectral density is defined as,

$$
S_{yz}(q,\omega) = \frac{1}{2\pi} \int_{-\infty}^{\infty} dt e^{-i\omega t} \langle \partial \alpha_{yz}^*(q,0)\partial \alpha_{yz}(q,t)\rangle
\tag{13.45}
$$

After carrying out integration and substituting in Eq. (13.45) we get

$$S_{yz}(q,\omega) = \pi^{-1}\,\mathrm{Re}\int_0^\infty dt\,e^{-i\omega t}\langle A_2^*(q,0)A_2(q,t)\rangle$$

$$= \pi^{-1}\langle|\alpha_{yz}(q)|^2\rangle\left\{\frac{\omega^2\Gamma_z(q)+q^2\Gamma_{11}''(q)\left[\omega_s^2(q)-\omega^2\right]}{\left[\omega_s^2(q)-\omega^2\right]^2+\omega^2\Gamma_s^2(q)}\right\}. \tag{13.46}$$

Similarly, we obtain the following expression for the xy component of spectral density,

$$S_{xy}(q,\omega) = \pi^{-1}\langle|\alpha_{xy}(q)|^2\rangle\left\{\frac{\Gamma_{22}^{xy}(q)}{\omega^2+\left[\Gamma_{22}^{xy}(q)\right]^2}\right\}. \tag{13.47}$$

In the limit $q\to 0$, $\Gamma_{22}^{xy}(q)$, which is involved in Eq. (13.47) goes to $\Gamma_{22}(q)$. It can be shown that for small q and $s\,\langle v_y^*(q)\hat{v}_y(q,s=0)\rangle = \langle|v_y|^2\rangle(q^2v_s)^{-1}$, where v_s represents kinematic shear viscosity. We substitute Eq. (13.46) in Eq. (13.47), we get in the limit $\dfrac{q^2v_s}{\Gamma_{22}}\ll 1$

$$I_{VH}(q,\omega) = \pi^{-1}\langle|\alpha_{yz}|^2\rangle\left\{\frac{\Gamma_{22}}{\omega^2+\Gamma_{22}^2}-R\left[\frac{v_sq^2}{\Gamma_{22}}\right]\frac{q^2v_s}{\omega^2+\left[q^2v_s\right]^2}\cos^2\frac{\theta}{2}\right\} \tag{13.48}$$

where $R = 1-\dfrac{\Gamma_{11}'}{v_s}$.

The above equation predicts the presence of two Lorentzians among which one is broader than the other. Therefore, the corresponding spectrum consists of a broad band of width Γ_{22} subtracted from a new Lorentzian that leads to a dip or splitting in the spectrum.

13.3.4 DERIVATION OF THE DIFFUSION EQUATION BY THE PO TECHNIQUE

The ordinary diffusion equation, $\dfrac{dP}{dt}=D\dfrac{d^2P}{dx^2}$, can be derived in many different ways. Here we derive the same but by using the projection operator technique, for practice and training purposes.

According to hydrodynamics calculation, the starting expression for diffusion is given by

$$\frac{\partial P(r,t)}{\partial t} = D\nabla_R^2 P(r,t). \tag{13.49}$$

Here $P(r,t)$ represent the probability of finding a particle at position r at time t starting some arbitrary position at time $t=0$. With this interpretation $P(r,t)$ can be defined as,

$$P(r,t) = \langle\delta(r-(r(t)-r(0)))\rangle \tag{13.50}$$

where $r(t)$ and $r(0)$ denote the initial and final position of the particle at time t and $t = 0$, respectively. In this definition, the angular bracket provides an average over equilibrium distribution of the particles. The Fourier transform of Eq. (13.50) yields,

$$\hat{P}(k,t) = \left\langle \int dr e^{ikr} \delta\left[r - \left(r(t) - r(0)\right)\right]\right\rangle$$
$$= \left\langle e^{ik(r(t)-r(0))}\right\rangle \tag{13.51}$$

The form of Eq. (13.51) suggests that one can express $\hat{P}(k,t)$ as a time correlation function of the form $\langle A \mid A(t)\rangle$ in which A is defined as $\mid A(t)\rangle = e^{ik.r(t)}$. Once we express $\hat{P}(k,t)$ as a function of time correlation function it should satisfy the Mori-Zwanzig formalism. Then we have

$$\frac{\partial \hat{P}(k,t)}{\partial t} = i\Omega\hat{P}(k,t) - \int_0^t dt_1 \hat{\Gamma}(k,t-t_1)\hat{P}(k,t_1). \tag{13.52}$$

Now, for a single variable

$$\Omega \propto \left\langle A \mid \dot{A}\right\rangle = 0. \tag{13.53}$$

With the help of Eq. (13.53), the above Eq. (13.52) transforms as,

$$\frac{\partial \hat{P}(k,t)}{\partial t} = -\int_0^t dt_1 \hat{\Gamma}(k,t-t_1)\hat{P}(k,t_1). \tag{13.54}$$

In this expression, memory function is given by

$$\hat{\Gamma}(k,t) = \langle A \mid A\rangle^{-1}\langle F \mid F(t)\rangle \tag{13.55}$$

where F represents the random force, which is given by,

$$\mid F(t)\rangle = e^{QiLt}QiL \mid A\rangle$$
$$= e^{QiLt}Qik.v(0)e^{ikr}. \tag{13.56}$$

Here, $v(0)$ denotes the initial velocity of the particle.
    Now, by definition, we can write,

$$Q \mid \dot{A}\rangle = \left(1 - \mid A\rangle\langle A \mid A\rangle^{-1} \mid A\rangle\right) \mid \dot{A}\rangle = \mid \dot{A}\rangle. \tag{13.57}$$

Then Eq. (13.56) becomes

$$\mid F(t)\rangle = e^{QiLt}ik.v(0)e^{ikr(0)}. \tag{13.58}$$

With this definition of random force, the memory function becomes,

$$\hat{\Gamma}(k,t) = \langle A \mid A\rangle^{-1}k.\left\langle e^{-ikr(0)}v(0)e^{QiLt}v(0)e^{ikr(0)}\right\rangle.k = k.\hat{D}(k,t).k. \tag{13.59}$$

Last equality arises from the fact that $\langle A \mid A \rangle = 1$. Now if we do Laplace transform of both sides of Eq. (13.54) we get

$$\tilde{\hat{P}}(k,s) = \frac{\hat{P}(k,0)}{s + \tilde{\hat{\Gamma}}(k,s)}$$

$$= \frac{\hat{P}(k,0)}{s + k.\tilde{\hat{D}}(k,s).k} \qquad (13.60)$$

where $\tilde{\hat{D}}(k,s)$ is the Fourier-Laplace transform of $D(k, t)$.

Now, if we directly perform Fourier-Laplace transform of both sides of Eq. (13.49), we have,

$$\tilde{\hat{P}}(k,s) = \frac{\hat{P}(k,0)}{s + k^2 D}. \qquad (13.61)$$

Comparing Eq. (13.61) and Eq. (13.60) one can conclude that the diffusion coefficient is a really limiting form of complicated object that depends on both wave vector and frequency. It can be shown that,

$$D = \lim_{k \to 0} \lim_{s \to 0} \frac{k.\tilde{\hat{D}}(k,s).k}{k^2}. \qquad (13.62)$$

The joint limit of low frequency (long time) and small wavenumber (large distance) is known as the hydrodynamic limit.

13.4   SUMMARY

The Mori-Zwanzig projection operator formalism is one of the most fundamental tools of nonequilibrium statistical mechanics. This formalism is useful in deriving the reduced, macroscopic equations of motions starting from the microscopic dynamics. Hence, it is a valuable ingredient of both basic and advanced courses in nonequilibrium statistical mechanics. However, an accessible description of this method is not easily available. In this chapter, we have tried to present a simple introduction to the Mori-Zwanzig formalism that will be helpful for the students in understanding the methodology in the form it is utilized in the modern day research. Several applications of the projection operator technique are also discussed in detail. In particular, we gathered here the derivation of the simple Langevin equation of motion when the mass of the tagged particle becomes very large. We also discussed Chandler-Yamamoto formulation of chemical reaction rate.

BIBLIOGRAPHY

1. Berne B., and R. Pecora. 1976, *Dynamic Light Scattering*. Hoboken: Wiley.
2. Zwanzig, R. W. 2001, *Non-Equilibrium Statistical Mechanics*. Oxford: Oxford University Press
3. Hansen, J. P., and I. R. McDonald. 1976, *Theory of Simple Liquids*. New York: Academic Press.
4. Chandler, D. 1978, Statistical mechanics of isomerization dynamics in liquids and the transition state approximation, *Journal of Chemical Physics*, *68*, 2959.

5.   Andersen, H. C., and R. Pecora. 1971, Kinetic equations for orientational and shear relaxation and depolarized light scattering in liquids, *Journal of Chemical Physics, 54*, 2584.

6.   Barron, L. D., and A. D. Buckingham. 1971, Rayleigh and Raman scattering from optically active molecules. *Molecular Physics, 20*(6), 1111–1119.

7.   Forster D. 1975. *Hydrodynamic Fluctuations, Broken Symmetry, and Correlation Functions.* Massachusetts: The Benjamin/Cummings Publishing Company, Inc. (Advanced Book Program Reading).

14 Moments and Cumulants

OVERVIEW

Together with the projection operator technique, moments and cumulants provide useful mathematical and theoretical tools in the study of statistical mechanics. Moments and cumulants are important because these quantities are often directly observable in experimental studies. Many random variables we encounter in statistical mechanics are Gaussian or close to Gaussian (due to the well-known central limit theorem). A useful theorem, often termed "the cumulant theorem", states that for a Gaussian random variable, all the cumulants of order larger than two are zero. Another important aspect of cumulants is that it naturally gives rise to correlation functions in the exponential, unlike moments that give rise to a series expansion. In recent applications, higher-order cumulants have been used to gauge the departure from the Gaussian behavior. Higher-order cumulants of mean square displacement are found to be nonzero in supercooled liquids. Their time dependence reveals interesting dynamical features. Cumulants have been widely used in spectroscopy, with initial formulation by Kubo's stochastic theory of line shape. This chapter is to be studied in conjunction with chapter 2 where we discussed probability distribution. But there are aspects that are technical and contained in this separate chapter.

14.1 INTRODUCTION

In Chapter 2, we introduced basic concepts of probability theory, with particular relevance to the theoretical formalism needed to understand time-dependent phenomena from a statistical approach. Along with Chapter 2 of our text on *Equilibrium Statistical Mechanics* (CRC Press), they provide the necessary mathematical background for this chapter.

We know that the first moment is the average of a random quantity over its probability distribution. We also know that the second moment gives the standard deviation, which provides us with thermodynamic response functions (like specific heat and compressibility) and also width of line shapes in spectroscopic experiments. *What we do not usually appreciate is the fact that the standard deviation is actually a cumulant.* Cumulants are not only important theoretical quantities but they often are directly observable through experiments as correlation functions. In addition, they are often used in the evaluation of time correlation functions in time-dependent statistical mechanics.

Moments and cumulants are intimately connected. Cumulants appear when we attempt to evaluate such averages as $\langle \exp(\xi X) \rangle$, where X is the random variable over which the average $\langle \ldots \rangle$ is to be performed, and ξ is a c-number or even a *function but not random.*

DOI: 10.1201/9781003157601-16

The importance of cumulants arise from the repeated or frequent occurrence of $\langle\exp(\xi X)\rangle$ in statistical mechanical calculations, as we discuss below.

In this chapter we shall first define moments. Subsequently, we define cumulants. After these introductions, we discuss applications of cumulants, leading to time correlation function representation.

14.2 MOMENTS

As discussed, several times already, experimental results are often explained in terms of the fluctuations of a physical observable quantity around its average. Fluctuations are random variables, and experimental observables are often moments of such random variables. For example, the thermodynamic quantities like specific heat C_V and C_P can be quantified by the fluctuation of energy and enthalpy, respectively. Similarly, the mechanical quantity compressibility is obtained from the fluctuation of volume. These fluctuations are defined in terms of the variance $\sigma^2 = \langle X^2\rangle - \langle X\rangle^2$ of a random variable X. For a random variable X, we need to find both the mean $<X>$ and the mean of the square $\langle X^2\rangle$, and sometimes also the mean of higher powers. Here the angular bracket represents the average over probability distribution $P(X)$ of and $\langle X^n\rangle$ with $n = 1, 2, 3\cdots$ called the nth order moment of the random variable X.

The moment of order n of a random variable X with probability distribution $P(X)$ is thus defined by,

$$\boxed{\mu_n = \int dX\, X^n\, P(X)}.$$
(14.1)

Here $P(X)$ is the probability distribution. In probability theory, we often work with *a moment generating function* $M(\xi)$, which generates these moments defined above. The generating function is defined as

$$M(\xi) = \langle\exp(\xi X)\rangle = \sum_{n=0}^{\infty} \mu_n \xi^n / n!.$$
(14.2)

Thus μ_n is the nth derivative of $M(\xi)$ with respect to ξ, evaluated at $\xi = 0$:

$$\mu_n = \left[\frac{d^n M(\xi)}{d\xi^n}\right]_{\xi=0}.$$
(14.3)

Eq. (14.3) is the reason why it is called the moment generating function – because it directly generates the moments by differentiation. In the TDSM, we find that the moments $\langle X^n\rangle$ themselves are not that helpful because the key quantity in the study of statistical mechanics is the fluctuations of the random variable ξ around its average value μ, i.e., fluctuation in $(X - \mu)$. Therefore, we define central moments $\langle (X-\mu)^n\rangle$ (moments about the mean value) of the random variable X, as $\langle(\delta X)^n\rangle = \langle(X-\mu)^n\rangle$. For example, its second moment is precisely the variance σ^2. We shall return to this issue when we discuss cumulants.

14.3 CHARACTERISTIC FUNCTION

A characteristic function of a random variable X is defined as the Fourier transform of the probability distribution $P(X)$

$$\tilde{p}(k) = \int dX\, P(X)\, e^{-ikX}.$$
(14.4)

This is just the expectation value of e^{-ikX} with the weight factor $P(X)$

$$\tilde{p}(k) = \langle e^{-ikX} \rangle. \qquad (14.5)$$

Clearly, the characteristic function $\tilde{p}(k)$ is related to the moment generating function $M(\xi)$ of the random variable. The relationship is given by

$$\tilde{p}(k) = \langle e^{-ikX} \rangle = M(-ik). \qquad (14.6)$$

Therefore, the characteristic function is another way of writing the moment generating function.

14.4   CALCULATION OF THE N^{TH} MOMENT OF GAUSSIAN (OR, NORMAL) DISTRIBUTION

To calculate the nth moment of the normal distribution, we start with the probability density for the Gaussian distribution of a continuous random variable X (where $-\infty < x < \infty$) given by,

$$\boxed{P(X) = \frac{1}{\sqrt{2\pi\sigma^2}} \exp\left(-\frac{(X-\mu)^2}{2\sigma^2}\right).} \qquad (14.7)$$

This is a normalized distribution, with mean μ and variance σ^2. A central moment is defined on $(X-\mu)$, not on X. The nth central moment of any random variable X is given by,

$$\zeta_n = \langle (X-\mu)^n \rangle, \qquad (14.8)$$

where X is a random variable and μ is its mean.

Now, we evaluate the following integral:

$$\zeta_n = \int_{-\infty}^{\infty} dX \frac{1}{\sqrt{2\pi\sigma^2}} (X-\mu)^n \exp\left(-\frac{(X-\mu)^2}{2\sigma^2}\right). \qquad (14.9)$$

We substitute $y = X - \mu$ to obtain

$$\zeta_n = \int_{-\infty}^{\infty} \frac{1}{\sqrt{2\pi\sigma^2}} y^n \exp\left(-\frac{y^2}{2\sigma^2}\right) dy. \qquad (14.10)$$

Due to the symmetry about the mean value, all the odd moments of the deviation from the mean value vanish, i.e., $\langle (X-\mu)^{2n+1} \rangle = 0$. On the other hand, in the case of even moments, we obtain

$$\boxed{\langle (X-\mu)^{2n} \rangle = \frac{(2n-1)!}{2^{n-1}(n-1)!} \sigma^{2n},} \qquad (14.11)$$

where $n = 0, 1, 2, \cdots$ *Thus, we can express any higher moments in terms of only the first and the second moments.* For a Gaussian distribution, the first four moments are given by

$$\begin{aligned}
\mu_1 &= \mu \\
\mu_2 &= \sigma^2 + \mu^2 \\
\mu_3 &= 3\mu\mu_2 - 2\mu^3 \\
\mu_4 &= 4\mu_3\mu + 3\mu_2^2 - 12\mu_2\mu^2 + 6\mu^4.
\end{aligned} \tag{14.12}$$

This is an important simplifying property of Gaussian distribution.

The characteristic function for normal distribution can be obtained by taking the Fourier transform of the distribution function (Fourier transform of a Gaussian also a Gaussian)

$$\tilde{p}(k) = \exp\left[-\left(ik\mu + \frac{k^2\sigma^2}{2}\right)\right]. \tag{14.13}$$

The moment-generating function for a Gaussian or normal distribution has the following form:

$$M(\xi) = \exp\left(\mu\xi + \frac{1}{2}\sigma^2\xi^2\right). \tag{14.14}$$

This exponential is a polynomial in ξ, which terminates at the quadratic level.

14.5   CUMULANTS

As already mentioned, in statistical mechanics we are often concerned with fluctuations of physical quantities that themselves are the sum over many individual contributions, for example, the interaction energy, from different molecules. As described above, second-order moments of these fluctuations are our familiar standard deviations. A standard deviation is a cumulant.

Cumulants have been well-known in the theory of probability theory, but it was made popular in statistical mechanics by Kubo in two classic papers. He pointed out that in most applications of statistical mechanics we need to evaluate average of a function that contains a random variable in the exponential, like $e^{\xi X}$. This is true both in equilibrium statistical mechanics (the Boltzmann factor) and TDSM (stochastic theory of line shapes).

Cumulants κ_n are defined by the following expression:

$$\boxed{\left\langle e^{\xi X} \right\rangle = \sum_n \frac{\xi^n}{n!}\mu_n = \exp(\sum_n \frac{\xi^n}{n!}\kappa_n) \equiv \exp(K(\xi)).} \tag{14.15}$$

The cumulants κ_n of a probability distribution extend the concept of variance to the higher moments of the distribution function.

The cumulant generating function is defined as

$$K(\xi) = \log_e M(\xi), \tag{14.16}$$

where $K(1) = 0$ because $M(0) = 1$. So in general,

$$K(\xi) = \sum_{n=1}^{\infty} \frac{\kappa_n \xi^n}{n!}, \tag{14.17}$$

where κ_n is the nth cumulant of the variable ξ. In the analogy with Eq. (14.17), cumulants can be written as

$$\kappa_n = \left[\frac{d^n}{d\xi^n} K(\xi) \right]_{\xi=0}. \tag{14.18}$$

From the equation, we note that both moments generating function $M(\xi)$ and the cumulant generating function $K(\xi)$ are related exponentially. Hence, the cumulants can be expressed in terms of the moments and vice versa.

The nth cumulants can be obtained by expanding the right-hand side of Eq. (14.15)and then using Eq. (14.17), and picking out the coefficients ξ^n in the power series. For the first cumulant κ_1, we get

$$\kappa_1 = \mu_1 = \mu, \tag{14.19}$$

which is the mean value of the random variable X. *The second cumulant turns out to be the central moment or variance itself of the random variable X* (as mentioned above)

$$\kappa_2 = \mu_2 - 2\langle X \rangle \mu + \mu^2 = \mu_2 - \mu^2, \tag{14.20}$$

The third cumulant is defined as

$$\kappa_3 = \mu_3 - 3\mu_2\mu + 2\mu^3. \tag{14.21}$$

Similarly, the fourth cumulant is given by

$$\kappa_4 = \mu_4 - 4\mu_3\mu - 3\mu_2^2 + 12\mu_2\mu^2 - 6\mu^4. \tag{14.22}$$

Therefore, one can get equation with n^{th} cumulant in terms of nth moment μ_n, $(n-1)^{th}$ cumulant in terms of $(n-1)^{th}$ moment μ_{n-1} and so on.

Kubo used an alternate (to κ_n) symbol for cumulants. He denoted them by $\langle .....\rangle_C$. Thus,

$$\begin{aligned} \langle X^2 \rangle_C &= \langle X^2 \rangle - \langle X \rangle^2, \\ \langle X^3 \rangle_C &= \langle X^3 \rangle - 3\langle X^2 \rangle \langle X \rangle + 2\langle X \rangle^3, \end{aligned} \tag{14.23}$$

and so on.

14.6   AN IMPORTANT THEOREM

One of the main reasons for the importance of cumulants is the theorem that states the following. *When the distribution $P(X)$ is Gaussian, then all the cumulants higher than second are identically equal to zero.* The student can easily appreciate the importance of this theorem *because many variables are indeed Gaussian because of the central limit theorem (CLT).*

There is an important corollary of this theorem. Since in many applications, as already emphasized, we require an average over an exponential of the random variable, the cumulant theorem proves especially useful. As a consequence of the above theorem, the cumulant expansion vastly simplifies, to

$$\left\langle e^{\xi X} \right\rangle = \exp(\sum_n \frac{\xi^n}{n!} \kappa_n) = \exp\left(\xi\kappa_1 + \frac{\xi^2}{2} \kappa_2 \right). \tag{14.24}$$

As we discussed earlier, the moment-generating function for a Gaussian distribution function depends only on the first and second moments [Eq. (14.14)]. The first cumulant is just the first moment, while the second moment is given by Eq. (14.11). Then the corresponding cumulant generating function becomes

$$K(\xi) = \mu\xi + \frac{1}{2}\sigma^2\xi^2.$$

(14.25)

We shall demonstrate the validity of this theorem by using Eq. (14.20) and Eq. (14.21). Towards this, we first assume that the distribution is centered around zero. That is, the mean μ is equal to zero. For a centrally placed Gaussian distribution, all the odd moments are zero. Thus, all the cumulant becomes 0 except the second cumulant like the following:

$$\begin{aligned}
\kappa_1 &= \mu \\
\kappa_2 &= \mu_2 - \mu^2 \\
\kappa_3 &= \mu_3 - 3\mu_2\mu + 2\mu^3 = \left[3\mu\mu_2 - 2\mu^3\right] - 3\mu_2\mu + 2\mu^3 = 0 \\
\kappa_4 &= \mu_4 - 4\mu_3\mu - 3\mu_2^2 + 12\mu_2\mu^2 - 6\mu^4 \\
&= \left[4\mu_3\mu + 3\mu_2^2 - 12\mu_2\mu^2 + 6\mu^4\right] - 4\mu_3\mu - 3\mu_2^2 + 12\mu_2\mu^2 - 6\mu^4 \\
&= 0
\end{aligned}$$

(14.26)

.....

With the aid of Eq. (14.12) we simplify Eq. (14.26) and find that for a Gaussian and only for a Gaussian distribution, all the cumulants higher than the second vanish identically.

The importance of the above expression can be appreciated if we notice that in the case of moment expressions, not only that all the moments are nonzero, but the expression is a series expansion that is cumbersome and the whose sum can be beyond reach.

This cumulant theorem is thus the main reason why cumulants are popular in statistical mechanics. We now discuss certain general aspects – more specific cases discussed later.

14.7 NON-GAUSSIAN PARAMETERS AND HETEROGENEITY

Since random variables are often found to be Gaussian (and justified by the central limit theorem), the Gaussian curve can be treated as a reference distribution in the study of the properties of any distribution, obviously, we turn to the third and fourth cumulants κ_3 and κ_4 to test the normality or the Gaussian-ness of a distribution function. The third cumulant κ_3 is called the *skewness* and is a measure of the asymmetry of the probability distribution about its mean. It can be positive, negative, or even zero. For the normal distribution, its value is zero. Similarly, the fourth cumulant κ_4 is termed as *kurtosis* of the distribution and tells the sharpness of the central peak around the mean. The kurtosis is also zero for normal distribution. The pictorial representation of these cumulants is shown in Figure 14.1, and Figure 14.2. So the conclusion is that the first four cumulants essentially tell us about the shape of the distribution.

Time-dependent cumulants have been used to quantify dynamic heterogeneity that emerges in supercooled glassy liquids. In these liquids, relaxation of density could become slow. As a result, a dynamic heterogeneity develops at intermediate times when the mean square displacement of different particles are different in different regions within the liquid. In some regions, diffusion of tagged particles may be fast while in some regions slow. Since the sum of Gaussian functions is not a Gaussian function, one can obtain an estimate of heterogeneous dynamics through a non-Gaussian parameter (NGP) for the displacement $\Delta r\,(t)$ by studying the fourth-order cumulant, defined as

$$\boxed{\alpha_2(t) = \frac{3}{5}[\langle(\Delta r)^4(t)\rangle > -\langle(\Delta r)^2(t)\rangle^2].}$$

(14.27)

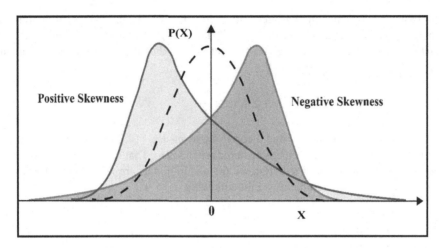

FIGURE 14.1 Pictorial representation of skewness for a distribution function $P(X)$ (Y-axis). The dashed line shows the normal distribution. The figure clearly demonstrates that skewness is a measure of the amount and direction of skew of the probability distribution of a random variable about its mean. The skewness value can be positive or negative, or even undefined. If skewness is 0, the data are perfectly symmetrical. If skewness is less than –1 or greater than 1, the distribution is skewed. If skewness is between –1 and –0.5 or between 0.5 and 1, the distribution is moderately skewed. If skewness is between –0.5 and 0.5, the distribution is approximately symmetric.

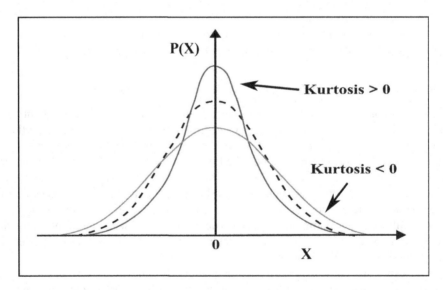

FIGURE 14.2 Pictorial representation of kurtosis for a distribution function $P(X)$. The dashed line shows the Gaussian distribution. Kurtosis is a measure of how differently shaped *the tails of distribution* are as compared to the tails of the normal distribution. Kurtosis tells you the height and sharpness of the central peak, relative to that of a standard bell curve. Kurtosis is sensitive to departures from normality on the tails. Kurtosis of distribution is positive (negative) if it has more (less) than a normal distribution. The skewness focuses on the overall shape, while Kurtosis focuses on the tail shape.

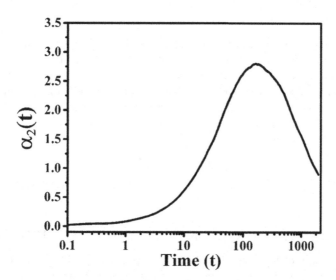

FIGURE 14.3 Time dependence of the non-Gaussian parameter (NGP), $\alpha_2(t)$, for a model glass forming liquid. In the long time, NGP goes to zero as the displacements become Gaussian.

NGP is zero if the distribution of displacement is Gaussian. In the short time, both the terms on the right-hand side is zero. In the long time, the system becomes ergodic, so NGP again goes to zero. However, it exhibits a peak in the intermediate time. The height of the peak provides a measure of heterogeneity in the system.

In Figure 14.3 we show a typical curve of NGP as a function of time.

The height of the peak provides a measure of the magnitude of heterogeneity while the time value where the NGP peaks provide a measure of the time scale of heterogeneity. This has become quite fashionable now as computer simulations have found wide use.

14.8 PROPERTIES OF CUMULANTS

14.8.1 ADDITIVITY

Suppose we have two independent random variables x_1 and x_2 having the probability densities $p(x_1)$ and $p(x_2)$, respectively. Now, we want to know what is the cumulant of a linear combination of these variables. This is given as follows:

$$
\begin{aligned}
\kappa_{x_1+x_2}(t) &= \log E\left[e^{t(x_1+x_2)}\right] \\
&= \log\left(E\left[e^{tx_1}\right]E\left[e^{tx_2}\right]\right) \\
&= \log E\left[e^{tx_1}\right] + \log E\left[e^{tx_2}\right] \\
&= \kappa_{x_1}(t) + \kappa_{x_2}(t).
\end{aligned}
\tag{14.28}
$$

The cumulant averages can be extended for many variables as,

$$
\left\langle \exp\left(\sum_i^N \xi_i x_i\right)\right\rangle = \exp\left\{\sum \prod \frac{\xi^{m_i}}{m_i!}\left\langle x_1^{m_1}\ldots x_N^{m_N}\right\rangle_c\right\}
\tag{14.29}
$$

Note the change of notation as it is easier to write this way. We note further that a cumulative average vanishes identically if any of the variables in it is statistically independent from the others.

14.8.2　Equivariance and Invariance

The first cumulant represents the mean of a probability distribution, which shifts equally with the constant scaling of the random variable. So the first cumulant is shift-equivariant. The second and higher moments show the properties of distribution. For example, the second cumulant tells us about the spreads, the third cumulant represents asymmetry from the normal distribution, etc., of the distribution function. So these higher moments are invariant under the shift of the random variable and called shift-invariant. Mathematically these properties can be shown as,

$$\kappa_1(\xi+c) = \kappa_1(\xi)+c$$
$$\kappa_n(\xi+c) = \kappa_n(\xi),$$

(14.30)

where $n \geq 2$.

14.9　AN IMPORTANT APPLICATION: KUBO THEORY OF LINE SHAPES

Kubo's stochastic theory of line shape employs cumulants and is widely used to understand both NMR and vibrational line shapes. The theory is formulated in terms of the coordinate $x(t)$ of a harmonic oscillator whose motion is defined in terms of its natural frequency ω, which is time-dependent and has the following form:

$$\dot{x}(t) = i\omega(t)x.$$

(14.31)

Here frequency $\omega(t)$ gets modulated randomly by the disturbances in the system. These fluctuations are responsible for the width. The line shape is given by the Fourier transform of $\langle x(0)x(t)\rangle$ or for fixed $x(0)$, just the $\langle x(t)\rangle$.

Eq. (14.31) is a first-order ordinary differential equation and its solution $x(t)$ can be obtained by integration as follows:

$$x(t) = \exp\left\{i\omega_0 t + \int_0^t \delta\omega(t')dt'\right\}.$$

(14.32)

Here ω_0 is the harmonic frequency of the system in the absence of any external disturbance and $\delta\omega(t)$ is the modulation or fluctuation in frequency is defined as $\delta\omega(t) = \omega(t) - \omega_0$. We shall make the assumptions of stationarity and ergodicity for the process $\delta\omega(t)$.

Thus for the line shape calculations we need to evaluate the relaxation function defined below

$$\varphi(t) = \left\langle \exp i\int_0^t \delta\omega(t')dt' \right\rangle,$$

(14.33)

where the angular bracket represents the ensemble average and due to ergodicity assumption can also be considered as a long-time average.

As discussed above, the spectral distribution function of an ensemble of these oscillators is given by

$$I(\omega-\omega_0) = \frac{1}{2\pi}\int_{-\infty}^{\infty} \phi(t)e^{-i(\omega-\omega_0)t}dt.$$

(14.34)

Since $\phi(t)$ is given by Eq. (14.33), it can be written in terms of cumulant function $K(t)$

$$\phi(t) = \exp K(t).$$

(14.35)

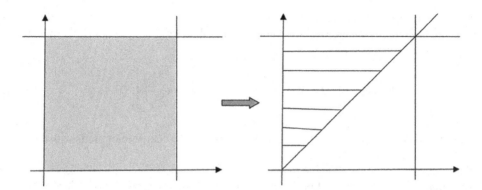

FIGURE 14.4 Pictorial representation of the region of integration from the contribution of the first term of Eq. (14.37). The figure illustrates how an integral over a square area is transformed into two times the integral over a triangle, with the limits on the internal integral changed. This transformation opens the door for approximation as discussed above.

And it admits a cumulant expansion

$$K(t) = i \int_0^t dt_1 \langle \delta\omega(t_1) \rangle + i^2 \frac{1}{2} \int_0^t dt_1 \int_0^t dt_2 \langle \delta\omega(t_1)\delta\omega(t_2) \rangle_c + \cdots \qquad (14.36)$$

The contribution from first term is zero as $\langle \delta\omega(t) \rangle = 0$. The next two terms in the expansion, can be written as

$$K(t) = -\int_0^t dt_1 \int_0^t dt_2 \langle \delta\omega(t_1)\delta\omega(t_2) \rangle_c + \cdots \qquad (14.37)$$

For the first term involves a two-dimensional integration, which can be simplified as follows. We first write as shown in the right panel of Figure 14.4 (the same can be done in three dimensions). Here the area under the curve is a square that can be split into two equal triangles that involve different times. Therefore, the above expansion has the following form:

$$K(t) = -\int_0^t dt_1 \int_0^{t_1} dt_2 \langle \delta\omega(t_1)\delta\omega(t_2) \rangle_c + \cdots \qquad (14.38)$$

If we neglect the second term on the right-hand side, and employ a change of variable, $\tau = t_1 - t_2$, the first term on the right-hand side can be written as

$$K(t) = -\int_0^t d\tau (t - \tau) \langle \delta\omega(0)\delta\omega(\tau) \rangle_c .$$

This can be evaluated by observing that the contribution to the integral comes from the small region of τ. Thus, the upper limit of the integral can be extended to infinity, as further discussed below.

Let us now consider a specific example – the line shape of a harmonic oscillator where Q is the normal coordinate and the Hamiltonian of is given by

$$H(Q) = H_v + V_B(Q) \qquad (14.39)$$

where H_v denotes the Hamiltonian for the vibrational degree of freedom and $V_B(Q)$ is the coupling energy of the normal coordinate with the bath. The coupling term can be evaluated by expanding $V_B(Q)$ in terms of Q as follows:

$$V_B(Q) = V_B(Q=0) + \left[\frac{\partial V_B(Q)}{\partial Q}\right]_0 Q + \frac{1}{2}\left[\frac{\partial^2 V_B(Q)}{\partial Q^2}\right]_0 Q^2 + \text{..} \tag{14.40}$$

For a harmonic coupling term, the first-order term is zero and the contribution starts with the second-order term.

In the simplest case closest to Kubo treatment, the experimental observable like isotropic Raman line shape function is given by the Fourier transform of normal coordinate correlation function

$$I(\omega) = \int_0^\infty \exp(i\omega t)\langle Q(0)Q(t)\rangle, \tag{14.41}$$

where ω is the Laplace frequency conjugate to time, t. If we take $x(t) = Q(t)$, the normal coordinate time correlation function can be related to frequency-modulation time correlation function by

$$\langle Q(t)Q(0)\rangle = \text{Re}\left[\exp(i\omega_0 t)\left\langle \exp\left[i\int_0^t dt' \Delta\omega_{mn}(t')\right]\right\rangle\right], \tag{14.42}$$

where ω_0 is the vibrational frequency and $\hbar\Delta\omega_{mn}(t) = V_{nn}(t) - V_{mm}(t)$ denotes the energy gap between vibrational levels m and n.

Now we shall consider a few things that are of interest. We decompose the frequency gap into two parts: one constant (time-independent) and one fluctuating: $\Delta\omega_{mn}(t) = \Delta\omega + \delta\omega(t)$ where the first term on the right-hand side gives the equilibrium shift of the spectrum from the gas phase to the solution while the second term gives the width.

We next perform cumulant expansion and truncate at the second order to obtain

$$\langle Q(0)Q(t)\rangle = \text{Re}\left[\exp\left(i\omega_0 t + i\Delta\omega t\right) \times \exp\left[-\frac{1}{2}\int_0^\infty dt'\int_0^{t'} dt'' \left\langle \delta\omega_{mn}(t')\delta\omega_{mn}(t'')\right\rangle\right]\right]. \tag{14.43}$$

We next simplify the equation to obtain

$$\langle Q(0)Q(t)\rangle = \text{Re}\left[\exp\left(i\omega_0 t + i\Delta\omega t\right) \times \exp\left[-\int_0^\infty dt'(t-t')\left\langle \delta\omega_{mn}(t')\delta\omega_{mn}(0)\right\rangle\right]\right]. \tag{14.44}$$

Eq. (14.44) can be further approximated in the limit of fast decay of frequency-modulation time correlation function as

$$\langle Q(0)Q(t)\rangle \approx \exp\left(-t/\tau_v\right) \tag{14.45}$$

where τ_v is the average dephasing time defined by

$$\boxed{\tau_v^{-1} = \int_0^\infty dt' \left\langle \delta\omega_{mn}(0)\delta\omega_{mn}(t')\right\rangle.} \tag{14.46}$$

The exponential decay described by Eq. (14.45) gives rise to a Lorentzian line width with a half-width at half maximum of τ_v^{-1}. Hence, Eq. (14.46) allows us to determine τ_v directly. However, $\langle Q(0)Q(t) \rangle$ might be nonexponential in other cases and may give rise to an overall subquadratic quantum number-dependent rate.

In many applications of time-dependent statistical mechanics, we are faced with the task of evaluating a term like $\langle \exp(X(t)) \rangle$ where $X(t)$ is a function that involves a random or stochastic variable. Sometimes $X(t)$ involves an integral. While the first choice involves expansion of the exponential leading to moments, it turns out that there is a better way to evaluate that involves cumulants, which are combinations of moments. While students are familiar with moments, cumulants are relatively less known.

14.10 TIME ORDERING IN QUANTUM MECHANICS

In classical time-dependent processes, like the ones considered to date, and in the Kubo's line shape and various TCF analyses, what matters is the time gap between the two variables, as exemplified in Figure 14.4. In quantum mechanics the situation changes. In most applications of dynamics, we prefer to work in the interaction representation and here the evolution operator depends on the order it is applied, time ordering is required because operators do not commute at different times.

14.11 SUMMARY

Along with projection operator technique, the method of moments and cumulants provides us with a powerful theoretical tool to address stochastic problems in natural and biological sciences. The method of cumulants has huge application in field of spectroscopy. We discussed above the elegant analysis of Kubo, which forms the "bread and butter" in theoretical analysis of line shapes. Unlike the projection operator technique, the method of cumulant is more intimately connected to experimental observation.

BIBLIOGRAPHY

1. Feller, W. 1968. *An Introduction Probability Theory and Applications*, Vol. 1., New York: John Wiley & Sons.
2. Chung, K. L. 1974. *A Course in Probability Theory*. New York: Academic Press.
3. Kubo, R. 1962. Generalized cumulant expansion method, *Journal of the Physical Society of Japan*. *17*, 1100.
4. Kubo. R. 1963. Stochastic Liouville equations, *Journal of Mathematical Physics 4*, 174.
5. Schiff, L. I. 1955. *Quantum Mechanics*. 2nd edition. Maidenhead: McGraw Hill Education.

Part III

Phenomenology

15 Brownian Motion and Langevin Equation

These motions were such as to satisfy me ... that they arose neither from currents in the fluid, nor from its gradual evaporation, but belonged to the particle itself.

Robert Brown (1828)

According to the molecular-kinetic theory of heat, bodies of a microscopically visible size suspended in liquids must, as a result of thermal molecular motions, perform motions of such magnitude that they can be easily observed with a microscope.

Albert Einstein (1905)

OVERVIEW

Brownian motion, Langevin equation, random walk, and the diffusion equation are all related theoretical tools of phenomenological time-dependent statistical mechanics. These are widely used in the study of relaxation phenomena. They in turn are deeply rooted in the linear response theory and fluctuation-dissipation theorem that we discussed in Chapter 11. Brownian motion is the incessant random movement of a large particle suspended in a liquid. The motion is due to the collisions of the tagged molecule with the molecules of the liquid medium, which themselves are in continuous motion due to the temperature. Brownian motion gives rise to the diffusion of the tagged particle. In 1905, Einstein demonstrated that the Brownian motion was a natural phenomenon, a consequence of the thermal motion of the liquid due to the temperature. Einstein's predictions were verified later by Perrin and co-workers. Brownian motion can be described by a Langevin equation, which is discussed here in detail. The Langevin equation is a stochastic equation with a random force term acting on the tagged particle whose motion is being followed. We discuss several solutions to the Langevin equation, including the important problem of the dynamics of a damped harmonically bound particle. We also discuss the generalized Langevin equation (GLE) that leads to the generalized Einstein relation between frequency-dependent diffusion and frequency-dependent friction. We discuss Zwanzig's derivation of a GLE from a Hamiltonian. *Interestingly, Brownian motion served to provide a convincing proof of the corpuscular nature of matter*

15.1 INTRODUCTION WITH HISTORICAL PERSPECTIVE

Robert Brown (1773–1858), a Scottish botanist, while studying fertilization processes in plants under a powerful optical microscope, observed an astounding phenomenon. He discovered that pollen grains exhibit *incessant rapid and irregular movements* while suspended on the surface of

DOI: 10.1201/9781003157601-18

water. At first, he speculated that the motions exist because of some kind of *vital force*; the particles were alive. Later he examined samples from other living and nonliving specimens and observed the presence of similar ceaseless motions. Such random movement of small particles suspended in liquids is named after him as "Brownian motion". Interestingly, Brown was not the first observer of Brownian motion. Before Brown's paper in 1828, Leeuwenhoek (1632–1723) reported similar observations. The theory of Brownian motion developed by Einstein played a crucial role in establishing that matter is made of atoms and molecules.

It was realized early that Brownian motion is of great importance in understanding a large number of nonequilibrium phenomena. As a result, the study of Brownian motion has received persistent attention from mathematicians, physicists, chemists, biologists, and also from other disciplines. In the late nineteenth century, Cantoni, Delsaulx, and Carbonelle independently speculated the origin of Brownian motion in terms of collisions of the tagged molecule with the surrounding solvent molecules. Guoy, Exner, and others studied the same phenomena and speculated on the underlying mechanism. Guoy's observations (1888) that supported a kinetic theory origin were as follows: (i) the movement was irregular and composed of translations and rotations; (ii) the motion became more active if the particles were smaller, the liquid was less viscous, and the temperature elevated.

It soon became clear that Brownian motion has a certain universal character and also that these motions are important manifestations of the natural thermal motions of liquids. However, the physics that governs Brownian motion had remained obscure for several decades. Einstein wrote several papers on Brownian motion that are collected in a Dover publication, although the 1905 paper remains the most influential one.

Contemporary Polish physicist Smoluchowski developed a theory on Brownian motion and published in 1906 after Einstein's paper. Later, Jean Perrin (*Nobel Prize 1926*) and co-workers experimentally tested Einstein's theory by preparing a colloidal suspension of gamboge (a type of gum) in water. They verified the theoretical expressions and obtained a good estimate of the Avogadro's number from it.[2]

It is to be noted that the Brownian motion of a tagged particle is a reduced description where the motion of other solvent molecules is hidden or eliminated. We shall see later that in the Langevin equation description of Brownian motion, the solvent contributions are included in the form of a random force term.

In this chapter, we follow Einstein's approach to derive and solve the diffusion equation. In the process, we develop the definition of the diffusion constant in terms of the mean square displacement of the tagged particle. We derive an expression for the mean square displacement from Einstein's solution. These two equations [given in Eq. (15.19) and Eq. (15.26)] form two pillars of nonequilibrium statistical mechanics.

We guide students toward alternative approaches to obtain the same relations. We also discuss and solve the equation of motion for Brownian particles, which is known as the Langevin equation. We discuss the generalized Langevin equation to include the memory effects. We shall discuss the derivation and solution of the Smoluchowski equation in the next chapter.

15.2 BROWNIAN MOTION THEORY *A LA* EINSTEIN

In Brownian dynamics, we are concerned with the perpetual irregular motions exhibited by small grains or particles of colloidal size immersed in a fluid made of molecules of *much smaller* size. First, we choose a time interval τ in which the particle under consideration changes its X-coordinate from x to $x+\Delta$ (Figure 15.1). Suppose that there are n numbers of colloidal particles. For each particle, the jump (Δ) along X-direction is random and different in amplitude. Next, we introduce a probability distribution $\varphi(\Delta)$ of this displacement. Let n be the total number of particles and dn be

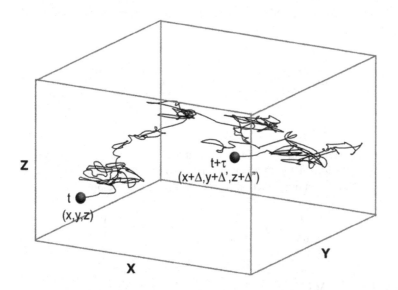

FIGURE 15.1 Schematic illustration that demonstrates the Brownian movement of a tagged point particle suspended in a liquid medium. The particle traverses Δ distance along X-axis, Δ' distance along Y-axis and Δ'' distance along Z-axis within a certain time interval Δ. During the displacement from $\{x, y, z\}$ to $\{x+\Delta, y+\Delta', z+\Delta''\}$, the particle undergoes several random displacements shown by the thin black trace.

the number of particles that exhibit displacement in the range of Δ to $\Delta + d\Delta$. Hence, one can write Eq. (15.1).

$$\frac{dn}{n} = \varphi(\Delta)d\Delta \tag{15.1}$$

where

$$\int\limits_{-\infty}^{+\infty} d\Delta\varphi(\Delta) = 1 \tag{15.2}$$

$$and \quad \varphi(\Delta) = \varphi(-\Delta). \tag{15.3}$$

Eq. (15.2) ensures normalization and Eq. (15.3) ensures symmetry; that is, *jumps along the positive X-axis are as probable as the jumps along the negative X-axis.* For simplicity, we shall treat the system in one dimension. However, the extension of these equations to higher dimensions is straightforward.

We next define a position and time-dependent concentration $c(x, t)$, that gives the number density of the colloidal particles at position x and time t. Therefore, the concentration of particles after a time interval τ is given by

$$c(x, t + \tau) = \int\limits_{-\infty}^{+\infty} d\Delta\varphi(\Delta)c(x + \Delta, t). \tag{15.4}$$

We expand both sides of Eq. (15.4) in Taylor series by considering both τ and Δ as infinitesimally small. The resultant equation is Eq. (15.5).

$$c(x,t) + \tau \frac{\partial}{\partial t} c(x,t) = c(x,t) \int_{-\infty}^{+\infty} d\Delta \varphi(\Delta) + \frac{\partial}{\partial x} c(x,t) \int_{-\infty}^{+\infty} d\Delta \varphi(\Delta) \Delta$$
$$+ \frac{\partial^2}{\partial x^2} c(x,t) \int_{-\infty}^{+\infty} d\Delta \varphi(\Delta) \left(\frac{\Delta^2}{2} \right). \qquad (15.5)$$

We realize that, on the right-hand side, the first integral is unity due to normalization (which arises because of the conservation of the number of particles) [see Eq. (15.2)] and the second integral vanishes as the integrand is an odd function. Upon simplification we obtain

$$\frac{\partial}{\partial t} c(x,t) = \frac{1}{\tau} \frac{\partial^2}{\partial x^2} c(x,t) \int_{-\infty}^{+\infty} d\Delta \varphi(\Delta) \left(\frac{\Delta^2}{2} \right). \qquad (15.6)$$

We next define $D = \frac{1}{\tau} \int_{-\infty}^{+\infty} d\Delta \varphi(\Delta) \left(\frac{\Delta^2}{2} \right)$. Then Eq. (15.6) becomes

$$\boxed{\frac{\partial}{\partial t} c(x,t) = D \frac{\partial^2}{\partial x^2} c(x,t)}. \qquad (15.7)$$

Eq. (15.7) is the diffusion equation and D is known as the diffusion coefficient. *It is related to the mean squared jump/displacement – a relation derived by Einstein for the first time.*

At a basic level, diffusion equation is a consequence of the conservation of mass and Fick's law as detailed below. According to the conservation law

$$\frac{\partial}{\partial t} c(x,t) = -\nabla_x . J_x. \qquad (15.8)$$

Here, ∇_x denotes the gradient operator and J_x is the flux along x-direction, which is given by Fick's law as follows:

$$J_x = -D\nabla_x c(x,t). \qquad (15.9)$$

One combines Eq. (15.8) and (15.9) to obtain

$$\boxed{\frac{\partial}{\partial t} c(x,t) = D\nabla_x^2 c(x,t)}, \qquad (15.10)$$

which is the same as Eq. (15.7). We note that the coefficient of diffusion (D) has a dimension of $[L^2T^{-1}]$.

As mentioned above, Einstein's derivation provides a *microscopic expression of diffusion in terms of the jump rate ($1/\tau$) and jump length.* This expression can be transformed into an integral over velocity time correlation function (TCF), as has been discussed in Chapter 10 on the velocity time correlation function.

The solution of the diffusion equation becomes quite simple in the Fourier space. Suppose, initially (some arbitrary time, $t = 0$) all the particles are concentrated at $x = 0$. Hence, one can represent $c(x = 0, t = 0)$ as a Dirac delta function as follows:

$$c(x, t = 0) = n\delta(x)$$
$$\begin{cases} \delta(x) = \infty \text{ if } x = 0 \\ \quad\quad = 0 \text{ if } x \neq 0 \end{cases}. \tag{15.11}$$

By borrowing a technique from mathematics to solve time and space-dependent partial differential equations, we perform a Fourier transformation in space on both sides of Eq. (15.7) with k as the conjugate variable,

$$\int_{-\infty}^{+\infty} dk \ e^{ikx} \frac{\partial}{\partial t} c(k,t) = D \int_{-\infty}^{+\infty} dk \ (-k^2) \ e^{ikx} c(k,t)$$

$$\text{or, } \boxed{\frac{\partial}{\partial t} c(k,t) = -Dk^2 c(k,t)}. \tag{15.12}$$

The solution of Eq. (15.12) yields

$$c(k,t) = c(k,t = 0) \exp\left(-Dk^2 t\right). \tag{15.13}$$

We now set $c(k, t = 0) = n$. Next, we Fourier transform $c(k,t)$ back into $c(x,t)$ by using Eq. (15.14)

$$c(x,t) = \frac{1}{2\pi} \int_{-\infty}^{+\infty} dk \ e^{ikx} c(k,t). \tag{15.14}$$

To obtain the following well-known expression:

$$c(x,t) = \frac{n}{(4\pi Dt)^{1/2}} e^{-\left(\frac{x^2}{4Dt}\right)}. \tag{15.15}$$

In this Gaussian form, the width of $c(x, t)$ increases with time while the peak height decreases. This indicates that the particles spread over the entire space and after a long time the system attains homogeneity. We demonstrate the solution graphically in Figure 15.2.

It is important to point out that the solution of Eq. (15.7) as given by Eq. (15.15), is a Green's function of the following partial differential equation [Eq. (15.16)].

$$\frac{\partial}{\partial t} G\left(x, x' \,|\, t, t'\right) - D \frac{\partial^2}{\partial x^2} G\left(x, x' \,|\, t, t'\right) = \delta(x - x')\delta(t - t'). \tag{15.16}$$

This equation finds wide use across many branches of natural science (for example, conduction of heat), even in quantum mechanics.

The solution given in Eq. (15.15) can be used to obtain expressions for the first and second moments of the probability/concentration distribution, defined as

$$\langle x \rangle = \frac{\int_{-\infty}^{+\infty} dx\, x\, c(x,t)}{\int_{-\infty}^{+\infty} dx\, c(x,t)} \ and \ \langle x^2 \rangle = \frac{\int_{-\infty}^{+\infty} dx\, x^2 c(x,t)}{\int_{-\infty}^{+\infty} dx\, c(x,t)}. \tag{15.17}$$

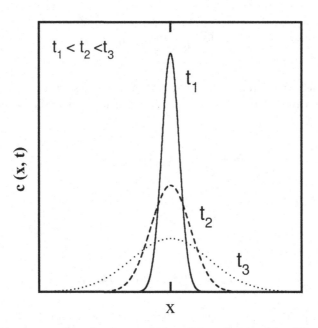

FIGURE 15.2 In this plot, we show the time evolution of the solution of the diffusion equation. At the beginning (time = t_1) the distribution is sharp that indicates the particles is concentrated at $x = 0$. As time progresses the distribution becomes flatter. This behavior is typical of a diffusive spread that gives rise to the establishment of a homogeneous distribution in the long-time limit.

As the integral in the expression of $\langle x \rangle$ contains the odd function in x, the result is zero. Hence, $\langle x \rangle = 0$. We now aim to derive expressions for the mean square displacement (MSD) of a Brownian particle from the position and time-dependent distribution $c(x,t)$.

The integral in the expression of $\langle x^2 \rangle$ in Eq. (15.17) is a simple Gaussian integral that can be easily solved. For a normalized distribution, the denominator becomes equal to the total number of particles n. Hence, the solution yields

$$\langle x^2 \rangle = \frac{1}{n} \int_{-\infty}^{+\infty} dx \, x^2 \frac{n}{(4\pi Dt)^{1/2}} e^{-\left(\frac{x^2}{4Dt}\right)}$$

$$= \frac{1}{(4\pi Dt)^{1/2}} \int_{-\infty}^{+\infty} dx \, x^2 e^{-\left(\frac{x^2}{4Dt}\right)} \tag{15.18}$$

$$= \frac{1}{(4\pi Dt)^{1/2}} \left[\frac{1}{2} \sqrt{\frac{\pi}{\left(\frac{1}{4Dt}\right)^3}} \right]; \text{ where we use } \int_{-\infty}^{+\infty} dx \, x^2 e^{-ax^2} = \frac{1}{2}\sqrt{\frac{\pi}{a^3}}$$

$$or, \quad \boxed{\langle x^2 \rangle = 2Dt}. \tag{15.19}$$

Hence, the root mean squared displacement ($\lambda_x = \sqrt{\langle x^2 \rangle}$) varies as $\sqrt{t}$. The relation between diffusion and MSD is Einstein's relation.

One can obtain the same expression in an alternative way without explicitly solving the integral. One needs to perform a time derivative on both sides of Eq. (15.17) and proceed as follows:

$$\frac{\partial}{\partial t}\langle x^2 \rangle = \frac{1}{n}\int_{-\infty}^{+\infty} dx x^2 \frac{\partial}{\partial t} c(x,t)$$

$$= \frac{D}{n}\int_{-\infty}^{+\infty} dx \, x^2 \frac{\partial^2}{\partial x^2} c(x,t)$$

$$= \frac{2D}{n}\int_{-\infty}^{+\infty} dx \, c(x,t) \quad [\text{integral by parts}] \qquad (15.20)$$

$$= 2D \quad \left[as \int_{-\infty}^{+\infty} dx \, c(x,t) = n \right]$$

Hence, $\langle x^2 \rangle = 2D\int_0^t dt = 2Dt$.

In a d-dimensional system the expression for MSD becomes $<\Delta r^2> = 2d\,Dt$. In three dimensions the diffusion equation becomes

$$\frac{\partial c}{\partial t} = D_x \frac{\partial^2 c}{\partial x^2} + D_y \frac{\partial^2 c}{\partial y^2} + D_z \frac{\partial^2 c}{\partial z^2}. \qquad (15.21)$$

Here, D_x, D_y, and D_z are the diffusion coefficients along the three different Cartesian axes. If diffusion is isotropic, then the three diffusion coefficients are equal in magnitude. c is a function of x, y, z, and t. One can employ the separation of variable technique to extract three independent one-dimensional diffusion equations from Eq. (15.21) and solve them separately with proper boundary conditions. We provide the final form of the distribution function in Eq. (15.22)

$$c(x,y,z,t) = \left\{ \frac{n}{(4\pi t)^{3/2}\sqrt{D_x D_y D_z}} \right\} \exp\left\{ -\frac{1}{4t}\left(\frac{x^2}{D_x} + \frac{y^2}{D_y} + \frac{z^2}{D_z} \right) \right\}. \qquad (15.22)$$

15.3   STOKES-EINSTEIN RELATION BETWEEN DIFFUSION AND VISCOSITY

Suppose that the movements of the suspended particles occur under the influence of an external force, F_{ext}. For example, if the particle falls through the liquid medium under the influence of gravity, this external force is the gravitational force (mg). When the viscous drag balances F_{ext}, the particle achieves a constant velocity known as *terminal velocity* (v). According to hydrodynamics (Stokes' law)

$$F_{ext} = \zeta v = C\eta r v. \qquad (15.23)$$

Here, ζ is the friction on the particle, η is the coefficient of viscosity of the medium, r is the radius of the suspended spherical particle, and C is a constant that is equal to 6π for *stick boundary condition* and 4π for *slip boundary condition*. Students should note that even though the particle is acted upon by a force, there is no acceleration; the particle moves with a steady velocity. This is because energy is getting dissipated. Friction provides a measure of this dissipation.

We note in passing that these are not valid for rotational motion. One can derive a similar expression for rotational friction with stick boundary condition, $\zeta_r = 8\pi\eta r^3$. We discuss hydrodynamics and molecular hydrodynamics in Chapter 5. In the subsequent discussions on translational motion, we use $C = 6\pi$ as was done by Einstein.

Suppose ρ is the number density of particles. Hence, the number of particles that pass through per unit area per unit time becomes $\rho v \left(= \rho F / 6\pi\eta r\right)$. Here, we assume that the external force is invariant with time and depends only on the position. On the other hand, because of diffusion, the number of particles that pass through per unit area per unit time is the flux as defined in Eq. (15.9). If one assumes a state of *dynamical equilibrium* in the Brownian system, one can write

$$\frac{\rho F_{ext}}{6\pi\eta r} = -D\frac{\partial\rho}{\partial x}. \tag{15.24}$$

We solve Eq. (15.24) by integration to obtain

$$\rho = \rho_0 \exp\left[-\frac{F_{ext}\left(x - x_0\right)}{6\pi\eta r D}\right]. \tag{15.25}$$

Eq. (15.25) shows the variation of density with x under the influence of an external force F_{ext}. One realizes that Eq. (15.25) is the Boltzmann distribution for density, under the influence of force F, and recognizes $6\pi\eta r D$ as $k_B T$ (from kinetic theory). Hence, one can write Eq. (15.26).

$$D = \frac{k_B T}{6\pi\eta r} = \frac{k_B T}{\zeta}. \tag{15.26}$$

Eq. (15.26) is known as the Stokes-Einstein relation for self-diffusion constant. The last relation between diffusion and friction is more fundamental, and separately known as Einstein relation, or Einstein equation.

In the previous section, we have derived an expression that connects diffusion constant with mean square displacement [Eq. (15.19)]. We now substitute the value of D from Eq. (15.26) and obtain the following relation:

$$\lambda_x = \sqrt{\langle x^2 \rangle} = \sqrt{2Dt} = \sqrt{\frac{RT}{3\pi\eta r N_A}}t, \tag{15.27}$$

where R is the universal gas constant and N_A is the Avogadro's number. Eq. (15.27) explains the observations made by Guoy in 1888 (see introduction). If we rearrange Eq. (15.27) we obtain

$$N_A = \frac{1}{\lambda_x^2}\frac{RT}{3\pi\eta r}. \tag{15.28}$$

Hence, one can validate this equation by determining mean square displacement from experiments and then by using this equation to obtain N_A. However, careful estimation of the radius of the spherical particles is required. This phenomenal task was achieved by Perrin and co-workers.[2] They observed quantitative agreement between the theory and experiment.

15.4 THE LANGEVIN EQUATION

The equation of motion that governs Brownian motion is known as the Langevin equation. *Langevin equation is a stochastic differential equation that contains both a systematic force and a random force.* Let us consider, for simplicity, a one-dimensional motion of a spherical particle of radius r

and mass m with velocity v suspended in a liquid and acted upon by a time-dependent force $F_{tot}(t)$, to be specified below. The liquid is characterized by a coefficient of viscosity η. We write Newton's equation of motion as follows:

$$m\dot{v} = F_{tot}(t). \tag{15.29}$$

As mentioned already, the total force on the tagged particle composed of two parts: a constant external force, which is kept on from infinite past and responsible for the steady velocity v, and a second part due to random collisions with the solvent molecules. At time $t = 0$, we switch off the external force. The velocity of the tagged particle is now time-dependent and shall approach zero following Stokes' law [Eq. (15.30)].

$$m\dot{v} = -\zeta v \tag{15.30}$$

The solution of Eq. (15.30) yields

$$v(t) = v(0)e^{-\left(\frac{\zeta}{m}\right)t}. \tag{15.31}$$

Thus, the velocity of the initially moving particle decays to zero. Why? This is because the effects of friction is due to the random collisions with the liquid molecules. This is the average behavior and easy to understand. Our problem is different. For the pollens moving in liquid, there is no external force! The motions observed are fluctuations. Although the average velocity of the particle is zero (in the absence of an external field), it has an instantaneous velocity due to the random force! This velocity gets damped by the friction. But how to describe this instantaneous velocity?

The deterministic equation alone is not enough to describe the equation of motion for the Brownian particle. The total force acting on a particle in liquid consists of two parts: (i) a systematic part (arises due to friction), and (ii) a fluctuating part $\delta F(t)$ (or noise) that arises due to the random kicks it receives from the surrounding medium. Both the friction and noise term comes from the interaction of the Brownian particle with its environment. Hence the modified equation becomes

$$\boxed{\begin{array}{c} m\dot{v} = -\zeta v + \delta F(t) \\ < \delta F(t) >= 0 \\ \left\langle \dfrac{1}{m}\delta F(0)\dfrac{1}{m}\delta F(t) \right\rangle = q\delta(t) \\ or, \\ \dot{v} = -\gamma v + \Gamma(t) \\ < \delta\Gamma(t) >= 0 \\ with, < \delta\Gamma(0)\delta\Gamma(t) >= q\delta(t) \end{array}} \tag{15.32}$$

The two alternativee representations are obviously connected by $\gamma = \zeta / m$ and $\Gamma(t) = \delta F(t) / m$. Eq. (15.32) is the Langevin equation for a Brownian particle. The force-force time correlation function is delta correlated with an amplitude q, which we determine below. This is a part of fluctuation-dissipation theorem.

However, the Langevin equation is a stochastic differential equation and cannot be analytically solved for most purposes. Hence, we eventually need Smoluchowski or Fokker-Planck equations, which are discussed in Chapter 17.

Now we emphasize the three important properties assumed for the fluctuating force below.

i.   The fluctuating force possesses a Gaussian distribution and the first moment of $\Gamma(t)$ is zero. $\langle \Gamma(t) \rangle = 0$.

ii. The second moment of $\Gamma(t)$ is delta correlated. $\langle \Gamma(t)\Gamma(t') \rangle = \underset{=}{q}\,\delta(t - t')$ where q is a measure of the strength of the fluctuating force. In general, $\underset{=}{q}$ is a tensor. However, for simplicity in the notations, we drop the double bar sign below q in the subsequent derivations. This condition depicts that the collisions are mostly uncorrelated.

iii. The power spectrum of $\Gamma(t)$ is constant. The spectral density is defined as, $S(\omega) = \int\limits_{-\infty}^{+\infty} dt\; e^{-i\omega t}\, \langle \Gamma(0)\Gamma(t) \rangle$. By definition, we have $\langle \Gamma(0)\Gamma(t) \rangle = q\,\delta(t)$. Hence, $S(\omega) = q$; that is, the spectral density is flat. This is known as *white noise*.

15.5   THE FLUCTUATION-DISSIPATION THEOREM FROM LANGEVIN EQUATION

We have discussed the fluctuation-dissipation theorem (FDT) as derived by Kubo in Chapter 11. However, long before Kubo, many aspects of FDT were understood in the context of the Langevin equation, and in fact was used by Einstein, although a detailed justification has to wait for Onsager and Kubo. In this section, we discuss the simpler approach specifically used in the context of Langevin equation.

The first fluctuation-dissipation theorem gives the relation between mobility and friction as already discussed in the context of Einstein's derivation. In this section, first, we provide a general solution of the Langevin equation [Eq. (15.32)]. Second, we aim to obtain an expression for the phenomenological constant q that is introduced as a measure of the strength of the fluctuating force.

We substitute $v(t) = e^{-\gamma t} y(t)$ in Eq. (15.32) and obtain the following equation:

$$\frac{dy(t)}{dt} = e^{\gamma t}\Gamma(t). \tag{15.33}$$

One can write the solution as

$$y(t) = v_0 + \int\limits_0^t dt'\; e^{\gamma t'}\Gamma(t'), \tag{15.34}$$

where v_0 is the value of the function $y(t)$ at $t = 0$. If we substitute $y(t)$ in terms of $v(t)$ in Eq. (15.34), we obtain

$$\boxed{v(t) = v_0 e^{-\gamma t} + \int\limits_0^t dt'\; e^{-\gamma(t-t')}\Gamma(t').} \tag{15.35}$$

In Eq. (15.35), the first term on the right-hand side shows the exponential decay whereas the second term provides an extra velocity produced because of the random noise term. We use this equation to obtain an expression for the mean squared velocity of a Brownian particle as follows. From Eq. (15.35) one can write

$$v(t)^2 = \underbrace{e^{-2\gamma t}v_0^2}_{term-1} + \underbrace{2v_0 e^{-\gamma t}\int\limits_0^t dt'\; e^{-\gamma(t-t')}\Gamma(t')}_{term-2} + \underbrace{\int\limits_0^t dt'\; e^{-\gamma(t-t')}\Gamma(t')\int\limits_0^t dt''\; e^{-\gamma(t-t'')}\Gamma(t'')}_{term-3}. \tag{15.36}$$

Now, the resultant expression possesses three terms named as term-1, 2, and 3 for convenience. Term-1 decays to zero in the long-time limit. Term-2 also becomes zero when we perform average over the noise term on both sides (property of the random noise). The only term that survives is term-3. We noise average term-3 and further analyze below.

$$\left\langle \int_0^t dt'\, e^{-\gamma(t-t')}\Gamma(t') \int_0^t dt''\, e^{-\gamma(t-t'')}\Gamma(t'') \right\rangle_{Noise}$$

$$= \int_0^t dt'\, e^{-\gamma(t-t')} \int_0^t dt''\, e^{-\gamma(t-t'')} \left\langle \Gamma(t')\Gamma(t'') \right\rangle \qquad (15.37)$$

$$= \int_0^t dt'\, e^{-\gamma(t-t')} \int_0^t dt''\, e^{-\gamma(t-t'')} q\delta(t'-t'') \text{ [by using property (ii) of } \Gamma(t)]$$

$$= \frac{q}{2\gamma}\left(1-e^{-2\gamma t}\right) \text{ [by using properties of delta function integrals]}$$

Hence, the expression for the mean squared velocity of a Brownian particle in the long-time limit becomes

$$\left\langle v(t)^2 \right\rangle = \frac{q}{2\gamma}. \qquad (15.38)$$

From the principle of equipartition of energy, one can write

$$\frac{1}{2}m\left\langle v(t)^2 \right\rangle = \frac{1}{2}k_B T$$

$$or, \ m\frac{q}{2\gamma} = k_B T. \qquad (15.39)$$

$$or, \ \boxed{q = \frac{2\gamma k_B T}{m}}$$

This final result is an example of a *fluctuation-dissipation theorem* that relates the strength (q) of the fluctuating force (random kicks from the solvent molecules) to the friction *(γ)* or dissipation. This also expresses the balance between the friction and noise that is required for a thermal equilibrium state at long times.

15.6 MEAN SQUARED DISPLACEMENT FROM FLUCTUATION-DISSIPATION THEOREM (FDT)

An important characteristic of the general solution of the Langevin equation is that it provides both the short- and the long-time behavior of the mean squared displacement of a Brownian particle. We express the displacement as

$$\Delta x(t) = \int_0^t dt'\, v(t'). \qquad (15.40)$$

We now substitute the expression for $v(t)$ from Eq. (15.35) and square both sides of Eq. (15.40).

$$
\begin{aligned}
\langle \Delta x(t)^2 \rangle &= \left\langle \int_0^t dt' v(t') \int_0^t dt'' v(t'') \right\rangle \\
&= \int_0^t dt' \int_0^t dt'' \langle v(t') v(t'') \rangle \\
&= \int_0^t dt' \int_0^t dt'' \left[\langle v_0^2 \rangle e^{-\gamma(t'+t'')} + \frac{q}{2\gamma} \left\{ e^{-\gamma(t'-t'')} - e^{-\gamma(t'+t'')} \right\} \right].
\end{aligned}
\tag{15.41}
$$

These exponential integrals are easily solvable and one obtains

$$
\boxed{ \langle \Delta x(t)^2 \rangle = \left(\langle v_0^2 \rangle - \frac{q}{2\gamma} \right) \frac{\left(1 - e^{-\gamma t}\right)^2}{\gamma^2} + \frac{q}{\gamma^2} t - \frac{q}{\gamma^3} \left(1 - e^{-\gamma t}\right). }
\tag{15.42}
$$

Eq. (15.42) is the general expression for the mean squared displacement (MSD). One can extract both short-time and long-time behavior of MSD from this equation. If one expands the exponential terms in the limit $t \to 0$, one finds a quadratic increase of MSD with time (t). This is the region where the effects of noise are not dominant and the motion is ballistic. However, in the long-time limit, we obtain the following expression:

$$
\langle \Delta x(t)^2 \rangle = \frac{q}{\gamma^2} t \quad \text{[by neglecting the constant term].}
\tag{15.43}
$$

We know that $\langle \Delta x(t)^2 \rangle = 2Dt$ [Eq. (15.19)]. We combine these two expressions to obtain

$$
2Dt = \frac{q}{\gamma^2} t = \frac{2 k_B T \gamma}{m} \frac{t}{\gamma^2} = \left(\frac{2 k_B T}{m \gamma} \right) t
$$

$$
\boxed{ \therefore D = \left(\frac{k_B T}{m \gamma} \right) = \left(\frac{k_B T}{\zeta} \right). }
\tag{15.44}
$$

This is the well-known Einstein's expression for self-diffusion. When we use Stokes' expression in terms of viscosity to determine ζ, this is known as Stokes-Einstein formula. In Chapter 10 we discuss how one obtains diffusion coefficient from velocity autocorrelation function – famously known as the Green-Kubo relation.

15.7 SOLUTION OF LANGEVIN'S EQUATION FOR HARMONICALLY BOUND PARTICLE

In section 15.4 we have described the Langevin equation for a *free Brownian particle* [Eq. (15.32)], where the force is divided into two parts – deterministic (frictional) and nondeterministic (random kicks from solvent). However, in many applications, particles are under the influence of other external potentials. In this section, we discuss and solve the equation of motion for a harmonically bound Brownian particle. For this system, the Langevin equation in one dimension is as follows:

$$
\frac{dv}{dt} = -\omega^2 x - \gamma v + \Gamma(t)
\tag{15.45}
$$

where ω is the frequency of the harmonic potential. We rewrite Eq. (15.45) in terms of the position of the particle x.

$$\frac{d^2x}{dt^2} + \gamma\frac{dx}{dt} + \omega^2 x = \Gamma(t) \tag{15.46}$$

We aim to solve this second-order stochastic differential equation [Eq. (15.46)] and obtain the probability distribution, $P(x, t \mid x_0, v_0)$ where x_0 and v_0 are the initial position and initial velocity of the Brownian particle. We can guess a solution for x as the linear combination of two exponential functions,

$$x = a_1 e^{\mu_1 t} + a_2 e^{\mu_2 t}, \tag{15.47}$$

where μ_1 and μ_2 can be described as the roots of the following quadratic equation:

$$\mu^2 + \gamma\mu + \omega^2 = 0$$
$$\text{hence, } \mu = \frac{1}{2}\left[-\gamma \pm \sqrt{\gamma^2 - 4\omega^2}\right], \tag{15.48}$$

a_1 and a_2 are time-dependent quantities that satisfy the following equation:

$$e^{\mu_1 t}\left(\frac{da_1}{dt}\right) + e^{\mu_2 t}\left(\frac{da_2}{dt}\right) = 0. \tag{15.49}$$

From Eq. (15.46), one can further derive

$$\mu_1 e^{\mu_1 t}\left(\frac{da_1}{dt}\right) + \mu_2 e^{\mu_2 t}\left(\frac{da_2}{dt}\right) = \Gamma(t). \tag{15.50}$$

We next solve Eqs. (15.49) and (15.50) for a_1 and a_2. We substitute $e^{\mu_2 t}\left(\dfrac{da_2}{dt}\right)$ from Eq. (15.49) in Eq. (15.50) to obtain

$$\mu_1\left(\frac{da_1}{dt}\right) - \mu_2\left(\frac{da_1}{dt}\right) = \Gamma(t)e^{-\mu_1 t} \tag{15.51}$$

$$\text{or, } a_1 = a_1(0) + \frac{1}{\mu_1 - \mu_2}\int_0^t ds\,\Gamma(s)e^{-\mu_1 s}. \tag{15.52}$$

Here, $a_1(0)$ is a constant term that can be calculated from the initial conditions: at $t = 0$; $x = x_0$ and $v = v_0$ and s is a dummy variable for time. From Eq. (15.51) we obtain

$$a_1(0) = -\frac{x_0\mu_2 - v_0}{\mu_1 - \mu_2}. \tag{15.53}$$

Similarly, the solution for a_2 becomes

$$a_2 = a_2(0) - \frac{1}{\mu_1 - \mu_2}\int_0^t ds\,\Gamma(s)e^{-\mu_2 s} \tag{15.54}$$

and

$$a_2(0) = \frac{x_0 \mu_1 - v_0}{\mu_1 - \mu_2}.$$ (15.55)

Hence, Eq. (15.47) becomes

$$x = \frac{1}{\mu_1 - \mu_2} \left\{ e^{\mu_1 t} \int_0^t ds \ e^{-\mu_1 s} \Gamma(s) - e^{\mu_2 t} \int_0^t ds \ e^{-\mu_2 s} \Gamma(s) \right\} + a_1(0) e^{\mu_1 t} + a_2(0) e^{\mu_2 t}.$$ (15.56)

This gives the time-dependent position in terms of fluctuating force. We perform time derivative on both sides of Eq. (15.56) in order to obtain an expression for velocity.

$$v = \frac{1}{\mu_1 - \mu_2} \left\{ \mu_1 e^{\mu_1 t} \int_0^t ds \ e^{-\mu_1 s} \Gamma(s) - \mu_2 e^{\mu_2 t} \int_0^t ds \ e^{-\mu_2 s} \Gamma(s) \right\} + \mu_1 a_1(0) e^{\mu_1 t} + \mu_2 a_2(0) e^{\mu_2 t}.$$ (15.57)

We note that the contribution that arises from the harmonic potential is embedded in μ_1 and μ_2. Our ultimate aim is to obtain a solution for the distribution function. The next part becomes quite rigorous in terms of mathematics and borrowed from Chandrasekhar's classic 1943 article. We present them here because of their elegance and generality. We introduce two lemmas without providing proofs. We guide interested students to refer [6] for the detailed proofs.

Lemma-I:
If $R = \int_0^t ds \ \psi(s)\Gamma(s)$, then the probability distribution of variable $\mathbf{R}$ can be expressed as

$$P(\mathbf{R}) = \left[4\pi q \int_0^t ds \ \psi(s)^2 \right]^{-1/2} \exp \left[-\left(\frac{|\mathbf{R}|^2}{4q \int_0^t ds \ \psi(s)^2} \right) \right]$$ (15.58)

Lemma-II:
If $R = \int_0^t ds \ \psi(s)\Gamma(s)$ and $S = \int_0^t ds \ \phi(s)\Gamma(s)$, then the joint probability distribution can be expressed as

$$P(\mathbf{R},\mathbf{S}) = \left[8\pi^3 \left(FG - H \right)^{-1/2} \right] \exp \left[-\left(\frac{G|\mathbf{R}|^2 - 2H\mathbf{R}.\mathbf{S} + F|\mathbf{S}|^2}{2\left(FG - H \right)^{1/2}} \right) \right].$$ (15.59)

where $F = 2q \int_0^t ds \ \psi(s)^2$; $G = 2q \int_0^t ds \phi(s)^2$ and $H = 2q \int_0^t ds \ \psi(s)\phi(s)$. ($q$ is the strength of fluctuating force that is obtained from the flu ctuation-dissipation theorem.)

Now, we substitute the values of $a_1(0)$ and $a_2(0)$ in Eqs. (15.56) and (15.57) to obtain

$$x = -\frac{1}{\mu_1 - \mu_2} \left[\left(x_0 \mu_2 - v_0 \right) e^{\mu_1 t} - \left(x_0 \mu_1 - v_0 \right) e^{\mu_2 t} \right] + \int_0^t ds \ \psi(s)\Gamma(s)$$ (15.60)

$$and, \ v = -\frac{1}{\mu_1 - \mu_2} \left[\mu_1 \left(x_0 \mu_2 - v_0 \right) e^{\mu_1 t} - \mu_2 \left(x_0 \mu_1 - v_0 \right) e^{\mu_2 t} \right] + \int_0^t ds \phi(s)\Gamma(s).$$ (15.61)

where $\psi(s)$ and $\phi(s)$ are defined as

$$\psi(s) = \frac{1}{\mu_1 - \mu_2}\left[e^{-\mu_1(t-s)} - e^{-\mu_2(t-s)}\right]$$

$$\phi(s) = \frac{1}{\mu_1 - \mu_2}\left[\mu_1 e^{-\mu_1(t-s)} - \mu_2 e^{-\mu_2(t-s)}\right]. \tag{15.62}$$

Now we substitute the explicit values of μ_1 and μ_2 from Eq. (15.48) in Eqs. (15.60)–(15.62) and evaluate three required integrals namely, $\int_0^t ds\,\psi(s)^2$, $\int_0^t ds\,\phi(s)^2$, and $\int_0^t ds\,\psi(s)\phi(s)$. The expressions, after one solves the integrals, become

$$\int_0^t ds\,\psi(s)^2 = \frac{1}{2\omega^2\gamma} - \frac{e^{-\gamma t}}{2\omega^2\beta^2\gamma}\left(2\gamma^2\sinh^2\left(\tfrac{\beta t}{2}\right) + \beta\gamma\sinh\left(\beta t\right) + \beta^2\right) \tag{15.63}$$

$$\int_0^t ds\,\phi(s)^2 = \frac{1}{2\gamma} - \frac{e^{-\gamma t}}{2\beta^2\gamma}\left(2\gamma^2\sinh^2\left(\tfrac{\beta t}{2}\right) - \beta\gamma\sinh\left(\beta t\right) + \beta^2\right) \tag{15.64}$$

$$\int_0^t ds\,\psi(s)\phi(s) = \frac{2e^{-\gamma t}}{\beta^2}\sinh^2\left(\tfrac{\beta t}{2}\right); \text{ here } \beta = \sqrt{\gamma^2 - 4\omega^2}. \tag{15.65}$$

The same substitutions in Eq. (15.60) and (15.61) leads to

$$x - x_0 e^{-\gamma t/2}\cosh\left(\tfrac{\beta t}{2}\right) - \frac{x_0\gamma + 2v_0}{\beta}e^{-\gamma t/2}\sinh\left(\tfrac{\beta t}{2}\right) = \int_0^t ds\,\psi(s)\Gamma(s) \tag{15.66}$$

$$v - v_0 e^{-\gamma t/2}\cosh\left(\tfrac{\beta t}{2}\right) - \frac{2x_0\omega^2 + \gamma v_0}{\beta}e^{-\gamma t/2}\sinh\left(\tfrac{\beta t}{2}\right) = \int_0^t ds\,\phi(s)\Gamma(s). \tag{15.67}$$

One uses *lemma-I* and *II* and substitute Eqs. (15.63)–(15.67) to obtain three useful distribution functions namely $P(x,t\,|\,x_0,v_0)$, $P(v,t\,|\,x_0,v_0)$, and $P(x,v,t\,|\,x_0,v_0)$.

$$P(x,t\,|\,x_0,v_0) = \left[\frac{m}{4\pi k_B T \int_0^t ds\,\psi(s)^2}\right]^{1/2}\exp\left[-\frac{\left\{x - x_0 e^{-\gamma t/2}\cosh\left(\tfrac{\beta t}{2}\right) - \frac{x_0\gamma + 2v_0}{\beta}e^{-\gamma t/2}\sinh\left(\tfrac{\beta t}{2}\right)\right\}^2}{\frac{2k_B T}{m\omega^2}\left\{1 - e^{-\gamma t}\left(\frac{2\gamma^2}{\beta^2}\sinh^2\left(\tfrac{\gamma t}{2}\right) + \frac{\gamma}{\beta}\sinh\left(\beta t\right) + 1\right)\right\}}\right] \tag{15.68}$$

$$P(v,t\,|\,x_0,v_0) = \left[\frac{m}{4\pi k_B T \int_0^t ds\,\phi(s)^2}\right]^{1/2}\exp\left[-\frac{\left\{v - v_0 e^{-\gamma t/2}\cosh\left(\tfrac{\beta t}{2}\right) - \frac{2x_0\omega^2 + \gamma v_0}{\beta}e^{-\gamma t/2}\sinh\left(\tfrac{\beta t}{2}\right)\right\}^2}{\frac{2k_B T}{m}\left\{1 - e^{-\gamma t}\left(\frac{2\gamma^2}{\beta^2}\sinh^2\left(\tfrac{\gamma t}{2}\right) - \frac{\gamma}{\beta}\sinh\left(\beta t\right) + 1\right)\right\}}\right] \tag{15.69}$$

In a similar fashion, by using Eq. (15.59), one can write down the expression for $P(x,v,t\,|\,x_0,v_0)$. If we compare Eq. (15.68) and Eq. (15.69) with general Gaussian forms, we find that the mean and standard deviation become dependent on time and the frequency of the harmonic oscillator.

However, we note that for all kinds of external potential analytic solution might not be possible. In those cases, one needs to employ molecular simulations. We discuss a few numerical techniques in Chapter 19.

15.8 GENERALIZED LANGEVIN EQUATION

The Langevin equation described above is called an ordinary Langevin equation. The random force that drives the Brownian motion has no memory of the immediate past. This is expressed mathematically by a delta function and is called either delta correlated noise or white noise. The resulting Langevin equation description is called Markovian. We know from Doob's theorem discussed in Chapter 2 (Probability Theory), a Gaussian Markov process gives rise to an exponential decay of the time correlation theorem. This is a surprisingly powerful theorem. In the case of the ordinary Langevin equation with Gaussian white noise, the velocity TCF is exponential, in agreement with Doob's theorem.

In many cases, however, the neglect of memory in the noise term is unacceptable. In such processes, the timescale separation assumed between the noise fluctuations and the variable of concern does not exist. Therefore, we need to include correlations between the fluctuating force term.

An example is the attempted use of ordinary Langevin equation description in activated barrier crossing dynamics. When the barrier height is large and sharp resulting in large barrier curvature, the reactant stays in the barrier top for a short time and the temporal correlation between noise is probed. This in a nutshell is the basis of a beautiful theory by Grote and Hynes, and has been discussed in great detail in Chapter 20. In the case of the generalized Langevin equation description of dynamics, Eq. (15.32) is replaced by equations with memory

$$\boxed{\begin{array}{l} m\dot{v} = -\int dt'\, \zeta(t-t')v(t') + \delta F(t) \\ < \delta F(0)\delta F(t) > = dk_B T \zeta(t) \end{array}}$$

(15.70)

where $\zeta(t)$ is the time-dependent friction and d is the dimension of the system. The second of the relations in the above equation is the fluctuation-dissipation theorem. This extended relation is sometimes referred to as the second fluctuation-dissipation theorem. The generalized Langevin equation has found wide use not just in chemical reaction dynamics, but also in vibrational relaxation and protein conformational dynamics.

15.9 AN EXACT DERIVATION OF THE GENERALIZED LANGEVIN EQUATION FROM A HAMILTONIAN USING A HARMONIC OSCILLATOR MODEL

The generalized Langevin equation is a phenomenological equation, with the value of the friction left undetermined. This is not quite a comfortable situation. Subsequent developments in this area closely followed the lead by Robert Zwanzig. who performed two seminal studies. First, Zwanzig provided a derivation of GLE by using the projection operator technique that we discussed in Chapters 12 and 13. In the latter, Zwanzig proposed the scheme to eliminate the irrevelant (or, the bath) degrees of freedom to obtain a formally exact, but somewhat intractactable equation of motion that has the same form as the GLE with an explicit expression for the frequency dependent friction.

Zwanzig realized the difficulty of implementing the exact equation. Subsequently, an exact derivation was presented of GLE by using an explicit and simpler model that allowed an explicit, and relatively easy, elimination of the bath degrees of freedom in lieu of a frequency-dependent friction. The bath was modeled as a system of oscillators linearly coupled to the system degrees of freedom and these oscillators are characterized by a density of states. The frequency-dependent friction is expressed as a function of the coupling elements and the density of states. This model has been used in barrier crossing dynamics.

Essentially, the same model but with a quantum degree of freedom coupled to a system of harmonic oscillators was studied by Caldeira and Legett and is often referred to in the physics literature as the Caldeira-Leggett Hamiltonian. Because of the earlier work of the same Hamiltonian by Zwanzig, we shall refer to it as the Zwanzig-Caldeira-Leggett (ZCL) model. Despite its apparent artificiality and unrealistic nature, the ZCL model has find several useful applications, a few are listed below.

(i)   Derivation of classical generalized Langevin equation (GLE) with an exact expression of the frequency dependent friction. This fulfils the basic premise of the projection operator technique.

(ii)   It provides a quantum analog of the Langevin equation and leads to a semi-classical Fokker-Planck equation that finds applications in spectroscopy.

(iii)   Forms the backbone of Kubo's stochastic Liouville equation discussed briefly later in this chapter.

The Zwanzig-Caldeira-Leggett (ZCL) Hamiltonian is given as

$$H = H_S + H_B + H_I + H_A \tag{15.71}$$

with

$$H_S = \frac{1}{2}M\dot{x}^2 + V(x); H_B = \sum_k \frac{1}{2}m_k\dot{q}_k^2 + \sum_k \frac{1}{2}m_k\omega^2_k q^2_k; H_I = x\sum_k C_k q_k. \tag{15.72}$$

And an auxiliary term

$$H_A = -x^2 \sum \frac{1}{2}\frac{C_k^2}{m_k\omega_k^2}, \tag{15.73}$$

which is introduced to facilitate future simplification. Here x is the system coordinate, and q_k is the bath coordinates.

Next, we obtain the equation of motion (from the Newton or Euler-Lagrange approach) for the coupled system and bath coordinates,

$$M\ddot{x} = -V'(x) + \sum_k C_k q_k - x\sum_k \frac{C_k^2}{m_k\omega_k^2} \tag{15.74}$$

$$m_k\ddot{q}_k = -m_k\omega^2_k q_k + C_k x. \tag{15.75}$$

The strategy now is to eliminate the bath degrees of freedom and obtain a closed form expression for the equation of the system degree of freedom, x. The basic philosophy of the following procedure is simple – it involves elimination of q_k in favor of $x(t)$, which is possible only when x and q_k are linearly coupled. Towards this goal, we Laplace transform Eq. (15.75) to obtain

$$\tilde{q}_k(s) = \frac{\dot{q}_k(0)}{s^2 + \omega^2_k} + \frac{sq_k(0)}{s^2 + \omega^2_k} + \frac{C_k\tilde{x}(s)}{m_k(s^2 + \omega_k^2)} \tag{15.76}$$

where s is the Laplace variable conjugate to time t, and $\tilde{q}_k(s)$ is the Laplace transformed bath coordinates. One next performs inverse Laplace transform of the above equations by introducing the Bromwich path to obtain the following equation of motion:

$$M\ddot{x} + V'(x) + x \sum_k \frac{C_k^2}{m_k \omega_k^2} = \frac{1}{2\pi i} \int_{\varepsilon - i\infty}^{\varepsilon + i\infty} \sum_k C_k \left\{ \frac{\dot{q}_k(0)}{s^2 + \omega_k^2} + \frac{s q_k(0)}{s^2 + \omega_k^2} \right\}$$

$$e^{st} ds + \sum_k \frac{C_k^2}{m_k} \frac{1}{2\pi i} \int_{\varepsilon - i\infty}^{\varepsilon + i\infty} \frac{\tilde{x}(s)}{s^2 + \omega_k^2} e^{st} ds \tag{15.77}$$

We next use the identity

$$\frac{1}{s^2 + \omega_k^2} = \frac{1}{\omega_k^2} \left\{ 1 - \frac{s^2}{s^2 + \omega_k^2} \right\}, \tag{15.78}$$

and rearrange the remaining terms to obtain the following expression:

$$M\ddot{x} + V'(x) + \sum_k \frac{C_k^2}{m_k \omega_k^2} \frac{1}{2\pi i} \int_{\varepsilon - i\infty}^{\varepsilon + i\infty} \frac{s^2 \tilde{x}(s)}{s^2 + \omega_k^2} e^{st} ds = \frac{1}{2\pi i} \int_{\varepsilon - i\infty}^{\varepsilon + i\infty} \sum_k C_k \left\{ \frac{\dot{q}_k(0)}{s^2 + \omega_k^2} + \frac{s q_k(0)}{s^2 + \omega_k^2} \right\} e^{st} ds. \tag{15.79}$$

The rest of the following derivation is rather interesting, but involves a bit of complex analysis. It is shown in the following that the last term on the left-hand side of the above equation leads to the time derivative of the system coordinate x, multiplied the dissipative term in the Langevin equation. The two terms on the right-hand side give the random force term, which is related to the dissipative term by the fluctuation-dissipation theorem.

The last term in the left-hand side of the above equation can be written as

$$\frac{d}{dt} \left\{ \sum_k \frac{C_k^2}{m_k \omega_k^2} \frac{1}{2\pi i} \int_{\varepsilon - i\infty}^{\varepsilon + i\infty} \frac{s \tilde{x}(s)}{s^2 + \omega_k^2} e^{st} ds \right\}. \tag{15.80}$$

We next use the convolution theorem (that the Laplace transform of a convolution integral is equal to the product of the LT of the terms)

$$\frac{d}{dt} \left\{ \sum_k \frac{C_k^2}{m_k \omega_k^2} \int_0^t \cos\left[\omega_k (t - t') \right] x(t') dt' \right\}. \tag{15.81}$$

The above expression suggests the introduction of a spectral function $J(\omega)$ in the following form:

$$J(\omega) = \frac{\pi}{2} \sum_k \frac{C_k^2}{m_k \omega_k} \delta(\omega - \omega_k). \tag{15.82}$$

This introduction finally allows us to express the sum in terms of the spectral function,

$$\sum_k \frac{C_k^2}{m_k \omega_k^2} \cos\left[\omega_k (t - t') \right] = \frac{2}{\pi} \int_0^\infty d\omega \frac{J(\omega)}{\omega} \cos\left[\omega(t - t') \right]. \tag{15.83}$$

Now next we assume a high-frequency cut-off

$$J(\omega) = \begin{cases} \eta\omega \text{ if } \omega{<}\Omega \\ 0 \text{ if } \omega{>}\Omega \end{cases}$$

And use the following simple identity $\dfrac{2}{\pi}\displaystyle\int_0^{\Omega} \eta\cos[\omega(t-t')] = 2\eta\delta(t-t')$ to obtain

$$\frac{d}{dt}\int_0^t 2\eta\delta(t-t')x(t')dt' = \eta\dot{x} + 2\eta\delta(t)x(0). \tag{15.84}$$

We have now achieved the desired simplification of the left-hand side of Eq. (15.79).

We next proceed to evaluate the right-hand side of Eq. (15.79) that contains the initial values of the bath coordinates. Towards this goal, we now evaluate the right-hand side by using the residue theorem to obtain

$$\frac{1}{2\pi i}\int_{\varepsilon-i\infty}^{\varepsilon+i\infty}\sum_k C_k\left\{\frac{\dot{q}_k(0)}{s^2+\omega_k^2} + \frac{sq_k(0)}{s^2+\omega_k^2}\right\}e^{st}ds = \sum_k C_k\left(\dot{q}_k(0)\text{Cos}(\omega_k t) + q_k(0)\omega_k\text{Sin}(\omega_k t)\right).$$

$$\tag{15.85}$$

The right-hand side can now be identified as the random force term $f(t)$. One can now show that the force-force time correlation function is given by

$$\langle f(t)f(t')\rangle = 2\eta k_B T\delta(t-t'). \tag{15.86}$$

With $\langle f(t)\rangle = 0$. Thus finally we get,

$$\boxed{M\ddot{x} + \eta\dot{x} + V'(q) = f(t)}. \tag{15.87}$$

In the derivation of the fluctuation-dissipation theorem, we need to make use of a few relations. They involve averages of each bath harmonic oscillator initially in thermodynamical equilibrium corresponding to the position $x(0)$. If we define $\Delta q_k(0) \equiv q_k(0) - \overline{q}_k(0)$ where

$$\overline{q}_k(0) \equiv \frac{C_k x(0)}{m_k \omega^2{}_k}. \tag{15.88}$$

One can also easily show

$$\langle q_k(0)\rangle = \frac{C_k x(0)}{m_k \omega^2{}_k} \text{ and } \langle \dot{q}_k(0)\rangle = 0 \tag{15.89}$$

$$\langle \dot{q}_k(0)\dot{q}_{k'}(0)\rangle = \frac{k_B T}{m_k}\delta_{kk'}. \tag{15.90}$$

This completes our proof of the Langevin equation for the ZCL model.

When the system is quantum mechanical, like a two-level system, one can derive a semi-classical Fokker-Planck equation.

The interaction of a quantum particle with a bath of harmonic oscillators has been further simplified by Kubo who introduced a quantum stochastic Liouville equation. Here the bath gives rise to stochastic fluctuations in the coupling matrix elements. The advantage is that one can alter the nature of the bath.

15.10   THE GENERALIZED EINSTEIN RELATION

The generalized Langevin equation leads to the generalized Einstein relation between diffusion and friction, both are frequency-dependent. This relation is given by

$$D(z) = \frac{k_B T}{m(z + \zeta(z))}. \qquad (15.91)$$

The above relation can be derived by employing the definition of the self-diffusion constant in terms of the velocity correlation function (discussed in Chapter 10) and the generalized Langevin equation presented above. The generalized Einstein relation finds use in describing intermediate time diffusion in the presence of memory. A dynamic process can be complete before the long-time diffusive limit could set in.

15.11   SUMMARY

In this chapter, we first introduce the then mysterious phenomenon of Brownian motion by providing a brief historical account from both experimental and theoretical viewpoints. We then discuss Einstein's theory of Brownian motion in great detail and derive the expression of mean squared displacement from first principles. This directly leads us to the determination of Avogadro's number from Brownian motion. This is a wonderful way to test the validity of Einstein's elegant theory.

Later, we discuss the Langevin equation, which is a stochastic differential equation of motion for Brownian particles. This equation successfully explains the origin of the observed incessant motion. Starting from the Langevin equation, we derived the expressions for the first fluctuation-dissipation theorem and mean squared displacement (MSD). This helps us to understand the short-time and long-time behavior of MSD.

Another important relationship is the Stokes-Einstein (SE) equation, which is phenomenal in many ways. We provide the derivation of SE relation using two different approaches: (i) from the barometric distribution law and (ii) form the first fluctuation-dissipation theorem. This beautiful equation relates friction with self-diffusion, which can be directly validated from molecular dynamics simulations. In the final section, following Chandrasekhar's classic review article,[6] we include the derivation of the probability distribution function for a Brownian particle bound in a harmonic potential.

BIBLIOGRAPHY AND SUGGESTED READING

1.   Einstein, A. 1905. Investigations on the theory of the Brownian movement. *Annalen der Physik.*
2.   Perrin, J. B. 1909. "Le Mouvement Brownien et la Réalité Moléculaire" appeared in the *Annales de Chimie et de Physique. 18*, 5–114; English translation, 2010, *Brownian Movement and Molecular Reality*, by Soddy, F. London: Taylor and Francis.
3.   Kubo, R., T. Morikazu, and H. Natsuki. 1991. *Statistical Physics II: Nonequilibrium Statistical Mechanics*. 2nd edition. New York: Springer Science & Business Media.

4. Zwanzig, R. 2001. *Nonequilibrium Statistical Mechanics*. Oxford: Oxford University Press.
5. Pathria, R. K. 1996. *Statistical Mechanics*. Oxford: Butterworth-Heinemann.
6. Chandrasekhar, S. 1943. Stochastic problems in physics and astronomy. *Reviews of Modern Physics, 15*, 1.
7. Brown, R. 1828. A brief account of microscopical observations on the particles contained in the pollen of plants and the general existence of active molecules in organic and inorganic bodies. *Edinburgh New Philosophical Journal, XXVII*, 358–371.

16 Random Walks

OVERVIEW

Random walk (RW) models provide a reduce description of dynamics where we consider only the motion of a tagged particle, and ignore the solvent molecules. Random walk models are widely used, along with the Brownian motion and Langevin equation, to describe dynamical processes in physical sciences and biology. It is in fact a well-known branch of mathematics and forms an essential branch of time-dependent statistical mechanics. For example, diffusion is intimately connected with random walk, so is the Langevin equation of motion. In this chapter, we first develop the concept and formalism of regular random walk, which is discrete in time and space. Then we address the effect of boundaries on the probability distribution function by following Chandrasekhar's solutions for absorptive and reflective boundary conditions. Subsequently, we address the generalization of discrete RW to include continuous time random walk. This formalism was developed by Montroll *et al.* and it greatly extends the reach of the RW models in treating realistic problems. We discuss several important applications of the RW models.

16.1 INTRODUCTION

Historically, the term "random walk" was first introduced, actually quite late (by the standard of the antiquity of mathematics), by Karl Pearson in 1905.[2] The underlying concepts closely resemble Einstein's treatment of the Brownian motion.[3] In fact, Brownian motion can be regarded as a random walk that is continuous in space and time. The derivation of the diffusion equation will be presented below. Associated discussions can be found in Chandrasekhar's well-known review article titled "Stochastic processes in physics and astronomy".[4] We should note that random walk was known earlier in the mathematics literature as the "gambler's ruin" problem.

We should remember that along with Brownian motion and Langevin equation, such reduced descriptions (as RW) ignore the molecularity of solvent molecules and focus is kept on the walk of a tagged molecule. Such descriptions are simple, elegant, and also have wide applicability; but depend on parameters (such as rate of hopping, and/or friction) as the input parameters, which need to be supplied from outside, and that absorbs the details of solute-solvent interactions. RW models are of value in modeling unusual phenomena like anomalous diffusion, and motion in a rugged energy landscape. Random walk has received attention from solid-state physicists as a model of transport and also popular in chemical reaction dynamics where reactants are assumed to execute a random walk over the reaction activation barrier, in the presence of random collisions due to viscosity.[5–7] Random walk models are used to describe configurations of a polymer chain where bonds from one carbon to another form a random walk in space due to the rotational degree of freedom.

DOI: 10.1201/9781003157601-19

Random walk is also closely related to the central limit theorem (CLT), which states that the probability distribution of the sum of weakly correlated random variables forms a Gaussian distribution. We have discussed the CLT in Chapter 2.

Because of the transparency of the physical picture behind the random walk, it has proven easy to extend the applicability to describe a wide range of phenomena, discussed later. This has given rise to many interesting variations of random walks. We list a few below, with brief explanations.

(i) Biased RW. This is often used in solid-state physics to describe diffusive transport in the presence of an external field on the charge particle (usually an electron or a polaron).

(ii) RW on a rugged surface. This has become popular in the recent times in biology, to describe such processes as protein folding, diffusion of proteins along a DNA.

(iii) Self-interacting RW. Polymer conformation is a good example of self-interacting random walk where it is named self-avoiding random walk, introduced first by Paul Flory.

(iv) Multiple interacting random walkers. A good example is provided by current conducting electrons in solids. The random walkers might interact and this introduces the model of interacting random walk.

(v) Continuous time RW (CTRW). This is perhaps the most recent and most advanced topic of RW. Historically, the random walk was considered a hopping between discrete sites at regular intervals. However, in most applications hopping from one site to another can occur at any point in time, with a waiting time distribution. This generalization has been termed as "continuous time random walk" or, CTRW. Another generalization is provided by "Levy flight" when the step sizes also follow a distribution.

We shall discuss them all in the subsequent sections. As can be appreciated from the above discussions, the theory of random walk is so general that one can use it to describe many natural phenomena in physics, chemistry, biology, materials science. It finds use to describe, for example, diffusion, protein folding, Brownian motion, dynamics of polymers, population dynamics, motility, etc., to name a few. It finds use even in economics and social sciences. It also finds extensive use in stock markets where it is studied as a part of "mathematical finance".

In this chapter, we first discuss the simple model of one-dimensional RW followed by RW in the presence of boundary conditions and finally continuous time RW with some applications.

16.2 RANDOM WALK IN ONE DIMENSION

For simplicity, let us first consider a particle/walker that moves along a straight line (by hopping) by steps of equal lengths. The particle can move to its left or right from the current position with equal probabilities, maybe decided by the toss of an unbiased coin. The one-dimensional chain ranges from $-N$ to $+N$. For simplicity let us assume that the particle starts from $x = 0$ and the step sizes are $x = \pm 1$ unit as shown in Figure 16.1. Below we follow the approach of Chandrasekhar,[4] to derive the probability distribution.

A popular, and widely discussed, an example of such a random walk is as follows. Suppose that a drunk man is walking in night (coming out straight from a tavern) along a one-dimensional path. He is confused (drunk) and takes steps in random directions. Sometimes, he moves right, sometimes left, and retains no memory of the last step. Let us now consider the random walker takes N number of steps. Our aim is to find the probability, $P(m, N)$, that the walker arrives at $x = m$ after N random displacements. As stated earlier, the probabilities of displacements along $+x$ or $-x$ direction are equal both equal to $1/2$; and the random walker does not possess any memory of its previous position/ direction. In this scenario, the probability of any given sequence of N steps is $(1/2)^N$. Therefore, the probability $P(m, N)$ is obtained by dividing the number of paths arrive at $x = m$ after N random displacements with the total number of possible paths 2^N.

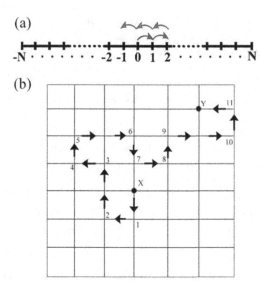

FIGURE 16.1 (a) Schematic representation of one-dimensional random walk with integral step size and integral chain length. A particle/random walker starts from the origin $x = 0$ and exhibits random displacements on either its right or left. In this figure, the particle suffers five random displacements and ends up at $x = -1$. (b) Schematic representation of two-dimensional random walks where the random walker starts from point X and reaches point Y via the sequential pathway marked as 1, 2, 3 …, 11.

$$P(m, N) = \frac{N!}{N_+! N_-!} \left(\frac{1}{2}\right)^N. \tag{16.1}$$

Here, N_+ and N_- are steps taken in the +ve and –ve directions, respectively. Since the walker (our drunkard man or a tagged particle) arrives at $x = m$ after N displacements then N_+ and N_- must obey the following two conditions [Eq. (16.2)].

$$N_+ + N_- = N$$
$$N_+ - N_- = m. \tag{16.2}$$

We note here that both m and N have to be either odd or even together, according to Eq. (16.2). Hence, Eq. (16.1) can now be rewritten as

$$P(m, N) = \frac{N!}{\left(\frac{N+m}{2}\right)! \left(\frac{N-m}{2}\right)!} \left(\frac{1}{2}\right)^N. \tag{16.3}$$

Eq. (16.3) is without any approximation. However, it becomes inconvenient to work with factorials of large numbers. Hence, Eq. (16.3) is further simplified by assuming $m \ll N$ and N is a large number (both turn out to be good approximations). This is, in practice, an important assumption because in statistical mechanics the volume of the sample space is huge. Hence, we apply Stirling's approximation to simplify the factorial terms. Stirling's approximation is given by Eq. (16.4) where n is a large number.

$$\log n! = \left(n + \tfrac{1}{2}\right)\log n - n + \tfrac{1}{2}\log 2\pi + O(\tfrac{1}{n}).$$ (16.4)

Application of Eq. (16.4) in Eq. (16.3) yields

$$\log P(m,N) \simeq \left(N + \tfrac{1}{2}\right)\log N - \tfrac{1}{2}(N+m+1)\log\left[\tfrac{N}{2}\left(1+\tfrac{m}{N}\right)\right]$$
$$- \tfrac{1}{2}(N-m+1)\log\left[\tfrac{N}{2}\left(1-\tfrac{m}{N}\right)\right].$$ (16.5)

As we have assumed $m \ll N$, Eq. (16.5) can further be simplified by employing a series expansion of the logarithmic terms (with m/N as a smallness parameter) as follows:

$$\log\left(1 \pm \tfrac{m}{N}\right) = \pm\tfrac{m}{N} - \tfrac{m^2}{2N} \pm \ldots \ldots \infty.$$ (16.6)

We use the above series expansion in Eq. (16.5) and retain the first two terms to obtain

$$\log P(m,N) \simeq -\frac{1}{2}\log N + \log 2 - \frac{1}{2}\log 2\pi - \frac{m^2}{2N}.$$ (16.7)

Hence, the probability of the walker to arrive at $x = m$ after N number of jumps acquires a Gaussian form as shown in Eq. (16.8).

$$P(m,N) = \sqrt{\frac{2}{\pi N}}\exp\left(-\frac{m^2}{2N}\right).$$ (16.8)

Eq. (16.8) can be regarded as the asymptotic formula. Surprisingly, however, even $N = 10\ or\ 12$ provides a good match with Eq. (16.3).

16.3   RELATIONSHIP OF RW WITH BROWNIAN MOTION AND DIFFUSION EQUATION

It is easy to appreciate the intimate connection between random walks and Brownian motion. Both give rise to the spread of an initially localized density of particles into the surrounding space. The main difference is that Brownian motion is a continuous and random walk, as presented till now, is discrete. There is really no need to assume that random walk is discrete in time, although discreteness in space remains the essence of random walk. In the limit of long length scale, meaning that net displacement (which is the sum of many small displacements) is larger than the size of the tagged particle, the restriction of discrete displacement in space also disappears. We now discuss how a random walk is equivalent to a diffusion equation, and all the three topics are in a broad sense are isomorphic.

In the chapter on Brownian motion, we have already derived and solved the diffusion equation. We have discussed in the previous section (section 16.2) the general problem of discrete space random walk and we discussed the resulting probability distribution in detail [Eq. (16.8)]. Below we relate the probability distribution of random walk (discrete in space) to the macroscopic phenomena of diffusion.

Let a be the step size of the walker and x be the net displacement such that $x = ma$ after m steps. Next, we aim to express the probability distribution in terms of x. If a is a real number (that is, not restricted to integers) we need to ask the following question: what is the probability of finding the walker between the interval x to $x+dx$ after N displacements? Here, N can be thought of as the time or a quantity directly proportional to time. We next use the gain-loss master equation for the change of probability $P(x, t)$ to obtain the following simple equation:

$$P(x,t+\Delta t)-P(x,t)=\frac{1}{2}k\Delta t\left[P(x-a,t)+P(x+a,t)-2P(x,t)\right],$$

which leads, in the limit of small Δt, to   (16.9)

$$\frac{\partial}{\partial t}P(x,t)=\frac{1}{2}k\left[P(x-a,t)+P(x+a,t)-2P(x,t)\right]$$

where, k is the rate constant associated with the step of size a at time t, and the factor $(1/2)$ accounts for the probability that during the time τ the particle can jump either to the left or right. Note that a RW continuous in time requires the use of a rate constant. Now, we employ the Taylor series expansion and retain up to the first two terms for $P(x-a,t)$ and $P(x+a,t)$ to obtain

$$\frac{\partial}{\partial t}P(x,t)=\frac{1}{2}ka^2\frac{\partial^2}{\partial x^2}P(x,t).$$ (16.10)

By comparing Eq. (16.10) with the diffusion equation as obtained from the equation of continuity and Fick's law, we can write $D=\frac{1}{2}ka^2$. Therefore, we recover the familiar diffusion equation

$$\frac{\partial}{\partial t}P(x,t)=D\frac{\partial^2}{\partial x^2}P(x,t).$$ (16.11)

The above simple derivation is illustrative and is a popular way to relate random walk with diffusion equation.

The solution of Eq. (16.11) is given by

$$\boxed{P(x,t)=\frac{1}{\sqrt{4\pi Dt}}\exp\left(-\frac{x^2}{4Dt}\right).}$$ (16.12)

This approach can be applied to derive the probability distribution in two or three dimensions given that the displacements in X direction do not affect the same in the Y direction. In that case, the probability is multiplicative, i.e., $P(x,y,z\,|\,t)=P(x,t)*P(y,t)*P(z,t)$. One can easily calculate the moments, namely, $\langle x\rangle$ and $\langle x^2\rangle$ of the probability distribution from random walk.

The relation between the diffusion constant and the parameters k and a in the form $D=\frac{1}{2}ka^2$ is a fairly general result. Also, note the relation of this expression to Einstein's formula that related the diffusion coefficient to the mean square displacement.

16.4 BOUNDARY CONDITIONS IN RANDOM WALK PROBLEMS

RW starts at one location in space and ends later at a different location. These conditions define a given problem, and just like in differential equations, serve as the boundary values that need to be satisfied. Fortunately, for several important cases of RW, the problem can be solved exactly, with the result expressed in a closed form. Therefore, it is important to understand the random walk in the presence of constraints and restrictions in order to build physically relevant models. In this section, we discuss two such important boundary conditions that are often used in physics

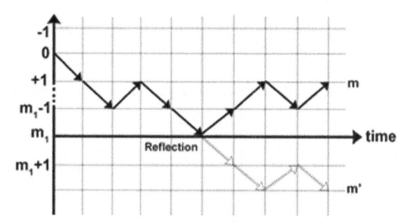

FIGURE 16.2 Schematic representation of the time trajectory of one-dimensional random walk in the presence of a reflecting barrier at m_1. The path leads the walker to a point m. However, in the absence of the reflector, the particle would have reached m' $(=2m_1-m)$, which is the image of m. Additionally, the particle can reach m without a single reflection. Hence, the probability of finding the particle at m after N steps is the sum of probabilities of reflected and unreflected paths.

and chemistry, namely: (a) the reflecting boundary condition and (b) the absorbing boundary condition.

Both these boundary conditions find extensive use in modeling problems of science, especially in chemical kinetics. These have been discussed in the chapter on "Reaction rate theory" (Chapter 20).

16.4.1 Reflecting Boundary Condition

A reflecting boundary restricts and confines the probability distribution of the random walker. In essence, it increases the probability in the confined region, as in a squash court where the ball must remain within the four walls, reflected repeatedly by the walls.

Let us assume that we have a reflecting barrier at $m = m_1$, as shown in Figure 16.2. The property of a reflecting barrier is such that after arriving at m_1 the particle, in the next step, returns to $(m_1 - 1)$ with unit probability (100% reflective wall). We can assume that in the absence of the reflecting barrier the particle would reach its image point $m_1 + 1$. This is demonstrated in Figure 16.2 by constructing a time trajectory of the random walker.

In Figure 16.2 a path is shown, which lead the particle to a point m after N steps by one reflection. In the absence of the reflecting barrier, the particle would have arrived at m', which is the image point of m. The coordinate of m' is given by $(2m_1 - m)$. Hence, the probability of arriving at m after N displacements becomes the sum of the probability of reaching m without any reflection and the probability of reaching the image point m' with reflections. The total probability is given by Eq. (16.13).

$$P(m, N \mid m_1) = P(m, N) + P(m', N)$$
$$= P(m, N) + P(2m_1 - m, N). \tag{16.13}$$

Note the addition sign. We can now use it for the probability of the unconstrained RW. The use of Eq. (16.8) yields,

$$P(m, N \mid m_1) = \sqrt{\frac{2}{\pi N}} \left[\exp\left(-\frac{m^2}{2N} \right) + \exp\left(-\frac{(2m_1 - m)^2}{2N} \right) \right]. \tag{16.14}$$

Let us convert this discrete distribution to a continuous one as demonstrated in section 16.2.

$$P(x, N \mid x_1) = \frac{1}{\sqrt{4\pi Dt}} \left[\exp\left(-\frac{x^2}{4Dt} \right) + \exp\left(-\frac{(2x_1 - x)^2}{4Dt} \right) \right]. \tag{16.15}$$

Mathematically the reflecting boundary condition is given by

$$\left(\frac{\partial P}{\partial x} \right)_{x = x_1} = 0. \tag{16.16}$$

16.4.2 Absorbing Boundary Condition

The absorbing boundary condition is employed when particles do not return after crossing a surface or a line or a point, depending on the dimension. Since particles are removed, this condition leads to a decrease in the population that eventually goes to zero unless supplemented from outside.

The formulation of this problem is similar to that of section 16.4.1, except that the particle after reaching at m_1 does not return. It gets absorbed or annihilated. In the presence of such an absorbing barrier, the probability of reaching m after N displacements decreases. Hence, instead of adding the probability of arriving its image point to $P(m, N)$, we now subtract the same to obtain the net probability, Eq. (16.17).

$$P(m, N \mid m_1) = P(m, N) - P(2m_1 - m, N)$$
$$= \sqrt{\frac{2}{\pi N}} \left[\exp\left(-\frac{m^2}{2N} \right) - \exp\left(-\frac{(2m_1 - m)^2}{2N} \right) \right]. \tag{16.17}$$

Again, by considering a continuous space we can transform variable m into x as,

$$P(x, N \mid x_1) = \frac{1}{\sqrt{4\pi Dt}} \left[\exp\left(-\frac{x^2}{4Dt} \right) - \exp\left(-\frac{(2x_1 - x)^2}{4Dt} \right) \right]. \tag{16.18}$$

Mathematically the absorbing boundary condition is given by,

$$P(x_1, N \mid x_1) = 0. \tag{16.19}$$

There are generalizations of the fully reflective and fully absorbing barriers in the form of partially reflective or partially absorbing barrier. We refer to the book of Feller for further study by the student.

16.5   CONTINUOUS TIME RANDOM WALK (CTRW)

Random walk can make the jump not just at regular intervals, but at any instant of time, which is determined by stochastic interactions with the surrounding medium. Let us consider that the random walker moves on a potential energy landscape with energy valleys, hills, and barriers. In Figure 16.3 schematic descriptions of realistic situations are shown.

The restriction of the "jump only at regular intervals" was removed by Montroll and Weiss[8, 9] who developed the mathematical formalism of the continuous time random walk (CTRW). It was later extended by Scher and Lax in order to explain the anomalous transit time distribution in current observed in photocopy machines.[10] In a discrete time random walk, the steps are taken periodically. In CTRW, in contrast, one assumes that the time taken between any two displacements is random but obeys a distribution. This time is known as the waiting time. Waiting times are assumed to be independent random numbers that follow a common distribution function known as the waiting time distribution (WTD), $\psi(t)$, which then gives the probability of a jump between $t = 0$ and $t = t$.

Suppose, n is the number of jumps/steps taken by the walker in the time interval of $(0, t)$. Thus, n is a random variable. For simplicity, we assume that the walker starts at origin at $t = 0$. We now define $\Psi_n(t)$ as a probability of making n jumps during a time interval $(0, t)$. In this context, we note that the probability of not making a jump in that time interval is given by the remaining or survival probability $\phi(t)$ as follows:

$$\phi(t) = 1 - \int_0^t dt' \, \psi(t') \tag{16.20}$$

where the integrand on the right-hand side gives the probability that a jump has occurred within time t. We now have the following sequence of equalities:

$$\Psi_0(t) = 1 - \int_0^t dt' \, \psi(t') = \int_t^\infty dt' \, \psi(t') = \phi(t)$$

$$\Psi_1(t) = \int_0^t dt' \, \psi(t')\phi(t - t') = \psi * \phi \ (convolution \ product) \tag{16.21}$$

$$\vdots$$

$$\Psi_n(t) = (\psi * \psi * \cdots n \, times) * \phi.$$

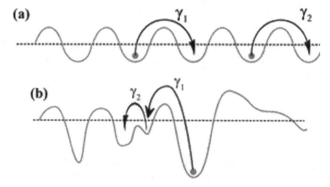

FIGURE 16.3   Schematic description of random walk on an energy landscape. (a) The energy barriers are equal. Hence, one can assume the same transition rate-constant/frequency (Υ) for every energy trap on the X-axis. (b) The energy barriers are not uniform. Accordingly, the transition rates differ for every energy trap, depending on the energy barrier the walker has to overcome. In this figure, for example, $\Upsilon_1 < \Upsilon_2$.

The second of the above series of equations simply state that no jump has occurred in the time interval t-t', followed by a jump afterwards. The rest of the equations follow similarly, and are not difficult to understand.

Since the Laplace transform of a convolution integral involving two functions is the product of the Laplace transformed functions [as shown in Eq. (16.22)], it is convenient to solve this problem in the Laplace plane.

$$L\left[\int_0^t dt'\,\psi(t')\phi(t-t')\right] = \psi(s).\phi(s) \tag{16.22}$$

where "L" denotes Laplace transform. Therefore, the Laplace transformation of the last line of Eq. (16.21) yields

$$L[\Psi_n(t)] = L[(\psi * \psi * \psi * ... n\ \text{terms}) * \phi] \tag{16.23}$$

Note that, in Eqs. (16.21) and (16.23) "*" denote convolution product. Hence, by the application of Eq. (16.22), we can write

$$\begin{aligned}
\Psi_n(s) &= \big(L[\psi].L[\psi].L[\psi]...n\ \text{terms}\big).L[\phi] \\
&= \big(L[\psi]\big)^n .L[\phi] \\
&= \psi^n(s).\phi(s) \\
&= \psi^n(s).L\left[1 - \int_0^t dt'\,\psi(t')\right] \\
&= \psi^n(s).\left[\frac{1}{s} - \frac{1}{s}\psi(s)\right]
\end{aligned} \tag{16.24}$$

$$or,\ \Psi_n(s) = \psi^n(s)\left[\frac{1-\psi(s)}{s}\right]. \tag{16.25}$$

Here, $\phi(s)$ is the probability, in the Laplace plane, for remaining at the original site. This in turn is the Laplace transform of Eq. (16.20),

$$\phi(s) = \frac{1}{s} - \frac{1}{s}\psi(s). \tag{16.26}$$

In a CTRW problem, we consider each step to be associated with a waiting time [that follows a distribution, $\psi(t)$] and displacement Δx. The position of the random walker, $x(t)$, after N jumps in the time interval 0 to t can be written as

$$x(t) = \sum_{n=0}^{N} \Delta x(n). \tag{16.27}$$

We now aim to obtain $P(x, t)$ that is, the probability that the walker arrives at x after executing N jumps in time t. If the spatial and temporal probabilities are separable, one can write

$$P(x,t) = \sum_{n=0}^{\infty} \Psi_n(t).P_n(x). \tag{16.28}$$

where $P_n(x)$ is the probability of arriving at x after n jumps and $\Psi_n(t)$ is the probability of executing N jumps within t. We shall now perform Fourier transform in space with k as the conjugate variable and Laplace transform in time. Since the waiting times and displacements are independent variables, we perform Fourier-Laplace transform of Eq. (16.28) to obtain,

$$P(k,s) = \sum_{n=0}^{\infty} \Psi_n(s) P_n(k),$$ (16.29)

$$or, \quad \boxed{P(k,s) = \frac{1-\psi(s)}{s}\left[\frac{1}{1-\psi(s)\lambda(k)}\right]}$$ (16.30)

where we have used the relation, derived below, $P_n(k) = [\lambda(k)]^n$, with $\lambda(k) = \cos(kl)$.

Eq. (16.29) is known as the Montroll-Weiss formula, which is widely used. Here, we adopt the method of induction to obtain an expression for $P_n(x)$ and hence, $P_n(k)$. If the step sizes are taken to be l, the probability of arriving at x after 1 step is given by $P_1(x)$ as

$$P_1(x) = \frac{1}{2}\left(\delta_{x,l} + \delta_{x,-l}\right).$$ (16.31)

Hence, the probability of arriving at x after 2, 3...n steps can be written as,

$$P_2(x) = \text{probability of arriving at } [x+l \text{ or } x-l] \text{ after 1step} \times P_1(x)$$
$$P_3(x) = \text{probability of arriving at } [x+2l \text{ or } x-2l] \text{ after 1step} \times P_2(x)$$ (16.32)
$$\vdots$$
$$P_n(x) = \text{probability of arriving at } [x+(n-1)l \text{ or } x-(n-1)l] \text{ after 1step} \times P_{n-1}(x).$$

Now we perform discrete Fourier transformations (DFT) on $P_n(x)$. The DFT of $P_1(x)$ can be evaluated as

$$P_1(k) = \frac{1}{2}\sum_x e^{-ikx}\left(\delta_{x,x+l} + \delta_{x,x-l}\right)P_0(x)$$
$$= \frac{1}{2}(e^{-ikl} + e^{ikl})\sum_x e^{-ikx}\delta_{x,0}$$ (16.33)
$$= \cos(kl) = \lambda(k).$$

Similarly, we can write

$$P_n(k) = \sum_x e^{-ikx}P_n(x) = \lambda(k)\sum_x e^{-ikx}P_{n-1}(x) = \lambda^2(k)\sum_x e^{-ikx}P_{n-2}(x) = \cdots = \lambda^n(k).$$ (16.34)

If one substitutes Eq. (16.25) and (16.34) in Eq. (16.29), the Montroll-Weiss formula can be derived by assuming, $\psi(s)\lambda(k) \ll 1$ as follows:

$$P(k,s) = \sum_{n=0}^{\infty} \Psi_n(s) P_n(k)$$
$$= \sum_{n=0}^{\infty} \psi^n(s)\left[\frac{1-\psi(s)}{s}\right].\lambda^n(k)$$
$$= \left[\frac{1-\psi(s)}{s}\right]\sum_{n=0}^{\infty} \psi^n(s).\lambda^n(k)$$ (16.35)
$$= \left[\frac{1-\psi(s)}{s}\right][1-\psi(s)\lambda(k)]^{-1}$$

with $\psi(s)\lambda(k)<1$. Next we consider two different cases: (i) exponential waiting time distribution and (ii) a power-law distribution of waiting times. Our goal is to understand the nature of Brownian motion and diffusion under these two specific conditions.

(i) Case 1: we assume $\psi(t) \sim \gamma \exp(-\gamma t)$. The Laplace transformation of $\psi(t)$ becomes

$$\psi(s) = \frac{\gamma}{s+\gamma}. \tag{16.36}$$

Hence, the remaining probability, according to Eq. (16.26), can be written as

$$\phi(s) = \frac{1}{s+\gamma}. \tag{16.37}$$

So, the probability distribution in the Fourier-Laplace plane, according to Eq. (16.29), is given by

$$P(k,s) = \left(\frac{1}{s+\gamma}\right)\frac{1}{1-\left(\frac{\gamma}{s+\gamma}\right)\cos(kl)} = \frac{1}{s+\gamma[1-\cos(kl)]}. \tag{16.38}$$

The inverse Laplace transform of Eq. (16.38) can be easily carried out to obtain

$$P(k,t) = \exp\left[-\gamma\{1-\cos(kl)\}t\right]. \tag{16.39}$$

Now, a series expansion of $\cos(kl)$ as $1 - \frac{(kl)^2}{2} + O(k^4 l^4)$ results in

$$P(k,t) \approx \exp\left[-\gamma t \frac{k^2 l^2}{2}\right].$$

This indicates a normal diffusion as also shown in the Brownian motion chapter as the solution of the diffusion equation in the Fourier space.

(ii) Case 2: we assume that the distribution is a power law, $\psi(t) \sim \left(\frac{t_0}{t}\right)^\alpha \frac{1}{t}$. The Laplace transformation of $\psi(t)$ becomes

$$\psi(s) = 1 - (t_0 s)^\alpha. \tag{16.40}$$

Hence, the probability of remaining at the original site can be written as

$$\phi(s) = \frac{1-\psi(s)}{s} = \frac{(t_0 s)^\alpha}{s}. \tag{16.41}$$

Now, one can follow the similar procedures as that of case 1 to obtain the probability distribution. That is, we use Eq. (16.34) to obtain $P(k, s)$, and then $P(x, t)$. This WTD leads to nondiffusive dynamics.

Let us now explore the nature of CTRW with the help of another example where the transition probability obeys an exponential distribution $\gamma(E) = \gamma_0 \exp(-E/k_B T)$ where E is the energy of the present state of the walker and γ_0 is the rate constant for the uniform energy landscape. The probability of exhibiting a jump in the time interval of $(0, t)$ is given by $\psi(t)$. Hence, the deeper the energy well, the lesser the transition rate constant [Figure 16.3(b)]. Thus, the form of $\psi(t)$ is exponential as $\psi(t) = \gamma(E)\exp(-\gamma(E)t)$.

In order to incorporate the effect due to degeneracy, suppose, the energy levels are associated with a density of states $g(E)$, which is exponential with a characteristic temperature T_C.

$$g(E) = \frac{1}{k_B T_C}\exp\left(-\frac{E}{k_B T_C}\right). \tag{16.42}$$

The probability to remain at the original site is now given by the average over the density of states. With these descriptions, we can derive an expression for the WTD as follows:

$$\psi(t) = \int_0^\infty dE\, g(E)\,\gamma(E)\exp[-\gamma(E)t]. \tag{16.43}$$

To solve this integral (integrating out E) we substitute $x = \gamma(E)t$. After this substitution, Eq. (16.43) becomes

$$\psi(t) = \left(\frac{T}{T_C}\right)\left(\frac{1}{t\gamma_0}\right)^{\frac{T}{T_C}}\frac{1}{t}\int_0^{t\gamma_0} dx\, x^{\frac{T}{T_C}}\exp(-x). \tag{16.44}$$

The integral in Eq. (16.44) is a lower incomplete gamma function, $\tilde{\gamma}\left(1+\frac{T}{T_C},t\gamma_0\right)$ [Abramowitz & Stegun (1983) Ch-6.5]. By considering $\alpha = T/T_C$ we rewrite Eq. (16.44) as

$$\psi(t) = \alpha\gamma_0^{-\alpha}t^{-(1+\alpha)}\tilde{\gamma}\left(1+\alpha,t\gamma_0\right). \tag{16.45}$$

The evaluation of the lower incomplete gamma function is nontrivial as it involves the hypergeometric functions and an exponential sum. In order to obtain a better insight, we direct the interested readers to [11]. However, the situation simplifies if one is interested in the long-time asymptotic limit $t \to \infty$, Eq. (16.45) can be written as

$$\boxed{\lim_{t\to\infty}\psi(t) = \left[\alpha\gamma_0^{-\alpha}\Gamma(1+\alpha)\right]t^{-(1+\alpha)}}. \tag{16.46}$$

From Eq. (16.46), we obtain power-law decay in the long-time limit. The theory of CTRW was first used to understand the charge transport in solids by Montroll *et al.*

16.5.1 Mittag-Leafller function of waiting time distribution

A popular and widely used form of waiting time distribution is the Mittag-Leafller function $\left(E_\beta(s)\right)$, which is given by

$$\psi_\beta^{ML}(s) = -\frac{d}{dt}E_\beta(-t^\beta)$$
$$\text{where}$$
$$E_\beta(s) = \sum_{n=0}^\infty \frac{s^n}{\Gamma(\beta n + 1)}$$

where $\Gamma(x)$ is a gamma integral and $0 < \beta \le 1$.

The merit of this function is that it interpolates between the exponential and the Gaussian waiting time distribution. It is popular among the practitioner of the fractional Brownian motion community. It can explain anomalous or fractional diffusion where the mean square displacement is sublinear, as witnessed in many natural phenomena.

16.6 APPLICATIONS

As stated in the introduction, the random walk model plays an important role in explaining many phenomena in physics, chemistry, and biology. Below we discuss some of the applications of RW and CTRW without going into mathematical complexities.

16.6.1 MODELING POLYMER CONFIGURATIONS

A polymer is a macromolecule formed by the repetition of several subunits known as monomers that are bonded with each other. The monomers can be similar (as in polythelene) or dissimilar (as in proteins). Interestingly, the random walk model plays an important role in the theoretical description of the conformational dynamics of polymers. There are several models, for example, freely jointed chain (FJC) model, worm-like chain (WLC) model, excluded volume random walk model, etc. In this section, we detail the FJC model as an example.

In the FJC model, the length of a segment of a polymer is assumed to be a random variable and the monomers are assumed to be linear (rod like). Due to the rotational degree of freedom, one of the ends of the above-mentioned segments is uniformly distributed on the surface of spheres whose radii are equal to the lengths of the linear segments (r). Say, $p_n(\underset{\sim}{r})$ is the probability density that the end-to-end distance is r after n steps. This can be expressed as

$$p_n(r) = \frac{1}{(2\pi)^3} \iiint d^3\underset{\sim}{\rho}\, C^n\left(\underset{\sim}{\rho}\right)\exp\left(-i\underset{\sim}{\rho}\underset{\sim}{r}\right) \qquad (16.47)$$

where $C\left(\underset{\sim}{\rho}\right) = \iiint d^3\underset{\sim}{r}\, p(\underset{\sim}{r})\exp\left(i\underset{\sim}{\rho}\underset{\sim}{r}\right)$. As we have previously assumed spherical symmetry around $\underset{\sim}{r}$ we can write

$$C(\rho) = \frac{4\pi}{\rho}\int_0^\infty dr\, r\, p(r)\sin(\rho r) \qquad (16.48)$$

where $p(r) = \frac{1}{4\pi l^2}\delta(r-l)$. $\qquad (16.49)$

Eq. (16.48) and (16.49) imply that

$$C(\rho) = \frac{\sin\rho l}{\rho l}. \qquad (16.50)$$

Hence one can write Eq. (16.47) as

$$p_n(r) = \frac{1}{2\pi^2 r}\int_0^\infty d\rho\, \rho\left(\frac{\sin\rho l}{\rho l}\right)^n \sin(\rho r). \qquad (16.51)$$

Here, l is known as the persistence length beyond which the monomer units become uncorrelated. The integral in Eq. (16.51) can be evaluated for large values of n. By employing the central limit theorem one can show

$$P_n(r) \sim \left(\frac{3}{2\pi n l^2}\right)^{3/2} \exp\left(-\frac{3r^2}{2nl^2}\right). \tag{16.52}$$

Hence, the end-to-end distance distribution of a chain polymer acquires a Gaussian form. This is a well-known model of polymer chemistry.

16.6.2 SWIMMING MICROORGANISMS

The motions of microorganisms (dimensions ~0.1 to 100 μm), for example, bacteria, archaea, and eukaryotes can be modeled in terms of random walks in two or three dimensions. These organisms swim by self-propulsion with the help of bodily extensions, such as flagella or cilia. The motors get power from either by ATP hydrolysis or by maintaining an ion flux through the cell membranes. The random walk that a bacterium exhibits mainly contains linear trajectories with occasional tumbling. Tumbling is a phenomenon where the bacteria chooses a new direction randomly by rotating at the same position. An exemplary trajectory of a bacterium is shown in Figure 16.4.

The motions of microorganisms are in general of many types, for example, chemotaxis, phototaxis, gravitaxis, etc. However, here we shall avoid including the intrinsic details of these movements and rather set up the problem from a general perspective with an external bias. The bias controls a bacterium in the following way. The time spent in a linear segment becomes elongated when a bacterium moves toward the bias (chemical, light, etc.) and shortened while moving away from it.

This is somewhat of a complex problem to be dealt in its entirety, and we shall present some of the main results. Students are advised to follow up the literature for further details.

We assume that the angles (θ) made by two successive linear segments are uniformly distributed between $-\pi$ and $+\pi$. We also assume that the motion along a particular linear segment possesses a constant (but randomly chosen for every linear segment, v_j) speed independent of the direction. The direction is denoted by φ as the angle between a linear segment and a fixed axis. The distribution of the time spent along a linear segment is given by

$$P_\varphi(t) = \frac{1}{T(\varphi)} \exp\left(-\frac{t}{T(\varphi)}\right). \tag{16.53}$$

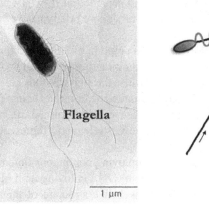
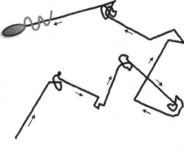

Flagella

1 μm

FIGURE 16.4 (Left) Microscopic view of a bacterium with flagella. (Right) A random walk model representation of a bacterium swimming trajectory that consists of linear segments and tumbling.

where, $T(\varphi)$ is the mean time. Suppose, $n(t)$ is the number of linear segments generated in time t. One can write

$$t = \sum_{i=1}^{n(t)} \tau_i \qquad (16.54)$$

where τ_i is the time taken to traverse the ith linear segment. Note that the tumbling times are ignored in this treatment. If the initial line form θ_0 angle with a fixed axis and the subsequent turn angles are $\theta_1, \theta_2, \ldots, \theta_{n(t)-1}$, then the cumulative angle traversed after $n(t)$ steps is given by

$$\xi_n = \sum_{i=0}^{n(t)-1} \theta_i. \qquad (16.55)$$

Hence, the displacement of the bacterium along x-direction after time t becomes

$$x_n(t) = \sum_{i=1}^{n} v_i \tau_i \cos \xi_i. \qquad (16.56)$$

Let us denote, the average of $x_n(t)$ with respect to v and τ as $E_{v,\tau}\{x_n(t)\}$. The Laplace transformation of this quantity is given by

$$\int_0^{\infty} dt\ e^{-st}\ E_{v,\tau}\{x_n(t)\} = \frac{\bar{v}}{s}\left(1 - F_n\right)\prod_{i=1}^{n-1} F_i \sum_{i=1}^{n} G_i \qquad (16.57)$$

where $F_i = \left[1 + sT(\xi_i)\right]^{-1}$ and $G_i = T(\xi_i)F_i \cos \xi_i$. However, it is impossible to analytically obtain a closed-form solution for $\langle x(t) \rangle$. In the long-time limit, it can be shown that $\langle x(t) \rangle \sim \bar{v}\dfrac{\alpha_1}{\alpha_0}t$, where $\alpha_j = \dfrac{1}{2\pi}\int_{-\pi}^{+\pi} d\varphi\, T(\varphi)\cos(j\varphi)$.

16.6.3 DNA GEL ELECTROPHORESIS

DNA gel electrophoresis is a useful experimental technique routinely used in chemical and biological analyses. It is primarily used to estimate the size and separate/purify DNA molecules (or other charged polymers). Let us understand the experimental procedure briefly before going into the theoretical description.

DNA molecules are negatively charged bio-polymers. In a solution of DNA (in a buffer) there is a distribution of size and net charge of the molecule. Depending on these two factors, the mobility of a particular DNA molecule changes when it is forced to diffuse through a porous gel matrix under the influence of an electric field. The DNAs are generally marked with intercalating dye ethidium bromide, which fluoresces under UV radiation. After the electric field is turned off, one observes distinct or smeared spots on the gel. In Figure 16.5 we show a schematic representation of the experimental setup and a model result.

The phenomena of electrophoresis can be modeled by employing continuous time random walk (CTRW) with a finite drift. This system is similar to a particle (DNA) moving on a rugged energy landscape full of entropic traps (owing to the porous and web-like structure of the gel). Hence, the process follows a waiting time distribution and becomes continuous in time. The longer the DNA (that is, more base pairs), the more is the waiting time in a particular trap. Following the

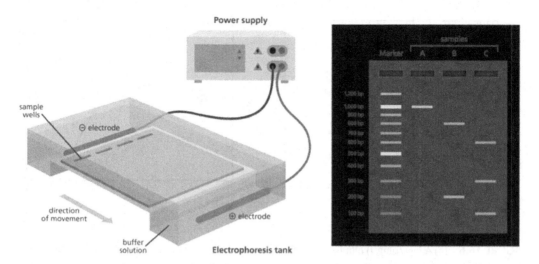

FIGURE 16.5 (Left) Schematic illustration of the gel electrophoresis experiment. The samples are put into the sample wells with a marker sample ladder. The ladder contains DNA molecules of several base pair lengths, which serve as an approximate scale. (Right) Exemplary picture of the result of the electrophoresis experiment. Sample A is pure, B is a mixture of two different sized DNAs, and C consists of DNA molecules of three different lengths.

formalisms discussed in section 16.5, it can be shown that the waiting time distribution becomes $\psi(t) \propto t^{-(1+\alpha)}$, where $1 < \alpha < 2$. However, in practice, because of the presence of inherent complexity in biomolecular systems, an analytic solution is not possible. One must opt for numerical solutions by employing the Brownian dynamics simulation approach.

16.6.4 POWER-LAW TIME DEPENDENCE OF PHOTOCURRENT

In order to explain the anomalous time dependence in the current, $I(t)$, observed in photocopy machines, Montroll, Scher, and Lax[10] employed continuous time random walk. In this problem, a transient current generated after the absorption of intense light by a solid plate. The current exhibits an interesting power-law decay with a cross-over in the exponent, as shown in Figure 16.6 below.

This rather peculiar result has been explained by using the results from CTRW. The main CTRW equation is similar to the one discussed above and is given by

$$\frac{d\tilde{P}(l,t)}{dt} = \int_0^t \phi(t-x) \sum_{l'} \left[p(l-l')\tilde{P}(l',x) - p(l'-l)P(l,x) \right] dx. \tag{16.58}$$

Here l signifies a cell number in a system, which is broken into equal-sized cells, and $P(l,t)$ is the probability that a cell l is occupied by a carrier at time t and $\phi(t)$ is the relaxation function.

16.6.5   PROTEIN'S SEARCH OF MUTATIONS ON DNA

In order to prevent long-term harmful mutations in DNA duplexes caused by pollutants, there are certain proteins in our body that move along the DNA and rectify the harmful chemical change. For example, HOGG1 repressor protein reduces oxoguanine back to its natural state guanine. The question that has been asked is the following: how does the protein find the defective nucleic acid?

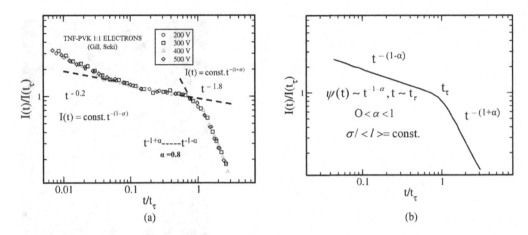

FIGURE 16.6 (a) A log(I)-log(t) plot for 1:1 TNF-PVK determined by Gill and taken with permission from a paper by Seki. The slopes of the dashed lines are −0.2, and −1.8, respectively. (b) A log(I)-log(t) plot indicating the current $I(t)$ associated with a packet of carriers moving, in an electric field, with a hoping-time distribution function $\psi(t) \sim t^{-1-\alpha}, 0 < \alpha < 1$, towards an absorbing barrier at the sample surface.

Source: Taken with permission from [16].

Models have been proposed that suggest a combination of one-dimensional sliding mixed with three-dimensional hop from one segment to another. Thus, a random walk model has been proposed to address the problem, in a novel recent application of the RW models in complex biological processes. We have discussed this problem in more detail in Chapter 21 on diffusion.

16.6.6 ROTATIONAL DIFFUSION MODELS

Rotational relaxation of water molecules presents an interesting problem. Here rotational motion is a combination of continuous diffusion and large amplitude (~60 degrees) jumps. The large amplitude jumps are treated in a model known as the Ivanov-Anderson model. The interesting complication here is the presence of a continuous diffusion in between two large amplitude jumps. Thus, the problem has been formulated as a combination of two continuous random walk problems where one mechanism consists of small jumps while the other mechanism consists of large amplitude jumps. Both are treated as CTRW with different waiting time distributions. These aspects have been discussed in the Chapter 22 on "Rotational Diffusion".

16.6.7 FINANCIAL MATHEMATICS AND STOCK MARKET

The combination of a continuous movement by small amplitude hops and a large amplitude jumps in the stock process have been modeled as two CTRW processes to predict stock prices. Such a model has been widely used in finances and forms the cornerstone of a major discipline called financial mathematics. Random walk is ideally suited to predict the evolution of stocks that are sensitive to many random events in the nation and world at large, such as wars, earthquakes, and weather change, to name a few, and thus are not fully predictable. This branch is also known as econophysics.

Other than the abovementioned examples, RW and CTRW models are used to understand interesting phenomena, for example, Ostwald's ripening, the effects of traps on transport properties, and many others. We direct interested readers to [12–15] for more details.

16.7 SUMMARY

Random walk (RW) presents an elegant method that is widely used to understand a large number of physical phenomena. It captures the randomness inherent in many natural processes, such as in Brownian motion and diffusion, chemical reactions and other transport phenomena. As a mathematical discipline, it is well developed. The discussion provided in this chapter is quite involved yet simple to understand. Below we summarize the content of this chapter.

(i)   We have detailed the theory of RW and derived the analytical expression for the spatio-temporal probability distribution in terms of the step size.

(ii) We have established a connection between RW and the Brownian motion that can be thought of as a continuous RW in space. This eventually led us to the microscopic explanation of diffusion.

(iii) We have solved for the RW probability distribution in the presence of two different boundary conditions, namely, absorbing and reflecting boundaries. Such boundaries are important to consider in real systems where repulsive/hard walls are present or a reaction is in progress.

(iv) To understand diffusion in a rugged landscape we have introduced the concept of continuous time random walk (CTRW). We have introduced an important quantity named waiting time distribution (WTD). By using the famous Montroll-Weiss formula, we have shown the emergence of power-law behavior of the WTD.

(v)   Scientists use RW theory or its variants to model several complex processes. Here we have discussed several applications of RW and CTRW. However, several other applications are there, including those outside the fields of physics/physical chemistry. We encourage readers to go through those applications to appreciate the beauty of the RW theory. A cartoon of widening initial distribution is depicted in Figure 16.7.

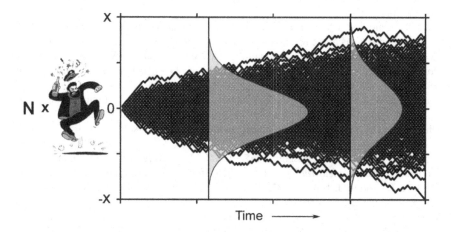

FIGURE 16.7 A simplistic cartoon representation of the random walk phenomena and the resulting distributions. If we let Captain Haddock (a bonafide drinker) start from $x = 0$ and let him walk towards an imaginary wall (several times) the probability distribution widens with increasing time of the walk.

PROBLEMS

(1) The probability of finding a random walker at a point $x = m$ after N steps in one-dimension is given by Eq. (16.3). For large values of N, the expression is approximated to Eq. (16.8). For any particular value of m, consider different values of N in an increasing order and plot the deviation of these two expressions against N. Comment on the result critically.

(2) Consider a one-dimensional system from $x = -10$ to $+10$ in integral steps [as in Figure 16.1(a)]. At $t = 0$, say 100 noninteracting random walkers are placed at $x = 0$. All the walkers have executed 10 steps. Write a computer code to obtain the population distribution along x. Increase the number of steps (and the system size accordingly) from 10 to 100 and 1000. Comment on the observed results.

(3) In problem 2, there were no boundaries. Now, consider an absorbing boundary at $x = +7$. Define a time-dependent survival probability $S(n)$ that denotes the fraction of walkers alive after n steps. Plot $S(n)$ vs n given that at $n = 0$, $S(n) = 1$. Increase the system size and total number of steps to observe the changing nature of the decay of $S(n)$.

(4) In section 16.6 we have discussed several examples where random walk models are applied. One such example is the motion of bacteria. Following the assumptions mentioned in that section for bacterial motions, produce a trajectory by writing a code. Consider them to be noninteracting and place fixed traps that denote chemical bias. Obtain the waiting time distribution for this system. (Consider a uniform random distribution of bacteria in space at $t = 0$.)

BIBLIOGRPAHY AND SUGGESTED READING

1. Feller, W. 1971. *An Introduction to Probability Theory and Its Applications*. Hoboken: John Wiley & Sons; Chung, K.-L. 1995. *Introduction to Probability Theory*. New York: Springer; van Kampen, N. G. 1997. *Stochastic Processes in Physics and Chemistry*. Amsterdam: North-Holland Personal Library.

2. Pearson, K. 1905. The problem of the random walk. *Nature, 72*(1867), 342.

3. Einstein, A. 1906. Investigations on the theory of the Brownian movement. *Annalens der Physik, 19*(4), 289–306.

4. Chandrasekhar, S. 1943. Stochastic problems in physics and astronomy. *Reviews of Modern Physics, 15*(1), 1.

5. Kramers, H. A. 1940. Brownian motion in a field of force and the diffusion model of chemical reactions. *Physica, 7*(4), 284–304.

6. Grote, R. F., and J. T. Hynes. 1980. The stable states picture of chemical reactions. II. Rate constants for condensed and gas phase reaction models. *The Journal of Chemical Physics, 73*(6), 2715–2732.

7. Bagchi, B. 2012. *Molecular Relaxation in Liquids*. New York: Oxford University Press

8. Montroll, E. W., and G. H. Weiss. 1965. Random walks on lattices. II. *Journal of Mathematical Physics, 6*(2), 167–181.

9. Montroll, E. W., and H. Scher. 1973. Random walks on lattices. IV. Continuous-time walks and influence of absorbing boundaries. *Journal of Statistical Physics, 9*(2), 101–135.

10. Scher, H., and M. Lax. 1973. Stochastic transport in a disordered solid. I. Theory. *Physical Review B, 7*(10), 4491.

11. Abramowitz, M., and I. A. Stegun. 1965. *Handbook of Mathematical Functions: With Formulas, Graphs, and Mathematical Tables*. Vol. 55. Chelmsford: Courier Corporation.

12. Weiss, G. H., and R. J. Rubin. 1983. Random walks: theory and selected applications. *Advances in Chemical Physics, 52*, 363–505.

13. Baumgärtner, A., and K. Binder. 1979. Monte Carlo studies on the freely jointed polymer chain with excluded volume interaction. *The Journal of Chemical Physics, 71*(6), 2541–2545.

14. Ariel, G., A. Rabani, S. Benisty, J. D. Partridge, R. M. Harshey, and A. Be'Er. 2015. Swarming bacteria migrate by Lévy walk. *Nature Communications, 6*, 8396.

15. Masoliver, J., M. Montero, and G. H. Weiss. 2003. Continuous-time random-walk model for financial distributions. *Physical Review E, 67*(2), 021112.

16. Scher, H., and Montroll, E.W., 1975. Anomalous transit-time dispersion in amorphous solids. *Physical Review B, 12*(6), 2455.

17 Fokker-Planck, Kramers, and Smoluchowski Equations and Their Solutions

OVERVIEW

Because of the presence of the random force in the Langevin equation, it cannot be solved analytically to obtain expressions for the time-dependent probability distribution functions. This is a hindrance for many practical purposes where it is necessary to obtain an analytical expression for the relevant probability distributions. This difficulty is removed by deriving a class of algebraic equations for the probability distributions. These are partial differential equations, which turn out to be as, or perhaps more, useful than the Langevin equation itself. They find use in many different areas, starting from astrophysics (as shown by Chandrasekhar), chemical reaction dynamics (Smoluchowski and Kramers), coagulation, Ostwald ripening, and many others. In this chapter we derive these equations and also present expressions for the probability distribution solutions for the free particle and when bound by a harmonic potential.

17.1 FROM STOCHASTIC EQUATION TO ALGEBRICIC DESCRIPTIONS

Langevin's equation, as discussed in the previous chapter, is a stochastic equation of motion of a continuous variable, like velocity, momentum, coordinate, etc. The random force present in this equation imparts stochasticity to the variable under concern (say A). As a result, the equation cannot be solved analytically for any probability distribution. It can however be solved analytically for time correlation functions and numerically, in computer simulations, as is indeed routinely done. We again mention that it is a reduced description where instead of following the trajectories of ($N+1$) particles, we do the same for only one tagged (special) particle.

Experimental observables are mostly obtained as averages over a probability distribution function. So, if we could have access to such a distribution function, theoretical descriptions of such observables are much easier and controlled than averaging over the trajectories, for every variable considered. Also, there arise situations, like the crossing over a high activation barrier that cannot be tackled by running trajectories alone.

We start with the usual form of the Langevin equation

$$\frac{dv(t)}{dt} = -\zeta v(t) + \delta F(t) \tag{17.1}$$

where $v(t)$ is the velocity of solute or tagged particle, ζ is the friction and $\delta F(t)$ is the random force term, which is irregular but its ensemble properties are specified. We have discussed this equation in Chapter 15. For the time being however we note that the structure of the Langevin equation is

DOI: 10.1201/9781003157601-20

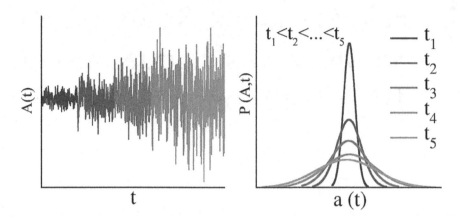

FIGURE 17.1 Schematic representation of the relationship between temporal evolution of a random variable *A* (Langevin equation) and its probability distribution (Fokker-Planck equation). This figure illustrates the transformation of a trajectory (left panel) to a distribution function (right panel) and thus serves to illustrate the conceptual process of transforming the Langevin equation to a Fokker-Planck equation.

more general than originally developed for velocity. This has been possible because of the *linear response theory* and the *fluctuation-dissipation theorem*, which provide the necessary support for the Langevin equation. These have been discussed in details in Chapter 11. The necessity for a general form of the Langevin equation becomes transparent from the treatment below. *The elegant treatment given below follows the procedure given in the books of Risken and Zwanzig.*

With a view to general applications, we start with the Langevin equation for a dynamical variable *A*.

$$\frac{dA}{dt} = M(A) + \delta F(t). \tag{17.2}$$

Here, $M(A)$ is the deterministic force and $\delta F(t)$ is the stochastic force. $M(A)$ is dependent on two terms; it contains a friction kernel and a systematic force, if any. The friction term is related to the $\delta F(t)$ by the *fluctuation-dissipation theorem*. In the absence of any systematic or external force, Eq. (17.2) takes the form of Eq. (17.1). However, in the presence of a force term, $M(A)$ is dependent not only on velocity, but also on position. In that case *A* becomes a vector and $M(A)$ should be treated as a tensor (matrix), which contains the force term. Specific examples and their derivations are discussed in section 17.3.

We now proceed to transform the stochastic equation [Eq. (17.2)] to an algebraic equation where the equation of motion would be in terms of a time-dependent probability distribution function $P(A,t)$ of the variable *A* which is defined as

$$P(A,t) = \langle \delta(A - A(t)) \rangle. \tag{17.3}$$

The angular brackets in Eq. (17.3) should be interpreted as follows. We follow a time trajectory of *A* and count the times $A(t)$ becomes equal to *A*. We can also consider many different trajectories starting from random values of $A(t)$. The length of the trajectory or the number of trajectories should provide enough "hits" to a given value to enable us to define a smooth $P(A,t)$. The physical essence of the above procedure is illustrated below in Figure 17.2. The probability distribution

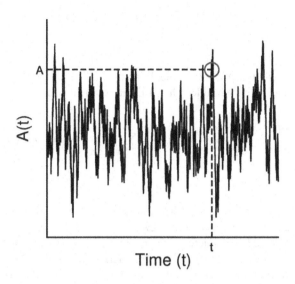

FIGURE 17.2 Construction of the distribution function from the trajectory of the time dependent continuous variable $A(t)$. The probability that at time t, $A(t) = A$, is given by the probability distribution function $\mathbf{P(A,t)} = \langle \delta(\mathbf{A} - \mathbf{A(t)}) \rangle$. The circle envelopes this point in the trajectory. One out of several such instances are shown.

such defined is time-dependent. The relationship between the time evolutions of $A(t)$ and $P(A,t)$ is shown in Figure 17.1.

Let us now consider the change in this distribution during a small time interval δt. We denote the distribution in this time interval by $P(A, t + \delta t)$. We can then write the following expression:

$$P(A, t + \delta t) - P(A,t) = \langle \delta(A - A(t+\delta t)) \rangle - \langle \delta(A - A(t)) \rangle. \tag{17.4}$$

Since A is a stochastic variable, and obeys the Langevin equation as a Brownian particle would, we have a stochastic equation of motion for $A(t)$ as described earlier. We shall show below that use of Langevin equation in Eq. (17.4) leads to the following general equation of motion for the probability distribution $P(A,t)$:

$$\boxed{\frac{\partial P(A,t)}{\partial t} = \left[-\frac{\partial}{\partial A} M(A) + \frac{\partial^2}{\partial a^2} D(A) \right] P(\mathbf{A},t).} \tag{17.5}$$

This form of equation is generally known as the *Fokker-Planck equation* (FPE). *Note the structure of the equation. First, it is an equation of motion in one variable, and generalization is required, and shall be discussed later.* Second, it is a second-order linear partial differential equation. Third, the functions $M(A)$ and $D(A)$ have definite meanings, as will be clarified in more detail later. We note for the time being that $M(A)$ and $D(A)$ stand for *drift* and *diffusion* coefficients, respectively. Note that in Eq. (17.5) both $M(A)$ and $D(A)$ are inside the derivatives. When $D(A)$ is independent of A and $M(A)$ is zero we get back the diffusion equation. Fourth, the solution of Eq. (17.5) provides $P(A,t)$, which, on integration, yields the average value of any function f that depends on the variable $A(f(A))$ [Eq. (17.6)].

$$\langle f[A(t)]\rangle \equiv f(t) = \int_{-\infty}^{\infty} dA f(A) P(A,t) \qquad (17.6)$$

where $f(t)$ can be an observable that depends on variable $A(t)$.

Eq. (17.5) is for only one dynamical variable. In many applications, there are multiple dynamical variables. For N such continuous variables, the FPE can be written as

$$\frac{\partial P(\mathbf{A},t)}{\partial t} = \left[-\sum_{i=1}^{N} \frac{\partial}{\partial A_i} M_i(\mathbf{A}) + \sum_{\substack{i,j=1 \\ j \geq i}}^{N} \frac{\partial^2}{\partial A_i \partial A_j} D_{ij}(\mathbf{A}) \right] P(\mathbf{A},t) \qquad (17.7)$$

where $\mathbf{A} = \{A_1, A_2, ..., A_N\}$.

The FPE is useful in describing a small subsystem of a macroscopic system, for example, velocity of a Brownian particle, electric field in a LASER, etc. If the subsystem is large enough, the stochasticity may be neglected, and a deterministic approach is sufficient to deal with it.

The rest of the chapter deals with two major parts: derivation of different kinds of FPE and subsequently, their solutions. In the next section we present a rigorous derivation of the noise averaged form of Eq. (17.7). This will be followed by specific examples of FPE and finally, their solutions.

17.2 DERIVATION OF GENERAL FOKKER-PLANCK EQUATION FROM LANGEVIN EQUATION

We now use the above method to present a compact and general derivation of the Fokker-Planck equation (FPE). We emphasize that the following derivation provides one of the simplest and most direct derivations of the FPE.

For the set of variables $\mathbf{A} = \{A_1, A_2, ..., A_N\}$, we can write a general Langevin equation of the form given in Eq. (17.8).

$$\frac{d\mathbf{A}}{dt} = \mathbf{M}(\mathbf{A}) + \delta\mathbf{F}(t). \qquad (17.8)$$

Here, $\mathbf{M}(\mathbf{A})$ is a function of the variables $\mathbf{A}$. The random force $\delta\mathbf{F}(t)$ is again a white noise. As mentioned previously, it is Gaussian with the first moment (mean) equal to zero and the second moment being delta-function correlated, is the fluctuation-dissipation theorem.

$$\begin{aligned} \langle \delta\mathbf{F}(t) \rangle &= 0 \\ \langle \delta\mathbf{F}(t).\delta\mathbf{F}(t') \rangle &= 2\mathbf{D}\delta(t-t'). \end{aligned} \qquad (17.9)$$

$\mathbf{D}$ is a tensor that measures the strength of the fluctuating force. It is given by the fluctuation dissipation theorem. As mentioned in the previous section, we seek the probability distribution of the variables $\mathbf{A}$ at time t [$P(\mathbf{A},t)$]. We note that this function is a conserved quantity. This means, it is a normalized function with respect to $\mathbf{A}$ for all t.

$$\int d\mathbf{A} P(\mathbf{A},t) = 1. \qquad (17.10)$$

For any conserved quantity, its time derivative is balanced by the divergence of a flux. This is equivalent to the Liouville equation description in classical mechanics. Hence, we may write the conservation equation as follows:

$$\frac{\partial P(\mathbf{A},t)}{\partial t} + \frac{\partial}{\partial \mathbf{A}} \cdot \left(\frac{\partial \mathbf{A}}{\partial t} P(\mathbf{A},t) \right) = 0. \tag{17.11}$$

Replacing the time derivative in Eq. (17.11) with Eq. (17.8) and rearranging, we obtain

$$\frac{\partial P(\mathbf{A},t)}{\partial t} = -\frac{\partial}{\partial \mathbf{A}} \cdot \left(\mathbf{M}(\mathbf{A})P(\mathbf{A},t) + \delta \mathbf{F}(t) P(\mathbf{A},t) \right). \tag{17.12}$$

For computational convenience, we define an operator L_s, such that

$$L_s P(\mathbf{A},t) = \frac{\partial}{\partial \mathbf{A}} \cdot \mathbf{M}(\mathbf{A}) P(\mathbf{A},t). \tag{17.13}$$

Hence, Eq. (17.12) is reduced to

$$\frac{\partial P(\mathbf{A},t)}{\partial t} = -\left[L_s - \frac{\partial}{\partial \mathbf{A}} \cdot \delta \mathbf{F}(t) \right] P(\mathbf{A},t). \tag{17.14}$$

This equation can be solved following the steps mentioned below.
  That involves rearrangements and an integration in the last stage.

$$\frac{\partial P(\mathbf{A},t)}{\partial t} + L_s P(\mathbf{A},t) = -\frac{\partial}{\partial \mathbf{A}} \cdot \delta \mathbf{F}(t) P(\mathbf{A},t)$$

$$\frac{\partial}{\partial t}\left(P(\mathbf{A},t)e^{tL_s} \right) = -e^{tL_s} \frac{\partial}{\partial \mathbf{A}} \cdot \delta \mathbf{F}(t) P(\mathbf{A},t)$$

$$P(\mathbf{A},t)e^{tL_s} - P(\mathbf{A},0) = -\int_0^t dt' e^{t'L_s} \frac{\partial}{\partial \mathbf{A}} \cdot \delta \mathbf{F}(t') P(\mathbf{A},t')$$

Hence, we get the probability distribution function of $\mathbf{A}$ at time t as

$$P(\mathbf{A},t) = e^{-tL_s} P(\mathbf{A},0) - \int_0^t dt' e^{-(t-t')L_s} \frac{\partial}{\partial \mathbf{A}} \cdot \delta \mathbf{F}(t') P(\mathbf{A},t'). \tag{17.15}$$

Substituting this back into the right-hand side of Eq. (17.14), we get

$$\frac{\partial P(\mathbf{A},t)}{\partial t} = -L_s P(\mathbf{A},t) - \frac{\partial}{\partial \mathbf{A}} \cdot \delta \mathbf{F}(t) e^{-tL_s} P(\mathbf{A},0)$$
$$+ \frac{\partial}{\partial \mathbf{A}} \cdot \delta \mathbf{F}(t) \cdot \int_0^t dt' e^{-(t-t')L_s} \frac{\partial}{\partial \mathbf{A}} \cdot \delta \mathbf{F}(t') P(\mathbf{A},t'). \tag{17.16}$$

Now, taking average of the noise,

$$\frac{\partial \langle P(\mathbf{A},t) \rangle}{\partial t} = -L_s \langle P(\mathbf{A},t) \rangle - \frac{\partial}{\partial \mathbf{A}} \cdot \langle \delta \mathbf{F}(t) \rangle e^{-tL_s} P(\mathbf{A},0)$$
$$+ \frac{\partial}{\partial \mathbf{A}} \cdot \int_0^t dt' e^{-(t-t')L_s} \langle \delta \mathbf{F}(t) \cdot \delta \mathbf{F}(t') \rangle \cdot \frac{\partial}{\partial \mathbf{A}} \langle P(\mathbf{A},t') \rangle. \tag{17.17}$$

The last decoupling term is delicate because **A** in $P(\mathbf{A},t')$ depends on the history of $\delta\mathbf{F}$. Using Eq. (17.9) and Eq. (17.13), we get

$$\frac{\partial\langle P(\mathbf{A},t)\rangle}{\partial t} = -\frac{\partial}{\partial\mathbf{A}}\cdot\mathbf{M}(\mathbf{A})\langle P(\mathbf{A},t)\rangle + \frac{\partial}{\partial\mathbf{A}}\cdot\int_0^t dt' e^{-(t-t')L_s} 2\mathbf{D}\delta(t-t')\cdot\frac{\partial}{\partial\mathbf{A}}\langle P(\mathbf{A},t')\rangle \quad (17.18)$$

where we have used Eq. (17.9). The delta function takes away the exponential operator term, yielding our final form of the equation

$$\boxed{\frac{\partial\langle P(\mathbf{A},t)\rangle}{\partial t} = \left[-\frac{\partial}{\partial\mathbf{A}}\cdot\mathbf{M}(\mathbf{A}) + \frac{\partial^2}{\partial\mathbf{A}^2}\mathbf{D}\right]\langle P(\mathbf{A},t)\rangle.} \quad (17.19)$$

Note that the factor 2 associated with **D** is used by the even nature of the δ function. We recognize that in Eq. (17.19) **M(A)** is the force term and **D** is the drift or the diffusion term.

Eq. (17.19) is the noise averaged generalized form of Eq. (17.7). Note that Eq. (17.19) holds only for a Markovian (memoryless) description. Consequently, this derivation will not work for a non-Markovian system.

What is the advantage of Eq. (17.19)? This form is so general that it can be easily and effortlessly applied to many situation to get the required equation of motion. In other words, by identifying the nature of the vectors and the tensor in this equation, we can study many interesting problems.

17.3 SPECIFIC EXAMPLES OF THE FOKKER-PLANCK EQUATION

In this section we shall show how Eq. (17.19) can be used in a straightforward fashion to derive several different algebraic equations, namely Fokker-Planck and Smoluchowski.

17.3.1 FP Equation in Velocity Space

In order to define the problem, start with the following Langevin equation:

$$\frac{dv}{dt} = -\zeta v + \delta F(t) \quad (17.20)$$

where the terms have their usual meanings. From fluctuation-dissipation theorem (discussed in the previous chapter) we can write

$$\langle\delta F(t)\delta F(t')\rangle = 2\zeta k_B T\delta(t-t'). \quad (17.21)$$

Hence, the vectors and tensor that go into FP equation [Eq. (17.19)] are

$$\mathbf{A} = (v) \quad (17.22)$$

$$\mathbf{M}(\mathbf{A}) = (-\zeta v) \quad (17.23)$$

$$\mathbf{D} = (\zeta k_B T) \quad (17.24)$$

Therefore, replacing the parameters in Eq. (17.19), we obtain,

$$\frac{\partial P}{\partial t} = \left(\zeta \frac{\partial}{\partial v} v + \zeta k_B T \frac{\partial^2}{\partial v^2} \right) P \qquad (17.25)$$

Such a process with linear drift coefficient $(B = -\zeta v)$ and constant diffusion coefficient $(D = \zeta k_B T)$ is called an *Ornstein-Uhlenbeck process*. For a vanishing drift coefficient, Eq. (17.25) reduces to an Einstein diffusion equation (derivation in previous chapter) with only the quadratic derivative term. This is known as a *Wiener process*.

17.3.1.1 An instructive alternative derivation (from S. Chandrasekhar)

This derivation uses the method of moments. Consider a time interval Δt, which is longer than the timescale of fluctuations of the Langevin force, but shorter than intervals in which the velocity of the Brownian particle changes appreciably. Then the probability distribution function $(P(v, t + \Delta t))$, that defines the possibility of velocity v at time $t + \Delta t$ is governed by transition probability $\psi(v; \Delta v)$, where, Δv is the increment in v in the interval Δt. Hence, we can write the following relation:

$$P(v, t + \Delta t) = \int d(\Delta v) P(v - \Delta v, t) \, \psi(v - \Delta v; \Delta v). \qquad (17.26)$$

In writing this relation we are assuming that the motion of the Brownian particle depends on the instantaneous system parameters only. This means that the history of the dynamics does not affect the instantaneous motions. This is known as the *Markovian approximation*.

Now we proceed by expanding the functions in Taylor series of respective variables. Hence, we have

$$P(v,t) + \frac{\partial P}{\partial t} \Delta t + O\left((\Delta t)^2 \right) = \int d(\Delta v) \left(P(v,t) - \frac{\partial P}{\partial v} \Delta v + \frac{1}{2} \frac{\partial^2 P}{\partial v^2} (\Delta v)^2 + O\left((\Delta v)^3 \right) \right)$$
$$\times \left(\psi(v; \Delta v) - \frac{\partial \psi}{\partial v} \Delta v + \frac{1}{2} \frac{\partial^2 \psi}{\partial v^2} (\Delta v)^2 + O\left((\Delta v)^3 \right) \right). \qquad (17.27)$$

Now, from normalization condition we have

$$\int d(\Delta v) \, \psi(v; \Delta v) = 1. \qquad (17.28)$$

We also have the following relations for the mean and variance of Δv:

$$\int d(\Delta v) \, \psi(v; \Delta v) \, \Delta v = \langle \Delta v \rangle$$
$$\int d(\Delta v) \, \psi(v; \Delta v) (\Delta v)^2 = \left\langle (\Delta v)^2 \right\rangle. \qquad (17.29)$$

Using Eq. (17.28) and (17.29) in Eq. (17.27), we have

$$P(v,t) + \frac{\partial P}{\partial t} \Delta t + O\left((\Delta t)^2 \right) = P(v,t) - \frac{\partial}{\partial v} \langle \Delta v \rangle P - P \frac{\partial}{\partial v} \langle \Delta v \rangle$$
$$+ \frac{1}{2} \frac{\partial^2}{\partial v^2} \left\langle (\Delta v)^2 \right\rangle P + \frac{\partial P}{\partial v} \frac{\partial}{\partial v} \left\langle (\Delta v)^2 \right\rangle + \frac{1}{2} \frac{\partial^2 P}{\partial v^2} \left\langle (\Delta v)^2 \right\rangle + O\left((\Delta v)^3 \right). \qquad (17.30)$$

By product formula of differentiation, this can be expressed in the following compact form:

$$\frac{\partial P}{\partial t}\Delta t = \left(-\frac{\partial}{\partial v}\langle\Delta v\rangle + \frac{\partial^2}{\partial v^2}\frac{\langle(\Delta v)^2\rangle}{2}\right)P. \tag{17.31}$$

This is the Fokker-Planck equation in velocity space with $\langle\Delta v\rangle$ and $\dfrac{\langle(\Delta v)^2\rangle}{2}$ as the drift and diffusion coefficients.

Now, from the Langevin Eq. (17.20), we have

$$\int\limits_{t}^{t+\Delta t} dt'v(t') = -\zeta v\Delta t + \int\limits_{t}^{t+\Delta t} dt'\,\delta F(t')$$

$$v(t+\Delta t) - v(t) = \Delta v = -\zeta v\Delta t + \int\limits_{t}^{t+\Delta t} dt'\,\delta F(t'). \tag{17.32}$$

On taking the average on both sides,

$$\langle\Delta v\rangle = -\zeta\langle v\rangle\Delta t + \int\limits_{t}^{t+\Delta t} dt'\,\langle\delta F(t')\rangle. \tag{17.33}$$

Since, the first moment of the fluctuating force is 0, we get

$$\langle\Delta v\rangle = -\zeta\langle v\rangle\Delta t. \tag{17.34}$$

Now, squaring both sides of Eq. (17.32),

$$(\Delta v)^2 = \zeta^2 v^2(\Delta t)^2 - 2\zeta v\Delta t\int\limits_{t}^{t+\Delta t} dt'\,\delta F(t') + \int\limits_{t}^{t+\Delta t} dt'\,\delta F(t')\int\limits_{t}^{t+\Delta t} dt''\,\delta F(t''). \tag{17.35}$$

Δt being a small quantity, the first term on the right-hand side can be neglected. Taking the average over noise, the second term becomes zero. Hence, we are left with

$$\langle(\Delta v)^2\rangle = \int\limits_{t}^{t+\Delta t} dt'\int\limits_{t}^{t+\Delta t} dt''\,\langle\delta F(t')\delta F(t'')\rangle$$

$$= \int\limits_{t}^{t+\Delta t} dt'\int\limits_{t}^{t+\Delta t} dt''\,2D\delta(t'-t'')$$

$$= 2D\int\limits_{t}^{t+\Delta t} dt'$$

Therefore, we get

$$\frac{\langle(\Delta v)^2\rangle}{2} = D\Delta t. \tag{17.36}$$

Using Eq. (17.34) and (17.36) in Eq. (17.31), we get

$$\frac{\partial P}{\partial t}\Delta t = \left(-\frac{\partial}{\partial v}(-\zeta v \Delta t) + \frac{\partial^2}{\partial v^2}D\Delta t \right)P$$

$$\boxed{\frac{\partial P}{\partial t} = \left(\zeta \frac{\partial}{\partial v}v + \zeta k_B T \frac{\partial^2}{\partial v^2} \right)P}.$$
(17.37)

Yet another method of derivation of Fokker-Planck-like equations can be achieved by using the method of moments. This method has been described in detail in the book by van Kampen.

17.3.2 KRAMERS EQUATION: FP EQUATION IN PHASE SPACE IN THE PRESENCE OF AN EXTERNAL FIELD

Here a neat derivation of a rather complex equation can be achieved by using the Risken-Zwanzig approach. Consider the phase space motion of a Brownian particle in a potential field $U(x)$. The Langevin equations for the two variables, namely x (coordinate) and p (linear momentum), are as follows:

$$\frac{dx}{dt} = \frac{p}{m}$$
(17.38)

$$\frac{dp}{dt} = -\zeta \frac{p}{m} - U'(x) + \delta F(t)$$
(17.39)

$U'(x)$ is the first derivative of $U(x)$ with respect to x, the negative of which gives the force experienced by the walker from the potential field. We note that the stochastic force is present only in the equation of linear momentum. The concerned vectors and tensors according to Eq. (17.19) are

$$\mathbf{A} = \begin{pmatrix} x \\ p \end{pmatrix}$$
(17.40)

$$\mathbf{M}(\mathbf{A}) = \begin{pmatrix} \dfrac{p}{m} \\ -\zeta \dfrac{p}{m} - U'(x) \end{pmatrix}$$
(17.41)

$$\mathbf{D} = \begin{pmatrix} 0 & 0 \\ 0 & \zeta k_B T \end{pmatrix}.$$
(17.42)

Note that there is no off-diagonal term in the $\mathbf{D}$ tensor. This means that the two variables x and p are orthogonal to each other. Hence, we can write the FPE as

$$\frac{\partial P}{\partial t} = \left(-\frac{p}{m}\frac{\partial}{\partial x} - \frac{\partial}{\partial p}\left(-U'(x) - \zeta \frac{p}{m} \right) + \zeta k_B T \frac{\partial^2}{\partial p^2} \right)P$$

$$\boxed{\frac{\partial P}{\partial t} = \left(-v\frac{\partial}{\partial x} + \frac{\partial}{\partial p}(U'(x) + \zeta v) + \zeta k_B T \frac{\partial^2}{\partial p^2} \right)P}.$$
(17.43)

Eq. (17.43) is known as *Kramers' equation*. It is the most general form of FP equation in phase space. This equation can be used to study several problems like rate of barrier crossing of a Brownian particle, molecular reorientation in liquids (with angle and angular momentum as the variables), etc. Note that in absence of noise and friction, Eq. (17.43) reduces to a Liouville equation. Hence, the Kramers equation is a generalized representation of the classical Liouville equation including Brownian noise.

Remember that the complete phase space description for a single Brownian particle requires three coordinates and three momenta components. In that case, replacing the first-order partial derivatives in Eq. (17.43) with gradients and the second-order partial derivative with Laplacian of the concerned variables gives the complete equation in a six-dimensional phase space [Eq. (17.44)].

$$\frac{\partial P}{\partial t} = \left(-\frac{\mathbf{p}}{m}\nabla_r - \nabla_p\left(-\mathbf{U}'(x) - \zeta\frac{\mathbf{p}}{m} \right) + \zeta k_B T \nabla_p^2 \right) P \tag{17.44}$$

where, $\nabla_r = \hat{x}\frac{\partial}{\partial x} + \hat{y}\frac{\partial}{\partial y} + \hat{z}\frac{\partial}{\partial z}$ and $\nabla_p = \hat{p}_x\frac{\partial}{\partial p_x} + \hat{p}_y\frac{\partial}{\partial p_y} + \hat{p}_z\frac{\partial}{\partial p_z}$. The capped variables denote unit vectors along the concerned vector direction. However, for the sake of convenience, we shall stick to the reduced dimensional form.

17.3.2.1 Alternative derivation from the master equation

This derivation follows the philosophy introduced earlier in the case of velocity space. In the presence of an external field, the extra force term $\mathbf{U}'(x)$ is present. Then, the deterministic increments Δx and Δp in position and momentum that the particle suffers in time Δt in the absence of Brownian motion is given by

$$\Delta x = x(t+\Delta t) - x(t) = \frac{p}{m}\Delta t \tag{17.45}$$

$$\Delta p = p(t+\Delta t) - p(t) = -U'(x)\Delta t. \tag{17.46}$$

Now, our system is in fact subject to Brownian forces. Considering the previously introduced Markovian approximation, we can write the probability distribution as

$$P(x,p,t+\Delta t) = \int d(\Delta x)d(\Delta p)P(x-\Delta x, p-\Delta p,t)\Psi(x-\Delta x, p-\Delta p; \Delta x, \Delta p) \tag{17.47}$$

where $\Psi(x,v;\Delta x,\Delta p)$ is the transition probability in phase space. This can be converted to a momentum space description [$\psi(x,p;\Delta p)$] according to Eq. (17.45). Using Dirac's δ-function notation, we can write

$$\Psi(x,p;\Delta x,\Delta p) = \psi(x,p;\Delta p)\delta\left(\Delta x - \frac{p}{m}\Delta t \right). \tag{17.48}$$

Hence, on integration over Δx, we get

$$P(x,p,t+\Delta t) = \int d(\Delta p)P\left(x-\frac{p}{m}\Delta t, p-\Delta p,t \right)\psi\left(x-\frac{p}{m}\Delta t, p-\Delta p; \Delta p \right). \tag{17.49}$$

Eq. (17.49) can be rewritten as

$$P\left(x+\frac{p}{m}\Delta t, p-U'(x)\Delta t, t+\Delta t\right)=\int d(\Delta p)P(x,p,t)\,\psi(x,p;\Delta p).\qquad(17.50)$$

This is possible by replacing the indices of x and p. Now, we expand each of the functions in a Taylor series of the respective variables and follow the same approach adopted in the previous derivation. This gives us the following equation:

$$P(x,p,t)+\frac{\partial P}{\partial x}\frac{p}{m}\Delta t-\frac{\partial P}{\partial p}U'(x)\Delta t+\frac{\partial P}{\partial t}\Delta t+O\big((\Delta t)^2\big)=\int d(\Delta p)$$
$$\times\left(P(x,p,t)-\frac{\partial P}{\partial p}\Delta p+\frac{1}{2}\frac{\partial^2 P}{\partial p^2}(\Delta p)^2+O\big((\Delta p)^3\big)\right)$$
$$\times\left(\psi(x,p;\Delta p)-\frac{\partial\psi}{\partial p}\Delta p+\frac{1}{2}\frac{\partial^2\psi}{\partial p^2}(\Delta p)^2+O\big((\Delta p)^3\big)\right).$$

Using the definitions of mean and variance previously introduced, we can write

$$\frac{\partial P}{\partial t}=-\frac{p}{m}\frac{\partial}{\partial x}P+\frac{\partial P}{\partial p}U'(x)-\frac{\partial}{\partial p}\big(\langle\Delta p\rangle P\big)+\frac{\partial^2}{\partial p^2}\left(\left\langle\frac{(\Delta p)^2}{2}\right\rangle P\right).\qquad(17.51)$$

Using Eq. (17.34) and (17.36), and employing minor algebraic manipulations, we can write

$$\boxed{\frac{\partial P}{\partial t}=\left(-v\frac{\partial}{\partial x}+\frac{\partial}{\partial p}\big(U'(x)+\zeta v\big)+\zeta k_B T\frac{\partial^2}{\partial p^2}\right)P.}\qquad(17.52)$$

This is the *Kramers equation*.

17.3.3 SMOLUCHOWSKI EQUATION: FP EQUATION IN THE OVERDAMPED LIMIT

In this example we consider the motion of a Brownian particle in the overdamped condition. This implies that the relaxation time $\tau=m/\zeta$ (ζ is the friction coefficient and m is the mass of the walker) much shorter than natural time scale of motion. The Langevin equation in this case has one variable velocity $\left(v=\dfrac{dx}{dt}\right)$, and can be written as

$$m\frac{d^2 x}{dt^2}=-\zeta\frac{dx}{dt}-U'(x)+\delta F(t).\qquad(17.53)$$

Because of the overdamped condition $\left(\dfrac{\partial^2 x}{\partial t^2}\ll\zeta\dfrac{\partial x}{\partial t}\right)$ we can drop the time derivative of velocity on the left-hand side of Eq. (17.53) and rearrange to get a Langevin equation in variable x.

$$\frac{dx}{dt}=-\frac{1}{\zeta}U'(x)+\frac{1}{\zeta}\delta F(t).\qquad(17.54)$$

Hence the parameters for the FPE are

$$\mathbf{A} = (x) \tag{17.55}$$

$$\mathbf{M(A)} = \left(-\frac{1}{\zeta}U'(x)\right) \tag{17.56}$$

$$\mathbf{D} = \left(\frac{k_B T}{\zeta}\right). \tag{17.57}$$

Note that here the stochastic force term is $\frac{1}{\zeta}\delta F(t)$ and not only $\delta F(t)$. Using the above three expressions, the FPE for the system becomes

$$\boxed{\frac{dP}{dt} = \frac{k_B T}{\zeta}\left(\frac{\partial}{\partial x}\beta U'(x) + \frac{\partial^2}{\partial x^2}\right)P.} \tag{17.58}$$

Eq. (17.58) is known as the *Smoluchowski equation*. It describes the motion of a Brownian particle in the high friction limit. Note that it can be expressed in the following way so as to describe the diffusion in an external potential in the form of the well-known diffusion equation.

$$\frac{dP}{dt} = D\frac{\partial}{\partial x}e^{-\frac{U(x)}{k_B T}}\frac{\partial}{\partial x}e^{\frac{U(x)}{k_B T}}P. \tag{17.59}$$

Here, D is the diffusion coefficient given by Eq. (17.57). Eq. (17.58) can be written in three-dimensional form following the convention introduced in Eq. (17.44).

17.4   SOLUTIONS OF THE FOKKER-PLANCK EQUATION

Analytical solution of the FP equation can be obtained for special conditions of drift and diffusion coefficients. The solution becomes nontrivial if separation of variables is not possible. Complications increase with the increase in the number of variables. Numerical solution, computer simulations can be used in such cases. In this section we discuss some of the analytical solutions of FPE.

17.4.1   STATIONARY SOLUTION

For a stationary process, the rate of change of probability density $P(a)$ is 0, that is, it is independent of time. Hence, the FPE [Eq. (17.5)] can be written as

$$\frac{d^2}{dA^2}(DP) = \frac{d}{dA}(BP). \tag{17.60}$$

Here, the drift coefficient (M), diffusion coefficient (D), and the probability density function (P) are functions of the variable A only. On integration of Eq. (17.60) with respect to A, we get

$$\frac{d}{dA}(DP) = MP + C$$

$$\frac{d(DP)}{DP} = \frac{M}{D}dA$$

C is the integration constant. On separation of variables and integration, we obtain

$$\ln\big(D(A)P(A)\big) = \int_A dA' \frac{M(A')}{D(A')}$$

$$\boxed{P(A) = \frac{N}{D(A)}\exp\left(\int_A dA' \frac{M(A')}{D(A')}\right)}. \qquad (17.61)$$

Here, N is a normalization constant that satisfies the following condition:

$$\int_A dA'P(A') = 1. \qquad (17.62)$$

Eq. (17.61) provides the solution of an FP equation describing a stationary state.

17.4.2 Solution for FPE in Velocity Space [Eq. (17.25)]

For convenience, we rewrite Eq. (17.25) here.

$$\frac{\partial P}{\partial t} = \left(\zeta \frac{\partial}{\partial v}v + \zeta k_B T \frac{\partial^2}{\partial v^2}\right)P. \qquad (17.63)$$

To start with the solution, we note that without the second-order differential term, Eq. (17.25) is a linear homogeneous first-order partial differential equation.

$$\frac{\partial P}{\partial t} - \zeta \frac{\partial}{\partial v}(vP) = 0$$
$$\frac{\partial P}{\partial t} = \zeta v \frac{\partial P}{\partial v} + \zeta P. \qquad (17.64)$$

Now, the solution of Eq. (17.64) involves the following equation (from Stokes law):

$$\frac{\partial v}{\partial t} = -\zeta v. \qquad (17.65)$$

The characteristic curve is determined by $-dt = dv/(\zeta v)$. The solution of Eq. (17.65) is

$$ve^{\zeta t} = v_0 = \text{constant} \qquad (17.66)$$

where $v_0 = v(t = 0)$. Hence, this the required first integral for the solution of Eq. (17.64). Accordingly, for solution of Eq. (17.25), we introduce a new variable

$$\rho = ve^{\zeta t}. \qquad (17.67)$$

Differentiating with respect to the variable v, we get

$$\frac{\partial \rho}{\partial v} = e^{\zeta t}. \tag{17.68}$$

Now, the second-order term in Eq. (17.61) can be written as

$$
\begin{aligned}
\frac{\partial^2 P}{\partial v^2} &= \frac{\partial}{\partial v}\left(\frac{\partial P}{\partial v}\right) \\
&= e^{\zeta t}\frac{\partial}{\partial \rho}\left(\frac{\partial \rho}{\partial v}\frac{\partial P}{\partial \rho}\right) \\
&= e^{2\zeta t}\frac{\partial^2 P}{\partial \rho^2}.
\end{aligned}
\tag{17.69}
$$

Hence, Eq. (17.61) reduces to

$$\frac{\partial P}{\partial t} = \zeta v\frac{\partial P}{\partial \rho} + \zeta P + \zeta k_B T e^{2\zeta t}\frac{\partial^2 P}{\partial \rho^2}. \tag{17.70}$$

The left-hand side of the above equation can be written as

$$\frac{dP}{dt} = \frac{\partial P}{\partial t} + \frac{\partial \rho}{\partial t}\cdot\frac{\partial P}{\partial \rho}. \tag{17.71}$$

On the other hand, we have

$$v\frac{\partial P}{\partial v} = v\frac{\partial \rho}{\partial v}\frac{\partial P}{\partial \rho}. \tag{17.72}$$

Now,

$$\frac{\partial \rho}{\partial t} = \zeta v\frac{\partial \rho}{\partial v} = \zeta \rho, \tag{17.73}$$

which should hold from Eq. (17.65). Then, we get,

$$\frac{\partial P}{\partial \rho} = \frac{\partial P}{\partial t}\left(\frac{\partial \rho}{\partial t}\right)^{-1} = \frac{\partial P}{\partial t}\left(\frac{\partial v}{\partial t}e^{\zeta t} + \zeta v e^{\zeta t}\right) = \frac{\partial P}{\partial t}\left(-\zeta v e^{\zeta t} + \zeta v e^{\zeta t}\right) = 0. \tag{17.74}$$

Plugging Eq. (17.74) into Eq. (17.70), we get

$$\frac{\partial P}{\partial t} = \zeta P + \zeta k_B T e^{2\zeta t}\frac{\partial^2 P}{\partial \rho^2}. \tag{17.75}$$

Bringing ζP to the left-hand side and multiplying both sides of Eq. (17.75) with $e^{-\zeta t}$, we get

$$\frac{\partial \left(Pe^{-\zeta t} \right)}{\partial t} = \zeta k_B Te^{\zeta t} \frac{\partial^2 P}{\partial \rho^2}$$

$$or, \frac{\partial \left(Pe^{-\zeta t} \right)}{\partial t} = \zeta k_B Te^{2\zeta t} \frac{\partial^2 \left(Pe^{-\zeta t} \right)}{\partial \rho^2}. \tag{17.76}$$

Now, we introduce

$$\frac{d\tau}{dt} = \zeta k_B Te^{2\zeta t}. \tag{17.77}$$

The solution of Eq. (17.77) is

$$\tau = \frac{k_B T}{2} e^{2\zeta t} - 1. \tag{17.78}$$

Hence, in terms of τ, we have

$$\frac{\partial \left(Pe^{-\zeta t} \right)}{\partial \tau} = \frac{\partial^2 \left(Pe^{-\zeta t} \right)}{\partial \rho^2}. \tag{17.79}$$

This is a diffusion equation for $Pe^{-\zeta t}$. Hence, using Eq. (17.67) we get the solution as

$$\boxed{P\left(v, t; v_0 \right) = \frac{1}{\sqrt{2\pi k_B T \left(1 - e^{-2\zeta t} \right)}} \exp\left(-\frac{\left(v - v_0 e^{-\zeta t} \right)^2}{2k_B T \left(1 - e^{-2\zeta t} \right)} \right).} \tag{17.80}$$

This is identical in form to the solution of the Smoluchowski equation in a harmonic surface. This is the classic solution of the FP equation in velocity space. A similar technique can be used to solve the FP equation in phase space (Kramers' equation) under special conditions of external field.

17.4.3 Solution of Kramers' Equation [Eq. (17.43)] for Field Free Case

In the absence of the potential field $U(x)$, the force term $U'(x)$ becomes zero. Hence, Eq. (17.43) becomes

$$\frac{\partial P}{\partial t} = \left(-v \frac{\partial}{\partial x} + \frac{\partial}{\partial v}(\zeta v) + \zeta k_B T \frac{\partial^2}{\partial v^2} \right) P$$

$$\frac{\partial P}{\partial t} = -v \frac{\partial P}{\partial x} + \zeta \frac{\partial}{\partial v}(vP) + \zeta k_B T \frac{\partial^2 P}{\partial v^2}$$

$$\frac{\partial P}{\partial t} = -v \frac{\partial P}{\partial x} + \zeta P + \zeta v \frac{\partial P}{\partial v} + \zeta k_B T \frac{\partial^2 P}{\partial v^2}. \tag{17.81}$$

For convenience, we consider $m = 1$, so that p (momentum) $= v$ (velocity). Before we proceed into the solution of Eq. (17.81), we note that without the second-order derivative term, the equation can be written as

$$\frac{\partial P}{\partial t} = -v\frac{\partial P}{\partial x} + \zeta P + \zeta v\frac{\partial P}{\partial v}. \tag{17.82}$$

The general solution Eq. (17.82) can be expressed in terms of the following equations:

$$\frac{\partial v}{\partial t} = -\zeta v \tag{17.83}$$

$$\frac{\partial x}{\partial t} = v \tag{17.84}$$

Hence, the first integrals for this system are $ve^{\zeta t}$ and $x + \dfrac{v}{\zeta}$.

Accordingly, for the solution of Eq. (17.81), we introduce the following new variables:

$$\rho = ve^{\zeta t} \tag{17.85}$$

$$\xi = x + \frac{v}{\zeta}. \tag{17.86}$$

Considering these variable transformations and minor algebraic manipulations, we get the following equations:

$$\frac{\partial P(x,v,t)}{\partial t} = \frac{\partial P(\rho,\xi,t)}{\partial t} + \zeta\rho\frac{\partial P}{\partial\rho} \tag{17.87}$$

$$\frac{\partial P}{\partial x} = \frac{\partial P}{\partial\xi} \tag{17.88}$$

$$\frac{\partial P}{\partial v} = e^{\zeta t}\frac{\partial P}{\partial\rho} + \frac{1}{\zeta}\frac{\partial P}{\partial\xi} \tag{17.89}$$

$$\frac{\partial^2 P}{\partial v^2} = e^{2\zeta t}\frac{\partial^2 P}{\partial\rho^2} + \frac{2}{\zeta}e^{\zeta t}\frac{\partial^2 P}{\partial\rho\partial\xi} + \frac{1}{\zeta^2}\frac{\partial^2 P}{\partial\xi^2}. \tag{17.90}$$

Substituting Eq. (17.87) to Eq. (17.90) in Eq. (17.81) we get,

$$\frac{dP}{dt} = -v\frac{\partial P}{\partial\xi} + \zeta v\left(\frac{\partial\rho}{\partial v}\frac{\partial P}{\partial\rho} + \frac{\partial\xi}{\partial v}\frac{\partial P}{\partial\xi}\right) + \zeta P + \zeta k_B T. \tag{17.91}$$

The left-hand side of the above equation can be expressed as

$$\frac{dP}{dt} = \frac{\partial\rho}{\partial t}\frac{\partial P}{\partial\rho} + \frac{\partial\xi}{\partial t}\frac{\partial P}{\partial\xi}. \tag{17.92}$$

Using Eq. (17.85) we can prove

$$\frac{\partial \rho}{\partial t} = \zeta v \frac{\partial \rho}{\partial v} = \zeta \rho. \tag{17.93}$$

Obviously, we have

$$\frac{\partial \xi}{\partial t} = 0 \tag{17.94}$$

and

$$-v = \zeta v \frac{\partial \xi}{\partial v}. \tag{17.95}$$

Using these and Eq. (17.86), we obtain

$$\frac{\partial P}{\partial t} = \zeta P + \zeta k_B T \left\{ e^{2\zeta t} \frac{\partial^2 P}{\partial \rho^2} + \frac{2}{\zeta} e^{\zeta t} \frac{\partial^2 P}{\partial \rho \partial \xi} + \frac{1}{\zeta^2} \frac{\partial^2 P}{\partial \xi^2} \right\}. \tag{17.96}$$

Now, we multiply both sides of Eq. (17.96) with $e^{-\zeta t}$ and define

$$\chi = P e^{-\zeta t}. \tag{17.97}$$

This reduces Eq. (17.96) to

$$\frac{\partial \chi}{\partial t} = \zeta k_B T \left\{ e^{2\zeta t} \frac{\partial^2 \chi}{\partial \rho^2} + \frac{2}{\zeta} e^{\zeta t} \frac{\partial^2 \chi}{\partial \rho \partial \xi} + \frac{1}{\zeta^2} \frac{\partial^2 \chi}{\partial \xi^2} \right\}. \tag{17.98}$$

The solution of Eq. (17.98) is achieved from the following lemma. Consider a differential equation for any two arbitrary variables of time, say $\phi(t)$ and $\psi(t)$.

$$\frac{\partial \chi}{\partial t} = \phi^2(t) \frac{\partial^2 \chi}{\partial \rho^2} + 2\phi(t) \psi(t) \frac{\partial^2 \chi}{\partial \rho \partial \xi} + \psi^2(t) \frac{\partial^2 \chi}{\partial \xi^2}. \tag{17.99}$$

$$\rho(t = 0) = \xi(t = 0) = 0$$

The solution of Eq. (17.99) is given by

$$\chi = \frac{1}{2\pi \Delta^{\frac{1}{2}}} \exp\left(-\frac{a\rho^2 + 2h\rho\xi + b\xi^2}{2\Delta} \right) \tag{17.100}$$

where

$$a = 2 \int_0^t dt' \psi^2(t') \tag{17.101}$$

$$b = 2 \int_0^t dt' \phi^2(t') \tag{17.102}$$

$$h = -2\int_0^t dt' \phi(t')\, \psi(t') \tag{17.103}$$

$$\Delta = ab - h^2 \tag{17.104}$$

This lemma can be easily verified by putting χ from Eq. (17.100) into Eq. (17.99) and comparing both sides of the resulting equation. Now, we obtain Eq. (17.98) from Eq. (17.99) by putting

$$\phi^2(t) = \zeta k_B T e^{2\zeta t} \tag{17.105}$$

$$\psi^2(t) = \frac{k_B T}{\zeta}. \tag{17.106}$$

Hence, the solution of Eq. (17.98) follows from Eq. (17.100) as

$$\chi = \frac{1}{2\pi \Delta^{1/2}} \exp\left(-\frac{a(\rho-\rho_0)^2 + 2h(\rho-\rho_0)(\xi-\xi_0) + b(\xi-\xi_0)^2}{2\Delta} \right) \tag{17.107}$$

where

$$a = \frac{2k_B T}{\zeta} t \tag{17.108}$$

$$b = 2\zeta k_B T \left(e^{2\zeta t} - 1\right) \tag{17.109}$$

$$h = -\frac{2k_B T}{\zeta}\left(e^{\zeta t} - 1\right) \tag{17.110}$$

$$\rho - \rho_0 = e^{\zeta t} v - v_0 \tag{17.111}$$

$$\xi - \xi_0 = x + \frac{v}{\zeta} - x_0 - \frac{v_0}{\zeta}. \tag{17.112}$$

Now, plugging Eq. (17.97) in Eq. (17.107), we get

$$P = \frac{e^{\zeta t}}{2\pi \Delta^{1/2}} \exp\left(-\frac{a(\rho-\rho_0)^2 + 2h(\rho-\rho_0)(\xi-\xi_0) + b(\xi-\xi_0)^2}{2\Delta} \right). \tag{17.113}$$

This is the solution of Kramers' equation for a field free situation.

17.4.4 SOLUTION OF KRAMERS' EQUATION [EQ. (17.43)] FOR A HARMONIC POTENTIAL FIELD

In this case the force from external field is given by

$$-U'(x) = -\omega^2 x \tag{17.114}$$

where ω is the harmonic frequency. Hence, Eq. (17.43) is modified to

$$\frac{\partial P}{\partial t} = -v\frac{\partial P}{\partial x} + \zeta v\frac{\partial P}{\partial v} + \zeta P + \omega^2 x\frac{\partial P}{\partial v} + \zeta k_B T\frac{\partial^2 P}{\partial v^2}. \tag{17.115}$$

We consider mass $m = 1$ for simplicity. Following the method developed in the previous section, the first integrals associated with this system are

$$\begin{aligned}\frac{dx}{dt} &= v\\\frac{dv}{dt} &= -\zeta v - \omega^2 x.\end{aligned} \tag{17.116}$$

This readily gives us the transformed variables as

$$\begin{aligned}\rho &= \left(x\mu_1 - v\right)e^{-\mu_2 t}\\\xi &= \left(x\mu_2 - v\right)e^{-\mu_1 t}.\end{aligned} \tag{17.117}$$

Here,

$$\begin{aligned}\mu_1 &= -\frac{\zeta}{2} + \sqrt{\frac{\zeta^2}{4} - \omega^2}\\\mu_2 &= -\frac{\zeta}{2} - \sqrt{\frac{\zeta^2}{4} - \omega^2}.\end{aligned} \tag{17.118}$$

The derivation of Eq. (17.118) is discussed in the previous chapter. Using these variables, Eq. (17.115) becomes

$$\frac{\partial P}{\partial t} = \zeta P + \zeta k_B T\left(e^{-2\mu_2 t}\frac{\partial^2 P}{\partial \rho^2} + 2e^{-(\mu_1+\mu_2)t}\frac{\partial^2 P}{\partial \rho \partial \xi} + e^{-2\mu_1 t}\frac{\partial^2 P}{\partial \xi^2}\right). \tag{17.119}$$

Multiplying both sides of this equation with $e^{\zeta t}$ and introducing a new variable

$$\chi = Pe^{-\zeta t} \tag{17.120}$$

we get

$$\frac{\partial \chi}{\partial t} = \zeta k_B T\left(e^{-2\mu_2 t}\frac{\partial^2 \chi}{\partial \rho^2} + 2e^{-(\mu_1+\mu_2)t}\frac{\partial^2 \chi}{\partial \rho \partial \xi} + e^{-2\mu_1 t}\frac{\partial^2 \chi}{\partial \xi^2}\right). \tag{17.121}$$

This equation has the same form as in the lemma introduced in Eq. (17.99). Accordingly, the solution of this equation is

$$\chi = \frac{1}{2\pi\Delta^{1/2}}\exp\left(-\frac{a\left(\rho-\rho_0\right)^2 + 2h\left(\rho-\rho_0\right)\left(\xi-\xi_0\right) + b\left(\xi-\xi_0\right)^2}{2\Delta}\right) \tag{17.122}$$

where

$$a = 2\zeta k_B T \int_0^t dt' e^{-2\mu_1 t'} = \frac{\zeta k_B T}{\mu_1}\left(1 - e^{-2\mu_1 t}\right)$$

$$b = 2\zeta k_B T \int_0^t dt' e^{-2\mu_2 t'} = \frac{\zeta k_B T}{\mu_2}\left(1 - e^{-2\mu_2 t}\right)$$

$$h = 2\zeta k_B T \int_0^t dt' e^{-(\mu_1 + \mu_2)t'} = \frac{2\zeta k_B T}{\mu_1 + \mu_2}\left(1 - e^{-(\mu_1 + \mu_2)t}\right)$$

$$\Delta = ab - h^2.$$

(17.123)

Plugging in Eq. (17.120) into Eq. (17.122), we get the solution of Kramers' equation for harmonic force field as

$$P = \frac{e^{\zeta t}}{2\pi\Delta^{1/2}} \exp\left(-\frac{a(\rho - \rho_0)^2 + 2h(\rho - \rho_0)(\xi - \xi_0) + b(\xi - \xi_0)^2}{2\Delta}\right).$$

(17.124)

It should be noted that the solutions of probability distribution functions derived in this chapter correspond to those derived in the previous chapter, starting from Langevin equations.

17.5 SUMMARY

The equations discussed here form the backbone of phenomenological time-dependent statistical mechanics. The robustness of these equations comes from the fact that they are based on conservation laws, and the fluctuation-dissipation theorem. All these equations obey number conservation. The velocity is dissipated but follows the F-D theorem. They all provide correct equilibrium distribution at long times.

These equations are closely related to the theory of random walk and to the Langevin equation. In the celebrated review of Chandrasekhar, the equivalence to random walk has been elegantly demonstrated.

In the following two chapters we discuss random walk and master equations.

BIBLIOGRAPHY

1. Chandrasekhar, S. 1943. Stochastic problems in physics and astronomy. *Reviews of Modern Physics, 15*, 1.
2. Risken, H. 1996. Fokker-Planck equation. In *The Fokker-Planck Equation*. Berlin: Springer. pp. 63–95.
3. Zwanzig, R. 2001 *Nonequilibrium Statistical Mechanics*. Oxford: Oxford University Press.
4. van Kampen, N. G. 1992. *Stochastic Processes in Physics and Chemistry*. Amsterdam: Elsevier.
5. Haken, H. 2004. *Synergetics: An Introduction*. Berlin: Springer. pp. 1–387.

18 Master Equations

OVERVIEW

Master equations (ME) provide a powerful yet intuitively appealing approach to describe time-dependent changes in an observable in terms of discrete changes in the microscopic state of the system. In many studies of nonequilibrium phenomena, time-dependent change in the property of interest involves the transfer of population from one state to the other. An example is the change of state in a chemical reaction. Another example is the transition from one energy level to another as in electronic and vibrational energy relaxation involving multiple eigenstates. The change in population in one state is determined not only by the population in that state but also by the transition rate from one state to the other. The equation of motion that describes such processes is called the master equation. It is also called the birth and death equation, for obvious reasons. Such a mathematical description of population dynamics is ubiquitous across many branches of physics, chemistry, biology, and even social sciences. Additionally, the master equation allows easy derivations of many phenomenological equations of time-dependent statistical mechanics and easy connection to random walk problems. The main goal of this chapter is to discuss interesting examples where the master equation has played an important role in explaining observed phenomena gathered from real-world science. The study of the master equation follows from the seminal work of Chapman and Kolmogorov. The master equation approach faces difficulty in the quantum world because of the presence of quantum coherences, which requires treatment of off-diagonal elements in a density matrix representation of the population dynamics. Some progress has been made in this direction in certain limits, and these have been discussed in this chapter.

18.1 INTRODUCTION

Master equations (ME) are phenomenological first-order differential equations, often employed to study the time-evolution of population distribution in a system characterized by multiple distinct states. The population distribution among the states changes due to repeated jumps or transitions from one state to another with time. As a result of these transitions, the macroscopic state of the system changes, as a function of time, from an initial nonequilibrium state. ME can provide a robust description, provided certain conditions are met. Because of its apparent simplicity, the master equation approach is widely used to model realistic systems and events, including population dynamics, energy transfer, diffusion, cell growth, bacterial colonization, epidemic spread, etc., to name a few.

In most examples of physics and chemistry, the master equation involves the rate of change of population distribution in a multistate or multilevel system. We denote the probability of

DOI: 10.1201/9781003157601-21

finding the system at a state $\mathbf{m}$ at a time t by $P(\mathbf{m}, t)$. Here $\mathbf{m}$ is a column vector with multiple components $\{m_1, m_2, m_3, m_4 \ldots\}$, which denotes occupation in different microscopic states. $P(\mathbf{m}, t)$ thus denotes the probability of finding the system in one of many states, characterized by population numbers $\{m_1, m_2, m_3, m_4 \ldots\}$ in microscopic states. Thus, in a three-state system, $P(1,0,0)$ denotes that the system is in the first state. The probability decreases when a transition occurs from $\mathbf{m}$ to $\mathbf{m}'$ and increases on transitions from $\mathbf{m}'$ to $\mathbf{m}$. Thus, it is also called the birth and death processes.

Let us explain this by considering a system of two coins. It has four distinct microscopic states (HH, HT, TH, TT), but three macroscopically distinct states because HT and TH are indistinguishable (the coins are identical), where H indicates the head and T indicates the tail of the coin. Now, we may start with the first state, HH. Then $m_1=1$, but all other m's are zero. Let us consider a single flip event. If we now flip one coin and get heads again, then there is no change. However, if we get tails, then we shall have $m_2 = 1$, and other populations zero. The important point to note is that one can reach HT or TH states in two ways. One can write a master equation to describe this process. In fact, Glauber's kinetic Ising model essentially models the same phenomenon but with spins instead of coins.

We now derive the master equation. Mathematically, the time-dependence of $P(\mathbf{m}, t)$ is given by an equation of motion of the following form that includes the sum if gain and loss of the population vector:

$$\frac{dP(\mathbf{m}, t)}{dt} = (rate\ in) - (rate\ out). \tag{18.1}$$

The "*rate in*" consists of all such processes for which the system makes a transition from $\mathbf{m}'$ to $\mathbf{m}$, already described above. Therefore, it is a sum over $\mathbf{m}'$. Each term of the sum is given by a product of the probability of the system at $\mathbf{m}'$ and the *probability* (w) of *transition* from $\mathbf{m}'$ to $\mathbf{m}$ [Eq. (18.2)].

$$rate\ in = \sum_{\mathbf{m}'} w(\mathbf{m}, \mathbf{m}') P(\mathbf{m}', t). \tag{18.2}$$

On the other hand, the "'*rate out*' term is given by Eq. (18.3) where the starting point is always $\mathbf{m}$, but going to all other $\mathbf{m}'$ states, so sum is over $\mathbf{m}'$,

$$rate\ out = P(\mathbf{m}, t) \sum_{\mathbf{m}'} w(\mathbf{m}', \mathbf{m}). \tag{18.3}$$

Therefore, one can write Eq. (18.1) as

$$\boxed{\frac{dP(\mathbf{m}, t)}{dt} = \sum_{\mathbf{m}'} w(\mathbf{m}, \mathbf{m}') P(\mathbf{m}', t) - P(\mathbf{m}, t) \sum_{\mathbf{m}'} w(\mathbf{m}', \mathbf{m})} \tag{18.4}$$

Eq. (18.4) is the generic form of a *master equation*. This form should seem obvious from the above arguments. *However, the crux of the problem lies in the determination of the transition probabilities (or, the rates), which requires a certain level of sophistication, as we shall see in the subsequent sections.* Note that Eq. (18.4) is instantaneous. That is, the change of population in a given state does not depend on the history of that state or the other states. Therefore, this equation is Markovian (no memory).

The simplest, a high school chemistry example of a master equation is provided by the two-state model of a chemical reaction, $A(1) \underset{k_{21}}{\overset{k_{12}}{\rightleftharpoons}} B(2)$, with states denoted by "1" and "2". The rate equation is given by

$$\frac{dP_1}{dt} = -k_{12}P_1 + k_{21}P_2 \qquad (18.5)$$

$$\frac{dP_2}{dt} = k_{12}P_1 - k_{21}P_2 \qquad (18.6)$$

where k_{ij} is the rate of transition from state "i" to the state "j", which is the same as "w". Note that $(P_1 + P_2)$ is conserved, that is, $\frac{d}{dt}(P_1 + P_2) = 0$.

In the subsequent discussions, we first introduce the Chapman-Kolmogorov (CK) equation and we outline the formal derivation of a master equation starting from the CK equation. We shall later discuss the ways to obtain the rates and transition probabilities from *Fermi's golden rule* and *transition state rate theory*. Subsequently, we elaborate (i) the Montroll-Schuler treatment of chemical kinetics; (ii) quantum master equations; (iii) methods of solutions of master equations, and (iv) the principle of detailed balance. We discuss two additional, modern applications of maser equations, namely, in the understanding of Levinthal's paradox in protein folding and the frequency-dependent heat capacity.

18.2   CHAPMAN-KOLMOGOROV (CK) EQUATION

This is one of the most important equations in the probabilistic (mathematical) description of dynamical events. The basic motivation of the CK equation is as follows. We often are not interested in the state/position of an individual particle in a system of many particles or the level of excitation in a donor molecule or the precise position of an individual walker. *Rather, it is the distribution, that is, the number of particles or excitations in different states, or walkers in different locations, that is required to determine the observed property.* Thus, a stochastic description becomes appropriate while modeling population dynamics in chemical kinetics or in biological processes.

In a Markov stochastic process transition probabilities do not depend on the past probabilities but only on the two states involved in the transition. Thus, in a sequence of events at times $t_1, t_2, \dots$ t_n, one can write the following condition (for any set of n successive times ($t_1 < t_2 < \dots < t_n$)

$$P\left(y_n, t_n \mid y_1, t_1; \dots; y_{n-1}, t_{n-1}\right) = W\left(y_n, t_n \mid y_{n-1}, t_{n-1}\right) \qquad (18.7)$$

Eq. (18.7) shows that *the conditional probability density at t_n, given the value at t_{n-1}, is uniquely determined and is not influenced by its earlier or past values.* In simple terms, one can call it a memory less process. The joint probability distribution can then be expressed as the starting probability and a sequence of transition probabilities. Such a decomposition is called a Markov chain, the hallmark of a Markov process. For example, for $n = 3$ one can write the joint probability distribution as a product of initial distribution and the transition probability distributions,

$$\begin{aligned}
P\left(y_1, t_1; y_2, t_2; y_3, t_3\right) &= P\left(y_1, t_1; y_2, t_2\right) P\left(y_3, t_3 \mid y_1, t_1; y_2, t_2\right) \\
&= P\left(y_1, t_1\right) W\left(y_2, t_2 \mid y_1, t_1\right) W\left(y_3, t_3 \mid y_2, t_2\right)
\end{aligned} \qquad (18.8)$$

The above is an interesting equation as it builds a transfer chain from one initial event. That is, the joint probability distribution of the whole sequence (the Markov chain connecting the states) can be decomposed into individual steps and can be constructed from the initial probability $P\left(y_1, t_1\right)$ and a sequence of transition probabilities $W\left(y_n, t_n \mid y_{n-1}, t_{n-1}\right)$.

Let us now eliminate the intermediate event "2" by integrating Eq. (18.8) with respect to y_2 and obtain the following equation:

$$P\left(y_1,t_1;y_3,t_3\right)=P\left(y_1,t_1\right)\int dy_2 W\left(y_2,t_2\mid y_1,t_1\right)W\left(y_3,t_3\mid y_2,t_2\right). \tag{18.9}$$

Now we use the definition of conditional probability to obtain a highly useful equation. Towards this, we divide both sides of Eq. (18.9) by $P_1\left(y_1,t_1\right)$ to obtain the conditional probability $P\left(y_3,t_3\mid y_1,t_1\right)$. We then recognize that the conditional probability is the same as the transition probability to arrive at Eq. (18.10). This is the Chapman-Kolmogorov (CK) equation.

$$\boxed{W\left(y_3,t_3\mid y_1,t_1\right)=\int dy_2 W\left(y_2,t_2\mid y_1,t_1\right)W\left(y_3,t_3\mid y_2,t_2\right)}. \tag{18.10}$$

The above equation is an identity that must be obeyed by the transition probabilities of a Markov process.

18.3   DISCRETE MASTER EQUATION FROM THE CK EQUATION

Some things in the probability theory are just intuitive, and quite simple but involves time translation invariance. This applies to the following derivation of the discrete master equation from the CK equation. Master equation is an equivalent but more convenient version of the CK equation.

$$\boxed{\frac{\partial}{\partial t}P_m\left(t\right)=\sum_{m'}\left\{W_{mm'}P_{m'}\left(t\right)-W_{m'm}P_m\left(t\right)\right\}}. \tag{18.11}$$

Eq. (18.11) is the best known and the most common form of master equation and is a discretized version of Eq. (18.10) with common notation in terms of probabilities. The first term on the right-hand side is the *gain term* that considers the transition to state m from other states (m'). On the other hand, the second terms are the *loss term* that considers the transition from state m to other states. We note that $m=m'$ does not contribute to the sum.

18.4   QUANTUM MASTER EQUATIONS

In systems involving multiple quantum energy states, the classical treatment in terms of static/site probabilities cannot provide a complete description. An important and elaborative example is the photosynthetic reaction center, which is usually regarded as made up of seven two-level systems (Figure 18.1). The issue of energy fluctuations in such systems is of great importance. The main interest here is to understand the absorption of photons of various kinds, often referred to as one of the most important phenomena in nature.

In order to understand the energy flow, one can write down a Pauli master equation where the rate $\left(W_{mn}\right)$ is obtained from Fermi's golden rule (FGR) [Eq. (18.12)].

$$W_{mn}=\frac{2\pi}{\hbar}\lambda^2\left|V_{mn}\right|^2\delta\left(E_n-E_m\right) \tag{18.12}$$

where the Hamiltonian of the system is given by $H=H_0+\lambda V$ and the energy eigenvalues for the unperturbed system are provided by $H_0\left|m\right\rangle=E_m\left|m\right\rangle$. However, in the presence of static coupling between the energy levels and the fluctuations, the Pauli description breaks down completely. In

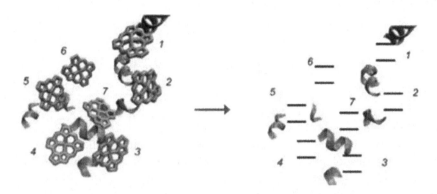

FIGURE 18.1 Schematic representation of the energy flows into the photosynthetic reaction center through the FMO (Fenna-Matthews-Olson) complex. Each of the seven sites, as highlighted on the right-hand side, can be considered as a collection of two-level systems.

Source: Taken with permission from [10].

order to study such systems, there are several existing approaches, which are elaborated in the *quantum dynamics* chapters. Here we provide a brief discussion.

A difficult job is to take care of the coherence, which is absent in classical treatments. In quantum master equations, the site probabilities are replaced by a density matrix, which contains the diagonal as well as off-diagonal terms. The Hamiltonian of such a system can be written as

$$H = H_S + H_B + V. \tag{18.13}$$

Here, H_S is the system Hamiltonian, H_B is the bath Hamiltonian, and V denotes the coupling between different states, which may or may not be influenced by the bath. In a photosynthetic complex, H_S gives the eigenvalues and eigenvectors. There are two possible descriptions as detailed below.

(i) *Site representation*: Here one assumes that V is small compared to system interactions. As a result, V contains a time-independent part and a time-dependent stochastic part.

(ii) *Excitonic representation*: Here one includes the static part of V into the Hamiltonian prior to diagonalization. This is a formidable approach as the coupling terms get influenced by the static terms. However, excitonic representation provides a better understanding of coherence.

In order to derive a quantum master equation, we need to integrate out the bath in a systematic fashion. Several approaches use perturbative approximation and limit up to V^2. Quantum master equations contain information of coherence, which is important not only for photosynthetic systems but also in electron transfer and NMR studies. The neglect of coherence allows us to have a description of rate constants. Several important master equations are given in Zwanzig's paper [6].

Below we give the final expression of the Nakajima-Zwanzig (non-Markovian) master equation [Eq. (18.14)]. The derivation is fairly complicated, which involves Zwanzig's projection operator technique and Dyson expansion.

$$\frac{d\rho_{mn}(t)}{dt} = -\int_0^t dt_1 \sum_{n \neq m} K_{mmnn}(t_1) \left[\rho_{nn}(t - t_1) - \rho_{mm}(t - t_1) \right] \tag{18.14}$$

where ρ_{mn} are the elements of the density matrix and the kernel (K) is explicitly given by Eq. (18.15).

$$K_{mmnn}(t) = \left[L_1 e^{-it(1-D)L} (1-D) L_1 \right]_{mmnn}. \tag{18.15}$$

Here the Liouville operator (L) is decomposed into unperturbed (L_0) part and perturbation (L_1). D is the projection operator.

There is another famous quantum master equation known as the Redfield equation, which is a Markovian master equation, widely applied in NMR studies. The underlying assumptions areas follows: (i) the system is weakly coupled to its environment, and (ii) the timescale of random fluctuations is much shorter than changes in the density operator. The Redfield equation is given by Eq. (18.16) where Γ is the relaxation super operator.

$$\boxed{\frac{\partial}{\partial t}\rho(t) = -i\left[\widehat{H}_0, \rho(t)\right] - \widehat{\widehat{\Gamma}}\left\{\rho(t) - \rho_{eq}\right\}.} \tag{18.16}$$

18.5 MASTER EQUATION DESCRIPTION OF THE RANDOM WALK MODEL AND DERIVATION OF DIFFUSION EQUATION

We have already discussed this topic in Chapter 16 and here we discuss the same here but from a master equation perspective, slightly more general. The master equation to describe change in probability distribution due a random walk is given by

$$P(m, n+1) = W(m \mid m-1) P(m-1, n) + W(m \mid m+1) P(m+1, n) \tag{18.17}$$

where $P(m, k)$ is the probability of being at m site after k steps and the transition probabilities as .....$W(m|m-1)$ and $W(m|m+1)$ are the transition rates to the left or right. For an unbiased RW, $W(m \mid m-1) = (1/2) = W(m \mid m+1)$. In a two-dimensional system, the right-hand side of Eq. (18.17) should contain four terms and each with a transition probability of $(1/4)$.

Next, we subtract $P(m, n)$ from both sides of Eq. (18.17) and divide by τ, the time taken to execute one step. Therefore, Eq. (18.17) becomes

$$\frac{P(m, n+1) - P(m, n)}{\tau} = \frac{1}{\tau}\begin{bmatrix} W(m \mid m-1) P(m-1, n) \\ + W(m \mid m+1) P(m+1, n) \\ -P(m, n) \end{bmatrix}. \tag{18.18}$$

We define the transition probability per unit time (rate constants) as $\tilde{w} = W / \tau$ and utilize the relation $W(m \mid m+1) + W(m \mid m-1) = 1$ to rewrite Eq. (18.18) as

$$\frac{P(m, n+1) - P(m, n)}{\tau} = \begin{bmatrix} \tilde{w}(m, m-1) P(m-1, n) \\ + \tilde{w}(m, m+1) P(m+1, n) \\ -\left\{\tilde{w}(m, m+1) + \tilde{w}(m, m-1)\right\} P(m, n) \end{bmatrix}. \tag{18.19}$$

Next we convert the discrete variable n to a time variable by using $t = n\tau$. Therefore,

$$\tilde{P}(m, t) \equiv P(m, t / \tau). \tag{18.20}$$

Hence, we can approximate the left-hand side of Eq. (18.19) as $\left(d\tilde{P} / dt \right)$ and write

$$\frac{d\tilde{P}(m,t)}{dt} = \begin{bmatrix} \tilde{w}(m,m-1)P(m-1,t) + \tilde{w}(m,m+1)P(m+1,t) \\ -\{\tilde{w}(m,m+1) + \tilde{w}(m,m-1)\}P(m,t) \end{bmatrix} \qquad (18.21)$$

Eq. (18.21) is the master equation for the one-dimensional random walk problem. Now we convert another discrete variable, m to a continuous variable by using $x = ma$, where x is the displacement after time t and a is the step size (or lattice spacing). Here, the walker is unbiased. Hence,

$$\tilde{w}(m,m-1) = \tilde{w}(m,m+1) = \frac{1}{2\tau}. \qquad (18.22)$$

Therefore, Eq. (18.21) can be written as

$$\frac{d\tilde{P}(m,t)}{dt} = \frac{1}{2\tau}\left[\tilde{P}(m-a,t) + \tilde{P}(m+a,t) - 2\tilde{P}(m,t) \right]. \qquad (18.23)$$

Now, we expand the right-hand side of Eq. (18.23) in a Taylor series with respect to x. After canceling the terms associated with a^0 and a^1 we get

$$\frac{\partial \tilde{P}(x,t)}{\partial t} = \frac{a^2}{2\tau}\frac{\partial^2 \tilde{P}(x,t)}{\partial x^2}. \qquad (18.24)$$

From Eq. (18.24) we finally arrive at the well-known form of diffusion equation where $D = \left(\frac{a^2}{2\tau}\right) = \frac{1}{2}ka^2$

$$\boxed{\frac{\partial \tilde{P}(x,t)}{\partial t} = D\frac{\partial^2 \tilde{P}(x,t)}{\partial x^2}}. \qquad (18.25)$$

The main merit, and also the outcome, of this derivation is the expression of D in terms of rate constant k and spacing a.

18.6 MONTROLL-SCHULER MASTER EQUATION TREATMENT OF CHEMICAL KINETICS

Montroll and Schuler developed a microscopic theory of chemical kinetics that can be thought of as a discrete quantum mechanical analogue of Kramers' Brownian motion model. In the following we summarize their work. Students should turn to the original article for details of the derivation.

Let us try to understand the rate of activation in unimolecular reactions of the following kind:

$$M + A(n) \underset{k_{-1}}{\overset{k_1}{\rightleftharpoons}} M + A^*(n')$$
$$A^*(n') \xrightarrow{k_D} \text{Products.} \qquad (18.26)$$

Here, $A(n)$ is the reactant (where n denotes a vibrational state), M is a molecule from the bath (of chemically inert gases) attached to the reaction system, $A^*(n')$ is the activated state (where $n'>n$),

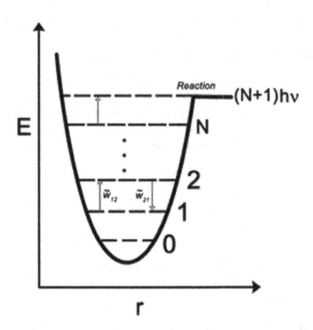

FIGURE 18.2 Schematic representation of the potential energy surface of a harmonic oscillator with nearest-neighbor transitions. The absorbing barrier is placed at $(N+1)^{th}$ level, which also serves as the activation energy required for dissociation.

which dissociates into products, k_1 is the rate of activation, k_{-1} is the rate of deactivation, and k_D is the rate of dissociation. This reaction scheme is the same as the Lindeman mechanism.

The ensemble of reactant molecules (A) possess quantized energy levels that are immersed in a large heat bath constituted of inert gas (M). The bath keeps the temperature of the reactive system fixed at T. The reactant molecules are excited by collision in a stepwise manner until they reach up to the $(N+1)^{th}$ energy level where the reactant molecules irreversibly form products. *The whole problem can be thought of as a one-dimensional random walk where the $(N+1)^{th}$ energy level acts as an absorbing barrier. The transition probabilities per unit time ($\tilde{w}_{mn}$) from energy level E_n to E_m depends on its distance from the n=0 level* (Figure 18.2).

Montroll and Schuler provided an elegant solution of the above problem to obtain a microscopic expression of the rate. The method of solution has been replicated in many other problems, including surface desorption. The approach of MS consists of two distinct steps. In the first step a master equation is written down for the time dependence of the population among the energy levels of the harmonic oscillator. In the second step, the detailed expression for the transition rates is obtained.

Let $F(t)$ be the fraction of reactant molecules that has not reached $n=N+1$ after time t. Hence,

$$F(t) = \sum_{n=0}^{N} x_n(t) \tag{18.27}$$

where $x_n(t)$ is the fraction of reactant molecules at the nth energy level after time t. The time evolution of $x_n(t)$ is given by the following master equation:

$$\frac{d}{dt} x_n(t) = \sum_{m \neq n} \left\{ \tilde{w}_{nm} x_m - \tilde{w}_{mn} x_n \right\}. \tag{18.28}$$

Now, in an infinitesimally small time interval (t to $t + \delta t$) the fraction of molecules that reaches the absorbing barrier is

$$-\left[F(t+\delta t) - F(t)\right] = -\left(\frac{dF}{dt}\right)\delta t. \tag{18.29}$$

We consider $P(t)$ as the distribution of fast passage times past level N. Hence, the number of molecules that pass the Nth level in the time interval of δt becomes $P(t)\,\delta t$. Therefore,

$$P(t) = -\left(\frac{dF(t)}{dt}\right) = -\frac{d}{dt}\sum_{n=0}^{N} x_n(t). \tag{18.30}$$

Hence, the mean first passage time (τ) can be calculated as

$$\tau = \int_0^\infty dt\, t\, P(t) = -\int_0^\infty dt\, t\, \frac{d}{dt}\sum_{n=0}^{N} x_n(t) = \int_0^\infty dt \sum_{n=0}^{N} x_n(t) \tag{18.31}$$

Now, for nearest-neighbor transitions (a selection rule) Eq. (18.28) can be written as

$$\frac{dx_0}{dt} = -\tilde{w}_{10}x_0 + \tilde{w}_{01}x_1 \tag{18.32}$$

$$\frac{dx_j}{dt} = \tilde{w}_{j,j-1}x_{j-1} - \left(\tilde{w}_{j-1,j} + \tilde{w}_{j+1,j}\right)x_j + \tilde{w}_{j,j+1}x_{j+1}\, ; j = 1, 2, \ldots, (N-1) \tag{18.33}$$

$$\frac{dx_N}{dt} = \tilde{w}_{N,N-1}x_{N-1} - \left(\tilde{w}_{N-1,N} + \tilde{w}_{N+1,N}\right)x_N. \tag{18.34}$$

It is quite formidable to obtain the collisional transition probabilities. However, with some simple assumptions, one can calculate $\tilde{w}_{mn}$. The assumptions are as follows:

(i) The reactant molecules are treated as simple harmonic oscillators (HOs).
(ii) There exist only weak interactions between the HOs and heat bath molecules.

Landau and Teller showed that if the transition probability per collision, W_{10} can be calculated for the transition $0 \rightarrow 1$, the linear perturbation of the vibrations of the diatomic molecules by the heat bath induce other transitions with probabilities

$$W_{mn} = \left[(m+1)\delta_{n-1,m} + m\delta_{n+1,m}\right]W_{10} = W_{nm}. \tag{18.35}$$

Here transitions are assumed to occur only between adjacent levels. Now, the transition probabilities per unit time, $\tilde{w}_{mn}$ are related to W_{mn} as follows:

$$\tilde{w}_{n+1,n} = ZN_b W_{n,n+1} \exp\left(h\nu/k_B T\right) \tag{18.36}$$

$$\tilde{w}_{n,n+1} = ZN_b W_{n,n+1}. \tag{18.37}$$

Here, Z is the collision number, that is, the number of collisions experienced by the HO per unit time when the gas density is unity and N_b is the concentration of the bath molecules. Eqs. (18.36) and (18.37) satisfy the principle of detailed balance as

$$\frac{\tilde{w}_{n+1,n}}{\tilde{w}_{n,n+1}} = \exp\left(\frac{h\nu}{k_B T}\right) \tag{18.38}$$

where $h\nu$ is the energy gap between any two adjacent levels of the HO. According to Eq. (18.30) and by using Eqs. (18.32)–(18.34), we can write

$$P(t) = -\frac{d}{dt}\sum_n x_n(t) = \tilde{w}_{N+1,N} x_N(t). \tag{18.39}$$

Therefore, the mean first passage time becomes [Eq. (18.31)]

$$\tau = \int_0^\infty dt\, t\, P(t) = \tilde{w}_{N+1,N} \int_0^\infty dt\, t\, x_N(t). \tag{18.40}$$

Montroll and Schuler showed that for the HO model of diatomic molecules, $x_n(t)$ can be written in terms of a linear combination as follows:

$$x_n(t) = \sum_{j=0}^N a_j \ell_n(\mu_j) \exp\left[\mu_j(e^{-\theta}-1)t\sigma\right] \tag{18.41}$$

where $\theta = \frac{h\nu}{k_B T}$. Here $\ell_n(\mu_j)$ denotes the nth order Gottlieb polynomial (which is related to the hypergeometric function) and σ is a parameter that depends on the coupling of reactant and the heat bath. By definition $\sigma = Z W_{10} N_b$. If one substitutes Eq. (18.41) into Eq. (18.40) for $N \to \infty$ limit, the final expression of the mean first passage time (MFPT) becomes

$$\boxed{\tau \sim \frac{\exp\left[(N+1)\theta\right]}{\sigma(N+1)(1-\exp[-\theta])^2}.} \tag{18.42}$$

Now we use the definition of σ, the relation between E_{act} and $(N+1)\theta$, and the partition function of HO $[Q_{HO} = (1-e^{-\theta})^{-1}]$ to rewrite Eq. (18.42) as

$$\tau = Q_{HO}^2 \left[Z W_{10} N_b (N+1)\right]^{-1} \exp\left[\beta E_{act}\right]. \tag{18.43}$$

As the rate is inversely proportional to the MFPT, we can write

$$\boxed{k = \frac{Z W_{10}(N+1)}{Q_{HO}^2} \exp\left[-\beta E_{act}\right].} \tag{18.44}$$

If one compares Eq. (18.44) with the celebrated Arrhenius expression $k = Ae^{-\beta E_{act}}$, one obtains an expression for the Arrhenius pre-factor $A = ZW_{10}(N+1)/Q_{HO}^2$.

The master equation approach was shown to be useful in developing models for barrierless reactions. Several of these models are exactly solvable, such as, Oster-Nishijima model, staircase model, and pinhole sink model. For further details and solutions we direct interested readers to [9] (Chapter 8).

18.7 PRINCIPLE OF DETAILED BALANCE

The principle of detailed balance is a principle of utmost importance in the description of nonequilibrium processes. It is derived from, and almost synonymous with the principle of microscopic reversibility. *It ensures that the correct equilibrium state is reached in the long-time limit.* At a quantitative level, it requires that there are as many transitions per second from state m to n as there are from n to m. That is, transitions between each pair of states, m and n, must balance. Hence, one can write

$$\boxed{W_{nm}P_m = W_{mn}P_n}.$$

(18.45)

Eq. (18.45) is the mathematical statement of the principle of detailed balance. We next prove Eq. (18.45), which holds for closed, isolated and finite physical systems with the following restrictions:

(i) the Hamiltonian of the system, $H(p, q)$ must be an even function of all momenta (p_k),
(ii) the set of macroscopic observables, $Y(p, q)$ must also be an even function of p_k.

Let us consider a system in a $2f$-dimensonal phase space Γ with positions q_k and momenta p_k $(k = 1, 2, 3 \dots f)$. Hamilton's equations of motions for this system are

$$\dot{q}_k = \frac{\partial H}{\partial p_k} \quad and \quad \dot{p}_k = -\frac{\partial H}{\partial q_k}.$$

(18.46)

If we substitute $t = -\bar{t}, q_k = \bar{q}_k$, and $p_k = -\bar{p}_k$ the equations of motions would remain invariant by virtue of condition (i). *Therefore, for a particular trajectory in the phase space Γ, one can obtain the reverse trajectory by reflecting the values of p_k which is also a solution of Eq. (18.46).* This tells us that, if a phase space point (q, p) goes to (q', p') after a time interval (τ) then the system shall, starting from $(q, -p)$, go to $(q', -p')$ in the same time interval. This is known as the time-reversal symmetry. The motion is fully deterministic in the phase space.

We shall hereafter use the following abbreviations for convenience: $x \equiv (q, p)$, $x^\tau \equiv (q', p')$ and $\bar{x} \equiv (q, -p)$. As mentioned earlier, the dynamic variables Y are even functions of momenta, we can write

$$Y_{\bar{x}}(0) = Y_x(0)$$
$$and, \ Y_{\bar{x}}(t) = Y_x(-t).$$

(18.47)

The equilibrium distributions are also an even function of the velocities (like Maxwell distribution) and we have

$$P^{eq}(\bar{x}) = P^{eq}(x).$$

(18.48)

Now, our primary is aim to show that $W(y_1,0\mid y_2,\tau)=W(y_2,0\mid y_1,\tau)$, which is the symmetry relation for P_2, by using the above relations. To this aim we write

$$
\begin{aligned}
W(y_1,0\mid y_2,\tau) &= \int dx\,\delta[y_1-Y_x(0)]\,\delta[y_2-Y_x(\tau)]P^{eq}(x)\\
&= \int d\bar{x}\,\delta[y_1-Y_{\bar{x}}(0)]\,\delta[y_2-Y_{\bar{x}}(\tau)]P^{eq}(\bar{x})\\
&= \int dx\,\delta[y_1-Y_x(0)]\,\delta[y_2-Y_x(-\tau)]P^{eq}(x)\\
&= W(y_1,0\mid y_2,-\tau)\\
&= W(y_2,0\mid y_1,\tau).
\end{aligned}
\tag{18.49}
$$

Eq. (18.49) can be written in terms of P as

$$
\boxed{W(y_2,\tau\mid y_1,0)\,P^e(y_1)=W(y_1,\tau\mid y_2,0)\,P^e(y_2).}
\tag{18.50}
$$

Eq. (18.50) is the same as Eq. (18.45), which is a rather simplified representation. It is noteworthy to mention that the principle of detailed balance is the same as the condition of chemical equilibrium for a reaction going back and forth between reactants and products. We would like to direct students to the textbook of van Kampen [2] for a detailed account of this derivation.

18.8   METHOD OF SOLUTION OF MASTER EQUATIONS

An important point is that master equations can often be solved by simple diagonalization. Of course, as shown by the examples in this chapter, the direct solution is only a minor part of the problem, but still it allows a simplification.

In the following we discuss how to solve a master equation to obtain the probability of the system being at a state i at time t [that is, $p_i(t)$], for a given set of equations. One of the most convenient ways is to form a transition probability matrix $(\mathbf{M})$ followed by obtaining its eigenvalues and eigenvectors. A set of coupled first-order differential equations can be written in the matrix form as follows [Eq. (18.51)]:

$$
\frac{d}{dt}\mathbf{P}=\mathbf{M.P}.
\tag{18.51}
$$

Here $\mathbf{P}$ is a column vector that contains the time-dependent population and $\mathbf{M}$ is the transition probability matrix that contains the rate constants. $\mathbf{M}$ is not necessarily symmetric and Hermitian. It becomes Hermitian only when $k_{ij}=k_{ji}$. In the symmetric case, $\mathbf{M}$ can be diagonalized by unitary transformation to obtain a diagonal matrix $\boldsymbol{\lambda}$.

$$
\boldsymbol{\lambda}=\mathbf{U^\dagger MU}.
\tag{18.52}
$$

Therefore, the solution of Eq. (18.51) can directly be written as

$$
\mathbf{P}(t)=\mathbf{U}\exp(\boldsymbol{\lambda}t)\mathbf{U^\dagger}\,\mathbf{P}(0)=\mathbf{T}(t)\mathbf{P}(0)
\tag{18.53}
$$

Here, $\mathbf{P}(0)$ is the population vector at time $t=0$ and the time-dependent coefficient matrix $\mathbf{T}(t)$ can be evaluated as, $T(t)=U\exp(\lambda t)U^\dagger$. According to Eq. (18.53), $p_i(t)$ is given by Eq. (18.54), where $T_{ij}(t)$ are the matrix elements that denote the conditional probabilities to find the system in a state i at time t provided that it was in state j at time $t=0$.

$$p_i(t) = \sum_j T_{ij}(t) p_j(0).$$ (18.54)

One can obtain the real eigenvalues (λ_α) and eigenvectors $(\mathbf{u}_\alpha)$ for a symmetric matrix, $\mathbf{M}$.

$$\mathbf{M}\mathbf{u}_\alpha = \lambda_\alpha \mathbf{u}_\alpha.$$ (18.55)

By using the eigenvalues and eigenvectors, the solution of a set of master equations can be written as the sum of exponentials as follows:

$$\mathbf{P}(t) = \sum_\alpha c_\alpha \exp(-\lambda_\alpha t) \mathbf{u}_\alpha.$$ (18.56)

The coefficients (c_α) are calculated from the initial conditions. Note that, if $k_{ij} \neq k_{ji}$, $\mathbf{M}$ becomes non-Hermitian and unitary transformation does not exist.

Numerical solutions of master equations can also be obtained by the method of propagation, Runge-Kutta method, cellular automata simulations, etc. We encourage students to follow this formalism and apply the same to the problem given in the exercise.

18.9 EIGENVALUE ANALYSIS IN CHEMICAL REACTION DYNAMICS

This analysis is due to Widom. It provides a basis for understanding the long-time nature of the reaction rate and allows us to separate the true rate from the transients.

Let us assume that each of the molecules is all the time in one of the well-defined microscopic states namely reactant (A) and product (B). The transition from one state to another state, which one molecule continuously experiences through collisions, can be represented by instantaneous jumps from one to another state. Mathematical formulation of such stochastic model can be represented by the following master equation:

$$\frac{dn_i}{dt} = \sum_j \left(W_{ji} n_j - W_{ij} n_i \right).$$ (18.57)

Here $n_i(t)$ is the number of molecules in state i and and $n_j(t)$ in state j at time t. W_{ij} is the transition probability per unit time from state i to state j. The sum of n_i overall reactant states is N_A and overall product states is N_B.

The solution of the above equation is given by

$$n_i(t) = n_i(\infty) + \sum_l c_{il} \exp(-\lambda_l t).$$ (18.58)

Here λ_l is the relaxation rate associated with l^{th} mode of relaxation that is a function of the transition probability W_{ij}. Mathematically one can construct λ_l from W_{ij} by finding the eigenvalues of the matrix of coefficients on the right-hand side of the coupled Eq. (18.57). One of these eigenvalues is necessarily zero, and is associated with the constant term $n_i(\infty)$ in Eq. (18.58), the negatives of the remaining eigenvalues (that are themselves negative) are relaxation rates λ_l.

At short time many modes of relaxation contribute, however, when elapsed time t will be very large, $(\lambda_2 - \lambda_1)t \gg 1$, then every term beyond for $l = 1$ in Eq. (18.58) will have negligible contributions compared to $l = 1$. So, for a larger value of t (when the transient faster modes of relaxation will be irrelevant, i.e., have negligible contribution) the relaxation rate is entirely governed by the slowest mode of relaxation.

18.10   A FEW IMPORTANT APPLICATIONS OF THE MASTER EQUATION

In this section we discuss three applications of master equations, although there are several others. These are (a) Levinthal's paradox; (b) frequency-dependent specific heat; and (c) epidemic modeling.

18.10.1   LEVINTHAL'S PARADOX

Levinthal's paradox is a conceptually interesting and extremely important topic in protein folding. Broadly there are two main problems in protein folding: (i) prediction of the three-dimensional folded structure of a protein from its amino acid sequences and (ii) the dynamics/timescale of protein folding. Levinthal's paradox (Cyrus Levinthal, 1969) is connected with the latter. Levinthal's paradox is that *finding the native folded state of a protein by a random search among all possible configurations (which is astronomically large) can take an enormously long time*. Nevertheless, proteins can fold in seconds or less. Let us understand this with the following numerical example.

We ask the following question: "*How long does it take for an unfolded polypeptide chain to fold to its unique native state?*" For example, let us consider a chain of 101 amino acid residues connected by 100 bonds. Each peptide bond may exhibit several conformations (say three). Therefore, there are $3^{100}(=5\times10^{47})$ possible conformations. If we assume that the protein explores 10^{13} configurations per second, it would take 10^{27} years to sample the full conformation space. Interestingly, protein folding happens in the milliseconds to seconds time scale. Here we follow the Zwanzig-Szabo-Bagchi (ZSB) formulation and address this paradox with the help of a master-equation-based kinetic model.

By following Richard Dawkins, ZSB presented an analysis exploring an analogy, which is later used as a clue to solve this problem, as discussed below. Dawkins framed the question differently: "How long does it take for a monkey to type Hamlet's remark '*Methinks it is like a weasel*' by random keystrokes?" Again the answer is an astronomically large number. However, if the monkey cannot change those letters that are already correctly in place, Hamlet's remark may be reached by a random search in only a few thousand keystrokes! That is, the introduction of a modest amount of bias might yield finite and believable timescales.

Let us consider a protein of $N+1$ residues connected by N number of bonds. The conformation around a bond is either "correct" (c) or "incorrect" (i) with respect to the native state. There can be several "i" states for a bond but only one unique "c" state. Therefore, the native state of the protein is "$ccccc....$(upto N terms)", starting from an arbitrary conformation of correct and incorrect bonds. Now we need to assign the rates of different transitions. A correct bond can become incorrect ($c\rightarrow i$) with the rate k_0 and an incorrect bond can become correct ($i\rightarrow c$) with the rate k_1. As a result, the number of incorrect bonds (S) changes over time and eventually reaches zero. So the protein is described by N peptide bonds and two rate constants. Here we aim to obtain an expression for the mean first passage time (MFPT), $\tau(S)$.

The kinetics can be described by the following equation, where $[c]+[i] = 1$:

$$\frac{d}{dt}[c] = -k_0[c]+k_1[i]. \tag{18.59}$$

And the equilibrium constant (K) can be expressed as

$$K = \frac{k_0}{k_1} = \frac{[i]_{eq}}{[c]_{eq}} = v\exp\left[-U/k_BT\right] \tag{18.60}$$

where v is the degeneracy of the incorrect bonds and $U (= E_i - E_c)$ is the energy difference between an incorrect and a correct bond. In other words, "U" is the penalty of making an incorrect bond. We

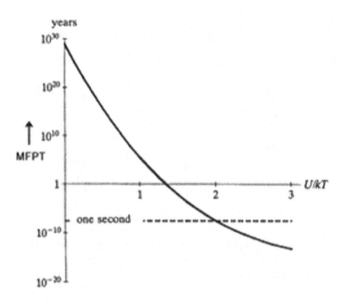

FIGURE 18.3 Mean first passage time of protein folding against the energy bias/penalty (U) as predicted by ZSB model. The model predicts biologically relevant timescales of protein folding for certain values of U.

denote the probability that there are "S" incorrect bonds at time t by $P(S, t)$. Therefore, the gain-loss master equation is given by

$$\frac{d}{dt}P(S,t) = (N-S+1)k_0 P(S-1,t) + (S+1)k_1 P(S+1,t)$$
$$- (N-S)k_0 P(S,t) - Sk_1 P(S,t). \tag{18.61}$$

With the following boundary conditions:

$$P(-1,t) = 0 = P(N+1,t). \tag{18.62}$$

Here we provide the final expression of the MFPT without the detailed derivation (given in [7]) and mathematical complexities. In the asymptotic limit of large N, we can write Eq. (18.63). The numerical results are given in Figure 18.3.

$$\tau(S) \approx \left(\frac{1}{Nk_0}\right)(1+K)^N \left[1 + \frac{1!}{NK} + \frac{2!}{N^2 K^2} + O(N^{-3})\right] \tag{18.63}$$

18.10.2 FREQUENCY-DEPENDENT SPECIFIC HEAT

We can measure the frequency-dependent specific heat [$C(\omega)$] by thermally perturbing the system slightly away from the equilibrium. We can also calculate $C(\omega)$ in terms of equilibrium energy fluctuations, as derived by Nielsen and Dyre for a system whose dynamics is described by a master equation. Below we give the expression of $C(\omega, T)$ in Eq.(18.64) where $s=i\omega$, $E(t, T)$ denotes the total energy of the system at time t and temperature T, and k_B is the Boltzmann constant.

$$C(\omega,T) = \frac{\langle E^2(T)\rangle}{k_B T^2} - \frac{s}{k_B T^2}\int_0^\infty dt\, e^{-st}\langle E(0,T)E(t,T)\rangle \tag{18.64}$$

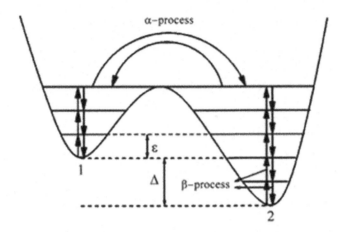

FIGURE 18.4 Schematic representation of the model under consideration. The horizontal lines within a well represent different excitation levels. The energy levels are in general degenerate, as they correspond to the sum of the energies of individual two-level systems in the collection.

Next we discuss a Green's function based formalism to obtain $C(\omega)$ by employing a kinetic model of glassy dynamics. Let us first briefly discuss the model. Here we consider a two-level systems (TLS) as shown in Figure 18.4, where the β-processes are associated with small-scale energy fluctuations and the α-processes are associated with large-scale energy fluctuations. The waiting time before a transition can occur from level i is assumed to follow a Poissonian for a β-transition, as shown in Eq. (18.65).

$$\psi_i(t) = \frac{1}{\tau_i}\exp(-t/\tau_i). \tag{18.65}$$

Here τ_i is the average residence time at the ith level. The condition for the detailed balance is given by Eq. (18.66), where ε is the energy difference between two levels.

$$K(T) = \frac{P_1(T)}{P_0(T)} = \frac{\tau_1(T)}{\tau_0(T)} = \exp\left[-\frac{\varepsilon}{k_B T}\right]. \tag{18.66}$$

Here $K(T)$ is the equilibrium constant. Now there are two metabasins, denoted by 1 and 2. $N_\beta^{(i)}$ is the number of noninteracting two level systems that characterize the ith metabasin. We define an occupation variable $\zeta_j^i(t)$ (for $j=1,2,\ldots,N_\beta^{(i)}$) that assumes a value "0" if the level 0 of the jth TLS is occupied, and "1" if otherwise. Thus one can define an order parameter, $Q_i(t)$ as follows:

$$Q_i(t) = \sum_{j=1}^{N_\beta^{(i)}} \zeta_j^i(t). \tag{18.67}$$

$Q_i(t)$ serves as an order parameter for dynamical changes involving metabasin i. Here an α-process is assumed to occur only when all the β-processes (TLSs) in a metabasin are simultaneously excited, i.e., when $Q_i = N_\beta^{(i)}$. There is a finite rate of transition k from each of the metabasins when this condition is satisfied. Therefore, the total energy of the system, as used in Eq. (18.64), can be expressed as

$$E(t,T) = \sum_{n=0}^{N_\beta^{(1)}} P_1(n;t,T)\left(N_\beta^{(1)} - N_\beta^{(2)} + n\right)\varepsilon + \sum_{n=0}^{N_\beta^{(2)}} P_2(n;t,T)n\varepsilon. \tag{18.68}$$

In Eq. (18.68), $P_i(n; t, T)$ is the probability that the order parameter Q_i assumes a value "n" in the ith well at time t and temperature T. The time evolution of these probabilities obeys the following master equation:

$$
\begin{aligned}
\frac{\partial}{\partial t} P_i(n; t, T) = & \left[\left(N_\beta^{(i)} - n + 1 \right) \middle/ \tau_0(T) \right] P_i(n-1; t, T) \\
& + \left[(n+1)/\tau_1(T) \right] P_i(n+1; t, T) \\
& - \left[\left(N_\beta^{(i)} - n \right) \middle/ \tau_0(T) \right] P_i(n; t, T) \\
& - \left[n/\tau_1(T) \right] P_i(n; t, T) - k \delta_{n, N_\beta^{(i)}} P_i(n; t, T) \\
& + k \delta_{n, N_\beta^{(i\pm1)}} \delta_{j, i\pm1} P_j(n; t, T).
\end{aligned}
\tag{18.69}
$$

In terms of the transition matrix $\mathbf{A}(T)$, we can write the master equations in the following compact form:

$$
\frac{\partial}{\partial t} \mathbf{P}(t, T) = \mathbf{A}(T) \mathbf{P}(t, T).
\tag{18.70}
$$

If $G_T(i, t \mid j, 0)$ is Green's function that gives the probability to be in state i at a later time t, given that the system is in the state j at time $t'=0$. Hence, Green's function also satisfies Eq.(18.70). Now one can express the energy-energy time correlation in terms of Green's function as follows:

$$
\langle E(0, T) E(t, T) \rangle = \sum_{i=1}^{N} \sum_{j=1}^{N} G_T(i, t \mid j, 0) E_i E_j P_{eq}(j, T)
\tag{18.71}
$$

where $P_{eq}(j, T)$ is the equilibrium probability of the jth state. If we define $\widehat{G}_T(i, s \mid j)$ to be the Laplace transform of $G_T(i, t \mid j, 0)$ then Eq. (18.64) can be expressed as

$$
C(\omega, T) = \frac{\langle E^2(T) \rangle}{k_B T^2} - \frac{s}{k_B T^2} \sum_{i=1}^{N} \sum_{j=1}^{N} \widehat{G}_T(i, s \mid j) E_i E_j P_{eq}(j, T).
\tag{18.72}
$$

In order to obtain $C(\omega, T)$, one needs to numerically solve Eq. (18.72) Interested readers are encouraged to go through [8] for a detailed encounter.

18.10.3 Epidemic Modeling

For a modern and important example, we now briefly discuss the master equation for the epidemic/pandemic progression. It is simple and also useful. This was popularized by Kermack and McKendrick in 1927. The master equation involved here is a classic birth-death equation. There are three states in this problem – the three human variants, susceptible (S), infected (I), and removed (R). The "R" category contains both *cured* and *deceased* population. The master equation takes the following form:

$$
\frac{dS}{dt} = -\beta SI
$$

$$
\frac{dI}{dt} = \beta SI - \gamma I
$$

$$
\frac{dR}{dt} = \gamma I.
\tag{18.73}
$$

The three coupled differential equations described in Eq. (18.73) are at the core of well-known *SIR* model of epidemic progression. Here β and γ are the rate constants. In mathematical epidemiology, the ratio of β to γ is known as the *basic reproduction ratio* (R_0), which indicates the potential of an infectious disease to cause an epidemic. Note that these equations are the same as the kinetic description of a consecutive chemical reaction. However, the simple *SIR* model is valid for a homogeneous distribution of population and requires several sophisticated modifications before its application to real scenarios.

18.11 SUMMARY

In this chapter we have discussed many interesting examples of dynamics where the master equation plays an important role in explaining the observed phenomena. These applications are taken from natural and biological sciences, in particular from chemistry and biology where, the macroscopic, thermodynamic state of a system is determined by occupation numbers of underlying microscopic states, like energy levels of a system. There are two aspects one should be aware of. (i) On one hand, the equilibrium fluctuations are determined by spontaneous change in the occupation number of the microscopic states. This has been illustrated in the example of specific heat calculation. (ii) On the other hand, the linear response of a system subsequent to a finite perturbation of the system is also determined by transitions among the microscopic states. Thus, relaxation of a nonequilibrium state to the equilibrium one requires reordering of the occupation probability among the microstates. Master equations can be used to describe both these situations.

The merit of the master equation approach is that it is intuitive and easy to formulate, and also easy to solve. Of course, certain conditions must be met before we can use master equation description, like the existence of well-defined microscopic states that can be used to describe the macroscopic state. It should be noted that the linearity of the master equations, encountered in most applications, is not a necessity. However, nonlinear equations are to be solved numerically by using quadrature as matrix inversion technique is no longer possible.

PROBLEMS

1. Consider the following scheme where a system possesses three states 1, 2, and 3. The transition probabilities, p and q, are shown with arrows.

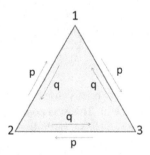

From the transition matrix, obtain its eigenvalues, and comment on the nature of the eigenvalues when $p \neq q$ and also when $p = q$. For the latter, find out the eigenvectors. Finally, plot the probability of being at state 3, that is, $P(3,t)$ against t by assuming that the initial state was 1. Try the same with any other values of p and q such that $p \neq q$ and $p + q = 1$.

2. Consider the following linear three site system where the transition from 1 to 3 (or the reverse) must occur via state 2.

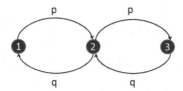

Suppose the system starts from $n = 1$. Find the equilibrium probability distribution. Plot $P(n, t)$ for $n = 1, 2$, and 3 considering $p = q$. What difference can you find compared to the system described in problem-1.

3. (a) Follow [7] and derive Eq. (18.63). (b) Follow [8] and derive Eq. (18.72).

4. Numerically solve the master equations given in Eq. (18.73) by assuming that at $t = 0$, 99% of the population is susceptible (S) and 1% is infected (I). Consider two situations: (i) $\beta \gg \gamma$ and (ii) $\beta \approx \gamma$. Plot the population of three different states, that is, S, I, and R against time.

BIBLIOGRAPHY AND SUGGESTED READING

1. Oppenheim I., K. Shuler and G. Weiss. 1977. *Master Equation*, Cambridge: MIT Press.
2. van Kampen, N. G. 1992. *Stochastic Processes in Physics and Chemistry*. Vol. 1. Amsterdam: Elsevier.
3. Haken, H. 2004. *Synergetics: An Introduction*. New York: Springer.
4. Montroll, E. W., and K. E. Shuler. 1957. The application of the theory of stochastic processes to chemical kinetics. *Advances in Chemical Physics*, *1*, 361–399.
5. Gillespie, D. T. 1992. A rigorous derivation of the chemical master equation. *Physica A: Statistical Mechanics and its Applications*, *188*(1–3), 404–425.
6. Zwanzig, R. On the identity of three generalized master equations. *Physica, 30*(no. 6), 1109–1123.
7. Zwanzig, R., A. Szabo, and B. Bagchi. Levinthal's paradox. 1992. *Proceedings of the National Academy of Sciences*, *89*(1), 20–22.
8. Chakrabarti, D., and B. Bagchi. 2005. Frequency dependent heat capacity within a kinetic model of glassy dynamics. *The Journal of Chemical Physics*, *122*(1), 014501.
9. Bagchi, B. 2012. *Molecular Relaxation in Liquids*. New York: Oxford University Press.
10. Dutta, R., and Bagchi, B. 2020. *The Journal of Physical Chemistry B*, *124*(22), 4551–4563.

19 Numerical Solution of Smoluchowski and Fokker-Planck Equations

OVERVIEW

As discussed in the last few chapters, the phenomenological descriptions of time-dependent statistical mechanics are based primarily on the Smoluchowski (SE) and the Fokker-Planck equations (FPE), in addition to the stochastic Langevin equation. The algebraic equations provide a better platform for mathematical formulations than the stochastic Langevin equation. Although they appear simple and physically transparent, SE and FPE are hard to solve analytically for realistic situations without making serious approximations. This is because even at a single-particle level of description, we need to solve second-order partial differential equations (PDE), which turn out to be prohibitively difficult in realistic situations where the external potential energy function is nonharmonic and involves multiple minima and maxima. These days it is a common practice to solve these equations numerically by generating stochastic trajectories on computer. Another important issue is that we often need to solve equations of motion that involve many-body potentials such that we need to deal with many-particle equations, which are far more difficult to deal with any analytical theory. Various numerical techniques have been developed that numerically simulate either the stochastic equations or propagate the algebraic equations directly. In this chapter we discuss a few such algorithms commonly used to solve these partial differential equations, with examples.

19.1 INTRODUCTION

In Chapter 17 we discussed the analytical solution of the Fokker-Planck equation for the special case of a particle constrained by a harmonic potential. *Even for such a simple system, the form of the analytical solution is rather complex.* This remains perhaps the only analytic solution for a somewhat nontrivial system.

The above statement underlines the difficulty. This partly was the reason that the field, despite its importance, remained stagnant for a long time. And as such there is no established or unique theoretical method to address the nonequilibrium problems. As mentioned above, in practical applications, the external potential acting on the particle is not harmonic. There are multiple reasons for the complexity that the student needs to appreciate. First, in liquids the motion of particles (atoms, molecules, colloidal particles) influenced by the interaction with other particles, and the force from these interactions is a complex function of positions of the surrounding particles. It is not possible to solve analytically even a Smoluchowski equation for such complex time-dependent many-body interactions. We have described earlier that for simple monatomic one-component liquids, one can use certain approximate treatment, initiated by de Gennes and further improved upon by using mode

DOI: 10.1201/9781003157601-22

coupling theory to obtain transport properties. However, most biological and chemical processes, the situations are too complex to be dealt by such simple analytical theories.

The alternative approach using computer simulations to directly solve these algebraic equations numerically by using various propagation techniques has proven highly effective. *The approach may involve two stages. (i) Computer simulations using advanced techniques (described in Chapter 25) can be used to obtain the free energy surface. (ii) Subsequently, the time evolution of the system is obtained by propagating the system on this free energy surface by using techniques to understand. This combined approach has proven useful in many cases, as discussed below.* Such a separation also has its limitations. This is currently an area of intense research.

What adds to the complexity is the fact that in nonequilibrium statistical mechanics, the dynamical variables of interest are not simple quantities like position of a particle or the angle of rotation, but collective variables like density, current density or magnetization. In many applications, the formulation of the appropriate collective variables require special attention, as discussed below. Thus, a probabilistic description can be involved as explained below with the important example of insulin dimer dissociation in water.

Briefly, insulin exists as a hexamer in pancreas and on exit from pancreas, it dissociates into three dimers. The two insulin monomers are bound together by hydrogen bonding and hydrophobic attractions. The biologically active species is the monomer. So, the dissociation of an insulin dimer into two monomers is an important biological process. However, how to quantitatively describe such a complex event?

We learned from our study of phase transitions (see Chapters 16–18 in the text "*Equilibrium Statistical Mechanics*", CRC Press, cited in the Bibliography) that collective phenomena can be successfully described by a set of order parameters. Order parameters are collective variables that take different values in the two stable states. We again explain with the example of insulin dimer dissociation. In this case, we need, at the simplest level, two order parameters, namely: (1) the distance separation, R, between the two monomers, and (2) the contact order parameter Q, which is the number of native contacts remaining during the dissociation. The description of the process of dissociation of the dimer has been described by use of two-dimensional free energy surface, which describes changes in free energy as the bound dimer separates by increasing R and decreasing Q. At the fully separated state R is larger than 6 nm and the value of Q is zero, as shown in Figure 19.1 we show the free energy landscape as a function of the two order parameters.

Thus, the goal is to calculate the rate of dissociation of the insulin dimer translates into calculation of the barrier crossing dynamics of the system on this free energy surface. Note that the free energy surface is dotted with multiple maxima (barriers) and minima. It is nontrivial to obtain this rate without serious approximations. In standard approaches of reaction rate calculation, discussed in our chapter on reaction rate theory (Chapter 20), we approximate the regions of interest (like the barrier region) as harmonic, and then attempt for a steady state solution, as pioneered by Kramers and Smoluchowski many years ago. However, a closed form analytical solution of a complex many-body problem is practically impossible. However, one can use computers to solve these, as discussed below.

The energy surface described in Figure 19.1 is already a vastly reduced description because we have absorbed many atomic degrees of freedom of proteins and water into only two order parameters.

Such cases are abundant in nature. We have discussed several examples of such processes in Chapters 20 and 25. In a recently discussed case, numerical solution of Brownian diffusion was employed to address the issue of rate enhancement in microdroplet where reactant executes random walk to reach the surface, which is the region of reaction [7]. Yet another example is the diffusion in a periodic potential where the barriers modify the motion of the tagged particle. This is relevant for surface diffusion.

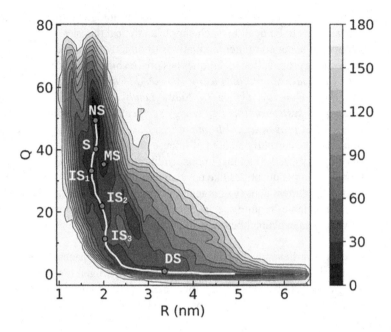

FIGURE 19.1 Two-dimensional free energy landscape (FEL) of insulin dimer dissociation in aqueous solution. The two order parameters that describe the FEL are center-of-mass distance between two monomers (R) and the number of contacts between the $C\alpha$ atoms of two monomers (Q). Starting from the native state, NS, several other metastable intermediate states are identified along the dissociation pathways. The structures of the dimer in this these states are shown below the FEL. Here, *IS* stands for intermediate state and MS stands for metastable state. The white line depicts the most probable minimum energy pathway (MEP).

Source: Taken with permission from [9].

In the following sections we describe numerical methods of solving two common probabilistic equations. The first one is a many-body Brownian diffusion, which simulates Smoluchowski equation of motion. The second one is a method of solving a one-body Fokker-Planck equation. There are several other approaches and techniques available, and here we focus on just two approaches with different attributes, to explain to students the scope of these methods.

19.2 SIMULATION OF *MANY-BODY* TRANSLATIONAL BROWNIAN DIFFUSION EQUATION

In this section, we discuss the method to numerically solve a many-body Smoluchowski equation given by

$$\frac{\partial P^{(N)}(r^{(N)},t)}{\partial t} = D\,\nabla^{(N)}.\left(\nabla^{(N)} - \beta \mathbf{F}^{(N)}\left(r^{(N)}\right)\right)P^{(N)}(r^{(N)},t). \tag{19.1}$$

Here, $P^{(N)}(r^{(N)},t)$ is the time-dependent joint probability distribution, which is equivalent to $P(\mathbf{r}_1,\mathbf{r}_2,...,\mathbf{r}_N)$, where $\mathbf{r}_i$ is the position vector of ith particle. $\nabla^{(N)}$ is N-dimensional gradient operator defined as

$$\nabla^{(N)} = \frac{\partial}{\partial \mathbf{r}_1} + \frac{\partial}{\partial \mathbf{r}_2} + \cdots + \frac{\partial}{\partial \mathbf{r}_N}. \tag{19.2}$$

$\frac{\partial}{\partial \mathbf{r}_i} = \hat{x}\frac{\partial}{\partial x_i} + \hat{y}\frac{\partial}{\partial y_i} + \hat{z}\frac{\partial}{\partial z_i}$. D is the diffusion constant. $F^{(N)}$ is the N-dimensional force vector, where each element is given by

$$\mathbf{F}_i^{(N)} = -\sum_{j \neq i} \nabla_{ji} U^{(N)}(\mathbf{r}_{ij}(t)). \tag{19.3}$$

That is, the summation of all the forces exerted on i^{th} particle by all other particles and $U^{(N)}\left(\mathbf{r}_{ij}(t)\right)$ is the potential energy of interaction.

As the particles move, the potential energy also acquires a time dependence, as indicated by the time in the parenthesis. The first term on the right-hand side is due to the presence of an (assumed) external force, such as an electric field on charged or dipole molecules of the systems). This can be set to zero in the absence of such an external force.

Note the difference from one-body Smoluchowski equation that is routinely used in chemical kinetics. The numerical method discussed below can be applied to the one-body problem also. However, in problems of interacting systems, one needs to use the many-body Smoluchowski equation. In a sense it is a simpler form of the BBGKY hierarchical equation of motion where the force acting on a tagged molecule contains the many-body effects. The equation of motion is given by the following Langevin equation type equation of motion,

$$\boxed{\frac{\partial \mathbf{r}_i}{\partial t} = \beta D_{tr} \mathbf{F}^{(Ext)}\left(\mathbf{r}_i(t)\right) + \beta D_{tr}\sum_j \mathbf{F}_{ij}\left(\mathbf{r}_{ij},t\right) + \eta_i(t), i,j = 1, N.} \tag{19.4}$$

Here $\eta_i(t)$ is the random velocity term with zero mean and satisfies the following fluctuation-dissipation relation

$$\left\langle \eta_i(0)\eta_j(t)\right\rangle = 2D_{tr}\delta_{ij}\,\delta(t). \tag{19.5}$$

The motion of the ith particle is connected to all other j particles through the force term, as in the BBGKY hierarchy. This equation needs to be written in finite difference form as

$$\boxed{\mathbf{r}_i(t+\delta t) = \mathbf{r}_i(t) + \left[\beta D \mathbf{F}^{(Ext)}\left(\mathbf{r}_i(t)\right) + \beta D_{tr}\sum_j \mathbf{F}_{ij}\left(\mathbf{r}_{ij},t\right) + \eta_i(t)\right]\delta t.} \tag{19.6}$$

The above equation resembles the Langevin equation, except the acceleration term has been dropped because we have neglected momentum relaxation, assumed to be fast.

Eqs (19.4)–(19.6) constitute the Brownian diffusion equation that can be propagated. The probability distribution, obtained from the trajectories so generated, obey the Smoluchowski equation. These equations have found wide use in dynamical studies of chemistry and biology. In the following we discuss a result obtained in the study of reactions in a microdroplet where one considers either a unimolecular or a bimolecular reaction. One of the reactants needs to diffuse to the surface. This is a slow reaction with time constant that range from ns to µs, hence cannot be easily studied by a molecular dynamic simulations, but can be investigated using Brownian dynamics simulations. In the example cited, the diffusion occurs through a noise term, thus we solve a one-body Smoluchowski equation.

Figure 19.2 gives the calculated survival probability obtained by solving the Smoluchowski equation.

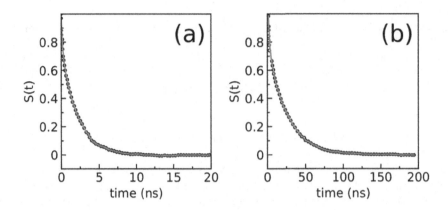

FIGURE 19.2 Survival probability ($S(t)$), which is defined as the fraction of particles unreacted at time t, for a bimolecular reaction occurring on a two-dimensional surface and obtained by solving Smoluchowski equation with reaction boundary conditions. (a) The reaction occurs in the bulk itself, and *(b)* reaction only on the circular ring, that is, the surface. The survival probability is plotted against time. It is important to note that the decay of $S(t)$ is nonexponential.

Source: Figures are taken with permission from [8].

19.3   ROTATIONAL BROWNIAN DIFFUSION EQUATION

We have discussed later in Chapter 22 that a single-particle rotational diffusion equation is given by the following pair of equations:

$$\frac{\partial}{\partial t}\rho(\Omega,t) = D_R \nabla_\Omega^2 \rho(\Omega,t) \tag{19.7}$$

$$\nabla_\Omega^2(\Omega) = \frac{1}{\sin(\theta)}\frac{\partial}{\partial\theta}\left[\sin(\theta)\frac{\partial}{\partial\theta}\right] + \frac{1}{\sin^2(\theta)}\frac{\partial^2}{\partial\varphi^2}. \tag{19.8}$$

However, this equation becomes inadequate when we need to study the many-body effects on rotational diffusion. In such a case we need a many-body rotational diffusion equation. This equation can be obtained by straightforwardly generalizing the above equation to obtain the following equation:

$$\frac{\partial \rho^{(N)}(\{\Omega_N\},t)}{\partial t} = D_R \sum_{n=1}^{N} \nabla_{\Omega n}\cdot\left\{\nabla_{\Omega n}\rho^{(N)} + \beta\rho^{(N)}\nabla_{\Omega n}U\right\}. \tag{19.9}$$

Here $\rho^{(N)}(\{\Omega_N\},t)$ is the N-particle orientational distribution with the set $\{\Omega_N\}$ denoting the orientation of N molecules. In Chapter 22 we consider a perturbative solution of the above complex equation for a system of point dipoles fixed on a simple cubic lattice. In Figure 19.3, we show a simple one-dimensional illustration to explain the scheme.

The Brownian dynamics equation of motion for the angle is given by

$$\dot{\theta}_i(t) = \left(D_R/k_B T\right)\nabla_\theta U_i(\theta,t) + N_i(t) \tag{19.10}$$

where $N_i(t)$ is the random angular velocity term with zero mean and satisfies the following fluctuation-dissipation relation:

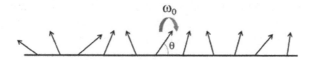

FIGURE 19.3 Schematic illustration of the model: spins arranged in a 1D lattice with the central spin rotating at a constant velocity of ω_0. The state of each spin is uniquely defined by an angle (θ) and a rotational velocity (ω). Coupling between the nearest-neighbor spins results in a gradient in the site-specific velocity profile.

$$\left\langle N_i(0)N_j(t)\right\rangle = 2D_R \delta_{ij}\,\delta(t). \tag{19.11}$$

And the corresponding finite difference equation is given by

$$\theta_i(t+\delta t) = \theta_i(t) + \left[\left(D_R/k_B T\right)\nabla_\theta U(\theta,t) + N_i(t)\right]\delta t. \tag{19.12}$$

As in the case of the translational motion case, rotational relaxation of the many-body diffusion equation provides a richer and complete description of the dynamics. As discussed in the course of Glauber spin dynamics covered in the chapter on the Ising model, the single-particle rotational dynamics can be obtained from the many-body equation.

In a detailed study, Zwanzig's dipolar lattice model was studied by Brownian dynamics simulations. The simplicity of the model allowed inquiry into several age-old problems like the validity of approximations made in Nee-Zwanzig dielectric relaxation theory.

19.4 FOKKER-PLANCK EQUATION

Although the Smoluchowski-equation-based description is widely used to model diffusion and reaction, inertial effects are important in many applications. Here we need to consider underdamped dynamics. In such cases, we need to use the full phase space Fokker-Plank equation (FPE) that we have discussed in Chapter 17. In a few simple cases analytical solution is still available, as discussed in Chandrasekhar's masterly review. However, analytical solution often eludes us in the presence of reactive event, such as reaction in the presence of a reaction sink on a harmonic surface. When the frequency ω of the harmonic potential is large ($\omega \geq \zeta/2$, ζ is the friction), the motion is considered underdamped.

The equation of motion for the probability distribution $P(x, v, t)$ is governed by the following full phase space FP equation

$$\frac{\partial P(x,v,t)}{\partial t} + v\frac{\partial P(x,v,t)}{\partial x} - \frac{\omega^2 x}{\mu}\frac{\partial P(x,v,t)}{\partial v} = \frac{\zeta}{\mu}\frac{\partial}{\partial v}\left(v + \frac{k_B T}{\mu}\frac{\partial}{\partial v}\right)$$

$$P(x,v,t) - k_{nr}S(x)P(x,v,t) - k_r P(x,v,t) \tag{19.13}$$

$$\left\langle R(t)R(t')\right\rangle = \left(2k_B T\beta/\mu\right)\delta(t-t') \tag{19.14}$$

where $\beta = \zeta/\mu$. ζ is the friction coefficient and μ is the effective mass of the particle. $S(x)$ is the sink function and k_{nr} is the magnitude of the radiationless rate at the origin of the sink where $S(x)$ is set to unity. k_r is the radiative rate constant.

We proceed as before and attempt to write the discretized equation of motion for propagation of the motion of a particle located at (x_{i-1}, v_{i-1}) at time t_{i-1}. In order to study the dynamics, one can obtain the following relation between the position of particles at times t_i and t_{i+1} in the phase space:

$$\begin{pmatrix} x_i \\ v_i \end{pmatrix} = \begin{pmatrix} 1 & \beta^{-1}\left[1 - \exp(-\beta\Delta t)\right] \\ 0 & \exp(-\beta\Delta t) \end{pmatrix} \begin{pmatrix} x_{i-1} \\ v_{i-1} \end{pmatrix} + \beta^{-1}K_{i-1}\begin{pmatrix} \Delta t - \beta^{-1}\left[1 - \exp(-\beta\Delta t)\right] \\ 1 - \exp(-\beta\Delta t) \end{pmatrix} + \begin{pmatrix} B_x(\Delta t) \\ B_v(\Delta t) \end{pmatrix}$$

(19.15)

where $K_{i-1} = -\mu^{-1}\partial v(x_{i-1})/\partial x$ denotes the acceleration of the particle at position x_{i-1}. $B_x(\Delta t)$ and $B_v(\Delta t)$ are the random variables with mean zero and variances given by

$$\langle B_x^2(\Delta t)\rangle = \left(k_B T/\mu\beta^2\right)\left\{2\beta\Delta t - 4\left[1 - \exp(-\beta\Delta t)\right] + \left[1 - \exp(-2\beta\Delta t)\right]\right\}$$ (19.16)

$$\langle B_v^2(\Delta t)\rangle = \left(k_B T/\mu\right)\left[1 - \exp(-2\beta\Delta t)\right].$$ (19.17)

Here the effects of the sink is introduced during time updates, with a decrease in the probability density. *Thus, the probability is no longer conserved.*

Fokker-Planck equation with a reactive sink has been solved numerically for the model problem of relaxation and reaction on a harmonic surface where a pinhole or Gaussian sink gives rise to the decay term. The results are shown in Figure 19.4.

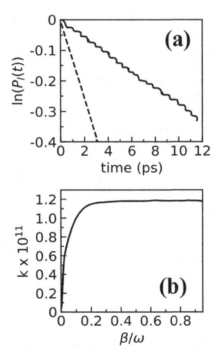

FIGURE 19.4 Example of numerical solution of a Fokker-Planck equation obtained by the procedure outlined above. (a) The decay of excited state population for a Gaussian sink at very low value of friction parameter β. The dashed line shows the solution for the overdamped Smoluchowski equation, which gives a much faster decay. (b) The rate constants of excited state population decay from a Gaussian sink plotted against the friction parameter β.

19.5 A NOTE ON NOMENCLATURE

We add this small section to clarify certain confusion in the literature regarding nomenclature. For example, we often find the Smoluchowski equation is referred to as the Fokker-Planck equation. Another subject of confusion concerns the Langevin equation. We note that the ordinary Langevin equation gets transformed into the Fokker-Planck equation. Only when we assume that the momentum relaxation is much faster than the positional relaxation do we recover the Smoluchowski equation. While the Smoluchowski equation is solved by using the discrete Brownian dynamics equation for the position coordinate, the Fokker-Planck equation is solved numerically by the Langevin equation. Alternatively, one can attempt to propagate the probability distributions directly as in quantum dynamics.

The Langevin equation has been discussed in great detail in Chapter 18, so we refrain from discussing it here, except to emphasize that algebraic equations are solved by going back to the discretized Langevin equation-like description. However, as discussed in the section above, we need a probabilistic formulation to account for reaction due to a sink term.

19.6 MACHINE LEARNING AND ARTIFICIAL-INTELLIGENCE-BASED STUDIES

Machine learning often means that one obtains by fitting or regression analysis the dependence of a critical quantity to environmental variation. For ML to work, we need a lot of *a priori* data or data generated during studies, of this dependence. This dependence, examples given below, is time consuming, yet needs to be known accurately. If one separate this part from the rest of the calculation where analysis can use this fitted data set, an enormous amount of simplification and saving of time results. The part which is determined *a priori* and called upon repeatedly can be called a core part. We need the information about the core part accurately

In such cases, one need not again calculate the core part in the large-scale analysis. So, considerable savings in computational effort and time can be achieved if we solve the core problem beforehand.

Let us discuss an example well-known in the spectroscopy literature. The vibrational relaxation rate of the -O-H stretch in water depends on the coupling of the -O-H bond to the fluctuation of the electric field acting on the bond from the surrounding water molecules. This fluctuation requires consideration of a system consisting of large number of water molecules, which can be treated classically. However, the response to the electric field is a quantum problem to be solved by detailed quantum calculation.

The electric field that is relevant is the one that acts on the bond (not on the molecule) and acts to modulate the frequency of the -O-H stretch.

In this problem, machine learning amounts to the generation of a master plot of the dependence of frequency ω on the magnitude of the electric field (E). This master curve of ω as a function of E can then be used to understand vibrational relaxation, where the fluctuation of the electric field can be obtained by classical simulations. The creation of the master curve requires extensive quantum chemical calculation of the variation of the frequency to the electric field acting on the said bond. In subsequent analysis, one generates a numerical fitting of this variation to an assumed form.

Thus, machine learning involves certain statistical procedures. In the above rather simple example, the fitting to the quantum calculation is to determine certain parameter values, which can be used repeatedly in classical, less demanding calculations.

However, the difficulty arises when the data set is very large and complex and continues to evolve continuously. Then the fitting parameters, even the form of the equation used to describe the core region, needs to evolve. This brings us to the domain of artificial intelligence.

As the name signifies, artificial intelligence (or AI) is an acquired intelligence. In computer simulations, it can facilitate theoretical studies by training certain aspects by using known data. This is already being used in coarse graining, where potential forms are obtained by fitting to known results.

Both machine learning and artificial intelligence involve fitting and regression analysis. What sets them apart is the ability of artificial intelligence programs to evolve continuously as new data set is obtained. This requires use of a very large number of fitting parameters.

While artificial intelligence can be quite useful in social sciences, and of course in essay writing on a given topic of interest, it finds difficulty in solving a specific problem. The success of alpha fold in predicting protein structure shows that this problem ca be circumvented if sufficient number of data is made available.

19.7   SUMMARY

In this chapter we have presented discussions on numerical solutions of two important stochastic equations, namely Smoluchowski and Fokker-Planck. For realistic external potentials, these stochastic equations are not amenable to analytical solutions. Thus, these need to be solved numerically. Fortunately, modern day computers allow us to solve many of these problems through propagation techniques.

The basic idea is to pursue a mixed approach where suitable techniques are first used to obtain the free energy surface, which is often multidimensional. In the second step one propagates stochastic equations to obtain the rate.

Any rate calculation remains a formidable challenge because we still need to obtain the diffusion constant(s) or the friction coefficient(s).

BIBLIOGRAPHY

1. Zwanzig, R. 2001. *Nonequilibrium Statistical Mechanics.* Oxford: Oxford University Press.
2. Zwanzig, R. 1963. Dielectric relaxation in a high-temperature dipole lattice. *Journal of Chemical Physics 38*, 2766.
3. Zhou, H., and B. Bagchi. 1992. Dielectric and orientational relaxation in a Brownian dipolar lattice. *Journal of Chemical Physics, 97*, 3610.
4. Nee, T., and R. Zwanzig. 1970. Theory of dielectric relaxation in polar liquids. *Journal of Chemical Physics, 52*, 6353.
5. Bagchi, B., S. Singer, and D. Oxtoby. 1983. Non-monotonic dependence of electronic relaxation rate and solvent viscocity. *Chemical Physics Letters, 99*(3), 225–231.
6. Zwanzig, R. 1969. Langevin theory of polymer dynamics in dilute solution. *Advances in Chemical Physics: Stochastic Processes in Chemical Physics, 15*, 325–331.
7. Chaudhari, P., *et al.* 2005. Equilibrium glassy phase in a polydisperse hard-sphere system. *Physical Review Letters, 95*, 248301.
8. Mondal, S., *et al.* 2018. Enhancement of reaction rate in small-sized droplets: A combined analytical and simulation study. *Journal of Chemical Physics, 148*, 244704.
9. Banarjee, P., S. Mondal, and B. Bagchi. 2018. Insulin dimer dissociation in aqueous solution: A computational study of free energy landscape and evolving microscopic structure along the reaction pathway. *Journal of Chemical Physics, 21*, 149(11), 114902.
10. Acharya, S., and B. Bagchi. 2022. Diffusion in the presence of correlated returns in a two-dimensional energy landscape and non-monotonic friction dependence: examination of simulation results by a random walk model. arXiv:2206.13565.
11. *Ermak, D., and J. A. McCammon. 1978. Brownian dynamics with hydrodynamic interaction. *Journal of Chemical Physics, 69*, 1352.
12. Jumper, J., R., Evans, A., Pritzel,T. Green, M. Figurnov,O. Ronneberger, ... and D. Hassabis. 2021. Highly accurate protein structure prediction with alpha fold. *Nature, 596*(7873), 583–589.
13. Keith, J. A., V. Vassilev-Galindo, B., Cheng, S., Chmiela, M., Gastegger, K. R. Müller, and A. Tkatchenko. 2021. Combining machine learning and computational chemistry for predictive insights into chemical systems. *Chemical Reviews, 121*(16), 9816–9872.
14. Noé, F., A. Tkatchenko, K. R. Müller, and C. Clementi. 2020. Machine learning for molecular simulation. *Annual Review of Physical Chemistry, 71*, 361–390.

20 Theory of Chemical Reaction Dynamics

OVERVIEW

The calculation of the rate of transition from one stable state to another over an activation barrier is of importance in chemistry and biology. Most common reactions occur in the liquid state where barrier crossing events face random environments. Hence statistical mechanics is called upon to explain the observed dependence of the yield and the rate of reactions on such variables as temperature and viscosity, solvent polarity, and composition. The theories of chemical kinetics were developed in stages. (i) The initial studies pioneered by Arrhenius and developed by Wigner and Eyring led to the formulation of the transition state theories (TST). These studies usually did not consider frictional effects but consider multidimensional reaction energy surface. (ii) On the other hand, the studies initiated by Smoluchowski in the early twentieth century, and followed by Kramers, focus on the dependence of the rate on the friction on one-dimensional reaction coordinate. (ii) However, these latter treatments largely consider a one-dimensional reaction energy surface. In an important work, Langer extended previous studies to treat multidimensional reaction surfaces with dissipation. (iv) Subsequent developments were motivated by reactions and have addressed complex issues of memory effects in the form of frequency-dependent friction. (v) More recent efforts have been directed to study memory effects in complex reactions that naturally involve multidimensional energy surfaces, as required in protein folding and protein association-dissociation reactions. We also discuss several interesting special cases, like barrierless reactions where the motion along the reaction coordinate from the reactant to the product does not face any sizeable barrier, enzyme kinetics, and electron transfer reactions that all need the methods of time-dependent statistical mechanics.

20.1 INTRODUCTION

The branch of physical chemistry that deals with chemical reaction dynamics is one of the most enduring, important, and exciting branches of chemistry. It is sometimes said that the study of chemistry is the study of chemical reactions, and calculation of rate forms an essential ingredient. This is also an important area of time-dependent statistical mechanics. We often need to calculate rates of processes that involve the crossing of an energy barrier known as the activation barrier. Sophisticated experimental techniques have been developed to investigate the dependence of the rate on temperature, viscosity, polarity, among other solvent properties. The development of laser-related techniques has given rise to a renaissance in this field since the 1980s. Time-dependent statistical mechanics, computer simulations and experimental techniques formed a powerful synergy during 1970–2000 to stimulate rapid developments in this field. Intense interest still continues.

DOI: 10.1201/9781003157601-23

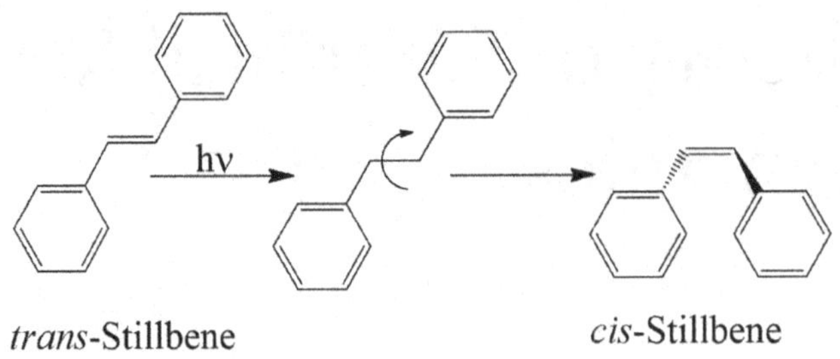

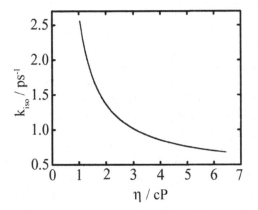

trans-Stillbene   **cis-Stillbene**

FIGURE 20.1 Photo-isomerization of trans-stilbene to cis-stilbene. This is an activated process and the reactive motion involves twist around the double bond in the middle of the two phenyl rings.

FIGURE 20.2 Viscosity dependence of the rate of trans-to-cis isomerization of stilbene in alkane solvents. **Source: Figure taken from [25].**

Let us consider a specific, well-known example, which is the trans-to-cis isomerization of stilbene. The reaction is shown below (Figure 20.1). This is a well-studied system where isomerization dynamics have been studied extensively, especially in the excited state.

This reaction faces a high activation energy barrier along the reaction coordinate (which is here the rotational motion along the backbone), the trans-to-cis reaction faces a slightly lower barrier than the reverse, cis-to-trans transformation. Importantly, the rate of this reaction is found to exhibit a strong viscosity dependence, which has been studied extensively as this serves as a prototype of chemical reaction in liquids, which is also of biological significance. The measured viscosity dependence is shown in Figure 20.2.

The viscosity dependence of a barrier crossing process was already in the mind of Smoluchowski. Einstein by then, had established the relation between diffusion and viscosity, through what came to be known as the Stokes-Einstein relation. The subsequent developments in this field were pioneered by Kramers, Wigner, and Eyring. The seminal review of Chandrasekhar that appeared in 1943 (published during the Second World War) does not mention the work of Kramers, which was published in 1940 but discussed the theory of Smoluchowski in an elegant fashion.

As mentioned, many reactions must cross a sizeable barrier. These are known as activated reactions, and they are abundant in nature. The height of the barrier offers great control on the rate because of the exponential dependence of the latter on the former, as given by $k = v e^{-\beta E_{act}}$, where E_{act} is the activation energy.

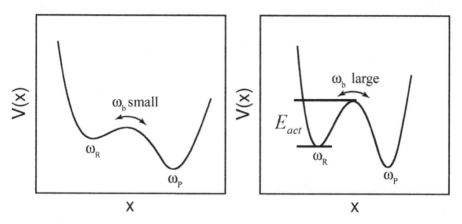

FIGURE 20.3 A bistable potential with two states R and P. k_f and k_b are the rates for forward, and backward reactions, respectively, and E_{act} is the corresponding activation energies. The theoretical formalisms we study in this chapter do not require the presence of the product minima. This is because the rates are calculated in terms of the flux over the barrier top and the absorptive wall is placed before the product well. ω_R, ω_b and ω_p are the reactant well, barrier top, and product well frequencies, respectively, giving the curvatures of the potential surface in the respective regions.

Much of the theoretical work has been devoted to the calculation of the rate in the activated reactions. This rate calculation involves two steps: (i) the calculation of the free energy or activation barrier, and (ii) calculation of the transmission coefficient (or, factor) which is often denoted by "v", as stated above, the calculation of the energy barrier was earlier left to quantum chemistry approaches. Of late, statistical mechanical approaches have been used to find the transition path, and hence the free energy barrier. We have discussed this aspect in Chapter 25.

The calculation of the pre-factor addresses the viscosity dependence of rate. It employs a stochastic description. We shall discuss this description in detail here. Figure 20.3 depicts bistable potential energy surfaces with two different barrier frequencies.

An elementary but important point to note in chemical kinetics is that the rate signifies *a steady-state flux from the reactant to the product.* If we consider that the entire population is in the reactant state, then the flow of the reaction from the reactant to the product occurs in three stages. (i) First, there is a transient time to build up the flow where the rate can be considered as time-dependent. (ii) This is followed by a steady flow characterized by a rate whose rate constant is the subject of the present chapter. (iii) The last stage is again a transient state when the flow dries up as the flow nears completion, and the system reaches equilibrium population distribution. These stages are shown in Figure 20.4, where we plotted the time dependence of rate as a function of time. The long plateau that sets in after a transient period is the steady-state rate.

The second important point to note is that for reactions in liquids, we often assume that a "dynamic equilibrium" exists between the population in the reactant well and the population at the barrier top and that the probability of finding the reactant at the barrier is given by the Boltzmann energy distribution. When activation energy is large, this probability at the barrier top is small, and this assumption holds. However, even after reaching the barrier, the reactant must cross the barrier. Thus, the rate can be given as a product of a transmission frequency, and the probability

$$k = v\exp\left(-E_{act} / k_B T\right). \tag{20.1}$$

This decomposition may appear to be *ad hoc*, but in reality, is fairly accurate when the barrier height is larger than $5k_B T$ or compared to the thermal energy available to the system, and it also follows

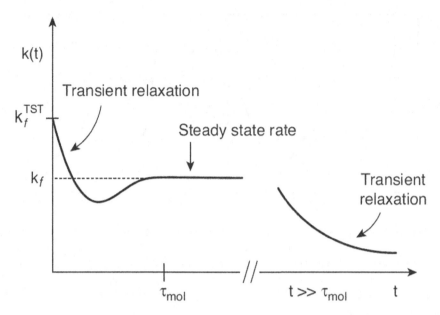

FIGURE 20.4 Characteristic behavior of the time dependence of rate. The steady-state plateau gives the long-time rate. It is the dominant stage when the barrier height is larger than ~$5k_BT$.

from a detailed steady-state calculation. The task then is to obtain an expression for the transmission coefficient, v. Many theoretical studies are centered around this calculation of v.

The final step of barrier crossing could be difficult because the already slow (at the relatively flat energy surface at the barrier top) forward speed could become further reduced at large viscosity due to collisions and as a result could be easily sent back to the reactant side by collisions with solvent molecules. This can happen even after the trajectory has crossed somewhat into the product side. Since the force acting on the reactant (or the product) near the barrier top is small (being a maximum in the reaction potential energy surface), several such re-crossings can occur as a result of collisions (and re-collisions) with the surrounding solvent molecules. Such crossings and re-crossings of barrier top lower the rate of the reaction. Such effects are called *dynamic solvent effects*. Such dynamic solvent effects are determined not only by the frequency of collisions but also by factors such as the curvature or the frequency of the reaction energy surface at the barrier top, denoted by ω_b. The latter determines the residence time at the barrier top.

Before we start on discussions of mathematical aspects, it should be pointed out that we shall also discuss later in the chapter several different kinds of reactions that do not involve a large activation barrier and where barrier crossing is not the rate-determining step. In particular, we shall discuss two types of chemical reactions:

(i) Barrierless chemical reactions, which are common in photo-isomerization processes, and
(ii) Bimolecular diffusion-controlled reactions are common in organic chemistry and biology, like enzyme kinetics and protein association. These reactions can be quite complex.

20.2 A SUCCINCT SUMMARY OF THE MAIN RESULTS OF RATE THEORY

It is important for students to know the main results as this is a developed and large area of ongoing study. This chapter thus aims to gather the main results to help students and researchers, as there are so *many different expressions that it tends to be confusing to students*. It also allows us to introduce the notations and nomenclature.

The most popular expression of the activated barrier crossing rate is given by the transition state theory (TST) developed by Wigner and Eyring. We provide the most popular and oft-quoted form of TST in Eq. (20.2).

$$k_f^{TST} = \frac{k_B T}{h} \frac{Q^{\neq}}{Q_R} \exp\left(-E_{act} / k_B T\right).$$

(20.2)

Here $Q^{\neq}$ and Q_R are the partition functions of the activated state (without the unstable vibrational mode) and the reactant, respectively. TST ignores dynamic solvent effects. It should be noted that Eq. (20.2) is a general expression valid for any dimension but it requires the existence of a well-defined transition state as a saddle surface separating the reactant and the product. We discuss later TST expression under harmonic approximations for the reactant well and the barrier partition functions.

(i) The frictional effects on the barrier top were considered in the elegant theories of Smoluchowski (1927) and Kramers (1940). Smoluchowki solved the barrier crossing problem in the high friction or overdamped limit, and Kramers' theory is valid in intermediate to high friction limit. The Kramers expression is given by Eq. (20.3) where ζ is the friction coefficient. This is a generalized form valid for all values of ζ.

$$k_f^K = \frac{\omega_R}{2\pi\omega_b}\left[\left(\frac{\zeta^2}{4} + \omega_b^2\right)^{\frac{1}{2}} - \frac{\zeta}{2}\right] \exp\left(-E_{act} / k_B T\right).$$

(20.3)

In the overdamped condition, that is, $\zeta \gg \omega_b$, we get the Smoluchowski limit, which is given by Eq. (20.4). Kramers' derivation was based on the phase-space Fokker-Planck equation, which has been discussed in Chapter 17. This gives us a viscosity-dependent rate expression and has been studied extensively.

$$k_f \equiv k_f^{SL} = \frac{\omega_R \omega_b}{2\pi\zeta} \exp(-E_{act} / k_B T).$$

(20.4)

In the opposite limit, that is, $\zeta \ll \omega_b$, one recovers the TST expression Eq. (20.5). *TST provides the upper limit of the rate constant.* One obtains the TST result when the barrier is high and sharp.

As a result, the system spends little time on the barrier top and feels almost no frictional forces.

$$k_f^{TST} = \frac{\omega_R}{2\pi} \exp\left(-E_{act} / k_B T\right).$$

(20.5)

However, in many cases where the barrier is large, and the viscosity is high, Kramers' theory breaks down and overestimates the effects of friction.

(ii) Later, in the early 1980s, Grote and Hynes generalized Kramers' formalism by removing the assumption of the white noise and incorporated non-Markovian effects. Grote and Hynes arrived at the following elegant expression given by Eq. (20.6).

$$k_f^{GH} = k_f^{TST}\left(\frac{\lambda_r}{\omega_b}\right)$$

(20.6)

where λ_r is given by a self-consistent relation $\lambda_r = \dfrac{\omega_b^2}{\lambda_r + \hat{\zeta}(\lambda_r)}$ and $\hat{\zeta}(\lambda_r)$ is the Laplace transform of the time-dependent friction, usually known as frequency-dependent friction. We note that one can arrive at Kramers' expression from Grote-Hynes. Later we present a combined derivation.

(iii) Landauer and Swanson (LS) (1961) extended Kramers' theory to multidimension. They showed that in the overdamped case, the reaction rate is reduced below the value derived from thermal equilibrium theory by the factor $\left(\omega_s / \eta\right)$ where ω_s is the angular frequency associated with the direction of steepest descent at the saddle point and η is the coefficient of viscosity (see section 20.7 for the detailed derivation). The expression derived by LS is given by Eq. (20.7).

$$k_f^{LS} = 2\eta \left(\frac{\gamma}{m}\right)^{1/2} \left[\left(\frac{\prod\limits_{i=1}^{N} v_i}{\prod\limits_{i=1}^{N-1} v_i'}\right) \exp\left\{-\frac{E_{act}}{k_B T}\right\}\right]. \tag{20.7}$$

The pre-factor is equal to ω_s / η, and the following part is the rate given by the equilibrium theory.

(iv) Langer (1969), in the context of the rate of nucleation and decay of metastable phases, formulated an elegant description of the rate in terms of the saddle frequency λ_s, which is also known as the unstable mode (see section 20.8 for the detailed derivation). The expression derived by Langer is provided in Eq. (20.83) where V_s is the volume of the saddle point subspace.

Note that the one-dimensional transition state rate provides an upper bound. Friction lowers the rate. At large viscosity, Kramers and Smoluchowski expressions converge, and the rate prediction forms the lower bound. Inclusion of the frequency dependence of friction can significantly increase the theoretical prediction of the rate.

Usually, different studies of chemical reaction dynamics have focused on various aspects of barrier crossing dynamics. From the 1970s onwards, stimulated by experiments with improved time resolution in laser spectroscopy, new theoretical techniques, and computer simulations, were developed. During this period, several fundamental aspects of reaction dynamics were addressed to and much new understanding evolved. In the subsequent sections, we outline the derivations of the same with detailed discussions.

Below we summarize all the expressions in Table 20.1.

20.3 TRANSITION STATE THEORY (TST)

A formally exact and elegant expression of the TST rate, first proposed by Wigner and Eyring, is given by

$$k_f^{TST} = \frac{1}{Q_R} \left\langle \mathbf{v}(0) H\left[\mathbf{v}(0)\right] \delta\left[X(0) - X_b\right]\right\rangle. \tag{20.8}$$

This rather formidable-looking expression has a simple meaning. First, the function is to be evaluated at the barrier top, at the dividing line between the reactant and the product through the term $\delta\left[X(0) - X_b\right]$. Next $H\left[\mathbf{v}(0)\right]$ is a Heaviside function selecting only the positive values of the initial velocity. In the above equation, the angular bracket $\langle...\rangle$ implies average over all the trajectories emanating from the reactant well and reaching the dividing line with a nonzero positive

TABLE 20.1
Expressions of Rate for Different Theoretical Approaches

Theoretical approach	Expressions
1D transition state theory	$k_f^{TST} = \dfrac{\omega_R}{2\pi} \exp\left(-\dfrac{E_{act}}{k_B T}\right)$
Kramers' theory	$k_f^K = \dfrac{1}{\omega_b}\left[\left(\dfrac{\zeta^2}{4}+\omega_b^2\right)^{\frac{1}{2}} - \dfrac{\zeta}{2}\right] k_f^{TST}$
Smoluchowski equation	$k_f^{SL} = \left(\dfrac{\omega_b}{\zeta}\right) k_f^{TST}$
Grote-Hynes theory	$k_f^{GH} = \left(\dfrac{\lambda_r}{\omega_b}\right) k_f^{TST}$
Multidimensional transition state theory	$k_f^{TST} = \dfrac{\prod_{i=1}^{N} v_i}{\prod_{i=1}^{N-1} v_i^b}\exp\left(-\beta E_{act}\right)$
Landauer-Swanson	$k_f^{LS} = 2\eta\left(\dfrac{\gamma}{m}\right)^{1/2}\left[\left(\dfrac{\prod_{i=1}^{N} v_i}{\prod_{i=1}^{N-1} v_i'}\right)\exp\left\{-\dfrac{E_{act}}{k_B T}\right\}\right]$
Langer	$k_f^L = |\kappa|\left(\dfrac{k_B T}{2\pi|\lambda_s|}\right)^{1/2} V_s \exp\left(-\dfrac{E_{act}}{k_B T}\right)$

Note: Here E_{act} is the activation energy, ω_R is the reactant well frequency, ω_b and the barrier frequency, ζ is the friction on the reaction coordinate, λ_r denotes the frequency with which the reactant molecule passes through the barrier.

velocity. Q_R in the denominator is the normalization constant over the initial equilibrium distribution. The average of the quantity in the bracket can be evaluated by following the method of Eyring. In the transition state, we can separate the degrees of freedom into two groups.

(i) The first group includes the motion along the reaction coordinate, as denoted by the Heaviside function in Eq. (20.8). If the reaction energy surface is available, it is straightforward to obtain the reactive motion and the associated frequency and also the friction.
(ii) The second group contains all the other degrees of freedom. We denote the associated partition function at the barrier top without the reactive motion by Q^*. We next evaluate the reactive velocity term.

If we regard this velocity term to originate from vibration with frequency v, then averaging over this unstable vibration gives the famous Eyring pre-fector $(k_B T/h)$. The combination of all these factors gives rise to the rate expression for the TST rate [see Eq. (20.2)].

Next, we present, following Eyring, a more rigorous derivation of the classical TST rate by assuming equilibrium between the reactant (R) and the transition state complex (or the activated complex, $X^{\neq}$).

$$R \rightleftharpoons X^{\neq} \to P. \tag{20.9}$$

Hence, the equilibrium constant of the first step $\left(K_c^{\neq}\right)$, according to statistical mechanical descriptions, can be expressed in terms of partition functions (Q_R and Q_*)

$$K_c^{\neq} = \frac{[X^{\neq}]}{[R]} = e^{-\Delta A/k_B T} = \left(\frac{Q_*}{Q_R}\right) e^{-\left(\frac{E_{act}}{k_B T}\right)} \tag{20.10}$$

where we have used the well-known relation between free energy and partition function, $A = -k_B T \ln Q$. Here ΔA is the free energy difference between the activated and the reactant state. Now we identify that particular vibrational degree of freedom, which is reactive, with no restoring force, out of the $3N$-6 (for nonlinear molecules) or $3N$-5 (for linear molecules) total vibrational degrees of freedom in the activated complex (this unique mode is also our reaction coordinate). That is, along that special direction $X^{\neq}$ can move to the product well without any restraint. Hence, we take that special vibrational part out of Q_* treat it as a pre-factor in the $\nu \to 0$ limit and write as [Eq. (20.11)]

$$Q_* = \left[\lim_{\nu \to 0} \frac{1}{\left(1 - e^{-h\nu/k_B T}\right)}\right]$$
$$or, Q_* \approx \frac{k_B T}{h\nu}. \tag{20.11}$$

We note that $Q^{\neq}$ possesses one less vibrational degree of freedom compared to Q_R. Now one can write the rate of the reaction as the product of the frequency (ν) and the concentration of the activated complex $[X^{\neq}]$

$$rate = \nu[X^{\neq}] = \left[\frac{k_B T}{h} \frac{Q^{\neq}}{Q_R} \exp\left\{-\left(\frac{E_{act}}{k_B T}\right)\right\}\right][R] = k_f^{TST}[R]$$
$$or, \boxed{k_f^{TST} = \frac{k_B T}{h} \frac{Q^{\neq}}{Q_R} \exp\left\{-\left(\frac{E_{act}}{k_B T}\right)\right\}}. \tag{20.12}$$

This form, however elegant, is not of much use in the condensed phases, as evaluation of partition functions is prohibitively difficult. It is thus converted into a more useful form [Eq. (20.13)] by using thermodynamic relation (*in the NPT or isothermal-isobaric ensemble*)

$$k_f^{TST} = \frac{k_B T}{h} exp(-\Delta G^{\neq} / k_B T), \tag{20.13}$$

where $\Delta G^{\neq}$ is the free energy difference between the activated state and the reactant and can be written as

$$\Delta G^{\neq} = \Delta H^{\neq} - T\Delta S^{\neq} \tag{20.14}$$

where $\Delta H^{\neq}$ and $\Delta S^{\neq}$ are the enthalpy of activation and entropy of activation, respectively. It is a standard procedure to relate the enthalpy of activation to the experimentally observed activation energy because $P\Delta V$ term is negligibly small at atmospheric pressure. The role of the entropy of activation is subtle. In many reactions, the activated state is more compact and $\Delta S^{\neq}$ is negative. This is often used as the reason to explain the observed decrease of the rate from the value given by $k_f^{TST} = \frac{k_B T}{h} exp(-E_{act} / k_B T)$. A major advantage of the thermodynamic TST expression is that one can include equilibrium effects of solute-solvent interactions through $\Delta H^{\neq}$ and $\Delta S^{\neq}$. This discussion gains more relevance for macromolecular hydration and biomolecular processes.

Quantitative experimental verification of TST was accomplished by Herschbach, Polanyi, and Lee in the late 1960s, for which they received Nobel Prize in Chemistry in 1986. *We strongly recommend students read the Nobel lecture of Dudley Herschbach to gain a good understanding of the efforts of that time.*

TST fails to account for the viscosity dependence of the rate observed in many chemical reaction rates (see section 20.1 for the detailed discussion). This is because re-crossing of trajectories is not allowed in TST. That is, at $X = X_b$, every trajectory with positive velocity at the barrier top is allowed to proceed to the product. In reality, however, a trajectory can cross back from the product side (near the barrier top) because of collisions with solvent molecules. Of course, it can again re-cross to the product side, and since the barrier top is flat, such crossings and re-crossings can happen several times before it finally relaxes to the product well. TST, however, would count each re-crossing as an extra reactive event, and thus, significantly over-estimate the actual rate, sometimes even by orders of magnitude. The reduction of the reaction rate by the solvent frictional forces experienced during the barrier crossing can be discussed using the well-known Smoluchowski-Kramers theory that provides a simple dependence of rate on friction (ζ) and predicts that in the limit of high friction, the rate should vary inversely with friction. If one further assumes that this friction is proportional to the zero-frequency shear viscosity (η) (via the Stokes-Einstein relation with a proper boundary condition), then one offers a simple relation connecting the rate to the solvent viscosity.

20.4   DERIVATION OF THE SMOLUCHOWSKI RATE EXPRESSION

This is the simplest theory to understand viscosity effects on chemical reactions. The rate, valid in the overdamped or high friction limit, is considered as diffusion of particles over the activation energy barrier [Eq. (20.4)]. We first consider a single particle moving in a one-dimensional potential energy surface, as shown in Figure 20.5. In this derivation, we consider the following assumptions. (i) The reaction is modeled as the motion of a solute over a barrier top, and the particle otherwise executes a Brownian diffusion. (ii) The barrier height is considerably larger than $k_B T$, and the number of particles at the barrier top $(X = X_b)$ is negligible. In other words, for barrier heights $>>k_B T$, and the probability of finding the particle at the barrier top is negligibly small. This makes particle

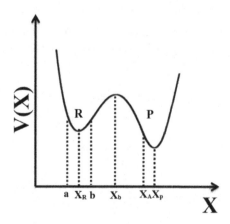

FIGURE 20.5   Schematic diagram of a potential energy barrier. Here X is the reaction coordinate. We assume there is a steady input of the particles into the neighborhood (a,b) of X_R. X_A represents an arbitrary location along the reaction coordinate near the product minimum X_P, and X_b is the position of the barrier.

flux across the barrier top also negligibly small (and does not significantly perturb the particle population in the reactant well), and ensures steady-state flow of particles from reactant to product energy basin on intermediate time scales, and (iii) equilibrium description of the population remains valid in the neighborhood of the reactant well, X_R.

We start with the Smoluchowski equation of motion for the probability distribution, $P(X, t)$

$$\frac{\partial P(X,t)}{\partial t} = D\frac{\partial}{\partial X}\left[\frac{\partial}{\partial X} - \beta F(X)\right]P(X,t). \tag{20.15}$$

Here D is the diffusion constant, which can be related to the friction of the solvent by the Stokes-Einstein relation as $D = k_B T/\zeta$ and $\beta = (k_B T)^{-1}$. $F(X)$ is the force term, given by $F(X) = -\frac{\partial V}{\partial X}$. Hence, Eq. (20.15) can be rewritten as Eq. (20.16).

$$\frac{\partial P(X,t)}{\partial t} = -\frac{\partial}{\partial X}\left[-\frac{k_B T}{\zeta}\left(\frac{\partial}{\partial X}\right) + \frac{F(X)}{\zeta}\right]P(X,t) = -\frac{\partial}{\partial X}J(X,t). \tag{20.16}$$

Now, our aim is to obtain an expression for the steady-state flux J at $X = X_b$. A steady state is defined by $\frac{\partial P(X,t)}{\partial t} = 0$. As Kramers pointed out, we need to find the state distinct from the equilibrium state, which also satisfies the time invariance but gives zero flux. We look for constant flux. This allows us to drop the time dependence of the population and write

$$\frac{d}{dX}P(X) - \frac{F(X)}{k_B T}P(X) + \frac{\zeta}{k_B T}J(X) = 0. \tag{20.17}$$

For a steady-state flow, J is independent of both t and X. We rewrite the above expression as

$$J = -\frac{k_B T}{\zeta}\exp(-\beta V(X))\frac{d}{dX}\exp(\beta V(X)P(X)),$$
or,
$$J = \exp(\beta V(X)) = \frac{k_B T}{\zeta}\frac{d}{dX}\exp(\beta V(X)P(X)). \tag{20.18}$$

One can solve Eq. (20.18) for J by integrating X from X_R to an arbitrary position X_A where $X_A > X_b$ such that there is no particle at and beyond that point, as shown in Figure 20.5. Then one obtains the following expression for the flux [Eq. (20.19)]:

$$J = \left(\frac{k_B T}{\zeta}\right)\frac{P(X_R)\exp\left\{V(X_R)/k_B T\right\}}{\int_{X_R}^{X_A} dX \exp\left\{V(X)/k_B T\right\}}. \tag{20.19}$$

At and around X_R the flux is zero. Hence, Eq. (20.17) yields

$$\frac{\partial P(X)}{\partial x} = -\frac{F(X)}{k_B T}P(X), \tag{20.20}$$

which leads to Eq. (20.21) where X is an arbitrary position in the vicinity of X_R

$$P(X) = P(X_R) \exp\left[\frac{-V(X) + V(X_R)}{k_B T} \right]. \tag{20.21}$$

The equilibrium population in the reactant well (n_R) can be defined as

$$n_R = \int_a^b dX \, P(X) = P(X_R) \int_a^b dX \, \exp\left[\frac{-V(X) + V(X_R)}{k_B T} \right]. \tag{20.21}$$

The rate of barrier escape is defined as $k = (J / n_R)$ (flux over the barrier/initial reactant population).

Note that analytical evaluation of the integrals is possible if we assume harmonic potential energy surface in the vicinity of X_R and X_b. In the harmonic limit (i.e., $a \to -\infty$ and $b \to \infty$), the reactant population becomes confined in the close vicinity of the potential energy basin minimum. After some simple algebraic manipulations, we arrive at Eq. (20.23), which is the Smoluchowski rate (k_f^{SL}) expression for activated barrier crossing.

$$k_f^{SL} = \frac{\omega_R \omega_b}{2\pi\zeta} \exp\left(-\frac{E_{act}}{k_B T} \right), \tag{20.23}$$

where ω_R is the reactant well frequency and ω_b is the frequency of the potential energy at the barrier top, often referred to as the barrier frequency. The above equation suggests that in the high friction limit, the rate decreases as $1/\zeta$. Since friction ζ is proportional to viscosity of the liquid medium, Smoluchowski rate theory predicts an inverse viscosity dependence of the rate.

In the above derivation, we have closely followed the treatment given by Chandrasekhar in his 1943 review.

20.5 FROM KRAMERS TO GROTE-HYNES THEORY: MEMORY EFFECTS WITHIN ONE-DIMENSIONAL FORMALISM

Smoluchowski rate expression is valid only in the overdamped regime. Kramers extended the limits of validity to arbitrary values of friction. The theory is based on an ordinary Langevin equation description along the reactive path. There the random force was considered as white noise, that is, totally uncorrelated in time, which produces frequency-independent friction. For some chemically activated processes in solution, there is a clear separation in the time scale between the correlation time of the interacting solvent molecules and the rate of the motion of the reactant molecule. So, the assumption of white noise holds, and Kramers' theory gives a reasonable description of the viscosity dependence of the rate constant.

However, there are situations where the use of frequency-independent friction can lead to erroneous results. For example, when the motion near the top of the barrier takes place on a picosecond or subpicosecond time scale, the solvent forces at two different instants (separated by a ps) can become correlated. For reactions in slow, viscous liquids, the use of Kramers' theory can lead to inaccurate results. Because these liquids with large viscosity derive a large fraction of viscosity from slow motions. Such slow motions cannot couple to barrier crossing dynamics. Thus, Kramers' theory highly overestimates the effect of friction on rate, as shown in Figure 20.6. Experimental evidence of this decoupling of the rate from solvent viscosity comes from the fractional dependence of the rate on solvent viscosity.

Grote and Hynes generalized Kramers' theory by removing the assumption of white noise. One interesting result of their theory is that the rate constant depends on the friction at a frequency

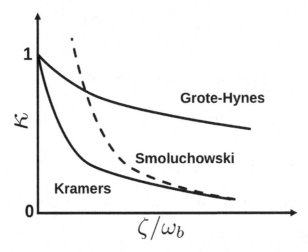

FIGURE 20.6 A comparative plot of friction dependence of the transmission coefficient κ obtained from Kramers, Smoluchowski, and Grote-Hynes theories. Note that the friction below which Smoluchowski rate sharply departs from Kramers' rate forms the boundary between the overdamped and underdamped limits. The rate is scaled by the transition state rate.

comparable to that of the barrier, ω_b. Since for many reactions, the barrier frequency, ω_b can be quite high, this friction can be much smaller than the zero frequency friction. Thus, Kramers' theory can vastly overestimate the effects of friction on those chemical reactions, which involve sharp, high-frequency barriers. To include non-Markovian effects, Grote and Hynes assumed the following generalized Langevin equation (GLE) for the dynamics along the reaction coordinateL

$$\mu\frac{dv}{dt} = F(X) - \int_0^t d\tau \zeta(\tau) v(t - \tau) + R(t) \qquad (20.24)$$

where the reaction coordinate is a coordinate connecting the two minimum points at the two isomeric forms (*cis* and *trans*) on the ground state potential surface through the transition state barrier between them. μ is the effective mass, v is the velocity along the reaction coordinate, $F(X)$ is the systematic force arising from the potential in the barrier region. $\zeta(t)$ is the time-dependent friction, and $R(t)$ is the random force from solvent assumed to be Gaussian. $F(X)$ is assumed to arise from a static potential, which is an inverted parabola, so that

$$F(X) = \mu\omega_b^2 X. \qquad (20.25)$$

In general, both $\zeta(t)$ and $R(t)$ appearing in Eq. (20.24) arise from microscopic motions of heat-bath (solvent) modes interacting with the reaction coordinate. For isomerization reactions in solvents, both of them arise from the microscopic motions of the solvent molecules interacting with the isomerizing moiety. So, they must be related to each other. They are related by the following relation (known as the fluctuation-dissipation theorem):

$$\zeta(t) = \beta\langle R(0)R(t)\rangle \qquad (20.26)$$

where β is the inverse of the Boltzmann constant (k_B) times the absolute temperature (T) and $\langle\cdots\rangle$ represents the statistical average over heat-bath modes at temperature T. By using the probability distribution from the generalized Fokker-Planck equation, Grote and Hynes obtained, after some formidable and lengthy analysis, the following simple and elegant expression for the rate constant, k_f^{GH},

$$\boxed{k_f^{GH} = k_f^{TST}(\lambda_r / \omega_b)},\tag{20.27}$$

where k_f^{TST} is the forward transition state rate constant given by,

$$k_f^{TST} = \frac{\omega_R}{2\pi}\exp(-E_{act}/k_BT).\tag{20.28}$$

Here ω_R is the frequency of motion in the reactant well, and E_{act} is the activation energy needed for the reactant molecule to surmount the barrier. The frequency with which the reactant molecule passes, by diffusive Brownian motions through the barrier region is λ_r and is given by the following *self-consistent* relation:

$$\lambda_r = \frac{\omega_b^2}{\lambda_r + \hat{\zeta}(\lambda_r)},\tag{20.29}$$

where $\hat{\zeta}(\lambda_r)$ is the Laplace transform of the time-dependent friction

$$\hat{\zeta}(\lambda_r) = \int_0^\infty dt e^{-\lambda_r t}\zeta(t).\tag{20.30}$$

The Grote-Hynes approach is quite involved. Fortunately, Eq. (20.27) can also be derived in simpler ways by following the treatments put forward by Pollak and separately by Munakata. We discuss below Pollak's approach.

Pollak's approach draws inspiration from an earlier work of Zwanzig, who showed that the equation of motion of a solute particle linearly coupled to a system of harmonic oscillators can be described exactly by a generalized Langevin equation, with an analytic expression for the frequency-dependent friction in terms of the density of states of the oscillator and the coupling constants. This approach can then be used to build a description based on a Hamiltonian, which is satisfactory. In the end, it leads exactly to the same expression derived by Grote and Hynes.

Thus, instead of starting directly with the generalized Langevin equation, Pollak modeled the stochastic interaction via a collection of the linearly coupled harmonic bath. The total Hamiltonian of the system and the bath can be written as

$$\begin{aligned}H &= H_{sys} + H_{bath}\\ &= \frac{p^2}{2M} + V(x) + H_{bath}\left(q_1,.....q_N, p_1,.....p_N; x\right)\end{aligned}\tag{20.31}$$

The first two terms on the right-hand side denote the kinetic energy and the potential energy of the system, respectively, where $V(x)$ is a double well potential. H_{bath} is the heat bath Hamiltonian described as

$$H_{bath} = \sum_{j=1}^N \left(\frac{p_i^2}{2m_j} + \frac{m_j}{2}\left[\omega_j q_j + \frac{C_j}{m_j \omega_j}x\right]^2\right).\tag{20.32}$$

Here $\left(p_j, q_j\right)$ are the momentum and coordinate of the jth bath oscillator whose mass and frequency are m_j, ω_j, respectively. C_j couples the bath oscillator to the system and measures the strength of the coupling of the system to the jth bath oscillator. Application of Zwanzig's scheme leads to the following generalized Langevin equation for the coordinate of the systems:

$$M \frac{d^2 x}{dt^2} = -\frac{dV}{dx} - M \int_0^t \zeta(t-\tau) v(\tau) d\tau + R(t) \qquad (20.33)$$

where the random force $R(t)$ is given in terms of the initial conditions of the bath variables $\left\{q_j(0), p_j(0)\right\}$ as

$$R(t) = \sum_{j=1}^{N} C_j \left[\left[q_j(0) + \frac{C_i}{m_j \omega_j^2} x(0) \right] \cos\left(\omega_j t\right) + \frac{p_j(0)}{m_j} \frac{\sin\left(\omega_j t\right)}{\omega_j} \right]. \qquad (20.34)$$

Random and frictional forces are relayed by the following fluctuation-dissipation theorem:

$$\left\langle R(t) R(s) \right\rangle = k_B TM \zeta(t-s). \qquad (20.35)$$

The time-dependent friction $\zeta(t)$ is given by

$$\zeta(t) = \frac{1}{M} \sum_{j=1}^{N} \frac{C_i^2}{m_j \omega_j^2} \cos\left(\omega_j t\right) \qquad (20.36)$$

The time-dependent friction can be expressed in terms of the spectral density

$$\zeta(t) = \frac{2}{\pi} \int_{-\infty}^{\infty} d\omega \frac{J(\omega)}{\omega} \cos\left(\omega t\right) \qquad (20.37)$$

where $J(\omega)$ is the spectral density of the bath, defined as

$$J(\omega) = \frac{\pi}{2} \sum_{j=1}^{N} \frac{C_j^2}{m_j \omega_j} \delta\left(\omega - \omega_j\right). \qquad (20.38)$$

20.5.1 NORMAL-MODE ANALYSIS

In the next steps, one evaluates the rate, and this treatment is elegant, although it is similar to the treatment adopted long ago by Langer in picking up the negative eigenvalue of a dynamical matrix to obtain the rate of barrier crossing. Let us assume that the potential $V(x)$ can be approximated at the barrier $x = x_b$ as

$$V(x) \approx E_{act} - \frac{1}{2} M \omega_b^2 \left(x - x_b\right)^2. \qquad (20.39)$$

Now, we can write the total Hamiltonian as a sum of $N+1$ independent harmonic oscillators in the vicinity of the barrier. This is achieved by first transforming to a mass-weighted coordinates

$$x' = \sqrt{M}x, \quad q_j' = \sqrt{m_j}q_j$$

and then diagonalizing the $(N+1)\times(N+1)$ force constant matrix $\mathbf{K}$ defined by the second derivatives of the total potential energy surface evaluated at the saddle point

$$x' = x_b, q_j' = -\frac{C_j}{\sqrt{m_j M}}\frac{1}{\omega_j^2}x_b'.$$

Let $\{\lambda^2\}$ denote the eigenvalues, then the solution of the secular equation, $\det(\mathrm{K}-\lambda^2 \mathrm{I}) = 0$, can be written as

$$\left[\prod_{i=1}^{N}\left(\omega_i^2 - \lambda^2\right)\right]\left[\omega_b^2\left(\Gamma^2 - 1\right) - \lambda^2\right] = \sum_{i=1}^{N}\frac{C_i^2}{m_i M}\prod_{\substack{j\neq i \\ j=1,N}}\left(\omega_j^2 - \lambda^2\right) \tag{20.40}$$

where $\Gamma^2 = \frac{1}{M\omega_b^2}\sum_{j=1}^{N}\frac{C_i^2}{m_i\omega_j^2}$. The coupling of the system to the bath modifies both the system and bath frequencies. We can rewrite Eq. (20.40) as

$$\lambda_r^2 = \frac{2}{1+(1/M)\sum_{i=1}^{N}\left(C_i^2/m_i\omega_i^2\right)\left[1/\left(\omega_i^2 + \lambda_r^2\right)\right]}. \tag{20.41}$$

We simplify Eq. (20.41) with the aid of Eq. (20.38) to obtain

$$\lambda_r^2 = \frac{\omega_b^2}{1+(2/\pi)(1/M)\int_{-\infty}^{\infty} d\omega\left(J(\omega)/\omega\right)\left(\omega^2 + \lambda_r^2\right)^{-1}}. \tag{20.42}$$

We next note that the Laplace transformation of $\zeta(t)$ defined in Eq. (20.37) is given by

$$\hat{\zeta}(z) = \int_0^{\infty} e^{-zt}\zeta(t)dt = \frac{2}{\pi}\int_{-\infty}^{\infty} d\omega\frac{J(\omega)}{\omega}\frac{z}{z^2 + \omega^2}. \tag{20.43}$$

Eq. (20.42) can now be simplified in terms of the frequency-dependent friction defined in Eq. (20.43) as

$$\boxed{\lambda_r = \frac{\omega_b^2}{\lambda_r + (1/M)\hat{\zeta}(\lambda_r)}.} \tag{20.44}$$

Eq. (20.44) is identical to Eq. (20.29) derived by Grote and Hynes. Therefore, we have proved that the reactive frequency of Grote-Hynes may be interpreted even in the continuum limit as an effective barrier frequency.

20.5.2 The Rate of Escape

The TST rate is given as the ratio of the partition function of all bath modes at the barrier to the partition function of the total system in the reactant well.

$$k_f = \left[\prod_{i=1}^{N} \left(\frac{\omega_i}{\lambda_i} \right) \right] \frac{\omega_R}{2\pi} \exp(-\beta E_{act}) \tag{20.45}$$

where λ_i^2 $(i = 1, N)$ are positive eigenvalues of the force constant matrix **K**. One can show

$$\prod_{i=1}^{N} \left(\frac{\omega_i}{\lambda_i} \right) = \frac{\lambda_r}{\omega_b}. \tag{20.46}$$

Eq. (20.45) can be written as

$$\boxed{ k_f = \frac{\lambda_r}{\omega_b} \frac{\omega_R}{2\pi} e^{-\beta E_{act}} = \frac{\lambda_r}{\omega_b} k_f^{TST} }. \tag{20.47}$$

Eq. (20.47) predicts the transition state result for very weak friction when $\lambda_r \sim \omega_b$, and Kramers' result for low barrier frequency (i.e., $\omega_b \to 0$) so that $\hat{\zeta}(\lambda_r)$ can be replaced by $\hat{\zeta}(0)$. If the barrier frequency is large ($\omega_b \geq 10^{13} s^{-1}$) and the friction is not negligible ($\hat{\zeta}(0)/\mu \sim \omega_b$) then the situation is not so straightforward. In this regime, which often turns out to be the relevant one experimentally, the *effective friction*, $\hat{\zeta}(\lambda_r)$ *can be quite small even if the zero frequency (i.e., the macroscopic) friction (proportional to viscosity) is very large. The non-Markovian effects can play a very important role in this intermediate regime.* One can also write the Grote-Hynes relation [Eq. (20.29)] as

$$\lambda_r^2 + \lambda_r . \hat{\zeta}(\lambda_r) = \omega_b^2. \tag{20.48}$$

The largest positive root of above quadratic equation is

$$\lambda_r = \left(\frac{\hat{\zeta}^2(\lambda_r)}{4} + \omega_b^2 \right)^{1/2} - \frac{\hat{\zeta}(\lambda_r)}{2}. \tag{20.49}$$

From Eqs. (20.27), (20.28), and (20.49) we can write

$$k_f^{GH} = \frac{\omega_R}{2\pi} \left[\frac{\left(\frac{\hat{\zeta}^2(\lambda_r)}{4} + \omega_b^2 \right)^{1/2} - \frac{\hat{\zeta}(\lambda_r)}{2}}{\omega_b} \right] \exp\left(-\beta E_{act}\right) \tag{20.50}$$

This is a landmark result in the activated rate theory. The above rate expression, i.e., Eq. (20.50) is suggesting that the transmission coefficient is determined by the frequency-dependent friction, not the zero-frequency (or, static friction) considered in the Kramers rate theory. Kramers' and transition state rate constants are special cases of Eq. (20.50). In order to apply the Grote-Hynes formula to realistic cases, we need a reliable expression for the frequency-dependent friction in terms of known quantities, especially as a function of viscosity.

20.6 MULTIDIMENSIONAL RATE THEORIES

Most applications of rate theory start and end with one-dimensional descriptions, especially in the liquid state. For example, the viscosity dependence of a rate has been extensively discussed by using Smoluchowski, Kramers, and, more recently, Grote-Hynes theories. However, for many complex reactions, the protein folding, and protein dimer dissociations, a one-dimensional description is incapable to capture the dynamics. It is important to realize that the multidimensionality of the reaction energy surface can alter our understanding of the reaction dynamics in a profound manner. We shall discuss one specific example in detail to illustrate this point.

In the following, we describe the existing theoretical formulations to treat the reaction in multidimensional reaction energy surface.

20.7 LANDAUER AND SWANSON TREATMENT

In a sense, Pollak's treatment of the rate involves a multidimensional reaction energy surface as the bath frequencies define additional dimensions. However, this view is somewhat limited because the bath coordinates do not experience any damping. However, these treatments are rather similar, as will be evident as we go through the remainder of the chapter.

The simplest approach, an extension of the Smoluchowski theory described above, was developed by Landauer and Swanson, who considered the case when the two potential minima had the same value and symmetrical about a line that passes through the saddle and perpendicular to the line connecting the two minima. These assumptions can also be generalized for the unsymmetrical potential wells, but here we restrict ourselves to this simple case, as described in Figure 20.7.

As in the earlier theories of reaction rate, the probability density, ρ, in the N-dimensional phase space can be decomposed as the time-independent thermal equilibrium part (ρ_t) and the deviation from the equilibrium (ρ_d)

$$\rho = \rho_{eq} + \delta\rho_d(t) \tag{20.51}$$

ρ_d decays with time as

$$\delta\rho_d(t) = \sum_i \rho_{di} \exp(-t/\tau_i). \tag{20.52}$$

Here it is assumed that there is a clear-cut separation of time scale between the motion within the potential well and between two wells. That is, the barrier is sufficiently large to allow the separation of time scale between the two wells, A and B. Our primary interest here is to find the time constant, τ_i, associated with the relaxation of the population difference.

The time dependence of the population difference (Δ) between the two wells is given by

$$\Delta(t) = \int_A \delta\rho_d(t)d\Gamma - \int_B \delta\rho_d(t)d\Gamma = \int_A \delta\rho_d(t)d\Gamma - \left(\int_{all\ space} \delta\rho_d(t)d\Gamma - \int_A \delta\rho_d(t)d\Gamma \right)$$
$$= 2\int_A \delta\rho_d(t)d\Gamma \tag{20.53}$$

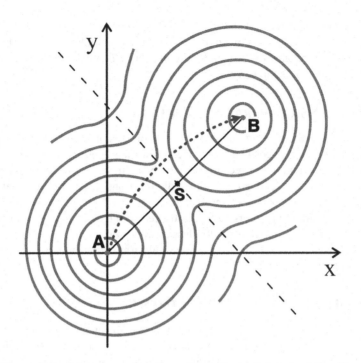

FIGURE 20.7 The equipotential contour in two dimensions is shown. Points A and B are the potential minima for the reactant and the product, respectively, and S is the saddle point.

Source: Taken with permission from [3].

(note that $\int\limits_{all\,space} \delta\rho_a d\Gamma = 0$).

The number of particles/systems crossing the saddle point per unit time ($\mathbf{J}$) is related to $\Delta(t)$ and can be given as

$$2\mathbf{J} = \frac{d\Delta}{dt} = \frac{\Delta}{\tau},\tag{20.54}$$

where τ is the time constant associated with the population difference. Our next aim is to find Δ and $\mathbf{J}$ to get the time constant τ in the overdamped case.

20.7.1 OVERDAMPED CASE

As we know in the overdamped case, the equation of motion of probability density of ensemble under the influence of external potential V obeys the following Smoluchowski equation,

$$\frac{\partial\rho}{\partial t} = -\nabla.\mathbf{J}\tag{20.55}$$

where,

$$\mathbf{J} = -D\left(\nabla\rho + \beta\nabla V\rho\right)\tag{20.56}$$

We assume that the diffusion constant D is independent of position. Near the saddle point, the potential V can be approximated as $V = V_s - \frac{1}{2}\gamma_b z^2 + \frac{1}{2}\sum_{i=1}^{N-1}\gamma_i x_i^2$, and near the reactant well A as $V = V_A + \frac{1}{2}\sum \delta_i x_i^2$. At equilibrium ρ satisfies the Boltzmann distribution, $\rho = c\exp(-\beta V)$, where c is the normalization constant. Here γ_b, γ_i are the harmonic force constants of the reactive mode and the nonreactive modes, respectively.

In the nonequilibrium case, we set

$$\rho = \alpha\exp(-\beta V), \tag{20.57}$$

where variation of α indicates the extent of deviation from equilibrium. Now we can write Eq. (20.56) as,

$$J = -D(\nabla\alpha)\exp(-\beta V). \tag{20.58}$$

From Eq. (20.58) it is clear that a large deviation from equilibrium can occur only at the saddle point because $\exp(-\beta V)$ is smallest at the saddle point. Let z be the coordinate perpendicular to the symmetry plane then we can integrate Eq. (20.58)

$$\alpha(z) = -D^{-1}\int_0^z j_z \times \exp\left[\beta\left(V_s - \frac{1}{2}\gamma z'^2 + \frac{1}{2}\sum\gamma_i x_i^2\right)\right]dz'. \tag{20.59}$$

The population difference is given as

$$\Delta = 2\int_A dx_1..dx_N \rho = 2\alpha(A)\exp(-\beta V_A)\prod_{i=1}^{N}(2\pi k_B T/\delta_i)^{\frac{1}{2}} \tag{20.60}$$

where $\alpha(A) = -D^{-1}\int_0^A j_z \times \exp\left[\beta\left(V_s - \frac{1}{2}\gamma z^2\right)\right]dz$ (j_z is the current at the saddle point and given by as $j_z = -D(\partial\alpha/\partial z)_{z=0}\exp(-\beta V_S)$).

Now the other quantity needed for calculation of τ is the total current J crossing the saddle point. The current density in the symmetry plane containing the saddle point is

$$j_z = -D\left(\frac{\partial\alpha}{\partial z}\right)_{z=0}\exp\left[\beta\left(-V_s - \frac{1}{2}\sum_{i=1}^{N-1}\gamma_i x_i^2\right)\right]. \tag{20.61}$$

We next follow the following steps.

1. Integration over N-1 transverse coordinates gives total current (J)

$$J = -D\left(\frac{\partial\alpha}{\partial z}\right)_{z=0}\exp(-\beta V_s)\prod_{i=1}^{N-1}\left(\frac{2\pi k_B T}{\gamma_i}\right)^{\frac{1}{2}}. \tag{20.62}$$

2. We use Eqs. (20.54), (20.60), and (20.62) to obtain the rate

$$\frac{1}{\tau} = 2De^{-\beta(V_s-V_A)} \times \prod_{i=1}^{N}\left(\frac{2\pi k_B T}{\gamma_i}\right)^{\frac{1}{2}} \bigg/ \left[\left(\frac{2\pi k_B T}{\gamma}\right)^{\frac{1}{2}}\prod_{i=1}^{N}\left(\frac{2\pi k_B T}{\delta_i}\right)^{\frac{1}{2}}\right]. \tag{20.63}$$

On more simplification we get,

$$\boxed{\frac{1}{\tau} = 2D\beta m \left(\frac{\gamma}{m}\right)^{\frac{1}{2}} \left[\left(\prod_1^N \nu_i \bigg/ \prod_1^{N-1} \nu_i'\right) e^{-\beta(V_s - V_A)}\right]}$$

(20.64)

where, $\left(\delta_i / m\right)^{\frac{1}{2}} = 2\pi\nu_i$, and $\left(\gamma_i / m\right)^{\frac{1}{2}} = 2\pi\nu_i'$.

As we know that the multidimensional TST rate $\left(k_f^{TST}\right)$ is $\left(\prod_{i=1}^N \nu_i \bigg/ \prod_{i=1}^{N-1} \nu_i'\right) e^{-\beta(V_s - V_A)}$, the above expression of rate can be written in terms of TST rate as

$$\frac{1}{\tau} = D\beta m \left(\frac{\gamma}{m}\right)^{\frac{1}{2}} k_f^{TST}.$$

(20.65)

The quantity $D\beta m$ is the friction ζ on the reaction coordinate, related to viscosity (η) by Stokes-Einstein relation ($\zeta = 6\pi\eta a$, with a as the effective radius of the reaction coordinate) and the quantity $\left(\gamma / m\right)^{1/2}$ is an angular frequency ω_b at the saddle point. The above equation can be written as

$$\boxed{\frac{1}{\tau} = \left(\frac{\omega_b}{\zeta}\right) k_f^{TST}.}$$

(20.66)

The factor $\left(\frac{\omega_b}{\zeta}\right)$ is same as in Kramers' expression of rate for one-dimensional overdamped case. So, the effects of multidimensionality enters only through k_f^{TST}.

20.8   LANGER'S EXPRESSION OF MULTIDIMENSIONAL RATE

The approach of Landauer and Swanson is limited by the assumptions that the friction affects only the reaction coordinate. This limitation was removed in an important paper by Langer. Interestingly Langer's goal was to calculate the rate of decay of a metastable phase that is relevant to a variety of thermally activated processes such as nucleation and growth. Langer generalized the prescription described by Landauer and Swanson using the classic nucleation theory of the rate of condensation of a supersaturated vapor. In this section, we outline the derivation of Eq. (20.83) as given in the introduction.

20.8.1   Equation of Motion for the Distribution Function

We consider a system with $2N$ degrees of freedom denoted by $\{\eta_i\}$, consisting of N coordinates and N momenta where $i = 1, N$ are for the position and the rest is for momentum. The system's dynamics is given by Eq. (20.67) where $E\{\eta\}$ is the Hamiltonian function.

$$\dot{\eta}_i = \sum_j A_{ij} \frac{\partial E\{\eta\}}{\partial \eta_j}$$

(20.67)

where A_{ij} are the elements of an asymmetric matrix defined below Eq. (20.68)

$$A_{ij} = \begin{cases} \delta_{i+N,j} & if\ i \leq N \\ -\delta_{i,j+N} & if\ j \leq N \\ 0 & otherwise \end{cases}.$$

(20.68)

The distribution function of the system $\left[\rho(\{\eta\},t)\right]$ that denotes the probability that the system can be found in the configuration $\{\eta\}$ at time t, can be decomposed into dynamical and fluctuating terms.

$$\frac{\partial \rho}{\partial t} = \left(\frac{\partial \rho}{\partial t}\right)_{dyn} + \left(\frac{\partial \rho}{\partial t}\right)_{fluc}. \quad (20.69)$$

Here the dynamical part determines the motion by the system's dynamical behavior can be described by the Liouville equation (or, conservation law), and the fluctuating part captures the rate at which ρ varies due to the system-bath interaction. Eq. (20.69) can be rewritten as a continuity equation in the η-space as follows [Eq. (20.70)]:

$$\frac{\partial \rho}{\partial t} = -\sum_i \frac{\partial J_i}{\partial \eta_i}, \quad (20.70)$$

where J_i is the ith component of the $2N$-dimensional flux or the probability current. Langer employed the following compact expression for this term [Eq. (20.71)]:

$$J_i = -\sum_j M_{ij} \left(\frac{\partial E}{\partial \eta_j} \rho + k_B T \frac{\partial \rho}{\partial \eta_j}\right) \quad (20.71)$$

where

$$M_{ij} = \left(\frac{\Gamma_i}{k_B T} \delta_{ij} - A_{ij}\right), \quad (20.72)$$

and Γ_i is constant that denoted the rate of variation of $\eta_i(t)$. Eq. (20.71) is just the flux in the Fokker-Planck equation, discussed earlier in Chapter 17 and discussed in Zwanzig's book using almost the same compact notation.

20.8.2 Rate of Decay of a Metastable State

One uses essentially the same procedure as used by Landauer and Swanson. The equilibrium solution of Eq. (20.70) is given by ($J_i = 0$ for all i)

$$\rho_0\{\eta\} \propto \exp\left(-E\{\eta\}/k_B T\right). \quad (20.73)$$

Near the position of local minima various stable and metastable configurations occur. Our interest would be such a transition where the system at a particular state $\{\eta\}$ makes a transition to another lower energy state. Such a transition passes across the lowest saddle point $\{\overline{\eta}\}$. *The saddle state possesses the same characteristics as that of the initial state except for the presence of a single localized fluctuation.* That is, when $\{\eta\}$ reaches $\{\overline{\eta}\}$ the system goes downhill in energy towards the stable region. Langer worked with the steady-state solution of Eq. (20.70) in the immediate neighborhood of the saddle (Figure 20.8) as opposed to the equilibrium solution Eq. (20.73).

Hence, the aim is to obtain an expression for the probability current (flux) across the saddle point. One can visualize J as a flowing straight line stream across the saddle (Figure 20.8)

In order to simplify the representation, we perform a coordinate transformation from η to ξ (origin at $\{\overline{\eta}\}$) via orthogonal transformation matrix $\boldsymbol{D}$.

$$\xi_n = \sum_i D_{ni}\left(\eta_i - \overline{\eta}_i\right). \quad (20.74)$$

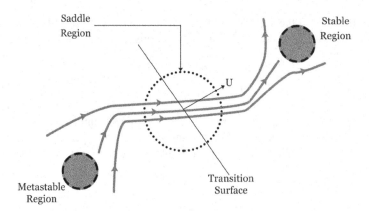

FIGURE 20.8   A schematic representation of the flow trajectories of the system going from a metastable to a stable region across a saddle. The saddle point approximation is assumed to be valid within the dotted circular region.

Source: Redrawn from [14].

Hence, in the immediate vicinity of the saddle point, that is, $\{\xi\} = \{0\}$ the energy function can be expanded as

$$E\{\eta\} = E\{\bar{\eta}\} + \frac{1}{2}\sum_{n=1}^{2N}\lambda_n \xi_n^2 + ... \tag{20.75}$$

In Eq. (20.75) λ_n are the eigenvalues of the Hessian matrix $\left[\partial^2 E/\partial \eta_i \partial \eta_j\right]$ evaluated at $\{\bar{\eta}\}$, where $\lambda_1 < 0$ (note only one eigenvalue of the Hessian matrix, say λ_1, is negative). Now one can define $J_n = \sum_i D_{ni}J_i$ and formulate the steady state version of Eq. (20.70) as follows:

$$\sum_n \frac{\partial J_n}{\partial \xi_n} = -\sum_{n,n'}\frac{\partial}{\partial \xi_n}M_{nn'}\left(\frac{\partial E}{\partial \xi_{n'}}\rho + k_B T\frac{\partial \rho}{\partial \xi_{n'}}\right) = 0. \tag{20.76}$$

If we transform $\rho = f(u)\exp\left[-E/k_B T\right]$ and define a new variable u as the linear combination of the ξ_n as $u = \sum_n U_n \xi_n$, we get the following equation:

$$\left(\sum_{n,n'}U_n \Gamma_{nn'}U_{n'}\right)\frac{\partial^2 f}{\partial u^2} - \left(\sum_{n,n'}\lambda_n \xi_n M_{nn'}U_{n'}\right)\frac{\partial f}{\partial u} = 0, \tag{20.77}$$

where the coefficient of $\dfrac{\partial f}{\partial u}$ must be proportional to u. That is,

$$\sum_{n,n'}\lambda_n \xi_n M_{nn'}U_{n'} = \kappa u = \kappa \sum_n U_n \xi_n. \tag{20.78}$$

In Eq. (20.78) the coefficients U_n are the solution of the eigenvalue problem

$$\lambda_n \sum_{n'}M_{nn'}U_{n'} = \kappa U_n. \tag{20.79}$$

Hence, Eq. (20.77) simplifies into Eq. (20.80)

$$\gamma \frac{\partial^2 f}{\partial u^2} - u \frac{\partial f}{\partial u} = 0; \; where \; \gamma = \frac{1}{\kappa} \left(\sum_{n,n'} U_n \Gamma_{nn'} U_{n'} \right). \tag{20.80}$$

Langer related κ to the growth rate of the unstable mode at the saddle point and showed that it is a unique negative eigenvalue of Eq. (20.79) that also makes γ negative. After several rigorous mathematical manipulations with an error function form for $f(u)$ Langer arrived at the following expression of the total rate (which we denote by k_f^L):

$$k_f^L = |\kappa| \left(\frac{k_B T}{2\pi|\lambda_1|} \right)^{1/2} V_s \frac{e^{-E\{\bar{\eta}\}/k_B T}}{Z_0} \prod_{n>1} \left(\frac{2k_B T}{\lambda_n} \right)^{1/2}, \tag{20.81}$$

where $Z_0 = e^{-E\{\eta_0\}/k_B T} \prod_l \left(\frac{2\pi k_B T}{\lambda_l^{(0)}} \right)^{1/2} = e^{-F\{\eta_0\}/k_B T}$ and $F\{\eta_0\}$ is the free energy of the metastable phase at $\{\eta_0\}$. Similarly one can relate the free energy of the saddle point $F\{\bar{\eta}\}$ with $\bar{Z}$ as follows:

$$\bar{Z} = e^{-E\{\bar{\eta}\}/k_B T} \prod_{n>1} \left(\frac{2\pi k_B T}{\lambda_n} \right)^{1/2} = e^{-F\{\bar{\eta}\}/k_B T}. \tag{20.82}$$

Hence, Eq. (20.81) can be rewritten as

$$\boxed{k_f^L = |\kappa| \left(\frac{k_B T}{2\pi|\lambda_1|} \right)^{1/2} V_s \exp\left(-\frac{E_{act}}{k_B T} \right),} \tag{20.83}$$

where V_s is the volume of the saddle subspace, a quantity difficult to calculate, and E_{act} is the free energy difference between the saddle and the metastable state. We can take harmonic approximation to calculate the volume of the saddle subspace. In this regard, we carry out the integration over volume element near the saddle to obtain $V_s = \prod_{n'=1}^{N-1} \left(\frac{2\pi k_B T}{\lambda_{n'}} \right)^{1/2}$. Here $\{\lambda_{n'}\}$ denote the eigenvalues of the Hessian matrix containing the second partial derivatives of the energy with respect to N-1 nonreactive modes at the saddle. We now solve the eigenvalue equation for κ to obtain $\kappa = \lambda_+ \sqrt{\frac{\det \mathbf{E}^R}{(2\pi k_B T)^N}}$. Here λ_+ is the only positive root of the equation $\det(\mathbf{A} - \lambda \mathbf{I}) = 0$.

$\mathbf{E}^R$ is the symmetric matrix containing the second partial derivatives of the energy at the reactant well. With the aid of simplified expressions of V_s, and κ, we rewrite Eq. (20.83) as

$$\boxed{k_f^L = \frac{\lambda_+}{2\pi} \left[\frac{\det \mathbf{E}^R}{|\det \mathbf{E}^s|} \right] \exp\left(-\frac{E_{act}}{k_B T} \right).}$$

Here $\mathbf{E}^s$ is the symmetric matrix containing the second partial derivatives of the energy with respect to N number of modes at the saddle point.

The above derivation of the rate by Langer is non-trivial and has become a classic.

20.8.3   An Application: Reduction of Langer's Expression in One Dimension

Owing to the compactness of the notation using in Langer's theory, the implementation of the theory could become a formidable task for a beginner. Hence, we demonstrate the application of the theory by considering a special case, i.e., reaction in one dimension where we of course recover *Kramer's formula*, but the derivation via Langer is instructive.

In this special case $\eta_1 = x$ and $\eta_2 = p$. One can write the noiseless Langevin equation as,

$$\frac{dx}{dt} = \frac{p}{m} ; \frac{dp}{dt} = -\frac{\zeta}{m} p - \frac{dV}{dx} \tag{20.84}$$

where V is the potential energy and ζ is the friction coefficient. We define the total energy as $E = \frac{p^2}{2m} + V(x)$. Using Hamilton's equation, we can write,

$$\dot{x} = \frac{p}{m} = \frac{\partial E}{\partial p} ; \dot{p} = -\zeta \frac{p}{m} - \frac{dV}{dx} = -\zeta \frac{\partial E}{\partial p} - \frac{\partial E}{\partial x}. \tag{20.85}$$

Following Langer we convert it to the following matrix form:

$$\begin{pmatrix} \dot{\eta}_1 \\ \dot{\eta}_2 \end{pmatrix} = \begin{pmatrix} 0 & 1 \\ -1 & -\zeta \end{pmatrix} \begin{pmatrix} \partial E / \partial \eta_1 \\ \partial E / \partial \eta_2 \end{pmatrix}. \tag{20.86}$$

Note that the transport matrix **M** is defined as negative of the matrix in Eq. (20.86).

$$M = \begin{pmatrix} 0 & -1 \\ 1 & \zeta \end{pmatrix}. \tag{20.87}$$

Let us consider the saddle point to almost zero ($\eta_1^s = 0$) and the momentum of the particle just escaping saddle point is almost zero ($\eta_1^s \approx 0$). One can then write the Taylor series expansion of energy around the saddle point and get the Hessian matrix at the saddle point ($\mathbf{E}^s$) and at the well minimum ($\mathbf{E}^R$) as

$$\mathbf{E}^s = \begin{pmatrix} m\omega_b^2 & 0 \\ 0 & -1/m \end{pmatrix} ; \mathbf{E}^R = \begin{pmatrix} m\omega_R^2 & 0 \\ 0 & 1/m \end{pmatrix} \tag{20.88}$$

where ω_b and ω_R are the frequency of the oscillators at the saddle point and well minimum, respectively. We can now linearize the matrix form and by using the Taylor expansion of energy at the saddle point

$$\begin{pmatrix} \dot{\eta}_1 \\ \dot{\eta}_2 \end{pmatrix} = \begin{pmatrix} 0 & 1/m \\ m\omega_b^2 & -\zeta/m \end{pmatrix} \begin{pmatrix} \eta_1 \\ \eta_2 \end{pmatrix} \tag{20.89}$$

where the matrix in Eq. (20.89) is defined as the transition matrix **A**. $\dot{\eta} = \mathbf{A}\eta$ with the eigen value problem as $\det(\mathbf{A} - \lambda\mathbf{I}) = 0$. We need the positive value of λ as that solution corresponds to the unstable barrier crossing mode. Upon solving the determinant equation, one obtains

$$\lambda_+ = \sqrt{\omega_b^2 + \left(\frac{\zeta}{2m}\right)^2} - \frac{\zeta}{2m}. \tag{20.90}$$

This leads to the same expression as Kramers

$$\boxed{k_K = \frac{\omega_R}{2\pi}\left[\sqrt{1 + \frac{\zeta^2}{4\omega_b^2}} - \frac{\zeta}{2\omega_b}\right]e^{-E_{act}/(k_B T)}.} \tag{20.91}$$

20.9 IMPLICATIONS OF GROTE-HYNES THEORY: FREQUENCY-DEPENDENT FRICTION

Many reactions in solution, for example, *cis* → *trans* isomerization in stilbene, involve large amplitude motion of a bulky group (isomerizing moiety) twisting around a molecular axis. Therefore, this twisting motion will certainly experience the effect of frictional forces exerted by the surrounding solvent molecules. The friction experienced by the moving moiety consists of two parts: one contribution comes from the translational drag by the environment, which we shall call the translational friction, ζ while the second could be the pure rotational friction. Generally, the rotational friction in most cases is smaller than its translational counterpart unless the bulky twisting group is either very large or carries a charge. So, the translational friction is the principal quantity that regulates the diffusive Brownian motion of the tagged moiety near the barrier region. Two approaches have been developed to estimate the frequency dependence of friction to be used in the Grote-Hynes formula. These are discussed below.

20.9.1 Frequency-Dependent Friction from Hydrodynamics

A hydrodynamic expression for the frequency-dependent friction can be obtained by solving the generalized Navier-Stokes equation. This calculation was carried out by Zwanzig and Bixon [24]. The resulting expression involves, among other things, frequency-dependent shear and bulk viscosity.

A schematic plot for the frequency dependence of the hydrodynamic friction is shown in Figure 20.9 for various values of shear viscosity. The frequency dependence becomes stronger as the shear viscosity increases. This implies that if we use zero-frequency viscosity then we vastly overestimate its effect, the effective friction at high viscosities is much smaller than its zero-frequency value. This is precisely the reason why the decrease of the rate slows down at high viscosities. Physically this means that many of the low-frequency motions that contribute to $\zeta(z = 0)$ do not affect the reactive motion across the barrier if the barrier frequency is sufficiently high. In the long chain alkanes, examples of low-frequency motions are those rotations around the backbone, which involve cooperative motions of several backbone atoms. These kinds of motions will not respond if the liquid is driven at a high frequency.

The above discussion of frequency-dependent friction based on the generalized hydrodynamic model becomes questionable when the barrier frequency is very high. The friction ζ on the tagged moiety in slow liquids is primarily determined by the molecular length scale processes, and the rate is, therefore, determined primarily by the *local* dynamics. The mode-coupling theory includes these aspects rigorously.

20.9.2 Frequency-Dependent Friction from Mode Coupling Theory

Generalized hydrodynamic description of friction becomes inadequate a short times or large frequencies. In chemical reactions, the barrier frequency is often in the excess of $10^{12} s^{-1}$, or so. At such large frequencies, even the viscoelastic response will be dominated by such processes as collision

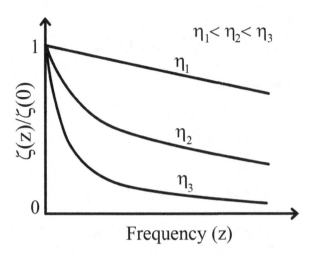

FIGURE 20.9 Schematic illustration of the generic behavior of the frequency-dependent friction $\zeta(z)$. Here values of $\zeta(z)/\zeta(0)$ are plotted against frequency z for various values of the zero-frequency shear viscosity. These values show the general trend in the frequency dependence when the liquid becomes more viscous. Similar functional dependence has been observed elsewhere.

Source: Redrawn from [17].

and inertia. Generalized hydrodynamics is grossly inadequate when such processes become dominant. However, we also need a self-consistent description at longer times because the viscosity of the solution is often determined by slow processes. Therefore, we have the difficult task of having a unified description valid both at short and long times. Fortunately, mode-coupling theory provides a reliable description in both the limits.

According to this theory, the frequency-dependent friction, $\zeta(z)$ is given by the following simple expression:

$$\frac{1}{\hat{\zeta}(z)} = \frac{1}{\hat{\zeta}_B(z) + \hat{\zeta}_{\rho\rho}(z)} + \hat{R}_{tt}(z) \tag{20.92}$$

where $\hat{\zeta}_B(z)$ is the bare (collisional) part of the total friction while $\hat{\zeta}_{\rho\rho}(z)$ is that component of the total friction which comes from the coupling of the reactive motion to the solvent density (ρ) fluctuation. $\hat{R}_{tt}(z)$ arises from the coupling to the transverse current mode of the solvent. Here $\hat{R}_{TT}(z)$ can be written as follows

$$\hat{R}_{TT}(z) = \frac{1}{\rho} \int_0^\infty dt e^{-zt} \int \left[dq'/(2\pi)^3 \right] \left[1 - \left(\hat{q}.\hat{q}' \right)^2 \right] \times \left[\gamma_{d12}(q') \right]^2 \omega_{012}^{-4} \left[F^s(q',t) - F^0(q',t) \right] C_{tt}(q',t) \tag{20.93}$$

We have discussed this in chapter 23 with great details. Note here that if one ignores both ζ_B and $\zeta_{\rho\rho}$ and retains only R_{tt}, then one recovers a generalized hydrodynamic type description for the friction experienced by a tagged molecule. However, in dense, slow liquids, the contributions from ζ_B and $\zeta_{\rho\rho}$ is far more important than R_{tt}, especially when the response is studied at high frequencies. The bare or the binary or the short-time friction, $\hat{\zeta}_B(z)$ is determined by equilibrium quantities. For example, for a hard sphere fluid, this can be approximated by the well-known Enskog friction (ζ_E) which is given by

$$\zeta_E = \frac{8}{3m}(2\pi\mu k_B T)^{1/2}\rho\sigma^2 g(\sigma) \tag{20.94}$$

where $g(\sigma)$ is the value of radial distribution function at contact. In writing the above equation, we have assumed for simplicity that both the solvent and the moving part of the reactant molecules are hard spheres of the same diameter (σ) with identical mass (m). It turns out that the Enskog expression is not very reliable, for the following reason. Reactions with high barrier frequency ω_b probe dynamics, which are in the collision frequency range in liquids. Thus, these reactions probe the details of the repulsive part of the potential and the hard sphere model gives too high a value at high frequency. Fortunately, it is not easy to overcome this problem because one can derive accurate short-time expansion for the time-dependent friction. The simplest is to assume a Gaussian form for $\zeta(t)$

$$\zeta(t) = \Omega_0^2 \exp\left(-(t/\tau_B)^2\right) \tag{20.95}$$

where τ_B is the time constant associated with the above Gaussian relaxation and Ω_0 is the Einstein frequency of the liquid. The value of the friction at zero time is given by the following expression:

$$\zeta(t=0) = \Omega_0^2 = \frac{4\pi\rho}{3m} \int dr r^2 g_{12}(r) \nabla_r^2 v_{12}(r) \tag{20.96}$$

where $g_{12}(r)$ is the solute-solvent radial distribution function and $v_{12}(r)$ is the interaction potential, ρ is the density of the solvent.

The collective contribution, $\hat{\zeta}_{\rho\rho}(z)$ is obtained from the mode coupling theory and is given by the following simple and elegant expression:

$$\hat{\zeta}_{\rho\rho}(z) = \frac{\rho k_B T}{6\pi^2 m} \int_0^\infty dt \exp(-zt) \int_0^\infty dq q^4 F_s(q,t)[c(q)]^2 S(q,t), \tag{20.97}$$

where $c(q)$ is the direct correlation function as a function of q. This function provides the coupling between the reactant and the solvent molecules. $F_s(q,t)$ is called the self-dynamic structure factor and it describes the self-motion of the solute (here reactant). In a homogeneous liquid, it is measured by incoherent neutron scattering. $S(q,t)$ describes the liquid (here the solvent) dynamic structure factor. Away from glass transition regime, these two can be approximated by simple expressions of the following forms:

$$F_s(q,t) = \exp[-Dq^2 t], \quad S(q,t) = S(q)\exp[-Dq^2 t/S(q)], \tag{20.98}$$

where D is the diffusion coefficient of a solvent molecule related to the translational friction, ζ by the relation, $D = k_B T/\zeta$.

In Figure 20.10, the total frequency-dependent friction, $\hat{\zeta}(z)$, is plotted against the Laplace frequency (z). In the same figure Enskog friction, ζ_E, and the binary contribution, $\zeta_B(z)$ are also shown. Figure 20.10 makes it clear that neither the hard sphere model nor hydrodynamics provides an accurate description of the frequency-dependent friction. The details of the interaction potential between the reactant and the solvent do matter in determining the rate of barrier crossing dynamics.

20.10 MULTIDIMENSIONAL AND NON-MARKOVIAN RATE THEORY

A general implementable non-Markovian rate theory for reactions on a multidimensional has not been developed. There have been several approximate approaches, the most notable being that of van der Zwan and Hynes, who considered a nonreactive mode coupled with a reactive mode. A process of adiabatic elimination was used to transform this into a one-dimensional

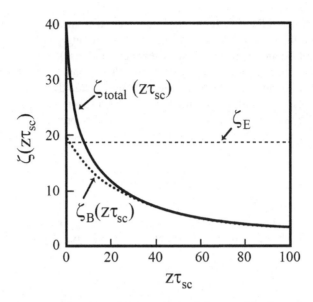

FIGURE 20.10 The frequency-dependent friction obtained from application of mode coupling theory, explained in Chapter 23. Here we plot the total friction, $\hat{\zeta}(z\tau_{sc})$ (the solid line) and the binary friction, $\hat{\zeta}_B(z\tau_{sc})$ (small dashed line) as a function of the Laplace frequency that enters the rate calculation in the Grote-Hynes theory. For comparison, Enskog friction, $\hat{\zeta}_E$, is also shown (large dashed line). The frequency-dependent friction, the Enskog friction, and the frequency are scaled by τ_{sc}^{-1} where $\tau_{sc} = \left[m\sigma^2 / k_B T \right]^{1/2}$.

Source: Taken with permission from [26].

description with resultant frequency-dependent friction. Ultimately, Grote-Hynes theory was used to obtain the rate.

The scheme of van der Zwan and Hynes can be extended to include multiple nonreactive modes by a straightforward generalization in a work by Acharya *et al*. An interested student is advised to read up on the literature. In a recent development, the issue of the interplay between memory effects and dimensionality on reaction dynamics in a multidimensional free energy surface was addressed. We refer to the literature for a detailed discussion. This study used a recently generated multidimensional free energy surface that we discuss below.

20.11 AN ILLUSTRATIVE EXAMPLE OF A MULTIDIMENSIONAL FREE ENERGY LANDSCAPE

In the previous sections, we have discussed several important theories and equations. Here we provide an illustrative example of their application on the free energy landscape (FEL) of a real system – that of insulin dimer dissociation. Here the order parameters often selected are (i) scaler distance (R) between the two monomers, and (ii) the number of the native state contacts during dissociation, Q. In the associated native state, distance R between the two centers of mass is close to 2 nm and the total number of contacts is close to 60. Both goes to zero in the fully dissociated state.

Such a free FEL is shown in 20.11(a), which is generated by enhanced sampling techniques, namely metadynamics and its variants which we have discussed in Chapter 25. However, other free energy calculation methods like umbrella sampling, replica exchange etc., can also be used depending on the system.

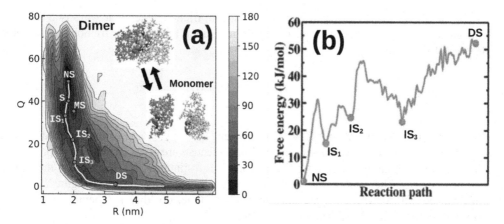

FIGURE 20.11   (a) Free energy contour of insulin dimer dissociation with respect to two collective variables namely, the distance between the center of mass of the monomers (R) and number of cross contacts between the C_α atoms of two monomers (Q). The white line represents the minimum energy pathway from the stable dimeric state (A) to the well-separated monomeric state (F) through the intermediate states B, D, and E. Here C is the metastable minimum. (b) The reduced one-dimensional rugged free energy landscape along the minimum energy pathway.

Source: Taken with permission from [19].

Figure 20.11 (a) shows that the native state of the associated dimeric species (well A) is a global minimum. The system goes from state A to state F, which corresponds to the dissociated state. The minimum energy pathway (MEP) is traced by the white dashed line. We note that the system may additionally follow some other pathways to reach well F. Along such a pathway the new collective variable (X) becomes a combination of R and Q.

Once we further reduce the dimensionality of the FEL into one dimension by considering the MEP, we can obtain the numerical values for the rate constant. If we traverse the MEP, the rugged one-dimensional FEL would look like Fig. 20.11(b). One can divide the whole process into several mutually uncorrelated steps such as $A{\to}B$, $B{\to}D$, $D{\to}E$ and $E{\to}F$. Then one needs to evaluate the rate constants of each step separately and apply Eq. (20.99).

$$\frac{1}{k} = \frac{1}{k_{A\to B}} + \frac{1}{k_{B\to D}} + \frac{1}{k_{D\to E}} + \frac{1}{k_{E\to F}}. \qquad (20.99)$$

We need the following quantities: (i) reaction well frequency ω_R; (ii) barrier top frequency ω_b; (iii) effective mass (μ) of the reaction path; (iv) the barrier height E_{act}; and (v) the coefficient of friction ζ. To obtain k_f^{TST} only (i) and (iv) are required. Evaluation of E_{act} can be easily achieved from the FEL graph as shown in Fig. 20.11(b). The well frequencies and the effective mass of the system along X can be calculated by using the following two equations:

$$\left\langle \delta \mathbf{X}(0)^2 \right\rangle = \frac{k_B T}{\mu \omega^2}; \;\; and \;\; \left\langle \delta \dot{\mathbf{X}}(0)^2 \right\rangle = \frac{k_B T}{\mu}. \qquad (20.100)$$

The frequency at the barrier top can also be calculated from Eq. (20.100) if one can generate a trajectory of statistically significant length at that configuration. Otherwise the same can be achieved by numerically fitting an inverted parabola.

The only nontrivial and difficult part is the calculation of friction. One of the ways is to calculate the force autocorrelation and then use the Green-Kubo relation to obtain the friction coefficient at $t = 0$ [Eq. (20.101)]

$$\zeta = \frac{1}{k_B T} \int_0^\infty dt \langle \delta F(0) \delta F(t) \rangle. \tag{20.101}$$

This is also known as Kirkwood's formula. Unfortunately, this only an approximate procedure. The time-dependent friction cannot be easily determined, and scientists are in search of a general procedure.

In recent work, insulin dimer dissociation rate was estimated using different theoretical prescriptions described here along the minimum energy pathway, as shown in Figure 20.9. The numerical values of the dissociation rate and find that both dimensionality and memory effects play an important role in influencing the dissociation rate.

20.12 BARRIERLESS CHEMICAL REACTIONS

As mentioned in the Introduction, there are many reactions in photochemistry and biology that do not face an activation barrier in their motion along the reaction coordinate from the reactant to the product. In physical chemistry, the best-known example is the electronic relaxation of triphenyl methane (TPM) dyes, like malachite green, crystal violet. These reactions exhibit surprisingly strong viscosity dependence, which has been the subject of several detailed theoretical studies. Another curious observation is the weak temperature dependence of the rate – the rate is found nearly independent of temperature so long as the temperature dependence of viscosity is removed from the dependence. The weak temperature dependence and the strong viscosity dependence, coupled with quantum chemical calculations, suggest a barrierless reaction.

It is found that in these reactions, the reaction coordinate is a rotation, like a propeller motion, around a -C-C- bond that connects the phenyl rings to the central carbon atom, as shown in Figure 20.12 below.

As mentioned above, many theoretical studies have been carried out to understand the unusually strong viscosity dependence of the rate. A simple model was developed by Bagchi, Fleming, and Oxtoby that modeled the reaction zone as a pin hole sink. See Figure 20.13 for a schematic illustration. This model (the pinhole sink) allows an analytical solution, for the solute particle starting

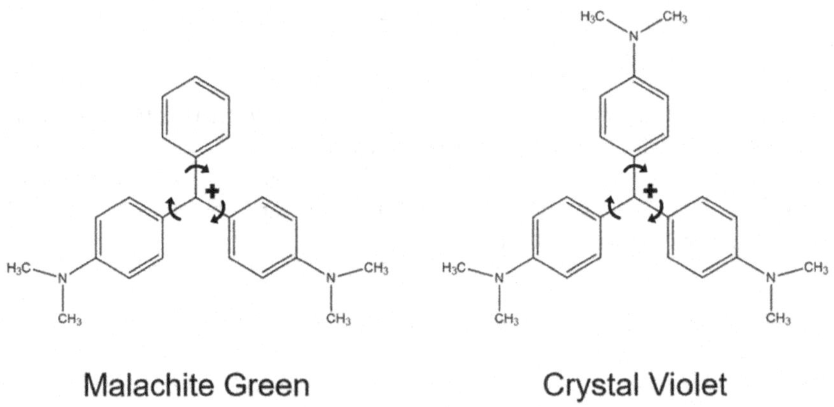

Malachite Green   Crystal Violet

FIGURE 20.12 TPM dyes, Malachite green (MG) and crystal violet (CV). Electronic relaxation of the excited state involves rotation around the central bonds. The rotations can be synchronous or asynchronous. The motion leading to the reaction is barrierless.

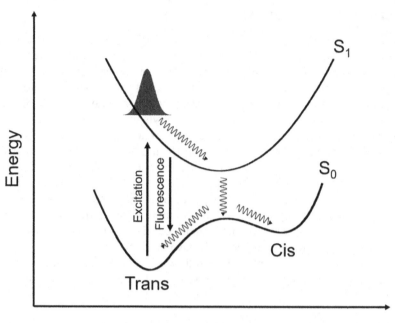

Reaction coordinate (rotation)

FIGURE 20.13 Barrierless reaction-free energy surface of triphenyl methane (TPM) dyes. The internal degree of the freedom relaxes on a harmonic surface towards a reaction window. The motion is resisted by frictional forces from the liquid.

at position x_0 at time $t = 0$, by using the method of images, described in probability theory books (W. Feller, Vol. I & II, Wiley). The details of the solution is available in the book "*Molecular Relaxation in Liquids*" (Bagchi, 2012).

The reaction free energy surface is rather uninteresting, as shown below.

20.13 DYNAMIC SOLVENT EFFECTS IN ELECTRON TRANSFER REACTIONS

Because electron is a charged particle, the rate of transfer of electron can be affected by the dielectric friction arising from solvent polarization relaxation in dipolar liquids like water, ethanol, acetonitrile. To understand these effects, one uses the reaction energy surface provided by Marcus. Specific application of these methods to electron transfer reactions has been discussed in the monograph "*Molecular Relaxation in Liquids*" cited in the Bibliography. Here we just mention a few interesting observations that have emerged from theoretical studies. First, Marcus' original reaction energy surface employs a collective coordinate, which is essentially the solvation energy. Thus, the reaction coordinate could be sensitive to solvent dynamics. Second, for a weakly adiabatic electron transfer reaction where the barrier frequency is large, ETR can couple only to the ultrafast solvation dynamics. This leads to a marked weakening of the solvent dynamic effects. Third, most of the reactions can be classified as weakly adiabatic when the coupling between the donor and acceptor surfaces is small. One can the use Fermi golden rule to obtain an expression of the rate. Usual electron transfer reactions are characterized by large activation energy, which can be obtained by approximating the reactant and product surfaces as harmonic where the reaction activation energy is determined both by the intersection of the reactant and product surfaces, which are separated by what is known as the "solvent reorganization energy". The details are available in the "*Molecular Relaxation in Liquids*" monograph, cited below.

20.14   UNDERSTANDING ENZYME KINETICS

Enzymes catalyze otherwise impossibly slow biological reactions by mechanisms whose molecular origins are often shrouded in mystery. The classical approach to understanding the rates of enzyme-catalyzed reactions start with Michelson-Menten theory that predicts a hyperbolic dependence of rate on substrate concentration. The derivation of MM rate expression followed Lindemann's theory of unimolecular reaction rate. This elegant expression however can be derived under a variety of approximations and appears to be insensitive to an underlying molecular mechanism.

Recent developments in this area has benefited from single molecule spectroscopic experiments that reveals that reaction energy surface is rugged, which is invoked to explain a broad distribution of reaction rates observed. This emphasizes the importance of conformational dynamics of the enzyme.

Another issue routinely discussed is the existence of dynamic effects in determining the rate of enzyme-catalyzed rates. The dynamic effects here imply the conformational fluctuations of the enzymes. There exist two opposite schools of thought, regarding the influence of solvent fluctuations.

Several detailed theories of the catalytic cycle have been proposed. We draw attention particularly to the work of Min *et al.* who employed a two-dimensional reaction surface to develop a model of the catalytic cycle. However, a detailed microscopic calculation still has remained outside the reach of theoreticians.

20.15   SUMMARY

Along with spectroscopy, chemical reaction dynamics is one of the most important areas of application of time-dependent statistical mechanics. This has remained so, now for more than one hundred years, since the initial studies of Smoluchowski, Arrhenius, and Ostwald. The derivation of the general expression of the transition state theory (TST) rate was an elegant application of statistical mechanics. Around the same time, Kramers presented his detailed, nontrivial expression of the effects of frictional forces on chemical reaction rate. Kramers' derivation and treatment remained the benchmark in this field for many years.

On the experimental side, things moved slowly. It was only in the 1970s and 1980s that detailed tests of the predictions of Kramers-Smoluchowski theory began to appear, facilitated by the rapid advance of laser spectroscopy during the same period. At the same time, computer simulation studies became possible allowing better interrogation of the merits of different theoretical approximations.

At the present time, the emphasis has shifted to understanding chemical dynamics in complex systems characterized by multidimensional free energy surface. The developments in this area started with Landauer and Swanson, then developed by Langer, and most recently by Szabo *et al.* We have discussed the above theories, except the study of Szabo *et al.* as that is too complex to discuss here in a textbook.

BIBLIOGRAPHY AND SUGGESTED READING

1.   Hynes, J. T. 1985. Chemical reaction dynamics in solution. *Annual Review of Physical Chemistry* *36*(no. 1), 573–597.
2.   Nitzan, A. 2006. *Chemical Dynamics in Condensed Phases: Relaxation, Transfer and Reactions in Condensed Molecular Systems*. Oxford: Oxford University Press.
3.   Bagchi, B. 2012. *Molecular Relaxation in Liquids*. New York: Oxford University Press.
4.   Laidler, K. J. 2008. *Chemical Kinetics*. (3rd edition). New York: Pearson Education.
5.   Eyring, H. 1935. The activated complex in chemical reactions. *Journal of Chemical Physics, 3*, 107.
6.   Wigner, E. 1938. The transition state method. *Transactions of the Faraday Society, 34*, 29.
7.   Smoluchowski, M. 1916. Three discourses on diffusion, Brownian movements, and the coagulation of colloid particles. *Phys. Z. Sowjet, 17*, 557.

8. Kramers, H. A. 1940. Brownian motion in a field of force and the diffusion model of chemical reactions. *Physica 7*, 284.

9. Hanggi, P., P. Talkner, and M. Borkovec. 1990. Reaction-rate theory: fifty years after Kramers. *Reviews of Modern Physics, 62*(2), 251–341.

10. Pollak, E. 1986. Theory of activated rate processes: a new derivation of Kramers' expression. *Journal of Chemical Physics, 85*, 865.

11. Chandler, D. 1978. Statistical mechanics of isomerization dynamics in liquids and the transition state approximation. *Journal of Chemical Physics, 68*, 2959.

12. Montgomery, Jr., A. John, D. Chandler, and B. J. Berne. 1979. Trajectory analysis of a kinetic theory for isomerization dynamics in condensed phases. *Journal of Chemical Physics, 70*, 4056.

13. Langer, J. S. 1967. Theory of the condensation point. *Annals of Physics (NY). 41*, 108–157.

14. Langer, J. S. 1969. Statistical theory of the decay of metastable states. *Annals of Physics 54*, 258.

15. Landauer, R., and J. A. Swanson. 1961. Frequency factors in the thermally activated process. *Physical Review, 121*, 1668.

16. Chandrasekhar, S. 1943. Stochastic problems in physics and astronomy. *Reviews of Modern Physics, 15*, 1.

17. Bagchi, B., and D. W. Oxtoby. 1983. The effect of frequency dependent friction on isomerization dynamics in solution. *Journal of Chemical Physics, 78*(5), 2735–2741.

18. Banerjee, P., and B. Bagchi. 2020. Dynamical control by water at a molecular level in protein dimer association and dissociation. *Proceedings of the National Academy of Sciences, 117*, 2302.

19. Acharya, S., S. Mondal, S. Mukherjee, and B. Bagchi. 2021. Rate of insulin dimer dissociation: interplay between memory effects and higher dimensionality. *Journal of Chemical Physics B, 125*(34), 9678–9691

20. Acharya, S., and B. Bagchi. 2022. Non-Markovian rate theory on a multidimensional reaction surface: complex interplay between enhanced configuration space and memory. *Journal of Chemical Physics, 156*, 134101.

21. Roy, S., and B. Bagchi. 1995. Effects of solvent polarization relaxation on nonadiabatic outersphere electron transfer reactions in ultrafast dipolar solvents. *Journal of Chemical Physics, 102*, 7937.

22. Bagchi, B., and N. Gayathri. 1999. Interplay between ultrafast polar solvation and vibrational dynamics in electron transfer reactions: Role of high-frequency vibrational modes. *Advances in Chemical Physics, 107*, 1.

23. Min, W., X. S. Xie, and B. Bagchi. 2008. Two dimensional reaction free energy surfaces of catalytic reaction: effects of protein conformational dynamics on enzyme catalysis. *Journal of Chemical Physics (Hynes Festschrift Issue), 112*, 454.

24. Zwanzig, R., and M. Bixon. 1970. Hydrodynamic theory of the velocity correlation function. *Physical Review A, 2*, 2005.

25. Kim, S. K., and G. R. Fleming. 1988. Reorientation and isomerization of trans-stilbene in alkane solutions. *Journal of Physical Chemistry, 92*, 2168–2172.

26. Murarka, R. K., S. Bhattacharyya, R. Biswas, and B. Bagchi. 1999. Isomerization dynamics in viscous liquids: Microscopic investigation of the coupling and decoupling of the rate to and from solvent viscosity and dependence on the intermolecular potential. *Journal of Physical Chemistry, 110*, 7365.

21 Diffusion on Flat and Rugged Energy Landscapes

OVERVIEW

Our modern understanding of diffusion began with the seminal paper of Einstein in the year 1905 on Brownian motion. In that paper he established two fundamental relations: (i) between the diffusion constant and the mean square displacement of a tagged large diffusing particle, and (ii) the relation between the diffusion and the friction acting on the moving particle. The former allowed us to link Brownian motion with random walk while the latter allowed the link with Langevin's equation, both associations enabled establishing the role of diffusion in molecular relaxation processes. These have already been discussed in Chapters 15 to 18. In this chapter, after discussion of certain elementary yet basic aspects of diffusion, we discuss several specific diffusion processes of importance in physics, chemistry, and biology, such as diffusion of a Brownian particle in a rugged energy landscape and diffusion of ions along a DNA chain. Subsequently, we discuss widely used relations between diffusion and entropy.

21.1 INTRODUCTION

Let us consider the following event. We place a small drop of blue ink at time $t = 0$ on the surface of water in a beaker. With time, we see the drop to disappear into the bulk of water. This happens due to diffusion of the dye molecules that comprise the ink drop into the bulk of water. At a microscopic level, we know that the diffusion is due to collisions with the surrounding water molecules. At a deeper level, diffusion as an irreversible process is driven by the entropic forces and is a consequence of the conservation of number density.

As discussed in the chapters on Brownian motion and random walk, diffusion is omnipresent in many natural processes. It is invoked to explain such diverse processes as the viscosity dependence of rate in chemical reactions, rotational and vibrational relaxation of molecules, transport processes in biological cells, protein-DNA interactions, crystal growth and nucleation of a new phase into an old parent phase, spinodal decomposition, dielectric relaxation in aqueous DNA solutions, and so on. We mentioned a few to give students the expanse of the field, and the advantage of gaining a good understanding of the diffusion process.

In a fundamental sense, the universality of diffusion arises from the basic conservation law of mass or number density. This is expressed in a quantitative form by the following continuity equation [Eq. (21.1)]:

$$\frac{\partial}{\partial t} c(\mathbf{r}, t) = -\nabla . J(\mathbf{r}, t). \tag{21.1}$$

DOI: 10.1201/9781003157601-24

where $c(\mathbf{r}, t)$ is the time (t) and space ($\mathbf{r}$)-dependent concentration or number density of the molecules under consideration. $J(\mathbf{r}, t)$ is the flux of particles passing through space-time point $(\mathbf{r}, t)$. The continuity equation (sometimes referred to as the continuity theorem) is the starting point in the study of hydrodynamics and in general, the same for the time-dependent statistical mechanics.

The continuity equation is exact but for completeness needs to be supplemented with an expression for the flux. Experimental observations suggest that mass moves from high density to low density (as in our example of spreading the dye in the ink). Thus, the flux (which is a rate) is proportional to the spatial gradient of concentration or density. This is described by the well-known Fick's law and is written as

$$J(\mathbf{r},t) = -D\nabla c(\mathbf{r},t) \tag{21.2}$$

This is the equation where we first encounter diffusion constant, D. While the continuity equation is exact, Fick's law is phenomenological. This is an example of a constitutive relation. When we combine the above two equations, we obtain the diffusion equation

$$\frac{\partial}{\partial t}c(\mathbf{r},t) = D\nabla^2 c(\mathbf{r},t) \tag{21.3}$$

As already remarked, this equation is nearly universal mainly because of the conservation law embodied in the continuity equation. Eq. (21.3) is the well-known diffusion equation that shall appear many times in the course of this book. The same equation holds for heat diffusion where heat energy is diffused with a thermal diffusion coefficient D_T.

In a liquid, the displacement of a tagged molecule occurs from several sources. First of all, molecules undergo incessant collisions with the surrounding molecules (as already mentioned), and each collision displaces the tagged molecule a little. Secondly, there are natural flows in the system. These are described as currents. *Currents are hydrodynamic modes and appear naturally as a consequence of momentum conservation.* Molecules get displaced or carried over because of these currents. Both these two mechanisms (collisions and currents) induce random walk in the tagged particle, but of different step lengths, and with different time intervals. The presence of two mechanisms needs to be remembered.

While the collision frequency of a tagged molecule is determined by the local density surrounding the molecule, the currents are controlled by viscosity. Thus, diffusion is related both to both density relaxation and current relaxation. These dependencies are the reasons that diffusion provides valuable information about the dynamics of the system. In turn, diffusion is used to describe the rates of many dynamical processes.

Diffusion is an irreversible process, as evident in the spread of the ink in the beaker of water. *It should be realized that diffusion is largely entropic in origin. Diffusion leads to an increase in entropy. Therefore, there have been attempts to relate diffusion to entropy.*

We next describe how a diffusion equation is solved to provide an expression for the position and time dependence of density. These expressions are often useful starting points.

21.2 SOLUTION OF THE SIMPLE DIFFUSION EQUATION

The diffusion equation [Eq. (21.3)] is a second-order partial diffusion equation that is to be solved as an initial and boundary value problem. In the example cited above, the dye molecules located within a narrow domain at time $t = 0$ spreads over the entire system. For simplicity, we adopt a simple one-dimensional description, whereby, the diffusion equation can be expressed as follows:

$$\frac{\partial}{\partial t}c(x,t) = D\frac{\partial^2}{\partial x^2}c(x,t). \tag{21.4}$$

This initial-boundary value problem (at $t = 0$) can be stated as,

$$c(x,0) = c_0 \delta(x) \tag{21.5}$$

where c_0 is the number of dye molecules. By definition, $\delta(x) = \infty$ when $x = 0$ and $\delta(x) = 0$ elsewhere. Hence the initial boundary value suggests that we consider the diffusion of the Brownian particles outward from a single point at $x = 0$. One should not be too worried about the singularity in the delta function – this local constrain disappears quite fast, as a consequence of the diffusion process.

One can solve Eq. (21.4) by using the method of Fourier transform. Introducing the conjugate Fourier variable k, we have the following relations:

$$\delta(x) = \frac{1}{2\pi} \int_{-\infty}^{\infty} dk e^{ikx} \tag{21.6}$$

$$c(x,t) = \frac{1}{2\pi} \int_{-\infty}^{\infty} dk e^{ikx} c(k,t) \tag{21.7}$$

Putting Eq. (21.7) into Eq. (21.4), we get

$$\int_{-\infty}^{\infty} dk e^{ikx} \frac{\partial c(k,t)}{\partial t} = D \int_{-\infty}^{\infty} dk e^{ikx} \left(-k^2\right) c(k,t). \tag{21.8}$$

Equating both sides of Eq. (21.8), we get

$$\frac{\partial}{\partial t} c(k,t) = -Dk^2 \, c(k,t). \tag{21.9}$$

Eq. (21.9) is a first-order differential equation, which can be easily solved to give

$$c(k,t) = c(k,0) e^{-k^2 Dt}. \tag{21.10}$$

Putting Eq. (21.10) into Eq. (21.7), we get

$$c(x,t) = \frac{1}{2\pi} \int_{-\infty}^{\infty} dk e^{ikx} c(k,0) e^{-k^2 Dt}. \tag{21.11}$$

Plugging in the initial boundary value [Eq. (21.5)] gives us

$$c(x,0) = c_0 \delta(x) = \frac{1}{2\pi} \int_{-\infty}^{\infty} dk e^{ikx} c(k,0). \tag{21.12}$$

Hence, comparing Eq. (21.12) and Eq. (21.6), we find

$$c(k,0) = c_0. \tag{21.13}$$

Therefore, from Eq. (21.13) and Eq. (21.11) we get

$$c(x,t) = \frac{c_0}{2\pi} \int_{-\infty}^{\infty} dk e^{ikx} e^{-k^2 Dt}.$$ (21.14)

The solution of Eq. (21.14) yields

$$\boxed{c(x,t) = \frac{c_0}{\sqrt{4\pi Dt}} e^{-\frac{x^2}{2Dt}}.}$$ (21.15)

[*Note: Students should study the dimensions here*]. The solution of the diffusion equation is a normal distribution, the width of which increases with time, with the value of the concentration decreasing at the origin but increasing at distant places, following Eq. (21.15), which shows how a drop of dye in our ink can spread throughout water with the progression of time and ultimately renders the system homogeneous.

For simplicity, we have worked out the solution in one dimension. The same method carries over to an arbitrary dimension d, for which, the expression becomes

$$c(\mathbf{R},t) = \frac{c_0}{(4\pi Dt)^{\frac{d}{2}}} e^{-\frac{\mathbf{R}^2}{2dDt}}.$$ (21.16)

We have discussed this solution in the chapter on Brownian motion where more details are given.

21.3 DIFFUSION COEFFICIENT FROM VELOCITY ACF

We have already discussed this subject in Chapter 10. Here, we present the same but with a somewhat different perspective.

In Chapter 15, we derived and discussed Einstein's equation [Eq. (21.17)]. This was the first analytical definition of Brownian motion in terms of self-diffusion coefficient of the tagged particles.

$$\frac{\partial}{\partial t} c(x,t) = D \frac{\partial^2}{\partial x^2} c(x,t).$$ (21.17)

Here, $C(x, t)$ is the space-time-dependent concentration of the diffusing particles. To find the mean square displacement of the particles, we can multiply both sides of the equation by x^2 and then integrate it over x. This gives us

$$\frac{\partial}{\partial t} \int_{-\infty}^{\infty} dx x^2 c(x,t) = D \int_{-\infty}^{\infty} dx x^2 \frac{\partial^2}{\partial x^2} c(x,t).$$ (21.18)

The left-hand side of Eq. (21.18) is the mean square displacement. Now, we note that the distribution $C(x, t)$ is Gaussian in nature with the following boundary conditions:

$$c(x,0) = c_0 \delta(x)$$

$$c(x,t)\big|_{x=\pm\infty} = 0$$

$$\frac{\partial}{\partial x} c(x,t)\big|_{x=\pm\infty} = 0$$ (21.19)

$$\int_{-\infty}^{\infty} dx c(x,t) = 1 \quad \text{(normalized)}.$$

With these boundary conditions, we integrate the right-hand side of Eq. (21.18) by parts and get the following expression:

$$\frac{\partial}{\partial t}\langle \delta x^2 \rangle = 2D. \tag{21.20}$$

On integrating over time, we obtain Einstein's diffusion equation [Eq. (21.21)] in one dimension.

$$\boxed{\langle \delta x^2 \rangle = 2Dt}. \tag{21.21}$$

This equation can be written in a general form for d-dimensions as

$$\langle \delta \mathbf{R}^2 \rangle = 2dDt. \tag{21.22}$$

Now, we can express the displacement of the diffusing particles in terms of their velocities as

$$x(t) = \int_0^t dt' v(t'). \tag{21.23}$$

Taking the square on both sides and averaging over initial conditions, we get

$$\langle \delta x^2 \rangle = \left\langle \int_0^t dt' v(t') \int_0^t dt'' v(t'') \right\rangle$$
$$= 2\int_0^t dt' \int_0^t dt'' \langle v(t')v(t'') \rangle. \tag{21.24}$$

The right-hand side of Eq. (21.24) contains the velocity autocorrelation function, which does not depend on any arbitrary time origin. Rather, it depends on the time difference $\tau = t'' - t'$.. Accordingly, by changing the variables of the integration in Eq. (21.24), already discussed in Chapter 10, we get

$$\langle \delta x^2 \rangle = 2t\int_0^t dt'' \langle v(0)v(t - t'') \rangle$$
$$= 2t\int_0^t d\tau \langle v(0)v(\tau) \rangle. \tag{21.25}$$

In the long-time limit, the left-hand side becomes $2Dt$, which translates into Eq. (21.25) as

$$\boxed{D = \int_0^\infty d\tau \langle v(0)v(\tau) \rangle}. \tag{21.26}$$

For a three-dimensional system, this equation becomes

$$\boxed{D = \frac{1}{3}\int_0^\infty d\tau \langle \mathbf{v}(0).\mathbf{v}(\tau) \rangle}. \tag{21.27}$$

As discussed in Chapters 8–10, this equation belongs to a class of equations known as the Green-Kubo relations. In these equations, a macroscopic dynamical property (here, diffusion coefficients) is expressed as the time integral of a microscopic time correlation function [here, that of velocity $v(t)$].

21.4 RELATION BETWEEN DIFFUSION COEFFICIENT AND SELF-DYNAMIC STRICTURE FACTOR, $F_s(k,t)$

Diffusion coefficient can be obtained from the self-dynamic structure factor, as discussed in Chapter 9, as these two are intimately correlated. By a cumulant expansion of the wavenumber-dependent diffusion of $F_s(k, t)$ we get the following relation:

$$F_S\left(k,t\right) = \exp\left(-\frac{k^2}{6}\left\langle (\mathbf{r}(t) - \mathbf{r}(t))^2 \right\rangle \right). \tag{21.28}$$

We know that the self-diffusion coefficient can be expressed by Einstein's relation as,

$$D = Lt_{t\to\infty}\frac{1}{6t}\left\langle \left| \mathbf{r}(t) - \mathbf{r}(0) \right|^2 \right\rangle. \tag{21.29}$$

We can then write $F_s(k,t)$ as $F_S\left(k,t\right) = \exp(-Dk^2 t)$. We can then find an alternate definition of D in terms of $F_s(k,t)$. *This definition can be formally `expressed as second derivative of $F_s(k,t)$ with respect to the wavenumber k and then taking the $k \to 0$ limit.* We have discussed this procedure in Chapter 9. Here we discussed them for the sake of continuity.

21.5 RELATION BETWEEN DIFFUSION AND VISCOSITY: THE STOKES-EINSTEIN RELATION

The relation between shear viscosity and the self-diffusion coefficient, known as the Stokes-Einstein relation (SE), was discussed in this text in Chapter 15. Einstein showed that we can express diffusion D as

$$D = \mu k_B T \tag{21.30}$$

where μ is the mobility, i.e., average velocity of the particle due to a constant external (valid in the linear limit), k_B is the Boltzmann constant and T denotes the temperature. This equation is an expression of linear response and at the same time is a manifestation of the fluctuation-dissipation theorem. Stokes showed that the mobility of the particle is related to the size of the particle, for a spherical solute, by the following relation:

$$\mu = \frac{1}{c\pi\eta\sigma}. \tag{21.31}$$

We then combine Eq. (21.30) and Eq. (21.31) to obtain the well-known Stokes-Einstein relation i.e.,

$$D = \frac{k_B T}{c\pi\eta\sigma}. \tag{21.32}$$

Stokes law gives $c = 4$ for slip and $c = 6$ for stick boundary conditions.

In Chapter 5, we have discussed a derivation of the Stokes expression from Navier-Stokes equation. Thus, a student should note the sequence of steps and approximations involved in deriving the oft-used Stokes-Einstein relation. The relation is expected to be valid when the size of the sphere is much larger than that of the solvent so that the solvent can be approximated by a continuum.

Several experimental and simulation results have revealed that the Stokes-Einstein relation works reasonably well even for tracer molecules of similar size to the solvent molecules. However, the SE relation is not exact, and deviations from it are consistently found. These deviations are reflected as follows:

The parameter c is used as a fitting parameter, with values between its slip and stick limits for different thermodynamic conditions. The local viscosity experienced by a tracer molecule may not be the same as the viscosity of the bulk liquid. Thus, viscosity is also used as a fitting parameter. The effective hydrodynamic diameter depends in a nontrivial way on intermolecular interactions.

21.6   RANDOM WALK AND DIFFUSION

Relationship between diffusion and random walk is so close that it is usually accepted that if a process admits of diffusive description, then it must also admit of a random walk description. Both embodies conservation of mass, although both can be extended to include "death" or decay of probability.

We have described this connection in Chapter 16 where we have presented a derivation of the diffusion equation from random walk description and considered related issues. Therefore, refrain from repeating here again.

21.7   ZWANZIG'S DERIVATION OF THE RELATION BETWEEN SELF-DIFFUSION AND VISCOSITY OF LIQUIDS

An interesting semi-microscopic derivation of the Stokes-Einstein relation between D and η from the Green-Kubo time correlation formula was presented by Zwanzig. This derivation has several insightful points, so we included it in this text. To perform these calculations Zwanzig introduced an elegant model, which assumes that the diffusion of a tagged particle takes place because the system jumps from one free energy minimum to another, causing the particle to displace, albeit by small amounts. This in turn was motivated by the discovery of inherent structures of liquids, which are local potential energy minima (we have discussed inherent structures in our text on "*Equilibrium Statistical Mechanics*"). Transitions of the system between these local minima cause small-amplitude displacements of individual particles. The model makes the following assumptions:

(i)   The volume of the entire system can be divided into a number of subvolumes each of which can undergo transitions between different states. Thus, the entire macroscopic system is divided into mosaics. Each mosaic can undergo transitions independently. We denote the volume of a mosaic by V^*.

(ii)   The configuration states in a mosaic are given by the inherent structures of the system.

(iii)   The waiting time distribution $\psi(t)$ for jumps between minima is $e^{-t/\tau}$.

(iv)   The mean time τ in a marked state γ is much larger than a Debye oscillation period τ_D.

(v)   The motions of the system in a particular state before the jump are uncorrelated with motions in a new state after the jump. That is, the system stays in a minimum long enough to forget where it came from.

Zwanzig started with the general definition of the self-diffusion coefficient as given by the Green-Kubo relation relating D to the velocity autocorrelation function of a tagged particle i,

$$D = \frac{1}{3}\int_0^\infty dt \langle v_i(t).v_i(0) \rangle = \frac{1}{3N}\int_0^\infty dt \sum_j \langle v_j(t).v_j(0) \rangle \tag{21.33}$$

where N is the number of particles in the system, the sum is over all the particles in the system, and we have assumed that all the N particles in the system are identical. The sum on the right-hand side of the last equation in Eq. (21.33) is now transformed into a sum over the normal modes, as in the Einstein-Debye picture of the specific heat of a solid. So, $3N$ denotes the total number of normal modes in the system. Thus, it is assumed that our tagged particle moves according to the normal modes, and superposition is taken. The displacements in the cell are harmonic in nature and we introduce normal mode frequency ω by diagonalizing the force constant matrix. With the above-mentioned approximations, the decay terms in velocity TCF become $\cos \omega t \exp\left(-\frac{t}{\tau}\right)$. The resulting expression for the self-diffusion coefficient becomes

$$D = (k_B T/3mN)\int_0^\infty dt \sum_\omega \cos\omega t \exp\left(-\frac{t}{\tau}\right) \tag{21.34}$$

where the sum is taken all over $3N$ normal modes. One can easily evaluate the above integral to obtain the expression for the diffusion coefficient,

$$D = (k_B T/3mN)\sum_\omega \tau/(1+\omega^2\tau^2).$$

Zwanzig used the Debye spectrum for the density of states to convert the sum into an integral, in a standard procedure. The cut-off is related to sound velocity, which brings in shear and bulk modulus.

In the next step, the sum in the above expression is expressed in terms of viscosity. This is accomplished in the following fashion. We know that the coefficient of viscosity is calculated by the time integrals of stress-stress time correlation function. These correlation functions are short-range in space, hence one needs to focus on the local viscosity of the region V^*. As long as the system remains in the neighborhood of a potential minimum, its motions are those of an elastic Debye continuum, and the local stress correlation function of the sub volume V^* gives a time-independent local elastic modulus; but when the system leaves this cell, all correlations in V^* are lost, and the stress correlation function of this region vanishes. Thus, the viscosity is the time integral of the constant local elastic modulus multiplied by the waiting time τ.

In Chapter 5, we already discussed a viscoelastic model for frequency-dependent viscosity. In this picture, shear viscosity is given by a product of shear modulus and relaxation time. A similar expression follows for bulk viscosity.

Finally, Zwanzig obtained the following expression for the self-diffusion coefficient:

$$\boxed{D = (k_B T/3\pi)(3N/4\pi V)^{1/3}(1/\eta_l + 2/\eta)} \tag{21.35}$$

where η_l and η denote the longitudinal viscosity and shear viscosity, respectively, and V is the volume of the system. It is found that the above expressions actually gives the coefficient in the diffusion-viscosity relation that is close to the prediction of Stokes-Einstein.

As pointed out by Zwanzig, the very real problem of estimating the lifetime for cell jumps has not been addressed in this work. Wolynes et al. later devised a method to determine the lifetime τ through the use of nucleation theory.

21.8   HYDRODYNAMIC EXPRESSIONS FOR DIFFUSION CONSTANTS OF NONSPHERICAL OBJECTS

In order to calculate diffusion through Einstein's equation $D = k_B T/\zeta$, we need the value of the friction, ζ. For spheres, this is given by Stokes and together they lead to Stokes-Einstein relation discussed above. However, the situation becomes complicated when the molecule (or the object) is not spherical. We do have some hydrodynamic expressions when the diffusing object can be approximated as a spheroidal, like prolates and oblates. In fact, diffusion of ellipsoids like prolates and oblates finds widespread use in the study of biopolymers and proteins. For such objects, Perrin has provided elegant expressions that we discuss here. In fact, Perrin's equations have been used to obtain a measure of the asymmetry in the shape of nonspherical protein molecules.

Rod-shaped molecules or prolates can diffuse preferentially in the direction of the long-axis, the disc-shaped molecules or oblates are likely to exhibit the saucer-like motion due to the expected tendency of the disc-shaped molecule to translate in the plane of the disc, though tumbling of its axis helps it in changing its direction of motion.

Microscopic theoretical approaches show that there is a strong coupling between the translational and rotational motions while studying the diffusion of nonspherical molecules in the suspension of spheres. A similar kind of coupling has been observed in the study of orientational relaxation of dipolar molecules.

The solution of the Navier-Stokes equation for nonspherical molecules is nontrivial. This nontrivial nature is reflected in the complexity of the expressions derived by Perrin. In fact, no analytical solution is available for rotational friction on spheroidals with the slip boundary conditions.

With the stick hydrodynamic prediction, it is known that for rod-shaped molecules, the translational diffusion coefficient in the direction parallel to its major axis (D_{II}) is equal to twice of the translational diffusion coefficient in the perpendicular direction ($D_\perp$). Rotation can change the direction of diffusion. This introduces a mechanism of translation-rotation coupling in ellipsoidal molecules, not present in cases of spherical molecules.

On the other hand, the slip hydrodynamic theory predicts that there could be decoupling between the perpendicular and parallel motion. The ratio between the diffusion coefficient parallel to that in the perpendicular direction approaches the value of aspect ratio (κ).

For ellipsoids, the stick hydrodynamic predictions for translational diffusion can be written as,

$$D_{II} = \frac{k_B T \left[(2a^2 - b^2)S - 2a \right]}{(a^2 - b^2)16\pi\eta} \tag{21.36}$$

and

$$D_\perp = \frac{k_B T \left[(2a^2 - 3b^2)S + 2a \right]}{(a^2 - b^2)32\pi\eta}, \tag{21.37}$$

where a and b are the lengths of the semi-major and semi-minor axes of the ellipsoid, respectively. For a prolate, S is given by,

$$S(P) = \frac{2}{(a^2 - b^2)^{1/2}} \log \frac{a + (a^2 - b^2)^{1/2}}{b}. \tag{21.38}$$

For an oblate, S is given by,

$$S(O) = \frac{2}{(b^2 - a^2)^{1/2}} \tan^{-1} \frac{(b^2 - a^2)^{1/2}}{a}. \tag{21.39}$$

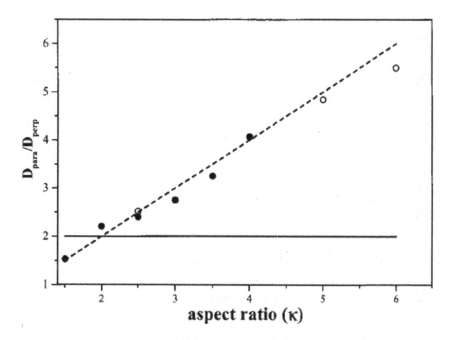

FIGURE 21.1 The plot of the ratio of diffusion coefficients, $D_{\parallel}/D_{\perp}$ with respect to the aspect ratio κ. The solid line represents the hydrodynamic prediction for stick boundary condition and the dashed line represents the prediction by slip boundary condition.

Source: This figure is taken with permission from [29].

These expressions reduce to the expression for the sphere when the aspect ratio approaches unity.

It is interesting to explore the components of diffusion in parallel and perpendicular directions for ellipsoidal molecules. The hydrodynamic stick boundary condition predicts that the ratio is constant with respect to the variation of the aspect ratio. However, the simulation shows that the ratio increases with the increase in the aspect ratio as shown in Figure 21.1. This is due to the fact that the motion in the perpendicular direction is weakly dependent on the aspect ratio.

21.9 BREAKDOWN OF THE STOKES-EINSTEIN RELATION

This is a "widely used" title of papers being published even today, with different authors addressing different aspects of the SE relation. As there are several aspects in this relation, different studies address different issues, with viscosity dependence being the central topic of attention.

21.9.1 POLLACK-ENYEART EXPERIMENTS

The relation between diffusion and viscosity is given by the celebrated Stokes-Einstein relation as discussed earlier.

$$D = \frac{k_B T}{C \pi \eta r} = \frac{\text{constant}}{\eta} \tag{21.40}$$

Here C is a constant dependent on the hydrodynamic boundary condition employed. Although this relation was derived for solute particles that are much larger in size than solvent molecules, it is

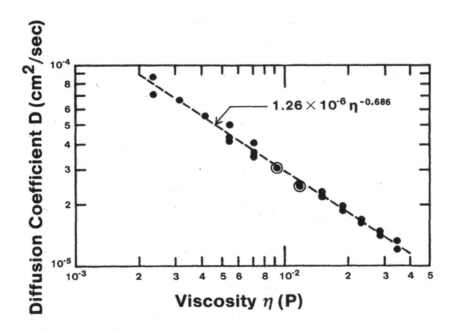

FIGURE 21.2 Translational diffusion of Xe is plotted against the viscosity of alkane solvents in Pollack-Enyeart experiments. Both the axes are represented in the log scale.

Source: The figure is taken with permission from [1].

frequently used for diffusion of solutes whose size is similar to those of the solvent molecules. Therefore, not surprisingly, in several systems, this relation is found to break down. In the 1980s, Pollack and Enyeart experimentally studied the translational diffusion of xenon (Xe^{133}) ($T = 20$ °C and $P = 1$ atm) in liquid alkane solvents in the homologous series starting from m-pentane to n-hexadecane. The only variable in their systems was the length of the alkane chain. With the increase in the chain length, the viscosity of the solvent increases. Hence the translational diffusion of Xe could be calculated as a function of the solvent viscosity. In Figure 21.2 we show the fractional viscosity dependence of diffusion observed in experiments.

From their experimental data, the following empirical relations could be retrieved.

$$D = 17.7 \times 10^{-5} e^{-0.169n}$$
$$\eta = 7.36 \times 10^{-4} e^{-0.247n}. \tag{21.41}$$

Here, D is the translational diffusion of Xe, η is the solvent viscosity and n is the number of carbon atoms in the alkane chain ($C_n H_{2n+2}$). Combining these two observed relations, we get a relation between diffusion and viscosity akin to the Stokes-Einstein equation.

$$D = \frac{1.27 \times 10^{-6}}{\eta^{0.686}} = \frac{\text{constant}}{\eta^p}. \tag{21.42}$$

The difference between Eqs. (21.42) and (21.40) is quite apparent. In the classical SE equation, the viscosity in the denominator has a power 1. However, from the experimental data, the exponent of η was found to be 0.686 ($p < 1$).

In the same year (1985), Zwanzig and Harrison gave a theoretical explanation of this observed deviation from the SE equation. Their theory was based on hydrodynamic considerations. Zwanzig and Harrison proceeded to explain the results of Pollack and Enyeart considering the classical SE equation to be correct. They explained the anomaly in terms of an *effective hydrodynamic radius* (EHR) of the diffusing particle. According to their argument, for the unaltered SE equation to hold, the effective hydrodynamic radius (EHR) a should also depend on the alkane chain length as in Eq. (21.41). This dependence could be represented as

$$a = 2.45 \times 10^{-8} e^{-0.0774n}. \tag{21.43}$$

Hence, the EHR is environment-dependent. The validity of the Zwanzig-Harrison theory is based on the following arguments.

Einstein diffusion equation $D = k_B T / \zeta$ is a direct consequence of linear response theory and is exact. It is independent of the mechanism of friction. However, a particle moving in a fluid medium produces a spatially nonuniform flow that results in viscous dissipation of momentum and energy. This dissipation is linearly dependent on viscosity according to classical hydrodynamics. A substantial fraction (say $1/\varepsilon$) of this energy dissipation occurs in a region beyond a sphere of radius εr from the center of the diffusing particle (r = radius of the particle) where hydrodynamics is expected to hold. Nonetheless, in the immediate vicinity of the moving particle, the effects of solute-solvent interactions are significant and may result in the breakdown of hydrodynamic laws. This is most significant in the absence of any other nonviscous dissipation mechanism. The EHR proposed in this theory is a measure of the strength of these complex interactions that define the coupling between solute and solvent.

It is interesting to note that a similar argument was proposed by Hynes, Kapral, and Weinberg quite some time back where they incorporated an additional contribution to friction to arise from collisions with the nearest neighbors. We discussed this work in Chapter 23.

21.10 JUMP DIFFUSION MODELS

Jump diffusion is a stochastic process that involves both jumps and continuous Brownian diffusion. The translational diffusion (and also rotational relaxation) sometimes occurs through large amplitude, infrequent jumps. In such cases, jumps are superimposed on natural continuous diffusive process.

It has been shown in recent computational studies on water and other viscous liquids, that the large amplitude motions may be more of a rule than an exception.

In some of the recent studies of linear molecules like carbon monoxide and nitric oxide, it has been shown that the translational diffusion of these molecules is strongly coupled to their own jump rotational motion, which in turn couples to the well-known jump dynamics of water molecules. To study this, a single CO or NO molecule trajectory and the trajectory of the nearest water are examined. Figure 21.3(a) and (b) show the trajectories of square displacement of CO and NO, respectively, are correlated with the nearest-neighboring water molecules from the initial time of trajectory. These figures demonstrate the coupled translational jump motion of these linear molecules and its nearest water molecules. The slaved motion of small solutes to the dynamics of water is also visible here.

The combined effects of jump and continuous diffusion has been described in theory by using the method of continuous time random walk, with two different waiting time distribution. We have discussed this aspect in the chapter on random walks (Chapter 16).

21.11 DIFFUSION IN SOLIDS AND ZEOLITES

Diffusion in zeolite provides a clear and interesting example of the application of diffusion phenomena to a specific industrial process, which is the separation of alkanes and other species in the

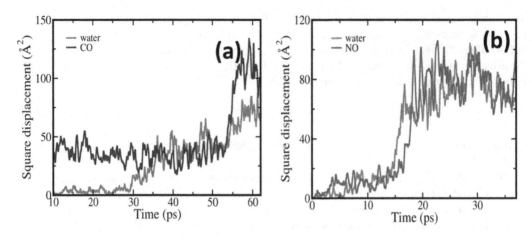

FIGURE 21.3 Square displacement trajectories of linear molecules (a) CO and (b) NO and the nearest neighboring water molecules. These figures show the coupled translational jump motion of the water molecules with the molecules CO and NO.

Source: These plots are taken with permission from [33].

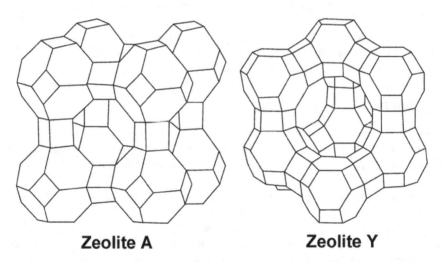

FIGURE 21.4 Two α cages of zeolite A and Y along with the interconnection window are shown here. Each cage in zeolite A is connected to six other cages placed octahedrally while each cage in zeolite Y is connected to four other cages in a tetrahedral fashion.

petroleum industry. Zeolites are also used in catalysis because of the large surface area. In both processes, the molecules move through the cages provided by the zeolites. We show a caged structure of zeolites. The important characteristic of many of the zeolites is that it has fairly big cages, which have separated from each other by a narrow neck and this structure allows the size selectivity in many ways the molecules inside zeolite cage behave as liquid and so one can imagine a dependence of diffusion on size as given by the Stokes-Einstein relation. However, the presence of the neck introduces a stringent selectivity allowing only molecules of a below certain size to pass through. Figure 21.4 shows a schematic diagram of two α cages of zeolite A and Y along with the interconnection window.

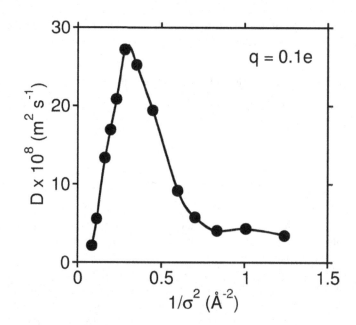

FIGURE 21.5   The plot of size-dependence of diffusion constant for ion with charge q= 0.1e shows the existence of a pronounced maximum at intermediate ion size. This has been termed as a levitation effect.

Source: Taken with permission from [26].

While it is easy to understand the slowing down or localization of big molecules, which are larger than the size of the neck, recent studies by Yashonath and coworkers have shown that there are certain specific sizes that exhibit larger diffusion in a total breakdown of Stokes-Einstein relation. In other words, the sizes of small molecules also diffuse slower than an intermediate size molecule and diffusion shows a size-dependent pronounced maximum as shown in Figure 21.5.

This has been termed as *levitation effect* and this is an interesting exercise in statistical mechanics. Although the discovery of levitation effect was made through computer simulation, it has subsequently been observed in experiments using isotopic dilution and it was possible to explain through a statistical mechanical model invoking free energy landscape.

21.12   DIFFUSION IN A RUGGED ENERGY LANDSCAPES

If we analyze the diffusion of a small tagged particle migrating between the cages of zeolites (discussed above), we find that the particle experiences different energies in different cages where the tagged particle experiences an energy landscape that contains multiple hills (maxima) and valleys (minima). In many practical situations, the energy landscape where diffusive dynamics or migration occurs, the heights of the maxima and the depths of the minima are random quantities. Such a landscape is termed as a rugged energy landscape (REL). In the case of zeolites the landscape is not random. So, it is not rugged.

In a rugged landscape, the diffusion gets modified. In fact, it can get drastically reduced, as we discuss below. The study of diffusion of a Brownian particle on a random energy landscape serves as an effective model in understanding different complex phenomena, such as diffusion in glassy systems and supercooled liquids, dynamics of molecular motors, diffusion of a protein along a DNA, enzyme kinetics, movements of a fluorescent labeled molecule inside the cell, etc., to name a few. An important application of this beautiful model is in protein folding where the transformation of the unfolded state is viewed as diffusion in the rugged conformation space.

21.12.1   THE ISSUE OF BROKEN ERGODICITY IN LOWER DIMENSIONS

The energy landscape is mostly multidimensional for complex systems and diffusion is considered via Einstein's definition as given in Eq. (21.44).

$$D(d) = \lim_{t \to \infty} \frac{\left\langle (\Delta r)^2 \right\rangle}{2dt}. \tag{21.44}$$

The above definition, for an *isotropic ergodic system*, shows that dimensionality is not an important issue. However, this simplicity is lost if the random walk and diffusion take place in the presence of the ruggedness of the energy landscape. Now, diffusion could show pronounced dependence on dimensionality, as discussed below.

In an elegant study, Newman and Stein addressed the issue of the dimensionality of diffusion in a rugged landscape. They treated diffusion as a percolation invasion problem and concluded that *diffusion in one dimension is pathological because the particle (in their language "water in a lake or river") can get trapped ("cannot flow to the sea") due to insurmountable barriers on both sides of exit* (Figure 21.6).

They observed that the height of the barriers faced by the diffusor/walker exhibits a logarithmic growth with time as $T \log(t)$, where T is the temperature. *Hence, the height of barriers increases with increasing time. Consequently, the probability of the walker getting reflected back and retracing the same path (in 1d) increases.* This leads to a sharp decrease of diffusion constant (even going to zero asymptotically). If one takes the limit of $t \to \infty$, at a fixed T, for systems with constant ruggedness, one raises the possibility of facing broken ergodicity.

However, at intermediate times, a well-defined value for the diffusion constant may exist. In higher dimensions, starting from $d = 2$, the walker can find practically an infinite number of escape

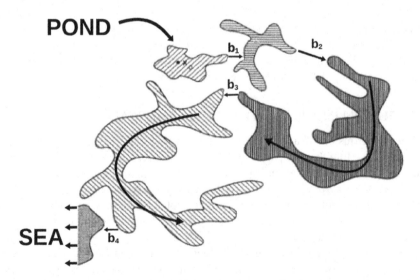

FIGURE 21.6   A rough sketch of the "ponds-and-outlets" picture illustrating the rugged landscape effect. The first pond contains the starting site x_o. Arrows indicate the large-scale direction of motion; once the process leaves a given pond, it does not return. The values of the b_n decrease as n increases; b_n controls the height of the minimal barriers confining the system to pond n.

routes. In the language of Newman and Stein, "*the water can flow to the sea*". Hence, this issue of broken ergodicity in a rugged landscape does not arise in higher dimensions.

21.12.2   ZWANZIG'S EXPRESSION OF DIFFUSION IN A RUGGED ENERGY LANDSCAPE

In a landmark paper, Zwanzig represented a rugged energy landscape potential [$U(x)$] as the sum of a background potential [$U_0(x)$] and a rugged potential [$U_1(x)$]. Hence, $U(x) = U_0(x) + U_1(x)$, where the latter term is random with a Gaussian distribution with a mean 0 and a standard deviation of ε.

$$P(U_1) = \frac{1}{\sqrt{2\pi\varepsilon^2}} \exp\left(-\frac{U_1^2}{2\varepsilon^2}\right). \tag{21.45}$$

Thus, ε is a measure of the ruggedness of the landscape. Zwanzig used the well-known method mean-first passage time to find that the effective diffusion coefficient is given by the following expression:

$$D_{eff} = \frac{D_0}{\left\langle e^{U_1/k_B T}\right\rangle\left\langle e^{-U_1/k_B T}\right\rangle}. \tag{21.46}$$

Here D_0 denotes the diffusion constant on the smooth potential, and $<...>$ stands for the spatial average over the rugged potential. The averages in the above expression can be evaluated by cumulant expansion (as the distribution of ruggedness is Gaussian) to obtain the following expression for the effective diffusion constant:

$$\boxed{D_{eff} = D_0 \exp\left[-\varepsilon^2 / (k_B T)^2\right]} \tag{21.47}$$

The derivation, however, employs a smoothening of the landscape by using a local average in order to smooth the rugged potential. This approximation breaks down in the presence of a configuration that involves one minimum flanked by two maxima as shown in Figure 21.7, where a

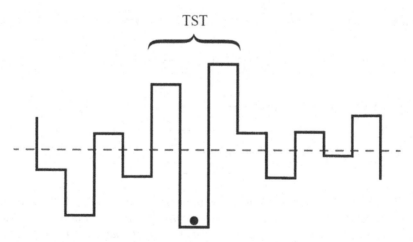

FIGURE 21.7   Schematic representation of a three site trap model (TST).

Source: Figure taken with permission from [34].

particle is trapped in a deep minimum (of negative energy) guarded by two maxima. The existence of such TSTs becomes more probable as we increase ε. This may give rise to long trapping and a subdiffusive growth of the mean first passage time (MFPT). In Zwanzig's treatment, such TSTs are coarse-grained and neglected.

Thus, the estimation of D by using the MFPT used by Zwanzig might not work here. This is because τ_{MFPT} (with respect to an initial and final position) is obtained from a coarse grained potential. Hence, this treatment implicitly invokes the relation.

Simulations showed that Zwanzig's expression is only semi-quantitatively reliable. In the following we present an improved expression that provides a quantitative description of diffusion without invoking too much complexity.

21.12.3 IMPROVED EXPRESSION OF EFFECTIVE DIFFUSION COEFFICIENT IN REL

In order to achieve the derivation, we consider a discrete random lattice where the site-energies are sampled from a Gaussian described in Eq. (21.45). According to the model we can have three different scenarios: (i) a trap with barriers on both sides (TST); (ii) a barrier with traps on both sides; and (iii) barrier on one side and trap on the other. We restrict the transitions to the nearest neighbors with the following description of the transition rate from site-i to site-j, Γ_{ij}.

$$\Gamma_{ij} = \begin{cases} \Gamma_0 & ; U_j < U_i \\ \Gamma_0 \exp\left[\left(U_j - U_i\right)/k_B T\right] ; U_j \geq U_i. \end{cases} \tag{21.48}$$

We solve this model numerically by simulating a Brownian particle and performing continuous time random walk (CTRW). D_{eff} can be obtained from Einstein's relation $\langle \Delta x^2 \rangle = 2D_{eff}t$ where $\langle \Delta x^2 \rangle$ is the mean-squared displacement of the random walker is. It is found that Zwanzig's prediction [Eq. (21.47)] systematically overestimates D_{eff}.

By using this model the expression of MFPT and D_{eff} can be derived analytically. The derivation is quite rigorous. Here we provide the final form in Eq. (21.49).

$$\tau_{MFPT} = \frac{N^2}{2D_0} \exp\left(\beta^2 \varepsilon^2\right)\left[1 + erf\left(\frac{\beta \varepsilon}{2}\right)\right]. \tag{21.49}$$

The relation between MFPT and D_{eff} is given by Eq. (21.50), where N is the number of sites

$$D_{eff} = \lim_{N \to \infty} \frac{N^2}{2\tau_{MFPT}}. \tag{21.50}$$

By using Eqs. (21.49) and (21.50) we can write

$$D_{eff} = D_0 \exp\left(-\beta^2 \varepsilon^2\right)\left[1 + erf\left(\frac{\beta \varepsilon}{2}\right)\right]^{-1}. \tag{21.51}$$

Eq. (21.51) is the modification of Zwanzig's expression given in Eq. (21.47). *This expression provides a substantially more accurate description of the dependence of diffusion on ruggedness, as shown in Figure 21.8. Note that Eq. (21.56) contains no additional parameters than temperature and ruggedness, ε.*

Diffusion on Flat and Rugged Energy Landscapes 353

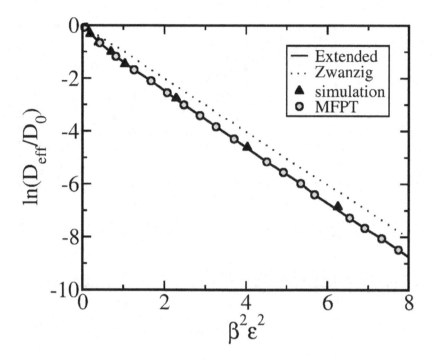

FIGURE 21.8 Plot of scaled diffusion coefficient (D_{eff}) against the squared ruggedness parameter (ε) for the Gaussian discrete random lattice. The black dotted line represents the theoretically predicted values given by Eq. (21.47). The solid triangle represents the results obtained from continuous time random walk simulation. The open circles represent the results obtained from the numerical evaluation of MFPT on the rugged potential energy surface. The solid black line indicates the results predicted by Eq. (21.51).

Source: Figure taken with permission from [34].

21.12.4 DIMENSIONALITY DEPENDENCE

A peculiarity of diffusion in a rugged energy landscape is that dimensionality has a strong influence on the value of the diffusion coefficient. For one dimension it depends on the order of limits, as discussed above. Next, the value of the diffusion coefficient increases by a factor of ~5–10 in going from one to two dimensions. In contrast, the changes in subsequent transitions (like two to three dimensions and so on) are far more modest, of the order of 10–20% only, see Figure 21.9.

21.13 EXAMPLES OF DIFFUSION ALONG RUGGED LANDSCAPE: DIFFUSION OF PROTEINS ALONG DNA

Proteins and nucleic acid enzymes diffuse along DNA to locate target sites. Because of the inherent heterogeneity of the DNA, the energy landscape experienced by the protein becomes rugged. Several earlier studies reveal that proteins can both hop and slide along a double-stranded DNA. Different models assume the pathway of the protein motion either linear (parallel to the DNA axis) or helical (by following a particular strand or groove). The protein can also rotate during the diffusive sliding. These complex coupled motions make this problem interesting.

A protein experiences three different frictional forces as it moves along the DNA: (i) friction on co-linear motion parallel to the DNA axis; (ii) the rotational friction for motion along the offset helical path due to circumnavigation of the DNA axis; and (iii) the additional rotational friction that

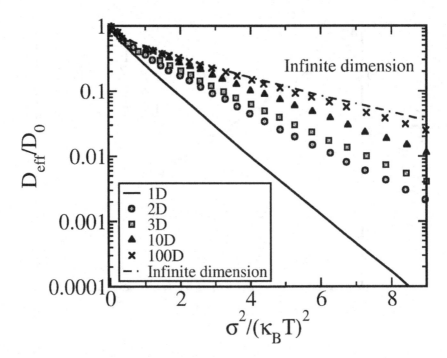

FIGURE 21.9   The plot of D_{eff}/D_0 against $\varepsilon^2/(k_BT)^2$ for one (solid line), two (open circle), three (open square), ten (filled triangle), and 100 (cross) dimensions. The black dotted line refers the result of infinite dimension.

Source: Redrawn from [7].

arises from the body-centric protein rotation. These frictional forces are proportional to the solvent viscosity and arise due to the collision of the protein with the solvent molecules. The presence of such frictional forces retards the motion of protein. In addition, the interaction between protein and DNA is responsible for slow diffusion.

If one considers a purely linear motion of the protein, only translational friction is applicable. In that case, the well-known Stokes-Einstein relation could predict $(1/R)$ size dependence of diffusion, where R is approximately the radius of a globular protein. On the other hand, if the protein traces out a helical path, it is bound to rotate. This may give rise to stronger size dependence such as $1/R^3$. A comparative study among different hydrodynamic models is given in Figure 21.10(b) and (c).

Single-molecule spectroscopic studies have quantified the effects of a rugged energy landscape on the diffusion of protein along a DNA. Such rough (and quasi-periodic) energy landscapes are generated due to the heterogeneity in the DNA sequence. The ruggedness is depicted in Figure 21.10(a). Here we briefly discuss the theoretical formalism that is available in the paper by Bagchi *et al*. listed in the Bibliography.

As shown in the previous section, rugged energy landscape, even with small barriers, can retard the diffusive motion of the particle. Let us denote this retardation factor by $F(\varepsilon)$. Following Zwanzig, one can write the effective diffusion constant as the product of the hydrodynamic diffusion constant and a factor $F(\varepsilon)$. Here, ε is the root mean squared variation of the fluctuations in the energy landscape. The theory used hydrodynamics to derive an expression of the hydrodynamic friction experienced by the sliding protein. The derivation considered both rotational and translational motion of the protein and arrived at the following interesting expression [Eq. (21.52)] for the diffusion constant of the protein:

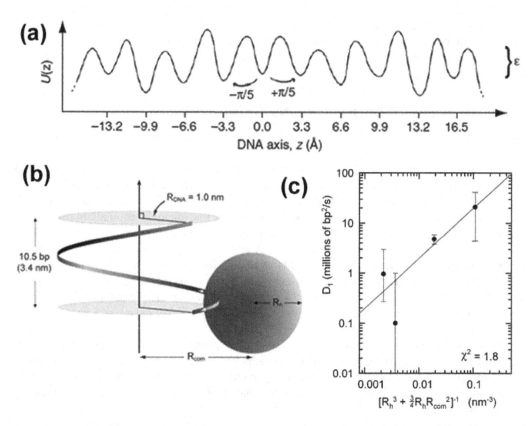

FIGURE 21.10 (a) Schematic illustration of a rugged free-energy landscape $U(z)$ experienced by sliding protein molecule according to the helical diffusion model. The root mean squared deviation of $U(z)$ is given by ε. (b), (c) Test of various hydrodynamic models with experimentally measured diffusion coefficients for four different proteins. Comparison of the data with the helical diffusion model which satisfactorily describes the experimentally observed values (bp = base pair).

Source: Taken from with permission [16].

$$D_{slide}^{helix} = b^2 \frac{k_B T}{\left[6\pi\eta Rb^2 + 8\pi\eta R^3 + 6\pi\eta R.R_{OC}^2 \right]} F(\varepsilon). \qquad (21.52)$$

In Eq. (21.52) R is the radius of the protein, b is the distance covered per full rotation (~10.5 base pairs or 3.4 nm), R_{OC} is the minimum distance between the protein's center of mass and the DNA axis, and η is the coefficient of viscosity of the solvent. Now, for a macromolecule like protein, rotational diffusion dominates over the translation and one can further approximate Eq. (21.52) as follows:

$$D_{slide}^{helix} \approx b^2 \frac{k_B T}{\left[8\pi\eta R^3 + 6\pi\eta R.R_{OC}^2 \right]} F(\varepsilon). \qquad (21.53)$$

If, for a particular DNA-protein complex $R \approx R_{OC}$, D scales as $1/R^3$. Single-molecule fluorescence tracking experiments obtained the values of the diffusion coefficient for labeled human oxo-guanine DNA glycosylase that slides along a double-stranded DNA. For this protein-DNA complex, $R = 3.2$ nm and $R_{OC} = 2.5$ nm. The calculated values of ε for the proteins range between $0.75k_B T$ to

$1.35k_BT$ with an average value of $1.1k_BT$. The results show that the tightly clustered small energy barriers facilitate the rapid exploration of the binding sites by the protein. Proteins that slide faster, have evolved to minimize the ruggedness parameter (ε). Of course, for different proteins the track of sliding can be different.

21.14   DIFFUSION OF COUNTERIONS ALONG DNA

We now discuss a complex diffusion from biophysics that has been widely discussed in the literature.

The long-time power-law behavior observed in solvation dynamics (SD) of aqueous DNA solutions has remained a puzzle for decades. Such long-time behavior is not observed in SD of aqueous proteins. To understand this anomalous behavior, here we follow an elegant theoretical description developed by Scher, Montroll, and Lax in order to explain the anomaly in transit time distribution in photocopy machines and experiments carried out on amorphous solid surfaces. They derived an expression for time-dependent current by approximating various forms of hopping time distribution. We find this description to be suitable to describe the above-mentioned phenomena.

The main idea behind the following discussion is that the power-law decay can arise from random walk of counter ions along the DNA chain. This random walk experiences a waiting time distribution, which is nonexponential due to the presence of correlations among the walkers imposed by the DNA chain.

The DNA backbone is a chain of negatively charged phosphate (PO_4^{2-}) groups that serve as localized traps where positively charged counterions can reside. There could be three types of transitions: (i) counterions arriving from bulk to a particular site; (ii) ions leaving a site and diffusing to the bulk solvent; and (iii) ions moving along the chain to the next site along the chain, if vacant. These processes give rise to a distribution of waiting times $[\psi(t)]$. We have already discussed the concept of waiting time and its distribution in Chapter 16. Montroll *et al.* considered the discontinuous flow of ions as propagation of Gaussian current packet. The propagation velocity of the mean position of the packet, $\langle l \rangle$, yields current.

Because of the fluctuation in the number of the counterions, water molecules, and the DNA itself, the probe (located at a major/minor groove or intercalated) experiences a continuous fluctuations of the dipolar/ionic electric field. Thus, the electric current $[I(t)]$ generated becomes proportional to the time derivative of the solvation energy $[E(t)]$.

$$I(t) \propto \frac{dE(t)}{dt}. \tag{21.54}$$

Now, the move from one site to another is controlled by several factors, namely, a potential barrier, availability of empty sites and separation distance between sites. Moreover, there is certain correlation in the location of empty sites and a potential barrier that depends on the local surroundings. Hence, one cannot approximate $\psi(\tau)$ as a rapidly decaying function with a single transition rate as often served by an exponential decay. We acquire evidence from our MD simulations that counterions hop along the DNA backbone with a *long-tailed distribution* of waiting times. Here, we approximate $\psi(\tau)$ as follows:

$$\psi(\tau) \sim \exp[\tau] \int_{\sqrt{\tau}}^{\infty} \left(z - \sqrt{\tau}\right)^2 \exp\left[-z^2\right] dz \tag{21.55}$$

such that the long-time tail acquires a $\tau^{-(1+\alpha)}$ dependence; where α is the exponent between 0 and 1. We now define $P(l, t)$, which is the probability of finding a random walker at l at time t. This is given by the following Laplace inversion

$$P(l,t) = \frac{1}{2\pi i} \int\limits_{c-i\infty}^{c+i\infty} dz \frac{\exp(zt)}{z} \left[1 - \psi^*(z)\right] P(l, \psi^*(z)) \tag{21.56}$$

where $\psi^*(z) = \int\limits_0^\infty \exp[-zt]\psi(t)\,dt$ and $P(l, z)$ is the random walk generating function defined as

$$P(l,z) = \frac{1}{N_1 N_2 N_3} \sum_{s_1=1}^{3} \sum_{s_2=1}^{3} \sum_{s_3=1}^{3} \frac{e^{i(\overline{\mathbf{l}}.\overline{\mathbf{k}})}}{1 - z\lambda(k)} \tag{21.57}$$

k is $(2\pi s/N)$ and $\lambda(k)$ is the hopping transition probability in Fourier space expressed as

$$\lambda(k) = \sum_l p(l)e^{i(\overline{\mathbf{l}}.\overline{\mathbf{k}})}; \ where \sum_l p(l) = 1. \tag{21.58}$$

Given the above probability and waiting time distribution, Montroll *et al.* showed that the electric current exhibits two different power laws in two limits as follows:

$$\begin{aligned} I(t) &\sim t^{-(1-\alpha)} \ if \ \langle l \rangle \ll L \\ I(t) &\sim t^{-(1+\alpha)} \ if \ \langle l \rangle \approx L \end{aligned} \tag{21.59}$$

where L is the total length of the backbone of DNA molecule. Now, from Eq. (21.54) it is clear that the solvation energy can also exhibit power law either $\sim t^\alpha$ or $\sim t^{-\alpha}$ depending on the two domains mentioned in Eq. (21.59).

While the above discussion may appear to be a bit abstruse, the pertinent point is that the power-law decay can arise from random walk of counter ions along the DNA chain.

21.15 DIFFUSION–ENTROPY RELATIONSHIPS

A relation between entropy (S) and diffusion (D) is fascinating at a fundamental level as it relates to two seemingly unrelated properties, namely, thermodynamics and dynamics. As we know, while entropy measures the volume of the accessible phase space, diffusion gives the rate of exploration of the configuration space. Therefore, a relationship between them is not fully unexpected

Two well-known diffusion-entropy relations are available in the literature, namely, the Rosenfeld scaling relation and the Adam-Gibbs relation. The first one was pioneered by Rosenfeld and is known as diffusion-entropy scaling relation. Rosenfeld plotted available transport coefficients of various systems like Lennard-Zones, hard spheres, etc., against their entropy to relate diffusion coefficient with entropy. Based on the figure shown in Figure 21.11, Rosenfeld proposed an exponential scaling relation between diffusion and entropy, which is given by [Eq. (21.60)]

$$D^* = a\exp(bS_{ex}). \tag{21.60}$$

Here S_{ex} represents the excess entropy of the system defined by $S_{ex} = \dfrac{S - S_{id}}{Nk_B}$. S and S_{id} denote the total entropy of the system in question and that of an ideal gas under the same thermodynamic conditions of temperature and density, respectively. N represents the total number of atoms/molecules present in the system. In the preceding equation, a and b are two constants that depend on the nature of the system. D^* denotes the reduced diffusion coefficient and is given by [Eq. (21.61)]

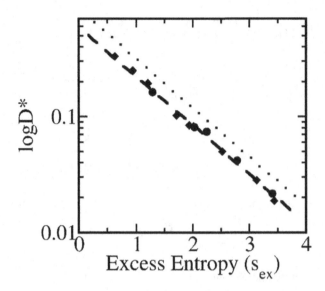

FIGURE 21.11 Demonstration of diffusion entropy scaling. Plot of reduced self-diffusion coefficient against excess entropy (S^E/k_B) for the soft sphere and LJ system as shown by the dotted line. The LJ shear-viscosity data along the saturated-vapor-pressure line is taken from Ashurst and Hoover. The LJ self-diffusion data is taken from Levesque and Verlet. The broken line compares the results for one-component plasma system with the results of soft sphere and LJ system.

Source: Redrawn from [7].

$$D^* = D\frac{\rho^{\frac{1}{3}}}{\left(k_B T/m\right)^{\frac{1}{2}}}. \tag{21.61}$$

Here $\rho, m,$ and D represent the number density, mass, and self-diffusion coefficient of the system. In the next section, we shall outline the derivation of the exponential relation between diffusion and entropy using basic principles of statistical mechanics.

Another celebrated relation, known as Adam-Gibbs relation, is given by [Eq. (21.62)]

$$D = Ae^{-\left(B/Ts_c\right)} \tag{21.62}$$

where s_c represents configurational entropy per particle of the system at a temperature T. A and B are two constants usually obtained by fitting to the experimental data. The configurational entropy is defined by $S=Nk_B \ln \Omega$, where Ω is the total number of configurations of the system and N is the total number of particles.

Adam-Gibbs proposed a heuristic derivation of the scaling relation using the concept of "cooperatively rearranging region (CRR)". *A CRR is defined as a region that contains at least two distinct configurational states such that relaxation can occur by transitions between them.* When the configurational entropy of the system decreases, the length of the region increases because we need to explore larger and larger region to find the two states, predicting a growing length scale in the system.

Later, Kirkpatrick, Thirumalai, and Wolynes (KTW) provided a more microscopic approach to establish the scaling relation. KTW employed the classical nucleation theory to calculate the nucleation barrier for the glass to liquid transition. If we assume that the entropy of glass is zero and enthalpy of liquid and glass phases are the same, then the free energy difference between the two phases is given by TS_c. The standard nucleation theory was modified by borrowing ideas from spin glass theory that gives a radius R dependent surface tension, $\sigma(R) = \dfrac{\sigma_0}{\sqrt{R}}$, where σ is the constant associated with the surface tension. The free energy cost for creating a nucleus of radius R is sum of two terms, a bulk term (-ve) and a surface term (+ve) and is given by

$$\Delta G(R) = -\frac{4\pi}{3} R^3 TS_c + 4\pi R^2 \sigma_0 / \sqrt{R} \tag{21.63}$$

We can find out the radius of the critical nucleus by setting the derivative $\dfrac{d\Delta G(R)}{dR}$ to zero to recover the nucleation free energy, which is found to be proportional to $(TS_c)^{-1}$.

This theory of KTW is known as the random first order phase transition theory, or, RFOT. It further predicts that a static correlation length emerges that scales as $1/s_c^{2/3}$, where s_c is the configuration entropy per particle.

Other than RFOT, we do not have any other statistical mechanical derivation of Adam-Gibbs diffusion-entropy scaling relation.

21.15.1   A MICROSCOPIC DERIVATION OF DIFFUSION-ENTROPY SCALING RELATION

This section discusses a derivation of diffusion-entropy exponential relation introduced by Rosenfeld employing the basic principles of statistical mechanics. Figure 21.12 shows a schematic representation of a bistable potential with two states. We are familiar with the underlying random walk behind the existence of diffusion in any system.

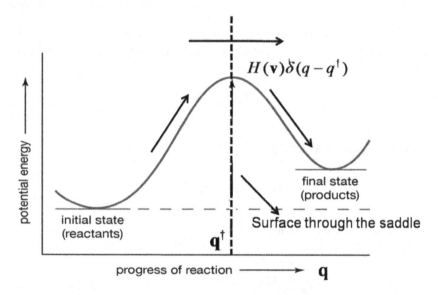

FIGURE 21.12   A schematic diagram of a bistable potential with two states initial state and product state. The dotted line represents the location of the saddle surface. The Heaviside function H takes care of the positive momentum of the reactant molecules and the δ function fixes the saddle point between the reactant and product.

On the other hand, the theory of regular random walk relates the rate constant (k) of crossing from one cell to the other to diffusion (D) through the following relation:

$$D = \frac{1}{2d} k a^2 \qquad (21.64)$$

where a is lattice constant, the distance between two adjacent cells, and d represents dimension of the system.

For such a reaction as in Figure 21.12, the exact rate constant is shown to be given in canonical ensemble by (see Chapter 20),

$$k_{P \to R} Q_R = \left\langle \left(\delta(q_1 - q_c) \frac{p_1}{m_1} H(p_1) \right) \right\rangle \qquad (21.65)$$

where "R" and "P" represent the reactant and product state, respectively. In Eq. (21.65), δ function specifies the location of the dividing surface between reactant and product and $k_{P \to R}$ represents the rate constant for the barrier crossing process. Q_R corresponds to partition function per unit volume for the reactant. The angular brackets $< ...>$ denote equilibrium average over the reactant state characterized by the Hamiltonian H_N is the total Hamiltonian of the system with N degrees of freedom and is defined as

$$H_N = \sum_{i=1}^{N} \frac{p_i^2}{2m} + V_N(q_1, ... q_N). \qquad (21.66)$$

We assume that one of the coordinates (say q_1) is perpendicular to the dividing surface, $\sigma(q_1 = q_c) = 0$ where q_c represents the critical value of q_1 at the transition point. With this condition, $\delta(q_1) \left(\frac{p_1}{m_1} \right)$ is the flux through the saddle surface where p_1 represents momentum operator conjugate to q_1. In Eq. (21.65), $h(p_1)$ takes care of the positive momentum at the dividing surface following the transition state theory approximation and is defined by

$$\begin{aligned} H(p_1) &= 1, \, p_1 > 0 \\ &= 0, \, p_1 < 0. \end{aligned} \qquad (21.67)$$

At fixed energy E, the classical density is given by,

$$\rho(E) = \frac{1}{h^N} \int dq^N dp^N \delta(H_N - E). \qquad (21.68)$$

At fixed E, Eq. (21.65) can be written as follows:

$$k(E) = \frac{\frac{1}{2} \int \frac{dq^N dp^N}{(2\pi\hbar)^N} \delta\left(H_N - E\right) \delta\left(q_1 - q_c\right) |\dot{q}_1|}{\int \frac{dq^N dp^N}{(2\pi\hbar)^N} \delta\left(H_N - E\right)}. \qquad (21.69)$$

We can simplify the denominator of Eq. (21.69) further as follows:

$$\rho(E) = \frac{d}{dE} \int \frac{dq^N dp^N}{(2\pi\hbar)^N} \Theta\left(H_N - E\right) = \frac{d\Omega(E)}{dE} \tag{21.70}$$

where $\Omega(E)$ represents the total number of states present in the system with energy less than or equal to E and Θ is a step function. Certain algebraic manipulations and use of Boltzmann's definition of entropy, i.e., $S = k_B \ln \Omega$, leads to the following expression for the rate:

$$k(E) = \frac{k_B}{2\pi\hbar \left(\dfrac{\partial S}{\partial E}\right)} \frac{\Omega^{tran}(E)}{\Omega(E)}. \tag{21.71}$$

Here $\Omega^{trans}(E)$ represents the total number of microstates at the transition state with energy less that E. The above expression is analogous to the Wigner and Eyring equation obtained from transition state theory approximation if we use the definition of temperature in a microcanonical ensemble, i.e.,

$$\left(\frac{\partial S}{\partial E}\right)_{N,V} = \frac{1}{T}. \tag{21.72}$$

Then, we substitute Eq. (21.71) for the rate in Eq. (21.64) and recall $S^\dagger = k_B \ln \Omega^{tran}$ (i.e., Boltzmann's formula of entropy) to recover the following scaling relation for the diffusion constant:

$$\boxed{D = \frac{a^2}{2d} \frac{k_B T}{2\pi\hbar} \exp\left(-\frac{S - S^\dagger}{k_B}\right).} \tag{21.73}$$

Eq. (21.73) is similar to the form of Rosenfeld but not identical. Here $S^\dagger$ is a scaling constant that serves the same role of S_{id} involved in the Rosenfeld scaling relation [Eq. (21.60)].

As mentioned earlier, the exponential form of diffusion-entropy scaling has been established via computer simulations.

21.16   SUMMARY

From the many applications discussed in this chapter, the reader should be able to acquire certain preliminary expertise that can help in further studies and research. Further skills, such as learning techniques to solve diffusion equations with complex boundary conditions, can be learned directly from specialized books/monographs like *"Probability Theory"* by Feller, the excellent book on *"Conduction of Heat in Solids"* by Carslaw and Jeager.

This chapter deals with basic theoretical aspects of diffusion phenomena. In developing this chapter, we have chartered a middle path where some complicated aspects have either been avoided or kept to a minimum but have emphasized applications so that students and researchers can appreciate the importance of the depth and breadth of the subject and the need for studying it in detail.

Translational (or, mass) diffusion is central to our understanding of nonequilibrium phenomena because many relaxation processes involve the transfer or migration of mass from one location to another, may it be over microscopic distances or much larger length scales. In addition to the great practical importance of this field, diffusion also is a beautiful and intellectually stimulating subject that has drawn the attention of many scientists, starting with Einstein. This field is intimately connected with phase transition, chemical reaction dynamics, biological processes like protein-DNA interaction or motion of counter ions along a DNA chain. It really amazes even a true practitioner of the field to find the number of different areas that the theory of diffusion, in some form or other, plays a central role.

Among the fundamental subtopics, we have discussed in detail the diffusion-entropy relationship, attempting to explain both Rosenfeld scaling and the Adam-Gibbs relation. We have discussed the interesting subject of diffusion in a rugged energy landscape. We have covered the use of diffusion in chemical kinetics, and also in complex systems like zeolites. The diffusion of ions has been discussed in a separate chapter.

BIBLIOGRAPHY

1. Pollack, G. L., and J. J. Enyeart. 1985. Atomic test of the Stokes-Einstein law. II. Diffusion of Xe through liquid hydrocarbons. *Physical Review A, 31*(2), 980.
2. Zwanzig, R. W., and A. K. Harrison. 1985. Modifications of the Stokes–Einstein formula. *The Journal of Chemical Physics, 83*(11), 5861–5862.
3. Hansen, J. P. and I. R. McDonald. 1976. *Theory of Simple Liquids*. New York: Academic Press.
4. Bagchi, B. 2012. *Molecular Relaxation in Liquids*. New York: Oxford University Press.
5. Zwanzig, R. W. 1988. Diffusion in a rough potential. *Proceedings of the National Academy of Sciences, 85*(7), 2029–2030.
6. Banerjee, S., R. Biswas, K. Seki, and B. Bagchi. 2014. Diffusion on a rugged energy landscape with spatial correlations. *The Journal of Chemical Physics, 141*(12), 09B617_1.
7. Seki, K., K. Bagchi, and B. Bagchi. 2016. Anomalous dimensionality dependence of diffusion in a rugged energy landscape: How pathological is one dimension? *The Journal of Chemical Physics, 144*(19), 194106.
8. Hu, C. M., and R. W. Zwanzig. 1974. *Rotational friction coefficients for spheroids with the slipping boundary condition. The Journal of Chemical Physics, 60*(11), 4354.
9. Fleming, G. R. 1986. *Chemical Applications of Ultrafast Spectroscopy*. New York: Oxford University Press.
10. Perrin, F. 1934. Mouvement brownien d'un ellipsoide - I. Dispersion diélectrique pour des molécules ellipsoidales *Journal de Physique Et le Radium, 5*(10), 497.
11. Berne, B., and R. Pecora. 1976. *Dynamic Light Scattering*. New York: Wiley.
12. Gennes, P. G. de, and J. Prost, 1995. *The Physics of Liquid Crystals*. Oxford: Clarendon.
13. Landau, L. D., and E. M. Lifshitz. 1966. *Fluid Mechanics*. Oxford: Pergamon Press.
14. Lamb, H. 1916. *Hydrodynamics*. Cambridge: Cambridge University Press.
15. Doi. M and S. Edwards. 1986. *The Theory of Polymer Dynamics*. Clarendon, Oxford.
16. Blainey, P. C., G. Luo, S. C. Kou, W. F. Mangel, G. L. Verdine, B. Bagchi, and S. X. Xie. 2009. Nonspecifically bound proteins spin while diffusing along DNA. *Nature Structural & Molecular Biology, 16*(no. 12), 124105.
17. Seki, K., B. Bagchi, and M. Tachiya. 2008. Orientational relaxation in a dispersive dynamic medium: Generalization of the Kubo-Ivanov-Anderson jump-diffusion model to include fractional environmental dynamics. *Physical Review E, 77*, 031505.
18. Boon, J. P., and S. Yip. 1991. *Molecular Hydrodynamics*. New York: Dover Publication.
19. Adam, G., and J. H. Gibbs. 1965. On the temperature dependence of cooperative relaxation properties in glass-forming liquids. *Journal of Chemical Physics, 43*, 139.
20. Rosenfeld, Y. 1977. Relation between the transport coefficients and the internal entropy of simple systems. *Physical Review A, 15*, 2545.
21. Xia, X., and P. G. Wolynes. 2001. Microscopic theory of heterogeneity and nonexponential relaxations in supercooled liquids. *Physical Review Letters, 86*, 5526.
22. Machta, J., and R. Zwanzig. 1983. Diffusion in a periodic Lorentz gas. *Physical Review Letters, 50*, 1959.

23. Bagchi, B., R. Zwanzig, and M. C. Marchetti. 1985. Diffusion in a two-dimensional periodic potential. *Physical Review A, 31*, 892.

24. Acharya, S., and B. Bagchi. 2020. Study of entropy–diffusion relation in deterministic Hamiltonian systems through microscopic analysis. *Journal of Chemical Physics,* 153, 184701.

25. Hummer, G., J. C. Rasaiah, and J. P. Noworyta. 2001. Water conduction through the hydrophobic channel of a carbon nanotube. *Nature, 414*(no. 6860), 188–190.

26. Yashonath, S., and P. K. Ghorai. 2008. Diffusion in nanoporous phases: size dependence and levitation effect. *Journal of Physical Chemistry B, 112*, 665–686.

27. Borah, B. J., P. K. Maiti, C. Chakravarty, and S. Yashonath. 2012. Transport in nanoporous zeolites: relationships between sorbate size, entropy, and diffusivity. *Journal of Chemical Physics, 136*, 174510.

28. Vasanthi, R., S. Bhattacharyya, and B. Bagchi. 2002. Anisotropic diffusion of spheroids in liquids: Slow orientational relaxation of the oblates. *Journal of Chemical Physics, 116*, 1092–1096.

29. Vasanthi, R., S. Ravichandran, and B. Bagchi. 2001. Needlelike motion of prolate ellipsoids in the sea of spheres. *Journal of Chemical Physics, 114*(18), 7989–7992.

30. Nandi, N., and Bagchi, B. 1997. Dielectric relaxation and solvation dynamics of water in complex chemical and biological systems. *Journal of Physical Chemistry B, 101*, 10954–10961.

31. Carslaw, H. S., and J. C. Jaeger. 1947. *Conduction of Heat in Solids*. New York: Springer.

32. Mukherjee, S., S. Mondal, S. Acharya, S., and B. Bagchi. 2018. DNA solvation dynamics. *The Journal of Physical Chemistry B, 122*(49), 11743–11761.

33. Nair, A. S., P. Banerjee, S. Sarkar, and B. Bagchi, 2019. Dynamics of linear molecules in water: translation-rotation coupling in jump motion driven diffusion. *Journal of Chemical Physics* 151, 034301.

34. Banerjee, S., R. Biswas, K. Seki, and B. Bagchi. 2014. Diffusion on a rugged energy landscape with spatial correlations. *Journal of Chemical Physics,* 141, 124105.

22 Rotational Diffusion
Study of Orientational Time Correlation Functions

OVERVIEW

Experimental investigations of rotational relaxation of molecules are numerous. These measurements are often direct, and easier to perform than its translational counterpart. This is partly the reason why theoretical calculations of rotational diffusion of molecules have been pursued vigorously. These studies constitute important parts of time-dependent statistical mechanics. Experimental techniques, like fluorescence depolarization, NMR, dielectric relaxation, IR spectroscopy, several nonlinear optical techniques, solvation dynamics (in part) can provide information about the rotational motion of molecules and each technique brings its own peculiarities. Computer simulations have also been used to great effects to understand this relaxation. Theoretically, both the kinetic theory of gases and hydrodynamics have been used to obtain predictions for the rates of rotational relaxation. Hydrodynamics predicts a zero rotational friction for the slip boundary condition on the surface of a sphere (that is, free, fast rotation) and a well-known $8\pi\eta R^3$ value for the stick boundary condition (R = radius of the sphere), which often over-estimates friction for small nonspherical molecules. Both experimental results and simulation predictions fall between the two. Among many other interesting aspects of rotational relaxation, the dependence of the rate on the shape of the molecule, and on translation-rotation coupling have drawn attention for many years. In this chapter we discuss various hydrodynamic and kinetic theory approaches, and also a nascent mode-coupling theory treatment to address the interesting translation-rotation coupling.

22.1 INTRODUCTION

Due to the intermolecular interactions such as collisions, the molecules in a liquid undergo rotational and translational motions continuously. If we study the detailed motion of a tagged nonspherical molecule, we find a random Brownian motion both in three-dimensional positional and orientational spaces. Moreover, motions of the nearest-neighbor molecules are often found to be correlated. This causes the departure from a simple Brownian motion description of relaxation. It is this departure that is a subject of much discussion as it offers valuable information.

In Chapter 3 we have discussed several techniques that measure rotational relaxation in liquids and gases. As discussed in that chapter, quantitative knowledge about the rate of orientational relaxation of anisotropic molecules is important to understand many other relaxation processes and chemical reactions. Between translational and rotational motions, there are certain features that are common while some are quite distinct and unique to each. In general, the time scales at which the orientational time correlation functions decays are faster than the corresponding time correlation functions of the linear motion.

DOI: 10.1201/9781003157601-25

This is an important point for a student to remember that the time scales of rotational and trans-lational diffusion are different, and they could couple to different aspects of solvent motion.

Another important characteristic of rotational relaxation is that the rotational diffusion depends rather strongly on the shape of the molecule. The dependence is more prominent in the case of rotational diffusion than translational diffusion. This is because the free rotation of a perfect sphere inside a liquid is possible, without causing any displacement to the liquid. But when the molecules is nonspherical or when the liquid surrounding a sphere sticks to it, the situation changes and the friction becomes significant.

The situation in which the fluid at the surface of the particle rotates along with the particle is often called the stick hydrodynamic boundary conditions. Instead of sticking to the surface, if the fluid is not displaced by the rotating body, it is known as the slip boundary conditions.

For a perfect sphere, the slip boundary condition gives zero friction, while the stick boundary condition gives too high a value. This presented a riddle. In an important paper, Hu and Zwanzig offered the resolution by studying the rotation of *nonspherical molecules*, and showed that good agreement with experimental results can be obtained when we apply the slip boundary condition to a nonspherical molecule. This work established a rather surprising result that hydrodynamics can be used even at a molecular level.

This chapter discusses the simple models of molecular reorientation, hydrodynamic theories for the rotational dynamics of ellipsoids, and also Evans kinetic theory of friction. Inspired by the series of work carried out by Evans on the rotation of nonspherical molecules, a mode-coupling theory analysis developed by us is also discussed here. In the subsequent sections, the translational-rotational coupling phenomena by taking examples of linear and polyatomic molecules are discussed.

Aqueous binary mixtures, especially water-alcohol mixtures are of great importance in chemistry and biology. It is well-known that the thermodynamic and transport properties such as diffusion and rotational correlation time of these mixtures show anomalies. In this chapter, we also discuss the rotational dynamics in aqueous binary mixtures like water-ethanol and also the rotational dynamics of water-ethanol binary mixture with linear molecules as solute molecules.

As already mentioned, Chapter 3 discusses several experimental techniques that probe rotational relaxation phenomena. In this chapter we shall focus on theoretical approaches.

22.2 ROTATIONAL DIFFUSION EQUATION

The rotational diffusion equation is analogous to translational diffusion, but quite different in internal content, as we describe here. For a spherical diffusor, where the orientation of the molecule can be described completely by the two angles – the polar angle θ and azimuthal angle φ [see Figure 22.1(a) for definition].

The rotational motion of some ions was found to be significantly coupled to the reorientational jump motion of its neighboring water molecules. Earlier, researchers could only find that the rotational motion of water has a large magnitude rotational jumps along with its diffusive nature. Recently, it was observed that the orientational motion of ions such as nitrate ions, which are hydrogen bonded with water, also shows jump rotations.

Figure 22.1(b) and (c) shows the trajectories capturing rotational jump motions of the water and nitrate ion that were hydrogen bonded initially. When the bond between the ion and water hydrogen breaks, both exhibit large amplitude angular jump motions and these jump motions are also coupled, which can occur simultaneously [Figure 22.1(a)] or after a short-time gap [Figure 22.1(b)].

For a spherical diffusor, the *rotational diffusion equation* can be given by,

$$\frac{\partial}{\partial t}\rho(\Omega) = D_R \nabla_\Omega^2 \rho(\Omega, t).$$

$$(22.1)$$

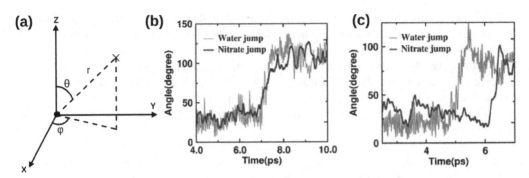

FIGURE 22.1   (a) The spherical coordinate system. A point in three-dimentional space is denoted by r (which is the distance from the origin), polar angle, θ (angle between line joining the point and origin or the radial line and the z axis) and the azimuthal angle φ (angle between the x axis and the projection of the radial line on the x-y plane). (b)-(c) Angular jump motion of the nitrate ions and the neighboring ions. It also demonstrates the coupling of the two rotational jump motions, which can occur simultaneously (b) or after a short time interval (c).

Source: Figures (b) and (c) are taken from [24].

Here $\rho\left(\Omega,t\right)$ is the probability density of finding a molecule with orientation Ω (determined by θ and φ) at time t, D_R is the rotational diffusion coefficient and ∇_Ω is the angular gradient operator. For a spherical diffusor (like a dipolar molecule or an ellipsoidal molecule), the Laplacian operator is given by

$$\nabla_\Omega^2(\Omega) = \frac{1}{\sin(\theta)}\frac{\partial}{\partial\theta}\left[\sin(\theta)\frac{\partial}{\partial\theta}\right] + \frac{1}{\sin^2(\theta)}\frac{\partial^2}{\partial\varphi^2}. \tag{22.2}$$

Therefore, to obtain rotational relaxation, we need to deal with this rather complex differential form, which the student should be familiar with from quantum mechanics where we solve this equation for the hydrogen molecule system.

In one limiting situation, the rotational diffusion equation can be solved rather easily. These are the limits of spherical diffusor or isotropic diffusor. The examples include a point dipole embedded at the center of a sphere. Many molecules like acetonitrile do come close to satisfying this condition, although this model is routinely used for many nonspherical molecules. These systems are characterized by only one diffusion constant. To obtain a solution in such cases we expand the orientational density in spherical harmonics as

$$\rho\left(\Omega,t\right) = \sum_{lm} A_{lm}(t) Y_{lm}(\Omega). \tag{22.3}$$

Since spherical harmonics form the basis set of the angular gradient operator, we have the following relation:

$$\nabla_\Omega^2 Y_{lm}(\Omega) = -l(l+1) Y_{lm}(\Omega). \tag{22.4}$$

When we substitute these expressions in the rotational diffusion equation, we obtain the following equation of motion:

$$\frac{\partial}{\partial t} A_{lm}(t) = -D_R l(l+1) A_{lm}(t). \tag{22.5}$$

where D_R is the rotational diffusion coefficient and t is the real time. This equation can be solved to obtain

$$A_{lm}(t) = A_{lm}(t=0)e^{\left[-l(l+1)D_R t\right]}.$$ (22.6)

Finally, the relation between the rotational diffusion coefficient and the rotational friction is given by

$$D_R = k_B T / \zeta_R$$ (22.7)

where ζ_R is the rotational friction. In the following sections, we present different calculation methods to obtain rotational friction. It is difficult to obtain reliable values for ζ_R. We shall discuss several approaches to obtain this friction later in this chapter.

We shall now discuss a few simple models of rotational relaxation, starting again with the rotational diffusion model.

22.3 THEORY: SIMPLE MODELS

22.3.1 THE DEBYE MODEL

The simple approach we described above [Eqs. (22.1)–(22.7)] is the Debye model, which is the oldest and the most known among the various theories of rotational relaxation. In 1929, Debye developed this model to account for the observed dielectric relaxation in dipolar liquids, and is described in the classic, ever-green monograph "*Polar Molecules*". Basically, one assumes that the reorientation processes through collisions occur frequently in a liquid such that a molecule can only rotate through a very small angle before undergoing another collision, thus essentially assuming rotational Brownian diffusion. As discussed above, for molecules reorienting in a liquid in small steps, the Debye model predicts an exponential decay of the lth rank orientational time correlation function, i.e.,

$$C_l(t) = \exp(-t/\tau_l)$$ (22.8)

where t is the real time and the corresponding orientational correlation time τ_l is given as

$$\tau_l = \frac{1}{l(l+1)D_R}$$ (22.9)

where D_R is the rotational diffusion coefficient in units of inverse time. This model predicts that the ratio of the first and second-order orientational correlation times, i.e., τ_1/τ_2 to be equal to 3.

So, the two basic characteristics of the Debye model are: (i) the single exponential decay of the rotational correlation time, τ_1, and that (ii) $\tau_1/\tau_2=3$.

Both these two predictions have been widely tested and are useful in various ways. The single exponential decay of the first rank correlation function gives rise to the well-known Debye model of dielectric relaxation, with the identification (made first by Debye) that $\tau_1=\tau_D$, where τ_D is known as the rotational diffusion constant. The second prediction that $\tau_1/\tau_2=3$ also finds wide use both in experiments and in theory, particularly in computer simulations. The departure of the ratio from 3 is attributed to several factors that include (a) the presence of memory effects; (b) the inadequacy of the Brownian diffusion picture due to the presence of large amplitude jumps.

In Figure 22.1, we show the calculated, through computer simulations, decay of the first and the second rank orientational relaxation functions for three cases: (i) of nitrate ion in water; (ii) of water

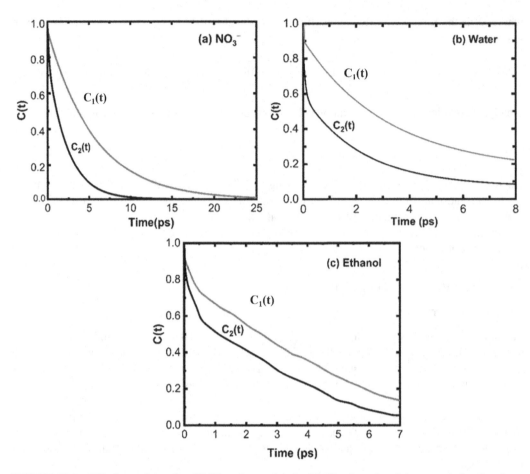

FIGURE 22.2 This figure shows first ($C_1(t)$) and second-order ($C_2(t)$) orientational time correlation functions, as defined above, obtained from simulations. Figure (a) shows the two TCFs for nitrate ion in water; (b) water in neat water; (c) liquid ethanol. The bond vectors used to calculate these are along NO bond in the case of NO_3^- and OH bond in the case of water and ethanol.

in neat water; and (iii) of ethanol in liquid ethanol. These correlation functions can be used to extract the relaxation times τ_1 and τ_2. For the nitrate ion in water, these are found to be equal to 6.01 and 2.35 ps, respectively. Thus, the τ_1/τ_2 ratio of the nitrate ion is 2.56, which can be compared with the τ_1/τ_2 =3 value of the Debye model. Also shown are the two other correlation functions in this figure, water and ethanol, the ratio of the decay times, τ_1/τ_2 is equal to 2.2 and 1.73, respectively.

Thus, the Debye model is only approximately valid in most of the cases. The magnitude of variation is often a clue to the underlying mechanism.

22.3.2 GORDON'S M AND J DIFFUSION MODELS

Debye provided the solution to the rotational diffusion equation in the case of spherical molecules and Perrin for asymmetric molecules. But small angular rotational diffusion models are found to be not sufficiently accurate to describe Raman and infrared spectra, especially in the gas phase, where the probe molecules rotate through large angles between collisions. And the small-angle diffusion model is inappropriate. To address this concern, in 1965 Gordon introduced two generalized rotational diffusion models to overcome this restriction of small angles. Both these models consider angular diffusion steps of arbitrary size. These models are known as *m-diffusion* and *J-diffusion* models. In the *m*-diffusion model, each collision randomizes the direction of the angular momentum, **J** while both the magnitude of angular momentum of rotation and the spatial orientation of the molecule are left unchanged by the collision. In the *J*-diffusion model, each collision randomizes both the direction and magnitude of the angular velocity. The magnitude is randomized onto a Boltzmann distribution by every collision.

The setting up of the problem is rather interesting. *Let* **U** *be the unit vector fixed to the molecule giving its orientation, and also the direction cosine (***U***(0).***U***(t)).* We first consider that there is only one collision during time t, at time t_1. In the first diffusional step, the molecule rotates with the angular velocity ω_1, hence $\mathbf{U}(0).\mathbf{U}(t)=\cos\omega_1 t$. If the first diffusional step ends at t_1 by the collision, then during the remaining period $(t-t_1)$, the final value of $\mathbf{U}(0).\mathbf{U}(t)$ becomes

$$\mathbf{U}(0).\mathbf{U}(t) = \cos\omega_1 t_1 \, \cos\omega_2(t-t_1) - \cos\alpha\sin\omega_1 t_1 \, \sin\omega_2(t-t_1) \qquad (22.10)$$

where ω_2 is the angular velocity of the molecule during the second step. α is the angle between the angular momentum of the molecule in the first and second rotational step. One now makes the crucial observation that various collision processes randomize the direction of **J**. So $\langle\cos\alpha\rangle = 0$, and thus, during the second step,

$$\langle\mathbf{U}(0).\mathbf{U}(t)\rangle_{(2)} = \cos\omega_2(t-t_1)\cos\omega_1 t_1. \qquad (22.11)$$

The collision, which ended the first step, could have also occurred with an equal probability at any time between 0 and t. So, it must be averaged over the intermediate time t_1 at which the first step ended:

$$\langle\mathbf{U}(0).\mathbf{U}(t)\rangle_{(2)} = \frac{1}{t}\int_0^t \cos\omega_2(t-t_1)\cos\omega_1 t_1 dt_1. \qquad (22.12)$$

With the contribution to the dipole correlation function from a molecule that has made $n+1$ angular diffusion steps, the generalized equation is given by

$$\langle\mathbf{U}(0).\mathbf{U}(t)\rangle_{(n+1)} = \frac{n!}{t^n}\int_0^t dt_n \cos\omega_{n+1}(t-t_n) \times \int_0^{t_n} dt_{n-1}\cos\omega_n(t_n-t_{n-1})...\int_0^{t_3} dt_2 \cos\omega_3(t_3-t_2)$$

$$\times \int_0^{t_2} dt_1 \cos\omega_2(t_2-t_1)\cos\omega_1 t_1 \qquad (22.13)$$

The above integrals represent average over the times where the first, second.....n^{th} diffusion steps have ended – that is, the time between two successive collisions. In terms of the rate of the rotational diffusion which is $1/\tau$, the number of molecules in the $n+1$ diffusion step (that is, after n collisions) at time t is given by the normalized Poisson distribution $P_n(t) = 1/n!(t/\tau)^n \exp(-t/\tau)$. So the total correlation function, which is the sum of contributions weighted by this distribution, is given as

$$\langle\mathbf{U}(0).\mathbf{U}(t)\rangle = \exp(-t/\tau)\sum_{n=0}^{\infty}\tau^{-n}\left[\begin{array}{c}\int_0^t dt_n \cos\omega_{n+1}(t-t_n)\times\int_0^{t_n} dt_{n-1}\cos\omega_n(t_n-t_{n-1})...\int_0^{t_3} dt_2 \cos\omega_3(t_3-t_2)\\ \times\int_0^{t_2} dt_1 \cos\omega_2(t_2-t_1)\cos\omega_1 t_1\end{array}\right].$$

$$(22.14)$$

Here, the nth term represents the total contribution to the dipole correlation function from molecules which are in their $n + 1^{th}$ diffusional step at time t. We still do not have information about the angular velocities, ω.

In the m-diffusion model, one sets $\omega_0 = \omega_1 = \cdots = \omega_n$ to be used in Eq. (22.14) and the result is averaged over a Boltzmann distribution of rotational frequencies: $\omega \exp\left(-\dfrac{1}{2}\omega^2\right)$ in the reduced units $\sqrt{k_B T / I}$, ω is the angular frequency. Then, we obtain the following expression:

$$\langle \mathbf{U}(0).\mathbf{U}(t)\rangle_m = \int_0^\infty d\omega\, \omega \exp\left(-\frac{1}{2}\omega^2\right) \times \exp(-t/\tau) \sum_{n=0}^\infty \tau^{-n} \int_0^t dt_n \cos\omega(t-t_n)$$

$$\times \int_0^{t_n} dt_{n-1} \cos\omega(t_n - t_{n-1}) ... \int_0^{t_3} dt_2 \cos\omega(t_3 - t_2) \int_0^{t_2} dt_1 \cos\omega(t_2 - t_1) \cos\omega t_1. \qquad (22.15)$$

In terms of the dipole correlation function for a free linear molecule,

$$F_0(t) = \langle \cos \omega t\rangle = \int_0^\infty \cos(\omega t)\, \omega \exp(-\frac{1}{2}\omega^2)\, d\omega \qquad (22.16)$$

$$F_0^{(n)}(t) = \langle P_n(\cos\omega t)\rangle = \int_0^\infty P_n(\cos\omega t)\, \omega e^{\left(-\frac{1}{2}\omega^2\right)} d\omega. \qquad (22.17)$$

For $n = 1$,

$$F_0(t) = \langle \cos \omega t\rangle = \int_0^\infty \cos(\omega t)\, \omega \exp(-\frac{1}{2}\omega^2)\, d\omega \qquad (22.18)$$

This integral can be computed from its power-series expansion, by expanding the cosine and integrating term by term,

$$F_0^{(1)}(t) = \sum_{l=0}^\infty (-1)^l t^{2l} / (2l-1)!! \qquad (22.19)$$

where the double factorial is defined as

$$(2l-1)!! = (2l)! / 2^l l! = 1 \cdot 3 \cdot 5 \cdots (2l-1). \qquad (22.20)$$

Using the recursion relation,

$$F_k(t) = -t^{-2}\left[F_{k-1}(t) - 2^{(k-1)}(k-1)! / (2k-2)!\right]. \qquad (22.21)$$

So, we combine the previous definitions to obtain,

$$\boxed{\langle \mathbf{U}(0).\mathbf{U}(t)\rangle_m = \exp(-t/\tau) \sum_{n=0}^\infty (n+1)(t/2\tau)^n \times \sum_{k=0}^{[(n+1)]/2} F_k(t) / 2^k k!(n+1-2k)!}. \qquad (22.22)$$

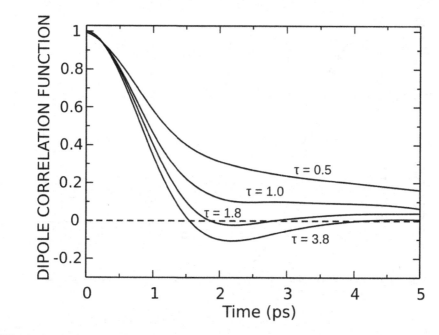

FIGURE 22.3   Dipole correlation functions for the *m*-diffusion model for the values of τ = 3.0, 1.8, 1.0, and 0.5 (from bottom to top). The *x*-axis time scale is in the reduced units $(I/k_B T)^{1/2}$.

Source: Redrawn from [3].

Eq. (22.22) allows us to evaluate the dipole moment time correlation function. Note that the only input is the collisional time τ that enters through the Poisson distribution.

In Figure 22.3, we plot the time dependence of the decay of the dipole moment time correlation function of the *m*-diffusion model for various values of τ. The curve sometimes becomes negative, which means that after a time the probability to find a dipole in the hemisphere opposite to that in which it began at a time t earlier, is more.

In the *J*-diffusion model, the rotation frequency ω_n and the direction of **J** during the nth step are considered independent of those in all other steps. The cosine function in Eq. (22.14) becomes a dipole correlation function for a free rotor. So,

$$\langle \mathbf{U}(0).\mathbf{U}(t)\rangle_J = \exp(-t/\tau)\sum_{n=0}^{\infty}\tau^{-n}\left[\int_0^t dt_n\, F_0(t-t_n)\times\int_0^{t_n} dt_{n-1}\, F_0(t_n - t_{n-1})...\right.$$

$$\left.\int_0^{t_3} dt_2\, F_0(t_3 - t_2)\int_0^{t_2} dt_1\, F_0(t_2 - t_1)F_0\, t_1\right] \qquad (22.23)$$

Eq. (22.23) can also be evaluated numerically using the expression of $F_0(t_n - t_{n-1})$.

We show a dipole moment time correlation function of the *J*-diffusion model for different values of τ in Figure 22.4. In this figure, the decay appears to be rapid at longer times.

Some of the most notable aspects of the above two figures are (i) the presence of a wide negative region in the time correlation function when the average collision interval time τ is large; (ii) cross-over from the negative to the positive exponential-like long-time decay behavior as the collision time τ is decreased; and (iii) a pronounced Gaussian decay at short times. Since the average collision time

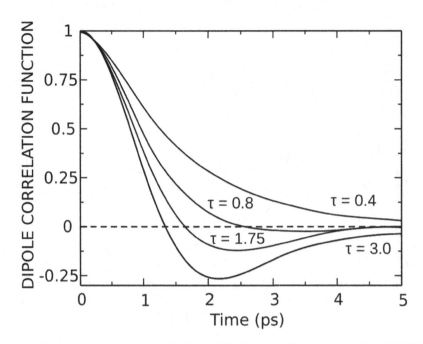

FIGURE 22.4   Dipole correlation functions for the *J*-diffusion model for the values of τ = 3.0, 1.75, 0.8, and 0.4 (from bottom to top). The *x*-axis time scale is in the reduced units $(I/k_B T)^{1/2}$.

Source: Redrawn from [3].

is inversely related to collision frequency and therefore inversely related to friction and viscosity. Figure 22.2 shows first $(C_1(t))$ and second-order $(C_2(t))$ orientational time correlation functions, as defined above, obtained from simulations. Figure 22.2(a) shows the two TCFs for nitrate ion in water; (b) water in neat water; (c) liquid ethanol. The bond vectors used to calculate these are along NO bond in the case of NO_3^- and OH bond in the case of water and ethanol. Figure 22.2 and Figure 22.3 are strikingly different from the predictions of the Debye model of single exponential TCF, especially at large τ, which we pointed out corresponds to low friction regime.

As mentioned earlier, these models are appropriate descriptions of rotational dynamics in the gas phase.

22.4   ORIENTATIONAL FOKKER-PLANCK EQUATION AND HUBBARD RELATIONS

Orientational motion is different from translational motion. Not only is the angular space limited, there are relations between angles that need to be taken care of. Although we speak of rotational diffusion constant D_R and use the Debye-Stokes-Einstein relation between D_R are viscosity, one needs to be careful about the validity of these relations. In fact, it is rather surprising, and a bit of a puzzle that such relations are often found to be quantitatively reliable. Gordon's *m*- and *J*-diffusion models discussed above should in principle be more appropriate for description of rotational TCF, but finds less use.

The usual treatment of rotational diffusion equation considers only the overdamped limit. However, just as in the case of translational diffusion, we often need to keep the inertial term, especially because rotational relaxation tends to contain a large inertial component. In this section, we describe an analogous problem of rotational Brownian motion that can be related to the calculation of the joint probability density of the angular velocity and orientation of a molecule in a fluid. As already mentioned, the rotational case is more complicated than the translational part. We need to introduce the rotational analog of Langevin's equation, based on Euler's equation, to derive the rotational Fokker-Planck equation. In 1972, Hubbard derived the Fokker-Planck equation for the joint probability density of the angular velocity and orientation of a molecule of general shape. We describe the steps below.

In classical mechanics, the rotational motion is ruled by the equivalent Newton's second law, which is given in terms of the external torque $\mathbf{N}$ and angular momentum $\mathbf{L}$, which is $\mathbf{N} = \left(\dfrac{d\mathbf{L}}{dt} \right)_{space}$. This is the equation in inertial space, i.e., fixed frame of reference. For convenience, we express the rotational motion in terms of the body-fixed frame of reference, where the inertia tensor of the rigid body is known. So, the rate of change of angular momentum in the body-fixed frame can be written by using the transformation from the space fixed inertial frame $\left(\hat{\mathbf{x}} \hat{\mathbf{y}} \hat{\mathbf{z}} \right)$ to the rotating frame $\left(\hat{\mathbf{e}}_1 \hat{\mathbf{e}}_2 \hat{\mathbf{e}}_3 \right)$ as

$$\mathbf{N} = \left(\frac{d\mathbf{L}}{dt} \right)_{space} = \left(\frac{d\mathbf{L}}{dt} \right)_{body} + \omega \times \mathbf{L}. \tag{22.24}$$

Choosing $\hat{\mathbf{e}}_i$ as the principal axis, we have

$$L_i = I_i \omega_i. \tag{22.25}$$

I_i is the principal moments of inertia. So the equation of motion using the body-fixed coordinate system is

$$\mathbf{N} = I_1 \dot{\omega}_1 \hat{\mathbf{e}}_1 + I_2 \dot{\omega}_2 \hat{\mathbf{e}}_2 + I_3 \dot{\omega}_3 \hat{\mathbf{e}}_3 + \begin{vmatrix} \hat{\mathbf{e}}_1 & \hat{\mathbf{e}}_2 & \hat{\mathbf{e}}_3 \\ \omega_1 & \omega_2 & \omega_3 \\ I_1 \omega_1 & I_2 \omega_2 & I_3 \omega_3 \end{vmatrix} \tag{22.26}$$

$$= \left(I_1 \dot{\omega}_1 - \left(I_2 - I_3 \right) \omega_2 \omega_3 \right) \hat{\mathbf{e}}_1 + \left(I_2 \dot{\omega}_2 - \left(I_3 - I_1 \right) \omega_3 \omega_1 \right) \hat{\mathbf{e}}_2 + \left(I_3 \dot{\omega}_3 - \left(I_1 - I_2 \right) \omega_1 \omega_2 \right) \hat{\mathbf{e}}_3 \tag{22.27}$$

So,

$$\begin{aligned} N_1 &= I_1 \dot{\omega}_1 - \left(I_2 - I_3 \right) \omega_2 \omega_3 \\ N_2 &= I_2 \dot{\omega}_2 - \left(I_3 - I_1 \right) \omega_3 \omega_1 \\ N_3 &= I_3 \dot{\omega}_3 - \left(I_1 - I_2 \right) \omega_1 \omega_2. \end{aligned} \tag{22.28}$$

So for a rigid body, the rotational motion can be expressed by Euler's equations as

$$\dot{\omega}_i = r_i \, \omega_j \omega_k + N_i / I_i \tag{22.29}$$

where i, j, and k are the cyclic permutation of 1, 2, and 3; $\dot{\omega}_i$ is the derivative of angular velocity ω_i with respect to time, which is along the ith principal body axis, I_i is the principal moment of inertia, N_i is the external torque and $r_i \equiv (I_j - I_k)/I_i$. Eq. (22.29) is known as Hubbard relations.

Similar to translational Brownian motion, it is assumed that the external torque N_i that arises due to the interaction of our tagged molecule with its surroundings is the sum of a viscous retarding torque proportional to the angular velocity ω_i and a torque $I_i A_i(t)$, which fluctuates at a rapid rate:

$$N_i / I_i = -B_i \omega_i + A_i(t) \tag{22.30}$$

where B_i is independent of t, $\vec{\omega}$ and the orientation of the body. By the fluctuation-dissipation theorem we find

$$\langle A_i(0) A_j(t) \rangle = \delta_{ij} \delta(t) k_B T B_i. \tag{22.31}$$

Use of Eq. (22.30) in the Euler equations Eq. (22.29) provides the following rotational Langevin equations

$$\dot{\omega}_i = r_i \omega_j \omega_k - B_i \omega_i + A_i(t). \tag{22.32}$$

We now proceed to transform the Langevin equation to Fokker-Planck equation, by following the steps we described earlier. We integrate Eq. (22.32) over a short time Δt to obtain

$$\Delta \omega_i = (r_i \omega_j \omega_k - B_i \omega_i) \Delta t + \int_t^{t+\Delta t} A_i(t') dt'. \tag{22.33}$$

As $A_i(t)$ fluctuates rapidly, an average over a coarse-grained time gives zero, so we set $\langle A_i(t) \rangle = 0$ in Eq. (22.33) to obtain the following equation:

$$\langle \Delta \omega_i \rangle = \left(r_i \omega_j \omega_k - B_i \omega_i \right) \Delta t. \tag{22.34}$$

Use of the above equations leads to the following equation for conditional probability density $P(\vec{\omega}, \Omega, t; \vec{\omega}_0, \Omega_0)$, where Ω is the orientation,

$$\boxed{\frac{\partial}{\partial t} P + \sum_{j=1}^{3} \left(i \omega_j J_j P - \frac{\partial}{\partial \omega_j} \left[E_j(\vec{\omega}) P \right] - a_j \frac{\partial^2}{\partial \omega_j^2} P \right) = 0} \tag{22.35}$$

where

$$E_i(\vec{\omega}) \equiv B_i \omega_i - r_i \omega_j \omega_k. \tag{22.36}$$

Eq. (22.35) is known as the *Fokker-Planck equation for rotational motion*. This is used extensively in NMR, where anisotropy is important.

22.5   HYDRODYNAMIC THEORIES OF ROTATIONAL FRICTION AND MOLECULAR SHAPE DEPENDENCE OF ROTATIONAL DIFFUSION

Solution of rotational diffusion equation provides expressions for rotational time correlation functions in terms of rotational diffusion coefficient, D_R. The latter is related to rotational friction through Einstein relation described above. So, we need an expression of the rotational friction. For rotational motion, the value of the friction depends strongly on the shape of the molecule, unlike that in translational friction. In this section, we present the hydrodynamic prediction for the rotational friction.

For a spherical molecule of radius R, and for the stick hydrodynamic boundary condition, the rotational friction is given by $\zeta_{DSE} = 8\pi\eta R^3$. The derivation of this elegant and simple expression is based on Navier-Stokes equation and can be obtained in [8]. The derivation is similar to Stokes' well-known derivation, also available in the same reference. The rotational diffusion constant is given by the Debye-Stokes-Einstein relation $D_R = \dfrac{k_B T}{8\pi\eta R^3}$ where η is the viscosity.

For the slip boundary condition, the rotational friction on a sphere is zero. This is an often quoted result and easily understandable.

The Debye-Stokes-Einstein (DSE) equation expresses the hydrodynamic predictions for the rotational friction, which depends on the viscosity of the solvent and size and shape of the solute. We have quoted below the well-known prediction for a sphere with stick boundary condition. However, the predictions depend strongly on the shape of the molecule. From the hydrodynamic theory, the rotational diffusion coefficient for a prolate molecule is given by

$$D_R = \frac{3}{2} \frac{\kappa\left[\left(2\kappa^2 -1\right)S - \kappa\right]}{\kappa^4 -1} D_s \tag{22.37}$$

where κ is the aspect ratio and

$$D_s = \frac{k_B T}{6V\eta} \tag{22.38}$$

where V is the volume of the prolate molecule given by $V = \dfrac{4}{3}\pi\, ab^2$; a is the semi-major axis and b is the semi-minor axis. S for prolate is defined as

$$S = \left(\kappa^2 -1\right)^{-1/2} \ln\left[\kappa + \left(\kappa^2 -1\right)^{1/2}\right]. \tag{22.39}$$

The same expression works for oblate but with a different expression for S, given by

$$S = \left(1 - k^2\right)^{-1/2} \tan^{-1}\left[\frac{\left(1 - k^2\right)^{1/2}}{k}\right]. \tag{22.40}$$

The above predictions are valid only for the stick boundary condition. One finds experimentally that small molecules rotate much faster in all solvents than the prediction of the hydrodynamic models with the stick boundary condition. The inadequacy of the hydrodynamic prediction may at first appear to be surprising when compared with the success of the former for translational diffusion. The resolution of the paradox turned out to be quite interesting, and somewhat unexpected.

In order to explain this apparent anomaly, Hu and Zwanzig observed that the main fault lies with the result that rotational friction is zero for the slip boundary condition for a sphere as no fluid displacement takes place. Hu and Zwanzig suggested that the slip boundary condition is not at fault, and should be more appropriate than the stick condition during the rotation of small solutes, and the fault lies with the assumption of perfect spherical shape of probe molecules which are hardly perfect spheres.

TABLE 22.1

The Table of Friction Coefficients Z^*, Defined by $\zeta^* = -\text{Torque}/(\text{max})^3\,\pi\mu_\omega$, Under Slip Boundary Conditions

	Prolate		Oblate	
κ	ζ^*	Slip/stick	ζ^*	Slip/stick
1.00	0.0	0.0	0.0	0.0
0.90	0.0468	0.00702	0.0505	0.0071
0.80	0.169	0.0307	0.199	0.0316
0.70	0.341	0.0751	0.441	0.0787
0.60	0.534	0.144	0.771	0.154
0.50	0.723	0.240	1.18	0.261
0.40	0.880	0.359	1.66	0.402
0.30	0.976	0.508	2.18	0.572
0.20	0.980	0.660	2.71	0.755
0.15	0.936	0.732	2.95	0.842
0.10	0.853	0.797	3.16	0.918
0.05	0.715	0.855	3.32	0.976
0.01	0.503	0.905	3.39	0.999

Source: This table is taken from [26].

Note: Here, M is the fluid viscosity and Ω is the angular velocity of the spheroid and K is the aspect ratio, defined as the ratio of semi-minor axis (min) to semi-major axis (max) of the spheroid. the slip/stick value is the ratio of friction coefficient with slip to the friction coefficient of the same spheroid with stick

Hu and Zwanzig numerically solved the Navier-Stokes equations to obtain the rotational friction coefficients for prolate and oblate under the slip boundary condition. This friction coefficients of Hu and Zwanzig can be used to obtain the rotational diffusion coefficient under slip boundary conditions. In Table 22.1, we present the results of friction calculation of Hu and Zwanzig. This table is taken with permission from [9]. Later, Youngren and Acrivos further extended this theory for particles of arbitrary shape.

In the case of translational diffusion, the difference in diffusion coefficient between stick and slip conditions is small, but in the case of rotational motion, a large difference exists between stick and slip boundary conditions.

22.6   KINETIC THEORY OF ROTATIONAL FRICTION

We have discussed two theoretical approaches above, Gordon's m- and J-diffusion models, and the hydrodynamic theory. Both provide methods to calculate the orientational TCF. However, both depend on several assumptions that can be removed only by a molecular approach. The important first step is to obtain an expression for the binary friction. This can play the same role as the Enskog friction for translational motion. In a certain sense, it can also be regarded as introducing local structural features to J- and m-diffusion models, just as Enskog introduced structure to Maxwell-Boltzmann kinetic theory of gases.

We model the solute – solvent system as an ellipsoid (E) and sphere (S), respectively. We use the kinetic theory approach of Evans *et al.* to determine the orientational correlation time of the ellipsoid in a sea of spheres. As already mentioned, the goal is to generalize Enskog kinetic theory of gases with local correlations to nonspherical molecules. Molecules are all considered as hard bodies to include short-range repulsive interactions. As in the Enskog theory, an important role is played

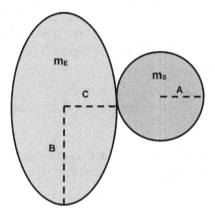

FIGURE 22.5 Schematic representation of a system of solute which is an ellipse and a solvent which is a sphere of mass m_E and m_S respectively. B and C represents the semi-major and semi-minor axes of the ellipse. A is the radius of the sphere.

by the radial distribution function (RDF) at contact. As we discussed in the kinetic theory of gases (Chapter 6), the friction in Enskog theory is proportional to $g(\sigma)$ where g is the RDF at contact. In the present case, we have an ellipsoid as the tagged molecule. Thus, we need to generalize the theory to include the angular dependence of an ellipsoid-sphere system.

For the interaction between an ellipsoid and a sphere, we consider only one angle or angle-like variable to specify the collision and let it be x_1. So we can write the radial distribution function at contact as a function of x_1 as

$$\begin{aligned} g^{E,S} &= g(x_1) \\ &= \sum g(n) P_n(x_1). \end{aligned} \tag{22.41}$$

Here the orientation dependence is expressed in terms of Legendre functions and the expansion coefficients $g(n)$. The shape anisotropy or eccentricity is given by $\varepsilon = (B^2 - C^2)/C^2$, where B is the semi-major axis and C is the semi-minor axis of the ellipsoid, which models the solute molecule, as shown in Figure 22.5.

Evans *et al.* carried out a detailed, and nontrivial, Enskog-type calculation that resulted in the following expression for the orientational correlation time for an ellipsoid-sphere system for the lth rank orientational correlation time [9].

$$\tau_l = \frac{g_l \rho C^3}{l(l+1)} \sqrt{\frac{8\pi I}{k_B T}} \left[\chi_1 \lambda_{E,S} \right] \varepsilon^2 \tag{22.42}$$

where χ_1 is the number fraction, g_l is the orientational pair correlation factor, ρ is the number density and the integral $\lambda_{E,S}$ is given by [8]

$$\lambda_{E,S} = \frac{1}{\sqrt{K_A}} \int_{-1}^{1} dx \, x^2 \, \frac{(1-x^2) g(x, r=0)}{\sqrt{(1+\varepsilon x^2)\left(1+\varepsilon x^2\left(1+\frac{\varepsilon}{K}\right)\right)}} \, \hat{S}(x). \tag{22.43}$$

Here, $K_A = I/\mu_A C^2$ where μ_A is the reduced mass of ellipsoid – sphere system which is given as $\mu_A = m_E m_S/(m_E + m_S)$, m_E: mass of ellipse and m_S: mass of sphere. In the above equation K is similar to the K used in the rough hard sphere kinetic theory calculations of Chandler, which is given as $K = 4I/md^2$, here I is the moment of inertia, m is the mass and d is the kinetic diameter of the molecule. $\hat{S}(x)$ is the surface area function of the ellipsoid-sphere system

We can write Eq. (22.42) as

$$\tau_l = \frac{g_l}{l(l+1)} \rho \frac{4\pi C^2 B}{3} \sqrt{\frac{I}{2\pi k_B T}} \left[\chi_1 \lambda_{E,S} \right] \varepsilon^2 \frac{3C}{B} \tag{22.44}$$

Using the reduced density, $\rho^* = ((4\pi C^2 B)/3)\rho$ and B/C as $\sqrt{1+\varepsilon}$ in Eq. (22.42), The final expression for the single particle orientational correlation function ($\chi_1 = 1$) becomes

$$\boxed{\tau_l = \frac{3\rho^* \varepsilon^2}{\sqrt{1+\varepsilon}} \sqrt{\frac{I}{2\pi k_B T}} \frac{\lambda_{E,S}}{l(l+1)}.} \tag{22.45}$$

So, the rotational diffusion D_R

$$D_R^{Bin} = \frac{\sqrt{1+\varepsilon}}{6\rho^* \varepsilon^2} \sqrt{\frac{8\pi k_B T}{I}} \frac{1}{\lambda_{E,S}}. \tag{22.46}$$

Substituting $\lambda_{E,S}$ and expanding the surface area function, we obtain the final expression for rotational binary friction as

$$\zeta_R^{Bin} = \sqrt{\frac{I k_B T}{8\pi}} \frac{6\rho^* \varepsilon^2}{\sqrt{1+\varepsilon}} \frac{1}{\sqrt{K_A}} \int_{-1}^{1} dx \frac{x^2 (1-x^2) g(x, r=0)}{\sqrt{(1+\varepsilon x^2)\left[1+\varepsilon x^2 \left(1+\frac{\varepsilon}{K}\right)\right]}} \cdot$$

$$\left\{ \begin{bmatrix} \left[A + \sqrt{C^2 + (B^2 - C^2)x^2} - \frac{C\varepsilon x^2}{\sqrt{1+\varepsilon x^2}} \right] \cdot \\ \left[\left[A + \sqrt{C^2 + (B^2 - C^2)x^2} - \frac{C\varepsilon x^2}{\sqrt{1+\varepsilon x^2}} \right] \\ + \left[C \left[\frac{\varepsilon}{\sqrt{1+\varepsilon x^2}} - \varepsilon^2 x^2 \left(1+\varepsilon x^2\right)^{-\frac{3}{2}} \right] \right] (1-x^2) \end{bmatrix} \right\}. \tag{22.47}$$

In Figure 22.6 we illustrate a simplified flow diagram to calculate the rotational binary friction.

Armed with Enskog's expression for the binary friction, one can now advance to the formulation of a mode coupling theory. MCT calculation would require derivation of expressions for density and current contributions.

22.7 MODE-COUPLING THEORY

In this section, we discuss a microscopic approach for the rotational friction due to interaction with the surrounding solvent molecules. This approach parallels the mode-coupling theory (MCT) for translational friction that has been developed in the last four decades, and has been applied with great success to calculate diffusion coefficients and other transport properties like viscosity. An essential ingredient in such theories is the dynamic structure factor, $F(k, t)$, which is determined

CALCULATION OF BINARY FRICTION TERM

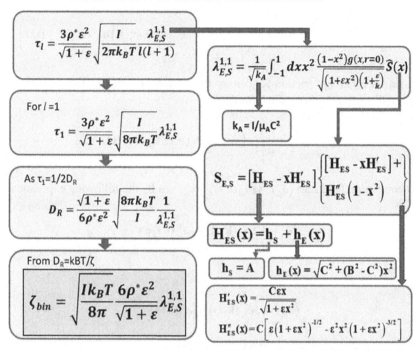

FIGURE 22.6 A flowchart to illustrating the step-by-step procedure to calculate the rotational binary friction of an ellipsoid-sphere system in the Evans scheme.

self-consistently with viscosity and diffusion coefficients. We have discussed the mode-coupling theory in detail in Chapter 23.

The basic idea is to obtain the torque-torque time correlation function. The torque on a molecule (be in nonpolar or dipolar) is a fluctuating quantity, which is due to the relative motions of the molecules, and *can be described, invoking concepts from hydrodynamics, as due to fluctuations in solvent density and solvent current fields.* Just as in the case of translational friction, we assume that the torque can be separated into a repulsive, short range part and a longer range part. *The short range part can in principle be described by binary term like an Enskog theory of the type developed by Evans and coworkers.* The longer range part of the torque can be related to space and density-dependent density fluctuations of the solvent. The latter is particularly important when the tagged molecule is dipolar or ionic so that polar long range interactions are important.

Since hydrodynamic theories neglect molecular aspects of the solute-solvent interactions, they cannot include many fascinating aspects of the dynamics that operate in dense liquids. For example, in a dense dipolar liquid, the orientational motion of the solute particle is strongly coupled to the translational and rotational motions of the solvent molecules. So, the equation of motion for the solute orientational density distribution is coupled to the equations of motion for the position and orientation-dependent solvent density distribution. An expression for torque on a point dipole due to density fluctuation is provided by classical density functional theory as

$$N(r,\Omega,t) = k_B T \nabla_\Omega \int dr' \int d\Omega' \, C_{S0}(r-r',\Omega,\Omega') \delta\rho_0 \left(r',\Omega',t\right) \tag{22.48}$$

where ∇_Ω is the angular gradient operator, $\delta\rho_0\left(r',\Omega',t\right)$ is the density fluctuation of the solvent and $C_{S0}(r-r',\Omega,\Omega')$ is the direct correlation function between the solute and the solvent molecules.

By expanding the density fluctuations in terms of spherical harmonics, we get

$$N(r,\Omega,t) = \frac{k_B T}{(2\pi)^3} \nabla_\Omega \int dk\, e^{-ikr} \left(\sum_{m_l=-l_l}^{l_l} (-1)^{m_l} C_{s0}\left(l_l, l, m_l; k\right) a_{lm_l}(k,t) Y_{l_l m_l}(\Omega) \right). \tag{22.49}$$

The expression for rotational friction for density fluctuation can be obtained as torque-torque time correlation function using Kirkwood's formula

$$\zeta^R = \frac{\beta}{2} \int_0^\infty dt\, \frac{1}{4\pi V} \int dr\, d\Omega \langle N(r,\Omega,0).N(r,\Omega,t) \rangle. \tag{22.50}$$

The solvent correlation functions $\langle a_{lm}(-k)\, a_{lm}(k,t) \rangle$ are given by the following expression:

$$\langle a_{10}(-\mathbf{k})\, a_{10}(\mathbf{k},t) \rangle = \langle a_{10}^2(\mathbf{k}) \rangle e^{\frac{-t}{\tau_L(k)}} \tag{22.51}$$

and

$$\langle a_{11}(-\mathbf{k})\, a_{11}(\mathbf{k},t) \rangle = \langle a_{11}^2(k) \rangle e^{\frac{-t}{\tau_T(k)}}. \tag{22.52}$$

Hence the final expression for rotational friction becomes

$$\zeta_{DF}^R = \frac{1}{2\beta(2\pi)^4} \int dk \left[2C_{s0}^2(111;k) \langle a_{11}^2(k) \rangle \tau_L(k) + c_{s0}^2(110,k) \langle a_{10}^2(k) \tau_T(k) \rangle \right]. \tag{22.53}$$

The new features of the above equation are first of all it is based on a reliable microscopic theory of solute-solvent interactions. Secondly, it includes the effects of molecular translations. An interesting point to note here is that the above expression incorporates contributions from all the wave vectors and we are aware that the dynamics of solvent relaxation is strongly dependent on k [11,12]. Also, we have contributions from both (110,111) components of the solvent polarization relaxation.

22.8   WATER REORIENTATION DYNAMICS: UNIQUE FEATURES

Even though the dynamics of water and the presence of jump motion were known, a sophisticated demonstration that such jump orientational motion is indeed a dominant relaxation mechanism, was put forward by Laage and Hynes. This mechanism has stimulated a great interest in exploring the jump motion behavior in more complex aqueous systems. Instead of considering the commonly accepted sequence of small diffusive steps, this mechanism of water reorientation involves large-amplitude angular jumps. They mapped out the detailed sequence of events, which leads to the jump motion of a central water molecule, in terms of motions of surrounding water molecules. To reorient spontaneously, it is generally assumed that a water molecule must break at least the H-bond involving the rotating hydrogen. Laage and Hynes focussed on the molecular mechanism which describes the migration of this H-bond donor site from one accepting molecule to another. They adopted the widely used geometric definition where the distance between the donor and the acceptor oxygen atoms is less than 3.5 Å and the angle between the OH bond and the OO vectors is less than 30°. Every time the hydrogen atom, which changes the acceptor oxygen with which the H-bond is formed is recorded. The H-bond switches, which is a common feature of all water molecules are also collected.

22.9 JUMP DIFFUSION MODELS

In this section, we discuss the jump diffusion model in more detail. Jump diffusion is a stochastic process, which involves both jumps and diffusion. In many cases, the jumps with large amplitudes are rare and they are superimposed on the frequently occurring small amplitude jumps. It is of great interest to understand the origin of such jumps quantitatively.

The existing theoretical studies on jump diffusion generally assume an exponential waiting time distribution for jumps. However, if the waiting time distribution is not Poissonian, then the difference between the Brownian diffusion and the large jumps becomes apparent. In the past, a most notable rotational jump diffusion model was proposed by Kubo, Ivanov, and Anderson.

22.9.1 Kubo-Ivanov-Anderson Rotational Jump Model

A most notable rotational jump diffusion model was proposed by Kubo in 1962 and Ivanov in 1964. In Kubo's model, the rotator was restricted to jump on a circle, thus is restricted to two dimension [Figure 22.7(a)]. In the more general Ivanov's jump model, the jumps were isotopically distributed in three dimensions, which is on the surface of a sphere [Figure 22.7(b)].

Ivanov's jump model is applied in many experimental studies. In this model, the waiting time distribution between jumps is an exponential distribution. So the decay is single exponential.

The first and second rank correlation functions are given by

$$
\begin{aligned}
C_1(t) &= e^{[-(1-\cos\Delta)t/\tau]} \\
C_2(t) &= e^{\left[-\frac{3}{2}(1-\cos^2\Delta)t/\tau\right]}
\end{aligned}
\tag{22.54}
$$

where τ is the average time interval between any two jumps and Δ is the constant amplitude of jump. A jump by $\Delta \sim \pi$ relaxes $C_1(t)$ and not $C_2(t)$. For small jumps, the equation behaves like the Debye model, which is $\tau_1/\tau_2=3$. But, the relaxation pattern is different for long jumps. So τ_1/τ_2 becomes much smaller than three and approaches zero.

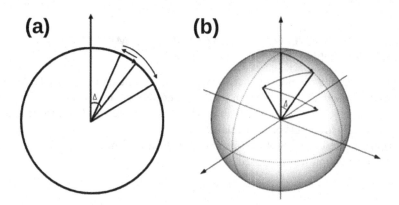

FIGURE 22.7 Schematics of two of the most notable models for rotational jump diffusions: (a) by Kubo and (b) Ivanov and Anderson. In the Kubo model, the jumps are restricted on a circle and in Ivanov-Anderson jump model, the jumps are isotopically distributed on a sphere.

Let us consider the situation where large and small amplitude jumps are simultaneous. Let $\psi_a(t)$ be the waiting distribution of large amplitude jumps and that of small amplitude jumps be $\psi_b(t)$. Let $\psi_a^0(t)$ be the waiting time distribution of large amplitude jump in the absence of small amplitude jump and $\psi_b^0(t)$ be the waiting time distribution of small amplitude jump in the absence of large amplitude jump. The waiting time in the presence of both the jumps is given by

$$\psi_a(t) = \psi_a^{(0)}(t)\left(1 - \int_0^t dt\psi_b^{(0)}(t)\right),$$

$$\psi_b(t) = \psi_b^{(0)}(t)\left(1 - \int_0^t dt\psi_a^{(0)}(t)\right).$$

$$(22.55)$$

When both amplitude jumps occur according to the exponential waiting time distribution, then

$$\psi_a(t) = \frac{1}{\tau_a}e^{(-t/\tau_a - t/\tau_b)}; \psi_b(t) = \frac{1}{\tau_b}e^{(-t/\tau_a - t/\tau_b)}.$$

$$(22.56)$$

The final expression for the correlation function is

$$C_1(t) = \exp\left[-(t/\tau_b)\left(1 - \frac{\cos(\Delta_b)\widehat{\psi}_b(0)}{1 - \cos(\Delta_a)\widehat{\psi}_a(0)}\right)\right],$$

$$C_2(t) = \exp\left[-3(t/\tau_b)\left(1 - \frac{\cos(2\Delta_a)\widehat{\psi}_a(0) - \cos(2\Delta_b)\widehat{\psi}_b(0)}{3 + \widehat{\psi}_b(0) - 3\cos(2\Delta_a)\widehat{\psi}_a(0)}\right)\right].$$

$$(22.57)$$

The exponential kinetics of these two expressions even under the algebraic waiting time distribution of the rare and large amplitude jumps is an important observation.

It has been shown in recent computational studies on water and other viscous liquids, that the large amplitude motions may be more of a rule than exception. Some of the earlier studies have shown that, in water, the translational diffusion is tightly coupled to the large-amplitude angular jumps of water molecules. At room temperature, the nonexponentiality in liquid water becomes weak but at low temperatures, the relaxation becomes progressively nonexponential.

22.10   RELATION BETWEEN MACROSCOPIC AND MICROSCOPIC ORIENTATIONAL RELAXATION TIMES

It is important to know the relationship between the single-particle orientational relaxation time τ_m and the many-body orientational relaxation time, τ_M. The single-particle rotational dynamics is not easily accessible to experiments. The available quantity is the Debye dielectric relaxation τ_D. But there is a need to obtain the single particle relaxation time, which is the quantity that enters most theoretical descriptions like Debye-Stokes-Einstein. So it is important to establish a relation between them. The first such relation was proposed by Debye

$$\tau_m = \frac{n^2 + 2}{\varepsilon_0 + 2}\tau_D$$

where ε_0 is the static dielectric constant and n is the refractive index. But this equation is inadequate for strongly polar solvents. Glarum modified this equation for strongly polar liquids and is given by

$$\tau_m = \frac{2\varepsilon_0 + \varepsilon_\infty}{3\varepsilon_0} \tau_D \tag{22.58}$$

ehere ε_∞ is the infinite frequency dielectric constant. As this expression is based on the static dielectric constant of Onsager, Eq. (22.58) is referred as the Onsager-Glarum expression. It is the limiting form of the more generalized relation of Powles, which is given as

$$\tau_m = \frac{2\varepsilon_0 + \varepsilon_\infty}{3\varepsilon_0} \left(\frac{\tau_D}{g} \right) \tag{22.59}$$

where g is Kirkwood's g factor (J.G. Kirkwood, in his classic paper [20] linked the macroscopic dielectric constant of polar liquids to the local orientation order which was measured by the g factor and was later named after him). This factor is a measure of correlations of short range in the dense dipolar liquid. He suggested that the corresponding dielectric constant at short range is effectively equal to the macroscopic value just after "a distance of molecular magnitude". An analytic expression for g can be given in terms of integration over the anisotropic part of the radial distribution function of the dipolar liquid. Madden and Kivelson showed that τ_D and τ_m are related by

$$\tau_m = \frac{\beta \mu^2 \rho_0}{3\varepsilon_0 (\varepsilon_0 - 1)} g' \tau_D \tag{22.60}$$

where $\beta = (k_B T)^{-1}$, k_B is Boltzmann's constant, T is the temperature, and μ is the magnitude of the dipole moment of liquid molecules, ρ_0 is the equilibrium density of the liquid and g' is the dynamic coupling parameter.

Later Chandra and Bagchi derived a microscopic relation between τ_m, τ_D, and τ_M of a dipolar liquid. The collective orientational correlation function is given by

$$C_M(t) = A_1(k = 0)e^{-t/\tau_{M1}} + A_2(k = 0)e^{-t/\tau_{M2}} \tag{22.61}$$

where the two time constants are given as

$$\tau_{M1} = \frac{1}{2D_R} \left\{ 1 + \frac{\rho_0}{4\pi} c(111,0) \right\}^{-1} \tag{22.62}$$

$$\tau_{M2} = \frac{1}{2D_R} \left\{ 1 - \frac{\rho_0}{4\pi} c(110,0) \right\}^{-1}. \tag{22.63}$$

τ_{M1} is the transverse polarization relaxation time constant and τ_{M2} is the longitudinal relaxation time. Thus, the following relations were established between the microscopic (τ_m) and macroscopic (τ_M) relaxation times

$$\tau_m = \frac{1}{2D_R} \tag{22.64}$$

$$\tau_M = \tau_m \left[1 + \frac{\rho_0}{4\pi} c(111, k = 0) \right]^{-1} \tag{22.65}$$

where ρ_0 is the average number density of the liquid and $c(111, k=0)$ is the Fourier transform of the (111) component in the spherical harmonic expansion of the two particle direct correlation function. The Nee and Zwanzig (NZ) predicts an l-independent expression for the dielectric friction. The zero frequency dielectric friction is given as

$$\zeta_{DF,l}^{NZ} = \frac{2k_B T}{\varepsilon_0} \frac{(\varepsilon_0 - 1)^2}{(2\varepsilon_0 + 1)} \tau_D \tag{22.66}$$

where ε_0 is the dielectric constant and τ_D is the dielectric relaxation time.

22.11   THE MANY-BODY ROTATIONAL DIFFUSION EQUATION

In the presence of interactions among spins (dipolar molecules), the most general and accurate description should involve the many-body distribution function, depending on the orientation of each molecule, at a time t. This can be developed similar to the one we consider in the case of translational diffusion.

The general many-body translational diffusion equation in the presence of external potential U is

$$\frac{\partial C}{\partial t} = D\nabla.\{\nabla C + \beta C\nabla U\} \tag{22.67}$$

where C is the concentration and D denotes the diffusion coefficient. Let, $r_1, r_2 ..., r_N$ and $C(r_1, r_2 ..., r_N)$ be the general coordinates of the configuration space, and the generalized distribution function in this space, respectively. Then the above diffusion equation can be generalized to the many-body form,

$$\frac{\partial C}{\partial t} = D\sum_n \nabla_{r_n}.\{\nabla_{r_n} C + \beta C\nabla_{r_n} U\}. \tag{22.68}$$

With this analogy, the generalized rotational diffusion equation can be written as

$$\frac{\partial g}{\partial t} = D\sum_n \Omega_n.\{\Omega_n g + \beta g\Omega_n U\}. \tag{22.69}$$

Here, Ω_n is the angular gradient operator and g represents the rotational analog of distribution function C. The operator $\wp$ introduced in the previous section is given explicitly by,

$$\wp f = D\sum_n \Omega_n.\{\Omega_n f + \beta f\Omega_n U\} \tag{22.70}$$

where f denotes some function of orientations. This many-body rotational diffusion was used in the study of dielectric relaxation of a dipolar lattice, with interesting results such as non-Debye dielectric relaxation and relation between single particle and collective relaxation times.

22.11   SUMMARY

In this long chapter, we discussed many aspects of rotational relaxation of nonspherical molecules, with emphasis on liquid state dynamics, although we did discuss Evans' nontrivial binary collision model and also Gordon's m- and J-diffusion models that are applicable to the gas phase. The importance of this chapter arises from the large number of experimental methods that are available and that are routinely used to study rotational dynamics. In fact from the beginning, study of rotational diffusion has remained intimately connected with our efforts to understand details of solute-solvent interactions.

An important aspect of the study of rotational diffusion is the dependence of diffusion on the shape of the molecule. Connected with the later, and also solute-solvent interaction, is the applicability of hydrodynamic boundary conditions. Just as in the case of translational diffusion, hydrodynamic predictions continue to remain popular and find enormous application because of the simplicity of expressions involved. Even in the case of slip boundary condition where no closed form expression for rotational friction is available, the results provided by Hu and Zwanzig on a nonspherical molecule prove useful.

In an interesting dichotomy, while rotational relaxation is relatively easy to access by experimental studies, microscopic theoretical study of rotational diffusion remains difficult and a formidable challenge. In this chapter we presented some of the modern analyses that address the plethora of factors that need to be included to understand rotational diffusion. Fortunately, computer simulations have played and still play an important role in filling the gap, and has begun to play a dominant role in our understanding of rotational motion. Students should note that this remains an acrive area of research, especially the subject of rotational relaxation in complex liquids and biological systems.

BIBLIOGRAPHY

1. Fleming., G. R. 1986. *Chemical Applications of Ultrafast Spectroscopy.* Oxford: Oxford University Press.
2. Berne, B., and Pecora, R. 1976 *Dynamic Light Scattering.* Hoboken: Wiley.
3. Gordon, R. G. 1966. On the rotational diffusion of molecules, *Journal of Chemical Physics, 44*, 1830.
4. Tang, S., and G. T. Evans. 1993. A critique of slip and stick hydrodynamics for ellipsoidal bodies, *Molecular Physics, 80*, 1443.
5. Bagchi, B. 2012. *Molecular Relaxation in Liquids.* Oxford: Oxford University Press.
6. Bagchi, B. 2018. *Statistical Mechanics for Chemistry and Materials Science.* Boca Raton: CRC Press.
7. O'Dell, J., and B. J. Berne, 1975. Orientational relaxation of a Fokker-Planck fluid of symmetric top molecules, *Journal of Chemical Physics, 63*, 2376.
8. Hubbard, P. S. 1972. Rotational Brownian motion, *Physical Review A, 6*, 2421.
9. Evans, G. T., R. G. Cole, and D. K. Hoffman. 1982. Kinetic theory of depolarized scattering R parameter. I. Formal theory, *Journal of Chemical Physics, 77*, 3209.
10. Lamb, H. 1994. *Hydrodynamics.* Cambridge: Cambridge University Press.
11. Hu, C., and R. W. Zwanzig. 1974. Rotational friction coefficients for spheroids with the slipping boundary condition, *Journal of Chemical Physics, 60*, 4354.
12. Perrin, F. 1934. Mouvement brownien d'un ellipsoide - I. Dispersion diélectrique pour des molécules ellipsoidales, *Journal of Physics Radium, 5*, 497.
13. Laage, D., and J. T. Hynes. 2008. On the molecular mechanism of water reorientation, *Journal of Physical Chemistry B, 112*, 14230.
14. Bagchi, B. 1991 Microscopic expression of dielectric friction on a moving ion, *Journal of Chemical Physics, 95*, 467.
15. Bagchi, B., and A. Chandra. 1991. Collective orientational relaxation in dense dipolar liquids, *Advances in Chemical Physics, 80*, 1.
16. Nair, A. S., S. Kumar, S. Acharya, and B. Bagchi. 2020. *Rotation of small diatomics in water-ethanol mixture, Journal of Chemical Physics, 153*, 014504.
17. Banerjee, P., S. Yashonath, and B. Bagchi. 2017. Rotation driven translation diffusion of polyatomic ions in water: A novel mechanism of breakdown of Stokes-Einstein relation, *Journal of Chemical Physics, 146*, 164502.

18. Nair, A. S., P. Banerjee, S. Sarkar, and B. Bagchi. 2019. Dynamics of linear molecules in water:Translation-rotation coupling in jump motion driven diffusion, *Journal of Chemical Physics, 151*, 034301.

19. Bagchi, B. 1991. Microscopic expression of dielectric friction on a moving ion, *Journal of Chemical Physics, 95*, 467.

20. Seki, K., B. Bagchi, and M. Tachiya. 2008. Orientational relaxation in a dispersive dynamic medium: Generalization of the Kubo-Ivanov-Anderson jump-diffusion model to include fractional environmental dynamics, *Physical Review E, 77*, 031505.

21. Kirkwood, J. G. 1939. The dielectric polarization of polar liquids, *Journal of Chemical Physics, 7*, 911.

22. Zhou, H. X., and B. Bagchi. 1992. Dielectric and orientational relaxation in a Brownian dipolar lattice. *Journal of Chemical Physics, 97*(5), 3610–3620.

23. Hu, C., and R. Zwanzig. 1974. Rotational friction coefficients for spheroids with the slipping boundary condition, *Journal of Chemical Physics, 60*, 4345.

24. Banerjee, P., and B. Bagchi. 2016. Role of local order in anomalous ion diffusion: Interrogation through tetrahhedral entropy of aqueous solvation shells, *Journal of Chemical Physics, 145*, 234502.

Part IV

Advanced Topics

23 Mode-Coupling Theory of Liquid State Dynamics

OVERVIEW

Mode-coupling theory (MCT) may be considered as the most sophisticated theory of liquid state dynamics. This successful theory evolved in stages, starting with the hydrodynamic five modes to describe relaxation of conserved variables like density, momentum, and energy. Important ingredients of MCT were borrowed from the kinetic theory of gases that provides a correct description of the short-time dynamics. Some of the notable early successes include treatment of critical slowing down of relaxation and the long-time tails of the velocity-time correlation function. Subsequently, MCT has been developed into a highly sophisticated theory to describe dynamics in dense atomic and molecular liquids. MCT is not just a theory of glass transition. It provides a theoretical tool to address diverse relaxation phenomena of dipolar liquids, electrolyte solutions, binary mixtures, polymers, and various relaxation processes.

Towards molecular level applications, hydrodynamics went through several developments like molecular hydrodynamic theory and generalized hydrodynamics. The latter is combined with the extended kinetic theory of gases developed during the 1980s. While hydrodynamics has been used to obtain time correlation functions, kinetic theory of gases has been applied to calculate the transport properties, like viscosity, thermal conductivity. The two approaches have been combined to obtain the present day MCT. The main characteristic of modern liquid state MCT is that it takes into account the microscopic structure present in dense liquids in a self-consistent fashion, which will be explained in this chapter. In a nutshell, self-consistency enters in the following fashion: the time correlation functions (TCFs), which are used to calculate the transport properties, are themselves determined by the time correlation functions. This is also called bootstrapping, as discussed below.

23.1 INTRODUCTION

Mode-coupling theory (MCT), as the name signifies, includes coupling between different modes. In most applications to liquid state dynamics, *these modes are the five hydrodynamic modes:* the heat mode, the two acoustic or sound modes, and two current modes, as discussed in Chapter 5. These modes control the decay of density and pressure fluctuations at large lengths and long time scales. The difficulty arises when we want to use these modes in the study of small length scale and short-time processes. We need to go beyond linearized hydrodynamics. The main merit of the theory is that it allows inclusion of the effects of microscopic structure (like static structure factor) in a self-consistent fashion.

In this chapter, we discuss many intertwined concepts originated from several different, seemingly unrelated areas, like the kinetic theory of gas, hydrodynamics, glass transition, etc. Some of these may surprise students, like the length-scale dependence of the conservation laws. But these are rich in physics and often a delight to delve into.

DOI: 10.1201/9781003157601-27

In linearized hydrodynamics, these modes are orthogonal to each other, as the vibrational normal modes of a molecule, or the normal modes of a solid. In the case of molecules, the mode-coupling effects enter through anharmonicity, which leads to a coupling between different vibrational modes, and change the vibrational relaxation substantially, sometimes enhancing the rate by many orders of magnitude, in the case of phase relaxation. In fact, the introduction of normal modes itself is an approximation, valid under idealized conditions.

Mode-coupling theory finds diverse applications, some of which are discussed here. They include long-time power-law decay of velocity TCF, anomalous dynamical exponents in critical phenomena, glass transition phenomena, transport properties of electrolyte solutions, and translation-rotation coupling in molecular liquids, to name a few.

The special character of hydrodynamic modes and their special role in time-dependent statistical mechanics arise because these, being the conserved variables, are the slowest modes in the system. The decay rate of these modes varies as q^2 where q is the wavenumber. The rate goes to zero in the long wave length (that is, $q \rightarrow 0$) limit. Historically, these slow modes were studied by light scattering experiments. It was only after the neutron scattering experiments became available that short wavelength phenomena could be studied, and limitations of hydrodynamic description became clear.

In terms of a physical picture, there are several ways to understand the coupling between modes and these shall become clearer as we go through the technical details later. At molecular length scales, liquid density does not relax as fast as it does at large length scales. At high density, a liquid molecule is "caged" by its neighbors, and hence its motion is coupled to the density of the surrounding molecules. Thus, we see that we need to involve two densities, the tagged particle density, and the solvent density. This is beyond Navier-Stokes hydrodynamics. It is important to realize that we are trying to build a reduced description, to replace an N-body problem by a much reduced few-body problem. We start with the hydrodynamic modes as the starting variables. They alone cannot describe the complex dynamics routinely observed in dense liquids, as they do not include microscopic structural information. One can then include the additional variables by forming products of these modes, at finite wavenumbers. This is done by maintaining the conservation of momentum. At the simplest level we form the binary products of the hydrodynamic variables. This in turn introduces coupling between different modes, like the current and the density modes.

A second way to understand mode coupling is to remember the derivation of the relaxation equation via the projection operator technique (see Chapter 12). Even at the cost of repetition, we write it down here to remind the reader the form of the relaxation equation for the dynamical variable $A(q, t)$ [Eq. (23.1)]

$$\frac{\partial \mathbf{A}(q,t)}{\partial t} = i\mathbf{\Omega}(q).A(q,t) - \int_0^t d\tau \mathbf{\Gamma}(q,\tau)\mathbf{A}(q,t-\tau) + \mathbf{F}(q,t). \qquad (23.1)$$

The derivation is provided in Chapter 12. $\mathbf{A}(q, t)$ is a set of wavenumber (q) and time-dependent dynamical variables. The quantity of interest here is the memory function $\mathbf{\Gamma}(q, \tau)$. Unlike the frequency matrix (which is often easily available), the memory function is determined by variables not included in the set $\mathbf{A}$. The construction of this set of variables orthogonal to A is tricky, and we have discussed this below. Mode-coupling theory can be essentially considered as a scheme to calculate the memory function. The name "mode coupling" originated from this step. In a nutshell, we select the product of hydrodynamic modes. Thus, mode coupling enters through the memory function.

As de Gennes' narrowing shows, density fluctuation slows down at intermediate wavenumbers, near the peak of the static structure factor. Hydrodynamic theory where density relaxation is governed by relaxation of the current mode cannot explain this slow down at intermediate-to-large wavenumbers. The point to realize that the current relaxation itself is governed by the density relaxation due the cage effect, which is described by a product of two density modes. These aspects have been discussed in detail in Chapter 5 and Chapter 9.

As mentioned in Chapter 5, at large wavenumber, neither the momentum nor the energy conservation is obeyed. However, number conservation must be maintained at all the length scales. Therefore, in the language of the field, *the heat modes degenerate into a self-diffusion mode. In the case of diffusion and related phenomenon in dense liquids, dynamics in controlled by the cage effect. At such small length scales, conservation of momentum and energy do not pose any constrain, and the number conservation is the dominant constrain. This simple fact serves to explain many of the expressions of MCT.*

There are three different situations where mode-coupling theory has been applied successfully. First, near the critical temperature, the heat mode undergoes a pronounced slowdown in relaxation. The width of the Rayleigh peak measures the relaxation rate. According to Navier-Stokes (N-S) hydrodynamics, this rate is given by $D_T k^2$, where D_T is the thermal diffusivity, and k is the wavenumber. The thermal diffusivity, in turn, is given by λ/C_p, where λ is thermal conductivity and C_p is the specific heat. As the critical temperature approaches, specific heat diverges, so the relaxation rate is predicted to go to zero. While the relaxation rate of the heat mode indeed goes to zero, the detailed prediction of N-S hydrodynamics completely breaks down, as a wrong exponent for the slowing down is predicted. The resolution of this paradoxical behavior was put forward first by Fixman and later by Kadanoff and Swift, who pointed out that the relaxation proceeds through a different channel that opens up due to the coupling of the heat mode to the current mode. This coupling can be captured by studying the decay of the heat mode into this product of two dynamical variables that form the orthogonal subspace. The success of MCT in describing the critical dynamics is really impressive and remained an intellectual triumph of time-dependent statistical mechanics.

The second well-known application of mode-coupling theory is to explain the long-time power-law decay observed in the velocity time correlation function discovered through computer simulations.

The third well-known application of MCT is to the glass transition phenomena. Here also, one sees a dramatic slowdown of relaxation as one approaches the glass transition temperature. However, physics is entirely different from that in critical phenomena. In dense liquids, at low temperatures (near or below the freezing temperature), because of the pronounced slowing down of the density relaxation at the intermediate wavenumbers (due to the cage effect), the nonlinear or the binary product terms, which make a negligible contribution in the dense gas or low-density liquid phase, become a significant contributor to molecular relaxation processes. MCT theory of glass transition was introduced in a pioneering work by Geszti who showed in 1983 that imposition of self-consistency in the mode-coupling theory expression of viscosity (in terms of the stress-stress time correlation function) could lead to the divergence of viscosity. Subsequently, Goetze, Leutheusser, and others have developed this theory further and made several interesting predictions. Many of these predictions have been verified, although the theory is found to break down in the deeply supercooled liquid, as discussed later.

23.2 HISTORICAL PERSPECTIVE

Mode-coupling theory (MCT) is formulated as a combined extension of two formidable theoretical approaches (as explained in Figure 23.1) that have been used to capture observed dynamics at two different time scales: kinetic theory of gases (for a short time) and hydrodynamics (for a long time). Starting from Maxwell and Boltzmann, and aided by Enskog, Chapman, Cowling, many scientists in the Dutch school, kinetic theory of gases have aimed at providing microscopic expressions for the transport properties, like viscosity, thermal conductivity, etc. Hydrodynamics, on the other hand,

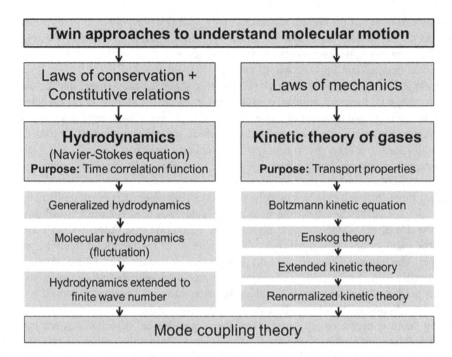

FIGURE 23.1 The two theoretical frameworks, namely, the classical hydrodynamics and the kinetic theory of gases, suitably extended, are combined to build the mode-coupling theory (MCT). The flowchart shows the origin of MCT and the ideas behind it. While the kinetic theory of gases offered methods to calculate the transport quantities (like diffusion and viscosity), the hydrodynamic theory showed how to obtain time correlation functions in terms of these transport properties.

dealt with inter-relationships between different transport properties (like those between friction and viscosity) and also expressions for time correlation functions (TCFs) (like density-density time correlation function and current-current correlation function, as in Rayleigh-Brillouin spectroscopy). *Hydrodynamics does not provide any expression for viscosity and thermal conductivity and they are used as input in the theory. However, the kinetic theory of gases provides expressions for these quantities, albeit for gases.*

We have discussed in Chapters 8–10 that the Green-Kubo formulae and the linear response theory provide the relations between transport properties and TCF. Transport properties are determined by the TCFs, which, in turn, determine the decay of the TCFs. This calls for self-consistency or boot-strapping.

Mode-coupling theory of liquids took many years to be developed, even after the seminal 1962 work of Fixman, because the realization that the nonlinear terms could play such an important role took time. When one found out that the standard kinetic theory of gases was inadequate to describe the long-time power-law tail in the tagged particle V-ACF, the natural attempt was to include it in correlated collisions, like ring collisions. This exercise was indeed successful to a great extent but failed at relatively high densities at ambient conditions. It became clear that including more and more correlated collisions would not solve the problem.

However, the hydrodynamic description is valid in intermediate-to-long time scales and breaks down in very short and very long time scales for different reasons. On the other hand, the renormalized kinetic theory works well in very short and very long-time responses. These two approaches are combined in MCT. None of these two approaches provides a self-consistent description that comes into the picture through the determination of the TCF of hydrodynamic modes in terms of transport coefficients or memory functions. Transport coefficients can be calculated as integrations over time and wavenumber of different time correlation functions (Green-Kubo relations) as shown in Eq. (23.14).

MCT formulation is based on the assumption of the separation of time scales among different dynamical events, such as fast collisional events and slower collective mode of relaxation. Therefore, the success of MCT largely depends on the validity of this separation of length and time scales.

23.3 BASIC PREREQUISITE

We next discuss the essential prerequisites of the mode-coupling theory.

23.3.1 WAVENUMBER-DEPENDENT HYDRODYNAMIC MODES

For an infinitely large system, there are several collective variables that relax slower than the rest. These slow modes are the hydrodynamic modes. They are conserved. These are five modes that come from the solution of linearized Navier-Stokes hydrodynamic equations: the heat, the two sound or acoustic modes, and two current modes.

The dynamic structure factor $S(k,\omega)$ (discussed in Chapter 9) is defined as the spatial and temporal correlation of the density fluctuations in wavenumber and Fourier frequency plane

$$S(k,\omega) = \int dt \exp(i\omega t)\langle \rho(k,t)\rho(-k,0)\rangle. \tag{23.2}$$

The spectrum of such density fluctuations can be observed in the light scattering experiment. This is popularly known as the *Rayleigh-Brillouin* spectrum. The derivation of the expression $S(k,\omega)$ is one of the great successes of the classical hydrodynamic theory.

The basic equations of fluid mechanics in the wavenumber (k) and Laplace frequency (z) plane are discussed in Chapter 5.

23.3.2 EXTENDED MOLECULAR HYDRODYNAMIC THEORY

Rayleigh-Brillouin spectrum discussed above is observed in a light scattering experiment involving wavelengths around 5000 Å. However, neutron scattering experiments of liquids capture completely different physics as the wavenumber in these experiments typically remains between 0.1 and 15 (Å)$^{-1}$. As already discussed earlier, the dynamic structure factor $S(k,\omega)$ has a three-peak structure at reduced wavenumbers $k\sigma \leq 1$ (where σ is the atomic diameter). The width of the central peak initially increases with wavenumber, k. However, it decreases significantly at a wavenumber close to the main peak ($k = k_m$). This effect is named after de Gennes, who observed that correlation effects between neighboring atoms play a predominant role [12]. This often leads to a strong narrowing of the density distribution function. At larger wavenumbers, the width of the central peak gets broadened again, and finally, it reaches its free-particle limit.

The extension of all five hydrodynamic modes (heat, sound, and viscous) to intermediate wavelengths was first done by de Schepper and Cohen in 1982 [13]. Their theory was based on a model kinetic equation for a hard sphere fluid (generalized Enskog theory). Another approach of the extended hydrodynamics to intermediate wavenumbers was provided by Kirkpatrick who included wavenumber and frequency-dependent transport coefficients in the generalized hydrodynamics [11]. Using the projection operator technique, he derived generalized hydrodynamic equations for the density and current modes for a hard sphere fluid. The final result is a closed set of generalized hydrodynamic equations for the time correlation functions of number density (ρ), longitudinal momentum density ($\mathbf{g_l}$), temperature T, and transverse momentum density t_i $(i = 1,2)$.

$G_{\alpha\beta}(k,z), (\alpha,\beta=\rho,l,T,t_i)$ denotes the normalized time correlation functions in Fourier-Laplace (z) space, given below

$$zG_{\rho\beta}(k,z) - \frac{ik}{\sqrt{\beta mS(k)}} G_{l\beta}(k,z) = \delta_{\rho\beta},\tag{23.3}$$

$$zG_{l\beta}(k,z) - \frac{ik}{\sqrt{\beta mS(k)}} G_{\rho\beta}(k,z) - ik\left[\frac{2}{3\beta m}\right]^{1/2}\left[1 + 2\pi\rho\sigma^3 g(\sigma)\frac{j_1(k\sigma)}{k\sigma}\right]G_{T\beta}(k,z)$$
$$+ \frac{2}{3t_E}\left[1 - j_0(k\sigma) + 2j_2(k\sigma)\right]G_{l\beta}(k,z) = \delta_{l\beta}\tag{23.4}$$

$$zG_{T\beta}(k,z) - ik\left[\frac{2}{3\beta m}\right]^{1/2}\left[1 + 2\pi\rho\sigma^3 g(\sigma)\frac{j_1(k\sigma)}{k\sigma}\right]G_{l\beta}(k,z)$$
$$+ \frac{2}{3t_E}\left[1 - j_0(k\sigma)\right]G_{T\beta}(k,z) = \delta_{T\beta},\tag{23.5}$$

and ($i = 1, 2$),

$$\left[z + \frac{2}{3t_E}\left[1 - j_0(k\sigma) - j_2(k\sigma)\right]\right]G_{t_i\beta}(k,z) = \delta_{t_i\beta},\tag{23.6}$$

where $j_l(k,\sigma)$ is the spherical Bessel function of order l, $g(\sigma)$ is the radial distribution function at contact and $t_E = \sqrt{\beta m\pi}/4\pi\rho\sigma^2 g(\sigma)$ is the Enskog mean free time between collisions. The transport coefficients in the above expressions are given only by their Enskog values (for hard sphere fluids), that is, only collisional contributions are retained. [14]. The eigenmodes can be obtained by solving the coupled Eqs. (23.3)–(23.6). The extended hydrodynamics reproduces the softening of the heat mode at intermediate wavenumbers. The theory of glass transition has also been developed based on the softening of the heat mode, as shown in Figure 23.2.

The complete solution of the heat mode eigenvalue $z_H(q)$ was found to be well approximated by the form obtained from the revised Enskog kinetic equation for not too small q. The latter is given by the following expression:

$$z_H(q) = -\frac{Dq^2}{S(q)}\left[1 - j_0(q\sigma) + 2j_2(q\sigma)\right]^{-1}.\tag{23.7}$$

From the above expression, one can say that the softening of the heat mode happens due to both, the peak in $S(q)$ near q_m and also the fact that the heat mode becomes essentially a density mode that decays via self-diffusion. An interesting way of stating the same is that the heat mode degenerates into the density mode. Physically, this occurs due to the obvious fact that at small length scales, only the number remains conserved – the conservation of momentum and energy poses no constrain, leading to the breakdown of Navier-Stokes hydrodynamics.

This slow mode is important in the theories that include mode-coupling effects. Such theories have been utilized to quantitatively understand the anomalous long-time tails of the stress-stress correlation function and the shear-dependent viscosity, observed in computer simulations.

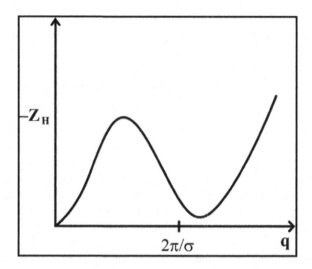

FIGURE 23.2 Wave-number (q) dependence of the eigenvalues (z_H) of the extended hydrodynamic modes. Note the small value of the frequency (or, rate) z_H at the intermediate value of q (when $q \simeq 2\pi/\rho$). Softening of density relaxation modes, also known as the heat mode (the nomenclature from the *Rayleigh-Brillouin* spectrum) is shown in this figure. This slow mode is important in the theories that include mode-coupling effects. Such theories have been utilized to quantitatively understand the anomalous long-time tails of the stress-stress correlation function and the shear-dependent viscosity, observed in computer simulations.

23.4 METHODS TO DEVELOP MCT EQUATIONS

There are multiple ways to develop an MCT of transport properties and/or time correlation functions. We shall discuss them in this chapter. A small introduction is provided below.

(a) As discussed above, a useful application of MCT arises in the evaluation of the memory function in the relaxation equations [Eq. (12.4)] derived by the projection operator technique. Here we evaluate the propagation operator by projecting it on a set of orthogonal dynamical variables. They are constructed using the product of hydrodynamic modes which lie in the orthogonal space. The propagator in the orthogonal subspace, $\exp(iQLT)$ is approximated as in Eq. (23.8). Here Q is $(1-P)$ where "P" is the projection operator that projects the dynamical variable to the relevant subspace, and Q projects to orthogonal subspace. That is, the approximation here remains essentially the same

$$\exp\left(iQLt\right) \approx P^{(2)}\exp\left(iLt\right)P^{(2)}, \tag{23.8}$$

This is an interesting approximation, widely used for memory function. The main advantage is that we get to deal with the full propagator $\exp(iLt)$. This advantage is made clear below.

(b) Sometimes we opt for a direct calculation of the time correlation function, $\langle x(0)x(t)\rangle = \langle x(0)\exp(iLt)x(0)\rangle$, by using the approximation

$$\exp\left(iLt\right) \approx P^{(2)}\exp\left(iLt\right)P^{(2)}, \tag{23.9}$$

We have already discussed in Chapter 12 this MCT procedure, involved in the calculation of the velocity TCF, first presented by Gaskell and Miller. $P^{(2)}$ is the projection operator. There are two main ideas that are in play here. First, we know the coupling between the dynamical quantity of interest and the modes in the orthogonal subspace. Second, we know the time evolution of the quantities in the subspace. One additional point is that the subspace need not be orthogonal.

(c) The next important step is the construction of orthogonal projection space. *It is very much like enhancing the basis set in quantum chemical calculations.* As already mentioned, the first and the most useful new members are the products of two hydrodynamic modes, like two density modes $(\rho_q \rho_{k-q})$, two current modes $(j_q j_{k-q})$, one current mode and one density mode $(\rho_q j_{k-q})$ and so on. There are often obvious reasons that lead to the proper choice. We shall see examples in the study of Gaskel and Miller, and also in the study of glass transition.

(d) In rare cases, we might need to consider three such products, like $\rho_q \rho_q \rho_{k-q-q'}$. As already mentioned, this is like augmenting the basis set in quantum chemistry calculations.

(e) Thus, in many applications of liquid phase dynamics, the bare and primary dynamical variables are the wavenumber and time-dependent hydrodynamic modes while the projection is on variables that are products of these hydrodynamic modes. In selecting the subspace, one takes into account the following points.

Let us consider the time and wavenumber dependent correlation function $C(k, t)$, which can now be expressed as,

$$\begin{aligned} C(q,t) &= \langle A(-q,0)\, A(q,t)\rangle = \langle A(-q,0)\, \exp(iLt)\, A(q,0)\rangle \\ &= \langle A(-q,0)\, P^{(2)}\, \exp(iLt)\, P^{(2)}\, A(q,0)\rangle. \end{aligned} \tag{23.10}$$

Next we consider the projection on a space spanned by B, so $P^{(2)} = |B\rangle\langle B|/\langle B\,|\,B\rangle$. Then Eq. (23.10) becomes,

$$\begin{aligned} C(q,t) &= \langle A(-q,0)\, A(q,t)\rangle \\ &= \langle A(-q,0)\, P^{(2)}\, \exp(iLt)\, P^{(2)}\, A(q,t=0)\rangle \\ &= \frac{\langle A(-q)|B\rangle\langle B|\exp(iLt)|B\rangle\langle B|\, A(q)\rangle}{\left[\langle B\,|\,B\rangle\right]^2}. \end{aligned} \tag{23.11}$$

$|B\rangle$ is a binary product of hydrodynamic modes. The above approach is particularly useful when we have an *a priori* idea about the hydrodynamic modes that would couple to $A(t)$. MCT allows a systematic method to include this coupling.

The MCT approach to the calculation of the time correlation function is useful but may become difficult sometimes because it shifts the onus on (i) the vertex function and (ii) new TCFs that emerge from this operation. It may fail to accurately describe the short-time dynamics. It is more suitable to study long time. In many applications, MCT is used to obtain long-time behavior. The above comments shall become transparent when we discuss the treatments of Gaskell and Miller below.

A common approach of mode-coupling theory is to start with the relaxation equation or memory function equation derived using Zwanzig's projection operator technique and given by.

$$\frac{\partial C(q,t)}{\partial t} = i\Omega(q)C(q,t) - \int_0^t d\tau\, \Gamma(q,\tau)C(q,t-\tau). \tag{23.12}$$

Eq. (23.12) is known as the generalized relaxation equation (GRE) or the memory function equation. In the chapter on the projection operator technique, we have derived this expression. Here $C(q, t)$ is the correlation matrix defined above. $\Omega(q)$ and $\Gamma(q,\tau)$ are the frequency and the memory function matrices, respectively.

As discussed above in point (c), *in MCT we attempt to develop an expression of the memory function* . This relaxation equation being exact provides powerful formalism to compute the time correlation functions. We have discussed this in detail in Chapter 10.

23.5 USEFUL TERMINOLOGY

There is certain terminology that the students should be aware of.

(a) The quantity $\langle B \mid A(q) \rangle$ in Eq. (23.11) above is called the vertex function. *This is an equilibrium quantity that carries weight at a given q.* This is often hard to calculate and contains detailed structural information at the molecular level, such as two- and three-particle distributions functions, both radial and direct correlation functions.

(b) Both "**q**" and "**k**" are used to denote wave vector. In order to use the wavenumber-dependent description successfully, the liquid must be isotropic. This allows such quantities as structure factor to be dependent on the magnitude of the wave vector. Second, the rule of combination of "**q**" and "**k**" is such that momentum is conserved as we vary one of them. Thus the combination "**q**" and "**k-q**" is used when "**q**" is summed or integrated over, as described above. This also follows from direct evaluation.

(c) Because B is a binary product, the time-dependent term $<B|e^{ilt}|B>$ involves four terms, a product of two time-dependent and two time independent. One uses the guidance of cumulant expansion (see Chapter 14) to decompose it into product of two terms. This is strictly valid when the dynamical variables obey Gaussian statistics. That is why it is called a Gaussian approximation. This has been a subject of many discussions. It is found to be generally valid in most applications. For example, the Gaussian approximation allows derivation of the limiting laws of viscosity, conductivity in electrolyte solutions, to give a few examples.

(d) MCT allows one to take into account of the cage effect, which is the effect of liquid structure, and as explained, this structure enters primarily through the vertex function. The relaxation of the cage plays an important determinant.

(e) This slowdown of motion due to the caging of a tagged molecule in liquid is called "cage effect". *This is not included in conventional hydrodynamics.* This cage effect eventually leads to a breakdown of both the Navier-Stokes hydrodynamics and also of the kinetic theory of gases. The emergence of slow density modes at intermediate wavenumbers (that is, at the nearest-neighbor distance) suggests that these modes need to be included in such processes as stress relaxation, as indeed was done by Geszti. We have discussed the extension of the hydrodynamic modes to include wavenumber dependence at intermediate wavenumbers.

(f) As discussed in Chapter 5, Navier-Stokes hydrodynamics provides analytical expressions for density relaxation in terms of several transport properties, like viscosity and thermal diffusivity. However, it is a theory valid only at long wavelengths and also long times – long compared to the diameter of a molecule and long compared to the collision time. Thus, hydrodynamics cannot describe the cage effect described above.

As was experienced by Boltzmann long ago, an extension of the kinetic theory of gases to include collisions and correlated collisions is exceedingly tricky. Thus, both the approaches (hydrodynamic and kinetic theory) have certain merits and certain limitations. MCT combines the two, as already mentioned. Of course, the success of MCT was first established in the explanation of slowing down of transport properties near the critical temperature and the long-time tail of velocity autocorrelation function. We shall discuss these, in detail, in the next section, where we discuss the explanation of three phenomena: long-time tail and power-law decay $t^{-d/2}$ of velocity autocorrelation function, critical phenomena, and glass transition. The details of the phenomena are written below *in three different sections.*

23.6   MCT AND POWER-LAW LONG-TIME TAIL IN VELOCITY TIME CORRELATION FUNCTION

The power-law tail is a well-known property of the velocity time correlation function (V-ACF) of a tagged particle. The latter is defined through,

$$Z(t) = \langle \mathbf{v}(0) \cdot \mathbf{v}(t) \rangle, \tag{23.13}$$

where angular bracket denotes the average over the initial condition. The autocorrelation is a microscopic property of an interacting system that can be related to the transport coefficient through the Green-Kubo formula. For example, the diffusion coefficient D (transport coefficient) is related to the time integration of the velocity autocorrelation function $Z(t)$ (time correlation function) in the form given below,

$$D = \frac{1}{3} \int_{0}^{\infty} dt Z(t). \tag{23.14}$$

Thus, the long-time decay of $C_V(t)$ acquires special importance.

Prior to the work of Rahman (1964) and of Alder and Wainwright in 1967 [5], it was believed since the time from Boltzmann, Enskog, and Langevin equations that the nonconserved hydrodynamic (e.g., current modes) variables decay exponentially in a long time.

This exponential decay of V-ACF is shown by all known solvable models like the Fokker-Planck equation. The explanation for this exponential decay can be understood from Figure 23.3.

As mentioned above, Rahman and later Alder and Wainwright carried out computer simulations on self-diffusion and velocity TCF with the models of Lennard-Jones and colliding hard disk and hard sphere. The results of the molecular dynamics came as a complete surprise, which indicated that the

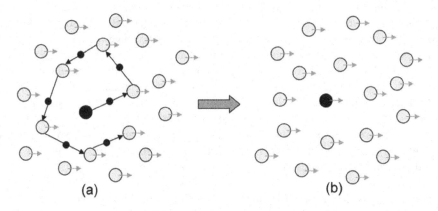

(a)   (b)

FIGURE 23.3   Here the tagged particle (filled black circle) gets repeatedly kicked around by the other particles (filled gray circles) moving in some positive direction. It was quite intuitive from these collisions' mechanism that this tagged particle loses memory of its initial velocity at some positive rate. The tagged particle couples ("feels") the collective hydrodynamic modes, like current and density, through these repeated collisions.

velocity autocorrelation function $\langle v(t) \cdot v(0) \rangle$ of a tagged particle decays as power law $t^{-d/2}$ in long time $t \to \infty$ where d indicates the dimensionality of the system that has been studied. This asymptotic power-law decay of $Z(t)$ in a fluid is called the long-time tail. Neutron scattering experiments also indicated a long-time tail in V-ACF for atomic liquids (liquid argon etc.).

Interestingly, for 3D LJ systems, both MD simulation and MCT predicts $t^{-\frac{3}{2}}$ the behavior of V-ACF over a wide range of density. For these systems, despite the existence of long-time tails of V-ACF, there exists a well-defined diffusion coefficient. In 2D LJ fluid, the presence of the long-time t^{-1} tail in V-ACF gives rise to the divergence of the diffusion in the long time. It should be clear from Eq. (23.14), if the integrand on the right-hand side of the Green-Kubo relation decays as $t^{-d/2}$ then the long-time diffusion coefficient does not for low dimensions. Therefore, it shows the conventional hydrodynamics breaks down in low-dimensional systems.

For the explanation of this long-time tail, consider a tagged particle in a sea of identical particles. As the tagged particle hits one of the surrounding molecules [shown in Figure 23.3(a)], it partially transfers its initial momentum to the surrounding liquid, which is moving collectively due to the presence of the hydrodynamic current. The fluid particles in turn, transfer some of their momentum to the tagged particle. Now, since there is an exchange of momentum due to repeated collisions, tagged particle inherits a bit of the solvent momentum and moves with the flow of surrounding particles [shown in Figure 23.3(b)]. This mechanism of coupling to the current mode results in a positive tail in the V-ACF in the long time. The particles themselves may move fast between collisions but the overall process is slow. Note these currents are the natural fluctuations in the system and the average current is zero. These fluctuations decay on a slow time scale as a consequence of the conservation of momentum. *This description of the coupling of the tagged particle's velocity to the naturally occurring current in the system and flow with them is called the mode coupling. This can be regarded as one of the developments that gave birth to the MCT.* The long-time behavior $Z(t)$ suggests that the self-diffusion process is a result of t several contributions: short time collisions that repeated collisions become coupled to the collective dynamical modes of the liquid. In fact, it has been shown that the diffusion coefficient can broadly be written as a sum of two terms

$$D = D_{Micr} + D_{Curr} \qquad (23.15)$$

where D_{Micr} and D_{Curr} are the two main components of particle migration. The first term on the right-hand side is due to collisions and density fluctuations while the second term is due to current flows in the system that carries the particle along. The above decomposition is approximate but semi-quantitatively reliable and captures the essence of diffusion in a neat way. MCT can be used to derive such insightful and elegant expressions.

Although the underlying picture is simple, the theory is not. Several theoretical approaches have been developed to derive the asymptotic form of V-ACF. The fluctuating hydrodynamics approach [7] explains the asymptotic behavior of V-ACF in terms of the current density. Starting with the continuity equation for the number density, $n_1(r,t)$ of tagged particles in a fluid

$$\frac{\partial}{\partial t} n_1(r,t) = -\nabla \cdot \mathbf{J}(r,t). \qquad (23.16)$$

The tagged particle current $J(r,t)$ can be expressed as

$$J(r,t) = -D_0 \nabla n_1(r,t) + \upsilon(r,t) n_1(r,t) + J_R, \qquad (23.17)$$

where the first term on the right-hand side is the diffusive component due to Fick's law and J_R is the fluctuating current. The convective current $\upsilon(r,t) n_1(r,t)$ involves the coupling between the tagged particle density, n_1 and the fluid velocity, $\upsilon(r,t)$ [7]. Later, it was pointed out that the coupling of the particle velocity to the transverse component of the current density determines the asymptotic behavior [8].

Thus, the tagged particle velocity TCF can be written as an integral over local current,

$$C_v(t) \approx \int dr \langle J(r,0).J(r,t) \rangle, \tag{23.18}$$

which can be written in the wavenumber plane as,

$$C_v(t) \approx (2\pi)^{-3} \int d\mathbf{k} \langle J(-\mathbf{k},0).J(\mathbf{k},t) \rangle. \tag{23.19}$$

One interesting aspect of the V-ACF is its strong dimensionally dependence on the long-time decay. It first came to light from the studies of Alder and Wainwright [5] on hard disk and hard sphere. The power-law tail can now be understood from the above equation by using the time dependence of the decay of the current-current correlation function. It decays as $e^{-\eta k^2 t}$. To explain the dimensional-dependent results, we take the $k-dependent$ integration of correlation function $e^{-\eta k^2 t}$, i.e., $\int dk \, k^{d-1} e^{-\eta k^2 t}$ where η is the kinematic shear viscosity.

For the two-dimensional scenario,

$$\int dk \, k \, e^{-\eta k^2 t} = \frac{1}{\eta t}. \tag{23.20}$$

For $d = 3$,

$$\int dk \, k^2 \, e^{-\eta k^2 t} = \frac{\sqrt{\pi}}{2} \frac{1}{(\eta t)^{3/2}}. \tag{23.21}$$

One can also show the dimension dependency $t^{-d/2}$ using an alternative approach, proposed by van Beijeren (1982), and based on the consideration that the motion of the tagged particle is diffusion-like. Therefore, suppose the particle (filled black circle) is at the origin in the diffusion regime. After time t the tagged particle travels $t^{1/2}$ distance in each dimension. It shows the tagged particle comes into contact (directly or indirectly) with the particles lying in the volume of order $t^{d/2}$. As a result of these collisions, the tagged particle shares momentum with other particles, and the velocity autocorrelation would be of order $t^{-d/2}$.

23.7 DYNAMICAL CRITICAL PHENOMENA

Dynamics near a thermodynamic critical point is characterized by large-scale density fluctuations, for the gas-liquid critical point, and in magnetization in magnetic transitions. Density and magnetization serve as the order parameter for the respective transitions. In the following we shall use both the systems in our discussion. The difference is that while density is a conserved order parameter, magnetization is a non-conserved variable. Isotropic-nematic liquid crustal transition also shows some of the critical anomaly, although it undergoes a weak first-order first transition.

Because of these fluctuations in the order parameters, many thermodynamic equilibrium properties like compressibility, specific heat show divergence near the critical point. The divergence is due to the appearance of correlated long wavelength density or magnetization fluctuations. Thus, there could be a large liquid-like region of high density surrounded by a large region of gas-like density. Or, in the Ising model picture, there would be large region of up spins surrounded by a region of mostly down spins.

An important point to note, that these large fluctuations survive for a long time, and contribute not just to divergence-like growth of equilibrium response functions but also to divergence of relaxation time. Like the static properties, the singularity exhibited by different transport properties also can be described by certain exponents. The singularities in the case of thermal diffusivity D_T or bulk viscosity ζ can be described as,

$$D_T \sim \left(T_C - T\right)^{-z_H} \tag{23.22}$$

or

$$\zeta(T) \sim \left(T_c - T\right)^{-z_\zeta} \tag{23.23}$$

where z_H and z_ζ are referred to as dynamical critical exponents. *Explanation of these anomalies was among one of the first successful applications of the mode-coupling theory.*

Let us try to motivate an understanding of this conceptually somewhat difficult problem. In Navier-Stokes hydrodynamics, we use a *linearized description* to obtain the expression for the Rayleigh-Brillouin spectrum (as discussed in Chapter 5). Such linearization leads to the neglect of nonlinear coupling between fluctuations in the density and momentum density (or, current) fields. When fluctuations become large, like the density fluctuation near the critical temperature, we have no justification in neglecting the nonlinear terms.

In fact, it was Fixman and Kawasaki who initiated the preliminary studies along these lines to treat the nonlinearity mentioned above [1, 2]. We rephrase Fixman's well-known work in the modern language. Viscosity is given by stress-stress time correlation function, which, as Geszti has shown, can be expressed in terms of current-current correlation function. The current in turn can have a normal part and a slowly decaying part, which can be described as a binary product of slow variables. In the case of viscosity anomaly of a binary mixture near the critical point, the binary product is taken as $\chi_{-q}\chi_{k+q}$, where χ_{-q} is the Fourier transform of the space-dependent composition fluctuation. As this is the dominant slow variable. Fixman showed that this decomposition can explain the rise in viscosity in terms of slow and large composition fluctuations in binary mixtures. An important point emphasized by Fixman was that the divergence is insensitive to the details of the intermolecular potential, which is indeed the hallmark of critical phenomena.

Subsequently, a systematic perturbation theory for the determination of the critical exponents was carried out by Kadanoff and Swift [3]. The language used by Kadanoff and Swift was interesting. In the presence of large fluctuations, the linearity breaks down and the modes interact with each other, and new decay mechanisms appear, which are responsible for the emergence of singularity. The latter is now described in terms of the decay of one transport mode into two other modes. In the present case of the decay of heat mode, a heat mode of wavelength q decays into a heat mode of wavenumber q and a viscous mode of wavenumber k-q. In this theoretical formalism, the critical exponents are predicted to be $z_H = 2/3$, $z_\zeta = 2$ at low frequency and $z_\zeta = 2/3$ at high frequency. The approach by Kadanoff and Swift not only was successful in explaining the singularity, but it also reproduced correct values of the dynamical critical exponent.

It is interesting to contrast the present situation with vibrational relaxation in molecules. In the latter the mode coupling effects enter through anharmonicity, which leads to a coupling between different modes and changes the vibrational relaxation substantially, sometimes enhancing the rate by many orders of magnitude. In the case of fluid dynamics, the situation is similar but the procedure is macroscopic. Now the fluctuations consist of a normal part, which is short ranged and a singular one, which is connected to slow, long wavelength density fluctuations. Viscosity, conductivity, and other transport properties show anomalies because of these density or composition fluctuations.

To summarize the physical picture, the relaxation proceeds through an additional channel, which opens up due to the coupling of the density mode to the current mode. In the last section (on the long-time tail), we have discussed that the tagged particle couples to the fluctuating current through repeated collision. Similarly, heat conductivity, which is the transfer of local momentum from one place to another, can couple to both density and current fluctuations. Near the critical temperature, large-scale density and current fluctuations appear in the system, which are also slow. Momentum density, which is a product of density and velocity, becomes coupled to them by the same mechanism described above. According to MCT, this coupling can be captured by studying the decay of the heat mode into the product of two dynamical modes that form the orthogonal subspace. The success of MCT in describing the critical dynamics is really impressive and remains an intellectual triumph of time-dependent statistical mechanics.

23.8 GLASS TRANSITION: THE MCT PICTURE

The third notable application of MCT is the microscopic description of the glass transition phenomena. Here also a dramatic slowdown of relaxation appears near the glass transition temperature. However, physics is entirely different from that in critical phenomena.

Glasses are amorphous solids that lack any long-range structural order. The most common route to a glassy state is by rapidly cooling a liquid. Almost every liquid, if cooled at a large enough cooling rate, enters a supercooled state bypassing the crystalline state below the melting point T_m, and eventually becomes a glass. The glass transition temperature T_g is defined as the point at where the structural relaxation time τ exceeds 100s or the viscosity of the liquid attains the value of 10^{13} poise. Basically, the liquid stops flowing on any practical time scale. The mystery behind molecular processes that lead to glass transition from a liquid has motivated a lot of research in recent decades, and a deep microscopic understanding has evolved.

To explain the liquid glass transition, MCT does not rely on any kind of phenomenological assumptions. A microscopic theory explains glassy behavior through the first principle, i.e., Newton's equations for classical particles. Explanation of glass transition using MCT is based on the idea of writing the mode coupling term solely in terms of the correlation functions (of slow modes) itself. This introduces the crucial concept of dynamic feedback mechanism, which is responsible for self-consistency in the theory. Subsequently, this feedback mechanism gives rise to the dramatic slowdown near the glass transition. The difference between the trajectories of particles in liquid and glassy phases is shown in Figure 23.4.

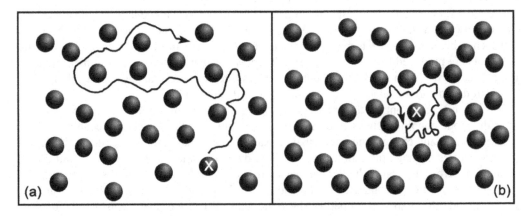

FIGURE 23.4   (a) The trajectory of a tagged particle in a low-density (normal) liquid phase. Here particles can move anywhere in the liquid without arresting in any cage formed by neighboring particles. (b) The trajectory of a tagged particle in a glassy phase where the particles become trapped in a cage formed by their neighbors.

This is not included in conventional hydrodynamics. This cage effect eventually leads to a breakdown of both Navier-Stokes hydrodynamics and also of the kinetic theory of gases. The emergence of slow density modes at intermediate wavenumbers (that is, at the nearest-neighbor distance) suggests that these modes need to be included in such processes as stress relaxation, as indeed was done by Geszti.

23.9   GESZTI-LEUNTHEUSSER-GOETZE THEORY OF GLASS TRANSITION

The self-consistent mode-coupling theory (MCT) provides a natural starting point for addressing the kinetic slowdown in the glass-forming simple liquid in terms of intermediate scattering function, $F(k,t)$. Time evolution of $F(k,t)$ upon cooling reveals the manner in which microscopic relaxation dynamics changes during liquid to the glass transition.

In this section, the explanation of glass transition is given by using the self-consistent MCT. The first self-consistent calculation of glass transition was presented by Geszti. He explained the viscosity trend near the glass-transition temperature. He suggested that the increase in viscosity results in a slower structural relaxation and the relaxation of the density mode. Due to this, the density mode contributes the most to the viscosity $\eta_{\rho\rho}$.

In the landmark work of Geszti, the MCT expression of viscosity was used to obtain the collective contribution to viscosity. Thus viscosity can be decomposed as,

$$\eta(t) = \eta_{bin}(t) + \eta_{\rho\rho}(t) \tag{23.24}$$

where $\eta_{bin}(t)$ is the short-time part originated from the binary interactions and can be approximated by a Gaussian function. However, $\eta_{\rho\rho}(t)$ is the long-time part originated from the density mode contribution. Using mode-coupling theory (MCT), one can obtain an analytical expression of $\eta_{\rho\rho}(t)$ as,

$$\eta_{\rho\rho} = \frac{k_B T}{60\pi^2} \int_0^\infty dt \int_0^\infty dk \, k^4 \left[S'(k)/S(k) \right]^2 \left[F(k,t)/S(k) \right]^2. \tag{23.25}$$

Here, $S(k)$ is the total static structure factor and $F(k, t)$ is the total intermediate scattering function.

The slow decay of the intermediate scattering function leads to an increase in the value of shear viscosity. In the above integration, we have assumed the lower limit of time integration as zero with minimal error.

The important observation made by Geszti was that the amplitude after the integration for MCT contribution to viscosity is itself proportional to viscosity. That is because the largest contribution derives from the intermediate wavenumbers where $F(k, t)$ decays as $S(k) \exp(-Dk^2 t/S(k))$. This is the same expression derived by de Gennes many years ago. Since we have a time integration and since $D \sim C/\eta$ where C is a constant. Therefore, we can write the final form as

$$\eta \approx A/(T - T_0) \tag{23.26}$$

and this equation is known as the Batchinski-Hildebrand law, which is found to be semi-quantitatively accurate in describing the growth of viscosity in the supercooled liquid regime as the temperature is lowered towards the glass transition. In Eq. (23.26), T_0 is the glass transition temperature in the Batchinski-Hildebrand law. Eq. (23.26) can be derived from the usual B-H law by carrying out a simple temperature expansion of the specific volume [28].

Although the liquid-glass transition phenomena have been studied in detail by different authors, the basic idea in all those approaches is the same that is the freezing of the density fluctuation near the wavenumber $q \simeq q_m$ where the static structure factor is sharply peaked. Here we discuss the approach of Leutheusser in detail [15].

Leutheusser derived the expression of the dynamic structure factor from the relaxation equation, which in this case is a nonlinear equation of motion for a damped oscillator:

$$\ddot{\Phi}(t) + \gamma \, \dot{\Phi}(t) + \Omega_0^2 \, \Phi(t) + 4\lambda \Omega_0^2 \int_0^t d\tau \, \Phi^2(\tau) \, \dot{\Phi}(t-\tau) = 0, \qquad (23.27)$$

where Ω_0 is the frequency of the free oscillator and γ is the damping constant. The initial conditions are $\Phi(t=0)=1$ and $\dot{\Phi}(t=0)=0$. In Eq. (23.27), the dot denotes the time derivative. The nonlinear term [fourth term on the left-hand side of Eq. (23.27)] resembles the memory kernel that depends on the past motion of the oscillator with a dimensionless coupling constant λ that varies from zero to infinity. The oscillator coordinate in this model represents the density correlation of a classical fluid. Introducing a Fourier-Laplace transform one can write,

$$\Phi(z) = \mathcal{L}\{\Phi(t)\} = i\int_0^\infty dt \, e^{izt} \, \Phi(t), \ \ \text{Im } z > 0. \qquad (23.28)$$

Eq. (23.27) can be written in the Fourier-Laplace space (z) as

$$\Phi(z) = -\frac{1}{z - \dfrac{\langle \omega_q^2 \rangle}{z + \eta_l(z)}}, \qquad (23.29)$$

where

$$\eta_l(z) = i\gamma + 4\lambda \langle \omega_q^2 \rangle \mathcal{L}\{\Phi^2(t)\}. \qquad (23.30)$$

Here $\eta_l(z)$ is the memory function in the Fourier-Laplace frequency plane and signified as the dynamical longitudinal viscosity. Here the expression of density correlation function involves $\eta_l(z)$ [Eq. (23.29)]. On the other hand, $\eta_l(z)$ itself depends on the density correlation function [Eq. (23.30)]. Thus the density correlation function calculation needs a self-consistent scheme.

For small coupling constant (zeroth order in λ),

$$\eta_l(z) = i\gamma. \qquad (23.31)$$

For this, the density correlation function $\Phi(z)$ has two simple poles in the lower complex half-plane, which are given below

$$\Phi(z) = -\frac{1}{z - \dfrac{1}{z + i\gamma}} = -\frac{a_1}{z + iv_1} - \frac{a_2}{z + iv_2}, \qquad (23.32)$$

where $v_{1/2} = \left[\gamma \mp (\gamma^2 - 4)^{1/2} \right]/2$ and $a_{1/2} = \left[1 \pm \gamma/(\gamma^2-4)^{1/2} \right]/2$. Using these and Eq. (23.30), the longitudinal viscosity in first order is obtained as

$$\eta_l(z) = i\gamma - \frac{4\lambda}{z - \dfrac{2}{z + i\gamma - \dfrac{2}{z + 2i\gamma}}}. \tag{23.33}$$

And the zero-frequency longitudinal viscosity in the first order in λ is given by

$$\eta_l = \gamma + 2\lambda(\gamma + 1/\gamma). \tag{23.34}$$

A comparison of Eq. (23.34) with Eq. (23.31) suggests that the zero-frequency value of the longitudinal viscosity in first order is found to be larger than its zeroth-order value. Thus in every loop of the self-consistent calculation, the zero-frequency longitudinal viscosity increases that might lead to a divergence of the zero-frequency value of $\eta_l(z)$ and $\Phi(z)$.

To understand the nature of the divergence, the following ansatz for $\Phi(z)$ was assumed by Leutheusser:

$$\Phi(z) = -\frac{f}{z} + (1 - f)\Phi_v(z). \tag{23.35}$$

Here $\Phi(z)$ is written as a combination of zero-frequency pole contribution with weight f and the remaining part $\Phi_v(z)$ with weight $(1 - f)$ where $\Phi_v(t = 0) = 1$.

So Eq. (23.30) can be rewritten as

$$\eta_l(z) = -4\lambda f^2/z + \eta_{lv}(z), \tag{23.36}$$

where

$$\eta_{lv}(z) = i\gamma + 8\lambda f(1 - f)\Phi_v(z) + 4\lambda(1 - f)^2 \big/ \mathcal{L}\{\Phi_v^2(z)\}. \tag{23.37}$$

Substituting it in Eq. (23.29) and comparing with the ansatz [Eq. (23.35)], the weight factor is obtained as

$$f = \left(1 + \sqrt{1 - 1/\lambda}\right)\big/ 2. \tag{23.38}$$

The expression of f suggests that for the value of λ larger than a critical value, $\lambda_c = 1$, the ansatz leads to an acceptable solution, and density fluctuations do not decay to zero for long times but decay to a finite value of f, that increases from $f = 1/2$ for $\lambda = 1$ to $f = 1$ for $\lambda \to \infty$. This is shown in Figure 23.5. Therefore, the density fluctuation spectrum exhibits a delta function peak with strength f, which is characteristic of the glassy phase. In the glass phase, the translational motion is frozen and the vibrational motion around the arrested position is described by $\Phi_v(z)$, which can be expressed in a simple way:

$$\Phi_v(z) = -\frac{1}{z - \dfrac{\Omega^2}{z + \eta_{lv}(z)}}, \tag{23.39}$$

where the oscillator frequency is defined by

$$\Omega^2 = 1 + 4\lambda f^2, \tag{23.40}$$

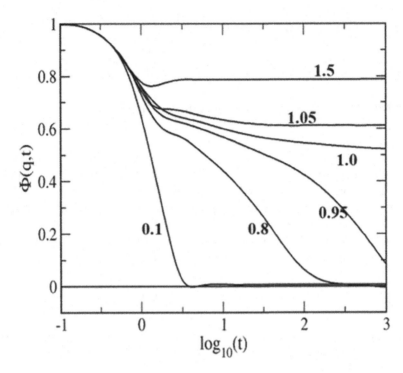

FIGURE 23.5 Time dependence of the density correlation function, $\Phi(t)$ for different coupling constants, λ (values are shown in the plot). Other parameters are $\Omega_0 = 1$, $\gamma = 1$, and $\lambda_c = 1$.

means Ω increases from $\Omega = \sqrt{2}\Omega_0$ at $\lambda = 1$ with increasing coupling constant.

This analysis by Leutheusser suggests that on approaching the glass transition, the slowing down of the density fluctuations is controlled by the increasing longitudinal viscosity, which in turn is coupled via a nonlinear feedback mechanism to the slowly decaying fluctuations. This leads to a divergence of the structural relaxation time at a certain critical coupling constant, λ_c.

23.10 MCT EXPRESSION FOR THE FRICTION TERM

We can use Zwanzig's projection operator technique (Chapter 12) to express friction on a tagged particle in terms of the time-dependent force-force autocorrelation function.

$$\zeta = \frac{1}{3k_B T}\int\limits_0^\infty dt \langle \mathbf{F}(0) \cdot \mathbf{F}(t) \rangle, \tag{23.41}$$

where $F(t)$ is assumed to be the force exerted on the ion at time t due to intermolecular interaction, k_B is the Boltzmann constant and T is the absolute temperature. There is an approximation in equating the orthogonal force in Zwanzig's approach to the actual force on a tagged particle, but it can also be shown to be valid under certain limiting conditions. The friction contains several contributions. The contribution of the binary collision to the total friction is calculated from kinetic theory; however, the long-time part involves the contribution of the solvent and the solute dynamics and the coupling between them. The coupling between solute and solvent dynamics is provided in terms of two particle direct correlation function. Now, the self-consistency comes into the picture as the self-dynamic structure factor itself depends on friction. The final expression of the friction is obtained as

$$\zeta_{\rho\rho}(\mathbf{k},z) = \frac{2k_B T \rho_0}{3(2\pi)^2} \int_0^\infty dt\, e^{-zt} \int_0^\infty dq\, q^4\, F_S(\mathbf{k}-q,t) |c(q)|^2 F(q,t). \qquad (23.42)$$

Here $F_S(\mathbf{k}\text{-}q,t)$ is the self-dynamic structure factor of the solute, $F(q,t)$ dynamic structure factor of the liquid and $c(q)$ is the two particle direct correlation function. ρ_0 is the average number density of the solvent. The limit $k \to 0$ and $z \to 0$ of the above equation defines the macroscopic friction.

One can model the bulk liquid or the solvent dynamic structure factor using the viscoelastic models (VEMs). This approximate, avoids the complication of self-consistency but is found to be fairly accurate at normal liquid density away from the glass transition temperature. Using the well-known Mori continued-fraction (truncated at second order), described in Chapter 13, the visco-elastic expression for $F(k, z)$ can be written as

$$F(k,z) = \cfrac{S(k)}{z + \cfrac{\langle \omega_k^2 \rangle}{z + \cfrac{\Delta_k}{z + \tau_k^{-1}}}} \qquad (23.43)$$

where $\langle \omega_k^2 \rangle = k_B T k^2 / m S(k)$ and $\tau_k^{-1} = 2\sqrt{\Delta_k / \pi}.\Delta_k = \omega_l^2(k) - \langle \omega_k^2 \rangle$, where $\omega_l^2(k)$ is the second moment of the longitudinal current correlation function, that is defined as

$$\omega_l^2(k) = 3k^2 \frac{k_B T}{m} + \omega_0^2 + \gamma_d^l(k). \qquad (23.44)$$

Here $\gamma_d^l(k)$ is the longitudinal component of the vertex function given later by Eq. (23.51).

The above approach allows a simple mode-coupling theory calculation of friction that has been widely used. In a more accurate description, the friction contains other contributions, which we describe below.

23.11 THE KINETIC THEORY MCT OF SJÖGREN AND SJÖLANDER

Sjögren and Sjölander developed a repeated ring collision kinetic theory that can describe various time correlation functions of liquid argon and rubidium fairly well. However, the numerical calculations needed structure of the liquid [like the radial distribution function and the static structure factor, $S(k)$] as input that could be obtained from computer simulation studies. The merit of this approach is that it provides detailed expression for required time correlation functions (like density-density and current-current) and also a detailed expression for the friction on a tagged particle.

Figure 23.6 shows the pictorial representation of three and four-particle repeated collisions, also referred to as ring-collision. These diagrams show that after the initial ($t = 0$) collision between two tagged particles "1" and "2", they undergo intermediate time collisions but again themselves re-collide at a later time, t. The intermediate collisions are to be treated as binary collisions, and they can be summed over. Based on these collisions, the dynamics of dense fluid is obtained as the sum of the contribution from binary collision and ring collisions.

In this approach, the exact microscopic expression of friction is written in terms of Green's function.

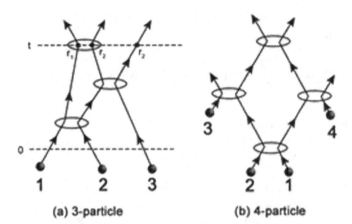

FIGURE 23.6 The sequence of binary collisions (two particles at a time) of a tagged particle (filled black circle) with the surrounding particles that form the cage. The collisions with particles labeled by 2 and 3 are uncorrelated. Tagged particle collides with particle 1 at point A, then with particle 4 at point B through intermediate collisions between 1 and 4. This is an example of ring collision.

$$\zeta(z) = \frac{1}{k_B TmV} \int dsds'dvdv' \left[\hat{\mathbf{k}} \cdot \nabla_{r_s} u(r_s - r_v) \right] G^s(sv;s'v',z) \left[\hat{\mathbf{k}} \cdot \nabla_{r_s'} u(r_s' - r_v') \right], \quad (23.45)$$

where the solute is denoted by subscript "s" and solvent is denoted by subscript "v". $u(r_s - r_v)$ is the interaction potential between the solute and the solvent molecules. $G^s(sv;s'v',z)$ is the Laplace transform of Green's function, $G^s(sv;s'v',t')$, which is the four-point correlation function that describes the correlated time evolution of the tagged solute (s) and the surrounding solvent molecules (v). It actually measures the probability that the solute moves from $(\mathbf{r}_s, \mathbf{p}_s)$ at a time t to $(\mathbf{r}_s', \mathbf{p}_s')$ at t' and the solvent particle located at $(\mathbf{r}_v', \mathbf{p}_v')$ at t' and that the same or another *solvent particle is found at* $(\mathbf{r}_v, \mathbf{p}_v)$ *at time t.*

For a system of charged solute in a polar solvent like water, there is one more contribution of the orientation of water dipole, Ω_v *that plays an important rol*e. Therefore, the equation becomes,

$$\zeta(z) = \frac{1}{k_B TmV} \int dsds'dvdv' \left[\hat{\mathbf{k}} \cdot \nabla_{r_s} u(r_s - r_v, \Omega_v) \right]$$

$$G^s(sv;s'v',z) \left[\hat{\mathbf{k}} \cdot \nabla_{r_s'} u(r_s' - r_v', \Omega_v') \right]. \quad (23.46)$$

Here $ds = dr_s dp_s$ and $dv = dr_v dp_v d\Omega_v$, Ω_v denotes orientation of solvent molecules, and all the other terms have their definitions discussed before. Figure 23.7 shows dipolar solvent molecules around a positively charged ion and dielectric relaxation of one of them from $(\mathbf{r}_v, \Omega_v)$ to $(\mathbf{r}_v', \Omega_v')$ when the ion is moving from position $\mathbf{r}_s$ to $\mathbf{r}_s'$.

We next focus on atomic liquids only, as orientational correlations remain to be developed. This theoretical formulation decomposes the total friction into a short-time part due to binary collisions and a long-time part. As already discussed, the long-time part arises due to correlated collisions

$$\zeta(z) = \zeta^B(z) + \zeta^R(z). \quad (23.47)$$

FIGURE 23.7 A spherical monatomic positively charged ion in dipolar solvent molecules; Orientational relaxation of solvent molecules (subscript: v) in the presence of ion (solute; subscript: s).

Source: Taken with permission from [35].

Here $\zeta^B(z)$ is the binary part of the friction. The correlated recollision term $\zeta^R(z)$ is obtained by expanding the total friction in the basis set of the eigenfunctions of the Liouville operator, i.e., the density, the longitudinal current mode, and the transverse current mode. As the contribution of longitudinal current mode to the total friction is not significant, the frequency-dependent friction can be written as given by

$$\frac{1}{\zeta(z)} = \frac{1}{\zeta^B(z) + R_{\rho\rho}(z)} + R_{TT}(z).$$

(23.48)

Here $\zeta^B(z)$ is the binary term, $R_{\rho\rho}(z)$ is the density term and $R_{TT}(z)$ is transverse current term. The binary term is assumed to be given by Gaussian approximation

$$\zeta^B(t) = \omega_0^2 \exp\left(-t^2/\tau_\zeta^2\right),$$

(23.49)

where τ_ζ is the relaxation time; the expression of this is derived as

$$\omega_0^2/\tau_\zeta^2 = (\rho/3m^2)\int d\mathbf{r}\left(\nabla^\alpha\nabla^\beta v(\mathbf{r})\right)g(\mathbf{r})\left(\nabla^\alpha\nabla^\beta v(\mathbf{r})\right)$$
$$+ (1/6\rho)\int\left[d\mathbf{k}/(2\pi)^3\right]\gamma_d^{\alpha\beta}(\mathbf{k})\left(S(k)-1\right)\gamma_d^{\alpha\beta}(\mathbf{k}).$$

(23.50)

In the above expression $S(k)$ is the structure factor and the expression for $\gamma_d^{\alpha\beta}(\mathbf{k})$ is written as a combination of the second moments of the longitudinal ($\gamma_d^l(\mathbf{k})$) and transverse current correlation functions ($\gamma_d^t(\mathbf{k})$).

$$\gamma_d^{\alpha\beta}(\mathbf{k}) = -(\rho/m)\int d\mathbf{r}\exp(-i\mathbf{k}\cdot\mathbf{r})g(\mathbf{r})\nabla^\alpha\nabla^\beta v(\mathbf{r})$$
$$= \hat{k}^\alpha\hat{k}^\beta\gamma_d^l(\mathbf{k}) + \left(\delta_{\alpha\beta} - \hat{k}^\alpha\hat{k}^\beta\right)\gamma_d^t(\mathbf{k}),$$

(23.51)

where $\gamma_d^l(\mathbf{k}) = \gamma_d^{zz}(\mathbf{k})$ and $\gamma_d^t(\mathbf{k}) = \gamma_d^{xx}(\mathbf{k})$.

Now, to derive the ring-collision term, Sjögren-Sjölander used the expression for the T-matrix, which is nontrivial. The final expression of the ring collision or the repeated collision can be written as

$$\zeta^R(z) = R_{\rho\rho}(z) - \zeta^B(z)R_{\rho L}(z) - R_{\rho L}(z)\zeta(z) - \zeta^B(z)\left[R_{LL}(z) + R_{TT}(z)\right]\zeta(z). \quad (23.52)$$

Here, the term $R_{\rho\rho}(z)$ includes the contribution of the coupling to density and is given by

$$\boxed{R_{\rho\rho}(t) = \frac{\rho}{m\beta}\int\left[d\mathbf{k}'/(2\pi)^3\right]\left(\hat{\mathbf{k}}\cdot\hat{\mathbf{k}}'\right)^2 k'^2\left[c(k')\right]^2\left[F^s(k',t)F(k',t) - F_0^s(k',t)F(k',t)\right],} \quad (23.53)$$

where $F^s = C_{00}^s$, $F_0^s = C_{000}^s$, $F = C_{00}/\rho$.

$R_{\rho L}(z)$ term contains the combined coupling to the density and the longitudinal current term and can be written as,

$$R_{\rho L}(t) = \int\left[d\mathbf{k}'/(2\pi)^3\right]\left(\hat{\mathbf{k}}\cdot\hat{\mathbf{k}}'\right)^2 c(k')\left[\gamma_d'(k') + \frac{\rho k'^2}{m\beta}c(k')\right]$$

$$\omega_0^{-2}\left[F^s(k',t)\frac{\partial}{\partial t}F(k',t) - F_0^s(k',t)\frac{\partial}{\partial t}F(k',t)\right]. \quad (23.54)$$

$R_{LL}(z)$ and $R_{TT}(z)$ provide the coupling to the longitudinal and transverse current modes, respectively

$$R_{LL}(t) = \frac{1}{\rho}\int\left[d\mathbf{k}'/(2\pi)^3\right]\left(\hat{\mathbf{k}}\cdot\hat{\mathbf{k}}'\right)^2\left[\gamma_d'(k') + \frac{\rho k'^2}{m\beta}c(k')\right]^2$$

$$\omega_0^{-4}\left[F^s(k',t)C_{ll}(k',t) - F_0^s(k',t)C_{ll0}(k',t)\right] \quad (23.55)$$

and

$$\boxed{R_{TT}(t) = \frac{1}{\rho}\int\left[d\mathbf{k}'/(2\pi)^3\right]\left[1 - \left(\hat{\mathbf{k}}\cdot\hat{\mathbf{k}}'\right)^2\right]\left[\gamma_{d12}'(k')\right]^2 }$$
$$\boxed{\omega_{012}^{-4}\left[F^s(k',t)C_{tt}(k',t) - F_0^s(k',t)C_{tt0}(k',t)\right]} \quad (23.56)$$

Here $C_{ll}(k,t)$ and $C_{tt}(k,t)$ are the longitudinal and the transverse current-current correlation functions, respectively, and $C_{ll0}(k,t)$ and $C_{tt0}(k,t)$ are the short-time parts of the same.

23.12   SELF-CONSISTENCY IN MCT EQUATIONS

A self-consistent calculation of friction becomes necessary due to the involvement of the self-dynamic structure factor, as discussed earlier. We now describe a simple scheme of self-consistent

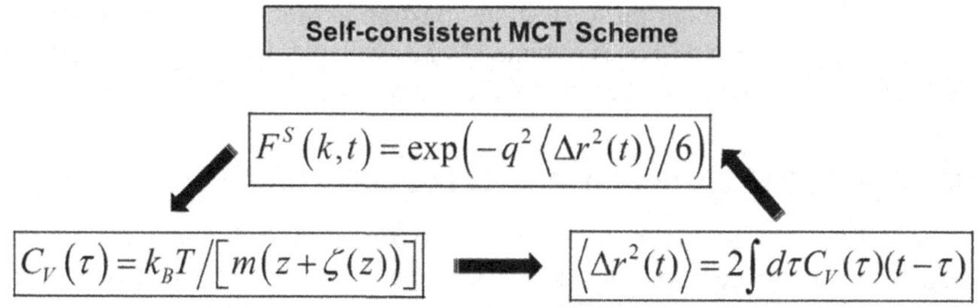

Self-consistent MCT Scheme

$$F^S(k,t) = \exp\left(-q^2 \langle \Delta r^2(t)\rangle/6\right)$$

$$C_V(\tau) = k_B T/\left[m\left(z+\zeta(z)\right)\right] \longrightarrow \langle \Delta r^2(t)\rangle = 2\int d\tau C_V(\tau)(t-\tau)$$

FIGURE 23.8 Scheme for self-consistent mode-coupling theory calculation.

calculation (the iterative scheme is shown in Figure 23.8). We stress that this is simplified and being presented to introduce self-consistency.

(1) The self-dynamic structure factor can be written as (using Gaussian approximation)

$$F_s(k,t) = \exp\left(-\frac{k^2\langle\Delta r^2(t)\rangle}{6}\right). \tag{23.57}$$

(2) $\langle\Delta r^2(t)\rangle$ is the mean-square displacement, which can be obtained from V-ACF using the Green-Kubo relation.

$$\langle\Delta r^2(t)\rangle = 2\int_0^t d\tau\, C_v(\tau)(t-\tau). \tag{23.58}$$

(3) The time-dependent V-ACF, $C_v(\tau)$ can be obtained from frequency-dependent V-ACF, $C_v(z)$ using Laplace inversion. Again, the Einstein relation relates $C_v(z)$ to frequency-dependent friction, $\zeta(z)$.

$$C_v(z) = \frac{k_B T}{m\left(z+\zeta(z)\right)} \tag{23.59}$$

(4) The friction is now expressed in the following fashion. We proceed from Eq. (27.58). In order to obtain the macroscopic friction that is measured experimentally, we need to set the $k = 0$ limit. This leads to the following expression:

$$\zeta(z) = \frac{2k_B T\rho_0}{3(2\pi)^2}\int_0^\infty dt\, e^{-zt}\int_0^\infty dq\, q^4\, F_s(q,t)\left|c_{id}(q)\right|^2 F(q,t).$$

The next steps are clear. We have to solve for $F_s(q, t)$ and $\zeta(z)$ self-consistently with each other, by iteration, as shown below.

The iteration process should be continued until the V-ACF obtained from two consecutive steps overlap. Finally, the diffusion coefficient is calculated using the relation

$$D = \frac{1}{3}\int_0^\infty d\tau\, C_v(\tau). \tag{23.60}$$

As stated above, this is a simple but fairly accurate illustration of a self-consistent MCT. Here the solvent dynamic structure factor $F(q, t)$ is assumed to be available. In the study of the glass transition described below, we need to include the dynamic structure factor in a self-consistent scheme.

23.13 GENERALIZED MODE-COUPLING THEORY (GMCT)

Even though the dynamical predictions of MCT for liquid dynamics are in general in good agreement with simulations, it fails to describe ultraslow relaxation near the glass transition in the deeply supercooled liquid regime. Most importantly, Leuthheusser-Goetze MCT predicts a value of critical transition density that is too low or, a value of the critical glass transition temperature T_c that is too high. In other words, the MCT predicted value is much higher than the temperatures T_g obtained from the experiments. This failure is attributed to the neglect of activated dynamics in MCT. This activated process is associated "hopping" of the particles out of their local cages to resist freezing. Therefore, MCT (in the ideal form) strongly overestimates the degree of caging. Along with MCT, this activated dynamics or hopping process above T_c may also be incorporated by random first-order transition theory.

Using nonlinear fluctuating hydrodynamic approach, Das and Mazenko in 1986 [20] introduced an extended MCT to cut off the sharp transition at T_c. Subsequently, in 1987 Götze and Sjögren [21] used projection-based formalism for the same purpose. Both theories show that the removing of sharp transition at T_c can be achieved through the activated dynamic or hopping process that restores ergodicity in the system below T_c. In a nutshell, these treatments lead to an additional decay term in the time dependence of the memory function.

In a similar but more transparent treatment, this additional decay term in the memory function was calculated by the random first-order theory (RFOT). RFOT provides a rate of melting of a glassy region in terms of the configuration entropy of the region. The configuration entropy approaches zero as the glass transition is approached. Relaxation of a region in RFOT is an activated event. When combined with MCT, it seems to capture many aspects of experimental results.

The q-independent (for a constant value of the wave vector) results for ideal MCT (where four-point correlation is written as a product of two-point correlation function) are shown in the application section. The results represented in this section are wave-vector-dependent and chosen value for $q = q_m$ corresponding to the inverse particle diameter, i.e., $k \approx 2\pi / \sigma$, where σ is the particle diameter. Similarly, in order to remove the sharp transition completely, a q-independent model for GMCT was introduced by Mayer et al. [23]. They reported analytically that the sharp transition can be completely removed when applying infinite-order GMCT. Finally, in 2015 Reichman [24] reported a q-dependent higher-order GMCT based on the unfactorized four-point intermediate scattering function. The calculation of this model is given in detail below.

Reichman et al. defined $2n$ points' density correlation function over n distinct values of wave vector k as

$$\Phi_n(\mathbf{k}_1, \cdots, \mathbf{k}_n, t)$$

$$\Phi_n(\mathbf{k}_1, \cdots, \mathbf{k}_n, t) = \frac{\langle \rho(-\mathbf{k}_1, 0), \cdots, \rho(-\mathbf{k}_n, 0)\rho(\mathbf{k}_1, t), \cdots, \rho(\mathbf{k}_n, t)\rangle}{\langle \rho(-\mathbf{k}_1, 0), \cdots, \rho(-\mathbf{k}_n, 0)\rho(\mathbf{k}_1, 0), \cdots, \rho(\mathbf{k}_n, 0)\rangle}. \tag{23.61}$$

The time evolution of these correlation functions is given by

$$\ddot{\Phi}_n(\mathbf{k}_1, \cdots, \mathbf{k}_n, t) + \Lambda \dot{\Phi}_n(\mathbf{k}_1, \cdots, \mathbf{k}_n, t) + \Omega_n^2(\mathbf{k}_1, \cdots, \mathbf{k}_n)\Phi_n(\mathbf{k}_1, \cdots, \mathbf{k}_n, t)$$
$$+ \int_0^t d\tau M_n(\mathbf{k}_1, \cdots, \mathbf{k}_n, \tau)\dot{\Phi}_n(\mathbf{k}_1, \cdots, \mathbf{k}_n, t - \tau) = 0, \tag{23.62}$$

where Λ represents the contribution from short time dynamics. The bare frequencies are given by

$$\Omega_n^2(\mathbf{k}_1,\cdots,\mathbf{k}_n) = \frac{k_B T}{m}\left(\frac{\mathbf{k}_1^2}{S(\mathbf{k}_1)} + \cdots + \frac{\mathbf{k}_n^2}{S(\mathbf{k}_n)}\right). \qquad (23.63)$$

The memory function M_n in Eq. (23.62) has the following form:

$$M_n(\mathbf{k}_1,\cdots,\mathbf{k}_n,t) = \frac{\rho k_B T}{16 m \pi^3}\sum_{i=1}^{n}\frac{\Omega_i^2(\mathbf{k}_i)}{\Omega_n^2(\mathbf{k}_1,\cdots,\mathbf{k}_n)}\int d\mathbf{q}\left|\tilde{V}(\mathbf{q},\mathbf{k}_i - \mathbf{q})\right|^2 S(q)S(|\mathbf{k}_i - \mathbf{q}|)$$
$$\times\,\Phi_{n+1}(\mathbf{q},|\mathbf{k}_1 - \mathbf{q}\delta_{i,1}|,\cdots,|\mathbf{k}_n - \mathbf{q}\delta_{i,n}|,t), \qquad (23.64)$$

where $\delta_{i,j}$ is the Kronecker delta function, ρ is the total density, and $\tilde{V}(\mathbf{q},\mathbf{k}_i - \mathbf{q})$ are static vertices given by

$$\tilde{V}(\mathbf{q},\mathbf{k}_i - \mathbf{q}) = (\hat{\mathbf{q}}_i.\mathbf{k})c(k) + \hat{\mathbf{q}}_i.(\mathbf{q}_i - \mathbf{k})c(|\mathbf{k}_i - \mathbf{q}|) \qquad (23.65)$$

where $S(k)$ is the static structure factor and $c(k)$ is the direct correlation function $c(k) \equiv [1 - 1/S(k)]/\rho$.

Eqs. (23.62) and (23.64) can be solved numerically as a coupled set of equations if a suitable expression is found for four-point (higher correlation) function in terms of two-point (lower order) correlation function. This higher-order GMCT shows MCT T_c higher than predicted by ideal MCT. It means GMCT captures at least some aspects of activated dynamics. Figure 23.9 depicts the different

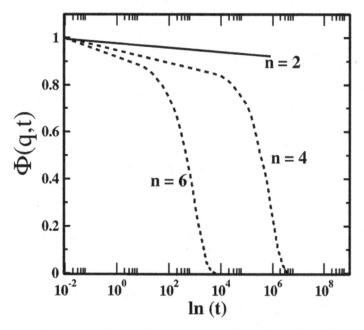

FIGURE 23.9 Time evolution of normalized density correlation function, $\Phi(q,t)$ at density 1.05. The values of n shown in the figure represent the order of GMCT at which theory closed and Eqs. (23.62)–(23.64) make a closed set of equations (details are given in Ref. [24]). The value of the wave vector corresponds to the first peak of the static structure factor $S(k)$. Here $n = 2$ is the case corresponding to the ideal MCT given by Eq. (23.27).

decay behaviour of the normalized dynamic structure factor at density 1.05, with increasing order of approximation.

23.14   MCT ANALYSIS OF THE THIRD-ORDER RAMAN SPECTRUM

A full microscopic MCT analysis for the third-order Raman spectrum of an atomic liquid Xe was presented by Denny and Reichman [34]. They have projected the dynamics onto bilinear pairs of fluctuating density variables. Classically third-order response function is expressed as

$$R^{(3)}_{\chi\lambda\mu\nu}(t) = -\theta(t)\beta \frac{d}{dt} C_{\chi\lambda\mu\nu}(t), \tag{23.66}$$

where $\theta(t)$ is the Heaviside theta function and χ, λ, μ, ν are the directions of the laboratory frame axes. The correlation function is defined as

$$C_{\chi\lambda\mu\nu}(t) = \left\langle \Pi_{\chi\lambda}(t)\Pi_{\mu\nu}(0) \right\rangle. \tag{23.67}$$

Here $\Pi_{\chi\lambda}$ are the specific tensor components of the total polarizability of the system that is expressed as

$$\Pi_{\chi\lambda}(t) = N\alpha^2\delta_{\chi\lambda} + \alpha^2 \sum_{i}\sum_{\substack{j>i \\ j}} T_{\chi\lambda}(\mathbf{r}_{ij}(t)), \tag{23.68}$$

where α is the atomic polarizability, and $T_{\chi\lambda}$ are the components of the dipole interaction tensor.

$$T_{\chi\lambda}(\mathbf{r}) = \frac{1}{r^3}\left(3\frac{\mathbf{r}_\chi \mathbf{r}_\lambda}{r^2} - \delta_{\chi\lambda} \right). \tag{23.69}$$

The correlation function of interest in this context is

$$C_{zzzz}(t) = \left\langle T_{zz}(t)T_{zz}(0) \right\rangle. \tag{23.70}$$

It can be written in the "interaction induced" form as

$$C_{zzzz}(t) = \int d\mathbf{r} \int d\mathbf{r}' A(\mathbf{r})A(\mathbf{r}') G(\mathbf{r},0;\mathbf{r}',t), \tag{23.71}$$

where

$$G(\mathbf{r},0;\mathbf{r}',t) = \frac{1}{N} \sum_{i}\sum_{\substack{j\neq i \\ j}}\sum_{k}\sum_{\substack{l\neq k \\ l}} \left\langle \delta\left[\mathbf{r}_{ij}(0) - \mathbf{r}\right] \delta\left[\mathbf{r}_{kl}(t) - \mathbf{r}'\right] \right\rangle. \tag{23.72}$$

For a simple dense liquid, the dynamics of the variable $T_{zz}(r)$ can be decomposed into the dynamics of fluctuating pairs of density variables

$$\sum_{i}\sum_{\substack{j\neq i \\ j}} T_{zz}(\mathbf{r}_{ij}) = \sum_{k} T_{zz}^{k}\left(\rho_{-k}\rho_{k} - N\right), \tag{23.73}$$

where T_{zz}^k is the Fourier transform of the dipole interaction tensor, and $\rho_k = \sum\limits_{i=1}^{N} \exp(i\mathbf{k} \cdot \mathbf{r}_i)$ is the Fourier transform of the position-dependent density.

Now, the propagator e^{iLt} can be replaced by a propagator acting in the subspace of the pairs of densities

$$\exp(iLt) \to \wp \exp(iLt) \wp, \tag{23.74}$$

where

$$\wp(A) = \lim_{q \to 0} \frac{1}{2N^2} \sum_k \frac{\hat{R}_{k,q-k}}{S(k)S(|q-k|)} \left\langle \hat{R}_{k,q-k}^{\dagger} A \right\rangle. \tag{23.75}$$

Here $\hat{R}_{k,q-k} = \rho_k \rho_{q-k}$, and $S(k)$ is the static structure factor at wave vector k.

Projection of $\sum_i \sum_j^{j \neq i} T_{zz}(\mathbf{r}_{ij})$ onto $\hat{R}_{k,-k}^{\dagger}$ divided by N will be denoted $V_{zz}(\mathbf{k})$ and is called the static vertex. Therefore, for a simple dense liquid, the correlation function can be expressed as

$$C_{zzzz}(t) = \frac{1}{4N^2} \sum_k \sum_{k'} \frac{V_{zz}(\mathbf{k})}{S(k)^2} \frac{V_{zz}(\mathbf{k}')}{S(k')^2} \left\langle \hat{R}_{k,-k}^{\dagger} \hat{R}_{k',-k'}(t) \right\rangle, \tag{23.76}$$

$$\left\langle \hat{R}_{k,-k}^{\dagger} \hat{R}_{k',-k'}(t) \right\rangle \approx N^2 \left(\delta_{k,k'} + \delta_{k,-k'} \right) F(k,t)^2, \tag{23.77}$$

where $F(k,t)$ is the intermediate scattering function defined as

$$F(k,t) = \frac{1}{N} \left\langle \rho_{-k} \rho_k(t) \right\rangle. \tag{23.78}$$

The expression of the vertex is given by,

$$V_{zz}(\mathbf{k}) = \frac{1}{N} \left\langle \sum_i \sum_{\substack{j>i \\ j}} T_{zz}(\mathbf{r}_{ij}) \hat{R}_{k,-k} \right\rangle. \tag{23.79}$$

It has the rotational symmetry

$$V_{zz}(\mathbf{k}) = \left(\frac{3\cos^2(\theta_k) - 1}{2} \right) \tilde{V}_{zz}(k). \tag{23.80}$$

Here $\tilde{V}_{zz}(k)$ is the static vertex for which k is assumed along the z-axis. Now, by performing angular integrals in the expression of $C_{zzzz}(t)$, one gets

$$C_{zzzz}(t) = \frac{N}{20\pi^2 \rho} \int dk\, k^2 \left(\frac{\tilde{V}_{zz}(k)}{S(k)^2} \right)^2 F(k,t)^2. \tag{23.81}$$

Here $F(k,t)$ is determined within the simple Lovesey approximation,

$$F(k,z) = \cfrac{S(k)}{z + \cfrac{\langle \omega_k^2 \rangle}{z + \cfrac{\Delta_k}{z + \left(\cfrac{1}{\tau_k}\right)}}}, \tag{23.82}$$

where $F(k,z)$ is the Laplace transform of $F(k,t)$, $\langle \omega_k^2 \rangle = \left[k_B T k^2 / m S(k) \right]$, $\Delta_k = \omega_L^2(k) - \langle \omega_k^2 \rangle$

$$\omega_L^2 = \frac{3k_B T k^2}{m} + \frac{\rho}{m} \int d\mathbf{r} \, \frac{\partial^2 \phi(r)}{\partial z^2} \left[1 - \exp\left(-ikz\right) \right] g(r) \tag{23.83}$$

and

$$1/\tau_k = 2\sqrt{\Delta_k/\pi}. \tag{23.84}$$

A comparison of Eq. (23.81) with computer simulation is shown in Figure 23.10. This theoretical formalism agrees with computer simulation reasonably well on all time scales. This also suggests that the short-time behavior of the nonresonant Raman response is governed by viscoelastic short-time response of the density modes as the cut-off, q_c does not change the short-time behaviour considerably.

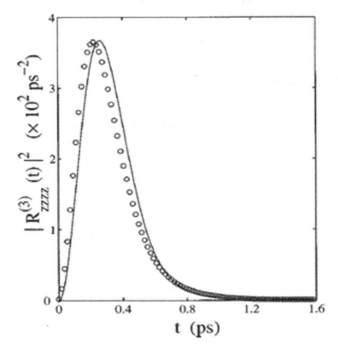

FIGURE 23.10 Plot of $\left| R_{zzzz}^{(3)}(t) \right|^2$ as a function of time for liquid Xe obtained from MCT formalism (solid line) and simulations (open circle). Other parameters are $q_c = 9.05\sigma^{-1}$, $\rho^* = 0.7$ and $T^* = 1.0$.

Source: The plot is taken with permission from [34].

23.15 MCT OF ELECTROCHEMICAL TRANSPORT

MCT has been used successfully to provide comprehensive derivations of the three famous class transport laws of electrochemistry, namely: (i) Debye-Huckel-Onsager limiting law of electrolyte conductance; (ii) Debye-Falkenhagen expression of frequency dependence of conductivity; and (iii) concentration dependence of viscosity expression that Onsager and Fuoss derived. In each case, MCT unerringly leads to the exact expression by using the basic principles as outlined above.

The new insight introduced by MCT in electrolyte transport is that in addition to usual slow variables' density and current, two additional slow variables need to be considered. These are (i) charge density and (ii) charge current density. The first one leads to ion atmosphere relaxation while the second one leads to the electrophoretic effect.

We refer the interested students to the work of Chandra and Bagchi who derived the classical laws from MCT and also showed that MCT provides a superior description.

However, a full mode-coupling theory of electrolyte dynamics needs to consider a three-component liquids mixture: water, cation and anion. Such a theory is yet to be developed.

23.16 MCT IN CHEMICAL REACTION DYNAMICS: BRIDGING THE GAP BETWEEN SHORT- AND LONG-TIME DYNAMICAL RESPONSE

In the Grote-Hynes theory of barrier crossing dynamics, we need values of the frequency-dependent friction on the reactive motion. As the rate is calculated self-consistently, we need the frequency dependence of friction over a wide range of frequencies. This necessitates a theoretical tool that can provide friction. While kinetic theory, like Enskog, provides reliable value at short times, hydrodynamics provide the same at long times. MCT allows us to bridge the gap accurately. The reason is that in many chemical reactions, the barrier frequency that describes the potential energy surface near the barrier is around 10^{13} s^{-1}. In that frequency range, friction derives contributions both from collisions and density relaxation.

The frequency dependence of friction provided by MCT has been used successfully to describe the viscosity dependence of the rate of isomerization at high pressures when the magnitude of the macroscopic viscosity is varied over many orders of magnitude. In fact, the viscosity dependence of rate can be explained only by using MCT friction [25,26], for reasons described above.

23.17 TRANSLATION-ROTATION COUPLING IN DIFFUSION

In many molecular liquids, translational and rotational modes are coupled. Rotational mode is not a hydrodynamic mode and can remain fast in the long wavelength limit. We have already discussed this topic in the chapter on "Rotational Diffusion". This interesting phenomenon has also been addressed by using MCT. Note that Navier-Stokes hydrodynamics cannot discuss the effects of the coupling.

In the MCT approach, coupling between translational and rotational motions enters through the respective frictions or memory functions. The time dependence of the respective memory function derives contributions from both translational and rotational modes. The binary product that we need to consider can be written as $\rho(-q)Y_{lm}(q)$ where Y_{lm} are the usual spherical harmonics with common symbols. Here we assumed molecules of spheroidal symmetry. In the four point correlation that emerges from use of Eq. (28.7) is then decomposed into products of density TCF and orientational TCF. Such expressions of memory function appeared in the monograph of Berne and Pecora, and also in the treatment of Chandra and Bagchi, available in the monograph *"Molecular Relaxation in Liquids"*. In addition to explicit coupling, further coupling enters through a self-consistent calculation.

Although both the density and the rotational relaxation are affected by the translation-rotation coupling, the relative impacts of the coupling are determined by the relative time scales and the shape-mediated interactions between molecules.

23.18   LIMITATIONS OF MODE-COUPLING THEORY

Despite the success of mode-coupling theory, it has been widely criticized for its failures, especially in glass where MCT predicts divergence of viscosity at lower than the experimentally observed temperature. In fact, experiments universally show a steady growth spread over 30–50° C while MCT predicts a transition to a nonergodic state at the start of the growth. There have been multiple attempts to rectify the situation, and some were briefly discussed above.

The main limitation of the MCT in describing glass transition is its failure to include the relaxation channels that indeed involve many-body nonhydrodynamic modes, like hopping. In deeply supercooled liquids when relaxation via usual hydrodynamic channels does not function, relaxation occurs by hopping from one free energy basin to another. An example could be a local melting of a quasi-crystalline region to a more disordered region. The number of particles involved could be of the order of hundreds, even thousands, and the transition could involve only small displacements of the particles. Such events cannot be included in the conventional MCT. However, one can attempt to develop a phenomenological approach to include hopping from another theory (RFOT, for example) inside the MCT framework in a self-consistent fashion.

The orthogonal subset that we project upon need not contain only binary products, but higher-order products too. There is no clear reason to stop only at the binary product when relaxation becomes slow.

A third limitation is a difficulty of implementing a self-consistent scheme because of the lack of knowledge on the relative importance of the different possible binary products. This is a difficulty we face when implementing MCT.

Nevertheless, with the time-dependent density functional theory, MCT provides a powerful approach to describe liquid state dynamics when inert-molecular correlations become important. This theory provides a method to explain, for example, translation-rotation coupling. MCT is also useful in predicting the diffusion coefficient and viscosity in the dense liquid regime away from the glass transition. We already discussed its success in explaining the long-time tail in velocity TCF and critical slowing down.

23.19   SUMMARY

Mode-coupling theory is often associated with glass transition phenomena, but this association is kind of unfortunate as it has cast doubt on the validity of MCT in general. MCT has been quantitatively successful in explaining a diverse range of relaxation phenomena, as discussed above.

In this rather long chapter, we have attempted to explain to students the basic physics behind the mode-coupling theory. We attempted to explain how developments in hydrodynamic theory and kinetic theory led to the self-consistent MCT. We pointed out the role of the extended hydrodynamic theory, which allowed expressions for the hydrodynamic modes and related eigenvalues to the intermediate and large wave numbers. This of course removes only one of the many limitations of the hydrodynamic approach. We discussed three important problems where the hydrodynamic approach was found to break down completely. The first was the power-law tail in the long-time decay of the velocity time correlation function. The second was the erroneous prediction of the width of the Rayleigh peak in the Rayleigh-Brillouin spectrum of a liquid near the gas-liquid critical temperature. The third was the prediction of the dynamic structure near the glass transition. In all three cases, the mode-coupling theory provided satisfactory explanations and made predictions that are in quantitative agreement with all the known results.

A less appreciated aspect of mode-coupling theory is its ability to include microscopic structural features in various relaxation phenomena. This success leads to successful applications of MCT to such areas as the transport laws of electrochemistry, translation-rotation coupling, vibrational spectroscopy, to name a few. We have discussed these applications in this chapter.

As is clear from discussions in this chapter, MCT is a difficult subject to master, both conceptually and technically. But the results are rewarding, especially in our efforts to understand computer simulations in complex systems including biology where these concepts and techniques find useful applications.

BIBLIOGRAPHY

1. Fixman, M. 1962. Viscosity of critical mixtures. *The Journal of Chemical Physics, 36*(2), 310–318.
2. Kawasaki, K. 1970. Kinetic equations and time correlation functions of critical fluctuations. *Annuals of Physics, 61*(1), 1–56.
3. Kadanoff, L. P., and J. Swift. 1968. Transport coefficients near the liquid-gas critical point. *Physical Review, 166*(1), 89.
4. Keyes, T., and I. Oppenheim. 1975. The self-diffusion constant for large heavy particles. *Physica A, 81*(2), 241–248.
5. Alder, B., and T. Wainwright. 1970. Decay of the velocity autocorrelation function. *Physical Review A, 1*(1), 18.
6. Rahman, A. 1964. Correlations in the motion of atoms in liquid argon, *Physical Review A, 136*, 405.
7. Pomeau, Y., and P. Resibois. 1975. Time-dependent correlation functions and mode-mode coupling theories. *Physics Reports, 19*(2), 63–139.
8. Gaskell, T., and S. Miller. 1978. Longitudinal modes, transverse modes and velocity correlations in liquids. I. *Journal of Physics C: Solid State Physics, 11*(18), 3749.
9. Mazenko, G. F. 1973. Fully renormalized kinetic theory. I. Self-diffusion. *Physical Review A, 7*(1), 209.
10. Sjogren, L., and A. Sjolander. 1979. Kinetic theory of self-motion in monatomic liquids. *Journal of Physics C: Solid State Physics, 12*(21), 4369.
11. Kirkpatrick, T. 1985. Short-wavelength collective modes and generalized hydrodynamic equations for hard-sphere particles. *Physical Review A, 32*(5), 3130.
12. De Gennes, P. 1959. Liquid dynamics and inelastic scattering of neutrons. *Physica, 25*, 825–839.
13. De Schepper, I., and E. Cohen. 1982. Very-short-wavelength collective modes in fluids. *Journal of Statistical Physics, 27*(2), 223–281.
14. Alley, W., B. Alder, and S. Yip. 1983. The neutron scattering function for hard spheres. *Physical Review A, 27*(6), 3174.
15. Leutheusser, E. 1984. Dynamical model of the liquid-glass transition. *Physical Review A, 29*(5), 2765.
16. Gotze, W., and F. Liquids. 1991. *The Glass Transition.* ed J. P. Haasma, D. Levesque, and J. Zinn-Justin. Amsterdam: North-Holland.
17. Geszti, T.1983. Pre-vitrification by viscosity feedback. *Journal of Physics C: Solid State Physics, 16*(30), 5805.
18. Farage, T. F. F., and J. M. Brader. 2014. Dynamics and rheology of active glasses. arXiv:1403.0928.
19. Ding, H., M. Feng, H. Jiang, and Z. Hou. 2015. Nonequilibrium glass transition in mixtures of active-passive particles. arXiv:1506.02754.
20. Das, S. P., and G. F. Mazenko. 1986. Fluctuating nonlinear hydrodynamics and the liquid glass transition. *Physical Review A, 34*, 2265–82.
21. Götze, W., and L. Sjögren. 1987.The glass transition singularity. *Zeitschrift für Physik B, 65*, 415–27.
22. Szamel, G. 2003. Colloidal glass transition: beyond mode-coupling theory. *Physical Review Letters, 90*, 228301.
23. Mayer, P., K. Miyazaki, and D. R. Reichman. 2006. Cooperativity beyond caging: generalized mode-coupling theory. *Physical Review Letters. 97*, 095702.
24. Janssen, L. M. C., and D. R. Reichman. 2015 Microscopic dynamics of supercooled liquids from first principles. *Physical Review Letters. 115*, 205701.
25. Biswas, R., and B. Bagchi. 1996. Activated barrier crossing dynamics in slow, viscous liquids. *Journal of Chemical Physics, 105*, 7543.

26. Murarka, R. K., S. Bhattacharyya, and B. Bagchi. 2002. Diffusion of small light particles in a solvent of large massive molecules. *Journal of Chemical Physics, 117*, 10730.

27. Keyes, T., and A. J. Masters. 1985. Tagged-particle motion in dense media: dynamics beyond the Boltzmann equation. *Advances in Chemical Physics, 58*, 1–53.

28. Ertl, H., and F. A. L. Dullien. 1973. Hildebrand's equations for viscosity and diffusivity. *Journal of Physical Chemistry, 77*(25), 3007–3011

29. Bagchi, B., and S. Bhattacharyya. 2001. Mode coupling theory approach to liquid state dynamics. *Advances in Chemical Physics, 116*, 67–222.

30. Chandra, A., and B. Bagchi. 1999. Ion conductance in electrolyte solutions. *Journal of Chemical Physics, 110*, 10024.

31. Chandra, A., and B. Bagchi. 2000. Frequency dependence of ionic conductivity of electrolyte solutions, *Journal of Chemical Physics, 112*, 1876.

32. Chandra, A., and B. Bagchi. 2000. Ionic contribution to the viscosity of dilute electrolyte solutions: Towards a microscopic theory, *Journal of Chemical Physics, 113*, 3226.

33. Chandra, A., and B. Bagchi. 2000. Beyond the classical transport laws of electrochemistry: new microscopic approach to ionic conductance and viscosity, *Journal of Physical Chemistry B, 104, 9067.*

34. Denny, R. A., and D. R. Reichman. 2002. Molecular hydrodynamic theory of nonresonant Raman spectra in liquids: Fifth-order spectra. *Journal of Chemical Physics, 116*, 1979.

35. Banerjee, P., and B. Bagchi. 2017. *Journal of Chemical Physics, 147*, 124502.

24 Irreversible Thermodynamics Revisited

OVERVIEW

Systems far from equilibrium are still far from understood. Bereft of the variational principles of equilibrium statistical mechanics, one is left only with a few guiding principles to understand nonequilibrium thermodynamics. As discussed already in Chapter 4, the quantitative study of irreversible thermodynamics was pioneered by Onsager through his classic paper in 1931. Subsequent developments in this area have been slow. In this chapter, we first present a proof of Onsager's reciprocal relations that makes extensive use of the properties of time correlation function formalism discussed in Chapters 8 to 10. We then discuss two recent developments of importance. The first one provides an estimate of free energy change in transforming the system from one state to another, even when the final state is a free energy minimum. The second development has become well-known as "active matter" where one develops models to explain the emergence of a steady flowing state when the system consists of driven particles. The latter means that each member of the assembly continuously spends energy to execute its own motion. We discuss the seminal work of Vicsek *et al.* who showed how *dynamic interactions* can give rise to an ordered state far from equilibrium. Many aspects of this emerging area are not only beautiful but also of far-reaching consequences, in the study of nonequilibrium phenomena.

24.1 INTRODUCTION

We have discussed elementary concepts of nonequilibrium thermodynamics in Chapter 4. We pointed out that in such systems our theoretical methods are severely limited by the absence of any extremum principle. In equilibrium statistical thermodynamics we use the minimization of free energy. Macroscopic properties are calculated at a free energy minimum. However, the postulates of statistical mechanics, which leads to the principle of free energy minimization, no longer hold when the system is away from equilibrium. We do however have Clausius' statement of the second law that asserts that the entropy of an isolated system increases till the equilibrium is reached where the entropy is maximum. This gave rise to the much-used phrase that *entropy can serve as an arrow of time*. The relation between irreversibility, entropy, and time was further extended by Boltzmann through his famous theorem and *H*-function discussed in Chapter 6. Boltzmann's *H*-theorem has motivated progress in our understanding of irreversible thermodynamics in more than one way. $H(t)$ serves as a relaxation function measuring the relaxation of the momentum space distribution, $f(p, t)$.

Onsager's regression hypothesis and Kubo's linear response theory are both based on the hypothesis that the response of the system can be described by properties at a nearby equilibrium state. At the basis of such an assumption is the expectation that the structure of the mildly displaced

DOI: 10.1201/9781003157601-28

nonequilibrium state remains similar to the equilibrium state. More fundamentally, as already mentioned, our assumptions of time average and ensemble average might not be applicable in systems away from equilibrium.

There are many states we find in nature that are maintained at a steady, time-invariant state but far away from equilibrium. Such a state is maintained by the flow of expenditure of energy, and/or matter, which could be generated internally or supplied from outside. The absence of any working principle for systems maintained far from equilibrium is indeed a large hindrance in our effort to calculate the properties of such systems.

We again start by considering a few questions with examples, as listed below.

(i)   How to determine properties, such as the viscosity, of a liquid flowing in a tube or between two parallel plates? The system is time-invariant because the liquid is flowing at a steady rate. A similar question can be asked when a temperature gradient exists in a liquid confined between two parallel plates. That is, the temperature of one plate is kept fixed at a significantly higher temperature than the other. How do we discuss either the transport properties or even the microscopic structure, such as the radial distribution function of such a liquid? As the system is maintained in a time-invariant state, such questions are valid, but we do not have satisfactory answers. There have been attempts to obtain these properties.

(ii) There have been several studies on the melting/freezing of a colloidal liquid under steady shear, such as in "Couette flow". Here a shear strain is induced at a steady rate by moving the two plates in the opposite directions. Under shear, the liquid confined between the two plates develops a flow where the direction of the velocities at the two surfaces is opposite but equal in magnitude. This is a nonequilibrium system. Here the conditions of phase transition we employ under equilibrium conditions are, strictly speaking, not valid. In fact, this has been a question of interest for some time now: how to describe a nonequilibrium phase transition when concepts of free energy are not available?

(iii) A classic example of nonequilibrium steady state is provided by oscillatory chemical reactions. These reactions have been studied extensively by the Prigogine school in Brussels. The best-known example is the Belousov-Zhabotinskii (BZ) reaction, which involves a complex set of reactions involving bromate ion, bromide ion, and hydroborate. The reactants exhibit oscillations at a certain concentration range but become chaotic outside the range. The study of this transition from oscillatory to chaotic behavior has drawn a lot of attention over the years.

(iv) We often see organized motions of animals, like flock of birds (as shown in Figure 24.1) and fishes, moving coherently. This behavior is found also among other living creatures, like in the movement of insects and bacteria. The notion of active matter is found to be valid in nonliving systems, for example, vibrating grains, self-organizing polymers etc.

Active systems are inherently out of thermal equilibrium where the active particles constantly absorb energy from the surroundings and use that for self-propulsion. What has intrigued people over generations is the emergence of coherent motion and stability in the sense of collective movement with nearly constant velocity in a given direction. Such motion of course has a beginning and an end. The emergence of such a *dynamically ordered state* has been a matter of intense investigation over the years. The seminal contributions toward a quantitative understanding were provided by the Vicsek model. Subsequently, a hydrodynamic picture emerged with the understanding that the advective force acting on a particle in the fluid state due to the motion of neighboring particles can interfere constructively to give rise to a dynamically ordered state as we see in the movement of fishes.

As mentioned in Chapter 4, Onsager's reciprocal relations serve as the starting point of nonequilibrium thermodynamics. We could not discuss its derivation as the present derivation makes extensive use of the time correlation function. In the following, a proof is presented.

FIGURE 24.1   An example of an active matter system: organization of a flock of birds.

Source: Released into the public domain by the photographer of this image, Christoffer A. Rasmussen.

24.2   ONSAGER'S RECIPROCAL RELATIONS

We discussed the statement of Onsager's reciprocal relations in Chapter 4. However, we could not derive this relation in that chapter as our proof makes use of several properties of time correlation functions. In the following, we accomplish this unfinished task. We start reminding the reader of the reciprocal relation

$$\mathbf{J} = \mathbf{L}.\mathbf{X} \tag{24.1}$$

where $\mathbf{J}$ is a row vector containing the fluxes of a set of n dynamical variables $y_1,\ y_2,\ y_3,...,y_n$. These functions $\{y\}$ are macroscopic variables that we shall use also to denote fluctuations. $\mathbf{X}$ is the external force vector driving the flux and time dependence of the set $\{y\}$. Thus, $\mathbf{X}$ is conjugate to $\{y\}$. For example, if $\{y\}$ is a set of concentrations, then $\mathbf{X}$ is the concentration gradient that drives the flow of $\{y\}$. L is a $n \times n$ matrix that gives the rate of change. Onsager's reciprocal relations state that $L_{ij} = L_{ji}$. We now proceed to prove this remarkable theorem. The proof is quite simple, although highly original and rich in ideas and concepts.

We start with the following definition of the time derivative of time correlation involving x_i and x_j:

$$\frac{1}{\tau}[\langle y_i(t+\tau)y_j(t)\rangle - \langle y_i(t)y_j(t)\rangle]$$
$$= \langle \dot{y}_i(t)y_j(t)\rangle \tag{24.2}$$

We next use time translation and time symmetry of the time correlation functions (see Chapter 8) to obtain the following identities:

$$\left\langle y_i(t+\tau) y_j(t) \right\rangle = \left\langle y_i(t-\tau) y_j(t) \right\rangle = \left\langle y_i(t) y_j(t+\tau) \right\rangle. \tag{24.3}$$

It now follows that

$$
\begin{aligned}
\left\langle \dot{y}_i(t) y_j(t) \right\rangle &= \frac{1}{\tau}\left[\left\langle y_i(t+\tau) y_j(t) \right\rangle - \left\langle y_i(t) y_j(t) \right\rangle \right] \\
&= \frac{1}{\tau}\left[\left\langle y_i(t) y_j(t+\tau) \right\rangle - \left\langle y_i(t) y_j(t) \right\rangle \right] \\
&= \left\langle y_i(t) \dot{y}_j(t) \right\rangle.
\end{aligned}
\tag{24.4}
$$

In the next steps, we use the macroscopic relation for the time evolution of the microscopic variables to obtain the following two relations:

$$\left\langle \dot{y}_i(t) y_j(t) \right\rangle = \sum_k \left\langle L_{ik} X_k(t) y_j(t) \right\rangle = \sum_k L_{ik} \left\langle X_k(t) y_j(t) \right\rangle = L_{ij}. \tag{24.5}$$

The last relation follows from the identity (to be discussed below) $\left\langle X_k(t) y_j(t) \right\rangle = \delta_{kj}$. Similarly,

$$\left\langle y_i(t) \dot{y}_j(t) \right\rangle = \sum_k \left\langle y_i(t) L_{jk} X_k \right\rangle = \sum_k L_{jk} \left\langle X_k y_i(t) \right\rangle = L_{ji}. \tag{24.6}$$

This completes our proof.

However, we have made use of the relation that $\left\langle y_i X_k \right\rangle = \delta_{ik}$, where δ_{ik} is a Kronecker delta function. This also requires proof. And this latter proof is quite involved. Onsager accomplished this proof by using the entropy fluctuation formalism introduced by Einstein. However, the main ideas are simple. Let us consider two forces X_1 and X_2 giving rise to fluxes J_1 and J_2. Let us assume further that the force X_2 is zero. Then $J_2 = L_{21} y_1$. Since $\langle J_2 X_2 \rangle$ is zero by symmetry (different time symmetry), we have $\langle y_2 X_1 \rangle = 0$. This completes our proof.

In the following, we present a further analysis. Boltzmann's entropy formula gives for the probability distribution function, $w = A \exp(S/k)$ where A = constant. This result follows as the probability of a given set of fluctuations $y_1, y_2, ..., y_n$ is proportional to the number of microstates that are available to execute that fluctuation. Assuming the fluctuations are small, the probability distribution function can be expressed through the second differential of the entropy.

$$
\begin{aligned}
P &= A e^{-\frac{1}{2}\beta_{ik} y_i y_k} \\
\beta_{ik} &= \beta_{ki} = -\frac{1}{k_B} \frac{\partial^2 S}{\partial y_i \partial y_k}
\end{aligned}
\tag{24.7}
$$

where we are using Einstein summation convention and β_{ik} is a positive definite symmetric matrix. We note that the coefficients β_{ik} can be regarded as force constants. We next assume that the system is only slightly disturbed from equilibrium so that we can write the following rate equation for the decay of the fluctuation x_i as a relaxation equation:

$$\dot{y}_i = -\lambda_{ik} y_k. \tag{24.8}$$

Thus λ_{ik} are the damping constants. Next, we define a thermodynamic force that is conjugate to a dynamical variable y_i in terms of the derivative of entropy

$$X_i = -\frac{1}{k_B}\frac{\partial S}{\partial y_i}. \tag{24.9}$$

We can regard X_i as the thermodynamic (or, system) force that gives rise to the fluctuation x_i. For small fluctuations, we can write down the following linear functions relationship:

$$X_i = \beta_{ik}y_k. \tag{24.10}$$

Thus, we can write

$$\dot{y}_i = -L_{ik}y_k \tag{24.11}$$

where $L_{ik} = \lambda_{il}\beta_{lk}^{-1}$ are called kinetic coefficients. The principle of symmetry of kinetic coefficients or Onsager's principle states that L is a symmetric matrix, that is $L_{ik} = L_{ki}$.

An important point to be stressed here is that we treated $\{y_i\}$ as fluctuations out of equilibrium so that we can use the harmonic assumption. In current or heat flow, we consider steady states. Thus, $\{y_i\}$ and $\{X_j\}$ are both nonzero. In such a situation, the fluctuations we discussed above are to be regarded as the fluctuations out of this nonequilibrium steady state. Onsager's treatment assumes that the relaxation of these fluctuations obeys the same law as in equilibrium. This is clearly an approximation.

24.3 RELATIONSHIP BETWEEN FREE ENERGY CHANGE AND WORK VIA FLUCTUATION THEOREMS

In the late 1990s, certain remarkable theorems were proved by Jarzynski and Crooks that allowed one (in principle) to obtain the free energy difference between two thermodynamic states by using an average over irreversible work. This work potentially opens the door towards using computer simulations to obtain a free energy gap between any two states, which is needed more than the absolute values of the free energy of the two states.

In general, however, the calculation of this difference is difficult because the thermodynamic relations involve work done by a reversible path (which is not practical in computer applications). The works of Jarzynski and Crooks allow one to circumvent this difficulty by establishing *a relationship between the work done over an ensemble of paths from the initial state (A) to the final state (B) – usually, both are sampled from the respective equilibrium ensemble.* Thus the usual reversible path approach is replaced by a set of paths that can be regarded as fluctuations around the "would be" reversible path in the conversion of the initial state to the final state. Along these fluctuation paths, the conversion is done at a finite rate. The latter freedom (from infinitesimal reversible path construction to change at a finite rate) allows us to implement such changes through computer simulations. The proof however seems to require the construction of the path by small changes, which are Markovian and obey microscopic reversibility in phase space. The proof of the theorem is not complicated but also not trivial. We direct the reader to the clear derivation provided by Crooks in 1998, cited in the Bibliography.

We know that the free energy difference between two thermodynamic states can be related to a combination of work done and heat released, by the first two laws of thermodynamics. The latter is related to entropy. As discussed above, we need to extend the traditional approach. Such an extension has become possible now because of the use of computer simulation techniques. Two such extensions, known as the Jarzynski equality (JE) and Crooks' fluctuation theorem (CFT), have been proposed and implemented recently. In a certain sense, they are quite remarkable because they offer exact relations between free energy change and the work done in a change carried out at a finite speed. In the Jarzynski equality (JE) the goal is to calculate the free energy difference (ΔF) between two thermodynamic states, A and B. Here the free energy difference between the two states is expressed in terms of the average of the exponential of the irreversible work done along an ensemble of trajectories that goes from one state (A) to the other (B). We first provide the final expression of JE in Eq. (24.12)

$$\boxed{\exp\left(-\beta\Delta F\right) = \overline{\exp\left(-\beta W\right)}} \tag{24.12}$$

where $\Delta F = F_B - F_A$, is the free energy difference between the final (B) and initial (A) states, W denotes the work done along a trajectory, and β is $\left(k_B T\right)^{-1}$. The overhead bar on the right-hand side of Eq. (24.12) denotes an average over W that is obtained for an ensemble of paths/trajectories joining A and B. Thus, with the help of JE, one can extract equilibrium information (ΔF) from an ensemble of nonequilibrium (finite-time) measurements. In a finite-step transformation from $A \rightarrow B$, $\overline{W} \geq \Delta F$. An important point to note is that the left-hand side of Eq. (32.12) is the ratio of two partition functions

$$\frac{Q_N(B,T)}{Q_N(A,T)} = \overline{\exp\left(-\beta W\right)}. \tag{24.13}$$

It is clear from the above description that one would need to make a transition from state A to state B. This can be carried out by the familiar Kirkwood charging process, discussed in our equilibrium statistical mechanics textbook (2018, CRC Press). We next outline the proof of Eq. (24.12).

The notation followed in the subsequent discussion is as follows. A and B denote two thermodynamic states that are free energy minima in the phase space of the system. "z" denotes a point in the phase space and z_0 denotes an arbitrary point in the basin A that denotes the start of the transformation at time $t = 0$ to a point z_τ in state B. $z(t)$ denotes a trajectory of the system in the phase space. The transformation takes time duration τ. One next assumes a control parameter λ that changes from λ_A (usually taken to be zero) to λ_B (unity) as the system moves from an initial to the final state. We imagine λ to be time-dependent. Thus, we can control the rate of transformation of A to B. Thus, time t goes from zero to τ as λ goes from zero to unity as the system goes from A to B.

Let $H_S(z,t)$ be the Hamiltonian that depends on time, with the initial state distribution of the system is given by,

$$\rho_{\lambda_A}(z_0) = \frac{e^{-\beta H_S(z_0)}}{Q_A} \tag{24.14}$$

Q_A being the initial partition function defined by

$$Q_A = \int dz_0 e^{-\beta H_S(z_0)}. \tag{24.15}$$

We already mentioned as z denotes symbolically a position (point) in the phase space of the system. As the system is made to evolve from state A to state B, it follows a trajectory denoted by $z(t)$ in the phase space of the system. The probability of a trajectory $z(t)$ in phase space is given by $P[z(t)]$. Since trajectory outlines a path, $P[z(t)]$ is a path probability.

Next, a time-reversed process is defined, which is described by the parameter $\lambda(\tau - t)$ acting on the system that is initially at equilibrium with the final value of the forward protocol. The reverse trajectory corresponding to the forward trajectory $z(t)$ is defined as $\tilde{z}(t) = z^*(\tau - t)$

Let us consider the following thought experiment. We consider a piston on a cylinder of gas. At the initial time ($t = 0$) the system at rest has a volume V_A and at a phase point z_0 (at $t = 0$). We then increase the pressure such that the system goes to volume V_B at time t to phase point z_F. While V_0 and V_B are macro variables, the phase points z_0 and z_F are microscopic. Thus, each such experiment starts with a different z_0 at $t = 0$ and ends at time $t = \tau$ at z_F. Each experiment is defined by a trajectory in the phase space. The reverse trajectory is defined in the same fashion. The only constraint here is that both z_0 and z_τ are sampled from equilibrium ensembles that define states A and B. We define the reverse trajectory by z_r where the process is repeated with different starting and ending phase points.

The initial state distribution of the system for the reverse process is then given by

$$\tilde{\rho}_{\lambda_B}(z_\tau) = \frac{e^{-\beta H_s(z_\tau^*)}}{Q_B}. \tag{24.16}$$

The probability of reverse trajectory in the reverse process is given by $\tilde{P}[\tilde{z}(t)]$. Note that this general notation is a bit different from the notation we used in the thought experiment, but should not be confusing.

If we take the ratio between $P[z(t)]$ and $\tilde{P}[\tilde{z}(t)]$, then we get,

$$\frac{P[z(t)]}{\tilde{P}[\tilde{z}(t)]} = \frac{\rho_{\lambda_A}(z_0)}{\tilde{\rho}_{\lambda_B}(z_\tau)} = e^{\beta[H_s(z_B) - H_s(z_A) - \Delta F]}. \tag{24.17}$$

This has an interesting, combined character, which appears because of the phase space positions given by z. It is important to note that probabilities in Eq. (32.16) are conditional probabilities – the trajectories start and end in two definite thermodynamic states. This condition gives rise to the ΔF term in Eq. (32.16). ΔF is free energy change of the system: $\Delta F = k_B T \ln \frac{Q_A}{Q_B}$, as mentioned above.

The total work W done on the system must be equal to the change in the Hamiltonian: $W = H_S(z_\tau) - H_S(z_0)$, we can write

$$\frac{P[z(t)]}{\tilde{P}[\tilde{z}(t)]} = e^{\beta[W - \Delta F]}. \tag{24.18}$$

Crooks' fluctuation theorem directly follows

$$\begin{aligned} P(W) &= \int D[z(t)]\delta(W - W[z(t)])P[z(t)] \\ &= \int D[z(t)]\delta(W - W[z(t)])\tilde{P}[\tilde{z}(t)]e^{\beta[W - \Delta F]} \\ &= e^{\beta[W - \Delta F]} \int D[\tilde{z}(t)]\delta(W + W[\tilde{z}(t)])\tilde{P}[\tilde{z}(t)] \\ &= e^{\beta[W - \Delta F]}\tilde{P}(-W). \end{aligned} \tag{24.19}$$

Thus, we have a remarkable relation

$$\boxed{\frac{P(W)}{\tilde{P}(-W)} = e^{\beta[W - \Delta F]}.} \tag{24.20}$$

We next multiply both sides of the above relation by $e^{-\beta W}$ and integrate over W and use normalization condition for $\tilde{P}(-W)$ to obtain

$$\boxed{\left\langle e^{-\beta W} \right\rangle = e^{-\beta \Delta F}}. \tag{24.21}$$

The Jarzynski equality can lead to an analog of the maximum work theorem as follows. We start with the inequality

$$\left\langle e^{r} \right\rangle \geq e^{\langle r \rangle} \tag{24.22}$$

when we use this in Eq. (24.21) we find that

$$e^{-\beta \Delta F} \geq e^{-\beta \langle W \rangle} \Rightarrow \left\langle W \right\rangle \geq \Delta F. \tag{24.23}$$

Multiplying both sides by -1 we have $-\left\langle W \right\rangle \leq -\Delta F$, which is an analog of the statement of the maximum work theorem.

The difference between $W_{A \to B}$ and ΔF denotes the work dissipated due to the forward process. The dissipated work becomes zero when both the forward and backward transformations are performed at an infinitely slow pace. Crooks' theorem is applicable to the following assumptions:

(i)   The system is deterministic and satisfies the time-reversal symmetry. This is ensured by Newtonian mechanics. Crooks pointed out that the transformation described above must also be Markovian.

(ii) The reverse trajectory is done following a reversed time schedule such that $W_{A \to B} = -W_{B \to A}$. This condition can be achieved in steered molecular dynamics simulations.

As mentioned before, the proof employs the notion of conditional probability and we elaborate on this a bit more. We start with the left-hand side of Eq. (32.20) and write the following equation:

$$\begin{aligned}
\frac{P_{A \to B}}{P_{B \to A}} &= \frac{e^{-\beta H_A}}{Z_A} \cdot \frac{Z_B}{e^{-\beta H_B}} \\
&= e^{-\beta\left(H_A - H_B\right)} e^{-\Delta F} = e^{\beta\left(W_{A \to B} - \Delta F\right)}.
\end{aligned} \tag{24.24}$$

Here one needs to use the fact that $\Delta F = -\ln\left(Z_B / Z_A\right)$ and $H_B - H_A = W_{A \to B}$ for an adiabatic trajectory.

These theorems relating work done over an ensemble of path to free energy has found considerable use in recent years. However, convergence still poses a serious problem.

24.4   ACTIVE MATTER: EMERGENCE OF SELF-ORGANIZATION

We discussed in the early part of the chapter that understanding the emergence of self-organization in the animal world, like the flock of birds (see Figure 24.1) has been a subject of intense interest over many years. Some notable progress has been made and these are discussed below. The modern era started with the Vicsek model.

The Vicsek model is a phenomenological and minimalistic model introduced by Vicsek and coworkers in 1995 to explain the emergence of self-ordered motion of active or driven particles. However, they did not use the term "active matter" and rather identified the constituents of such systems as *particles with biologically motivated interactions*. Vicsek *et al.* carried out the calculations in two dimensions that can easily be extended into three dimensions.

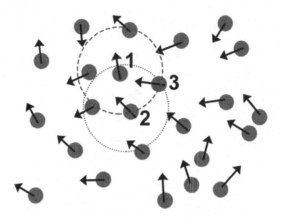

FIGURE 24.2 Schematic representation of the Vicsek model where the "*boids*" [bird like objects] are represented by gray circles and the instantaneous velocities are shown by black arrows. Each particle have a "*circle of influence*" around it, shown by circles with dashed (for particle #1) and dotted (for particle #2) peripheries. According to Eq. (24.26), the average direction of the particles within the circle of influence controls the future direction of the particle in concern. In a hydrodynamic system, the influence enters through the advective term. As it is a dynamical system, the neighbors change with time. Some particles (like particle #3) can influence multiple particles.

We first provide the equations that govern the dynamics of the system studied. The position of the ith particle at time $(t + \Delta t)$ is given by Eq. (24.25). Here (x_i, y_i) are the coordinates of the ith particle and (v_i^x, v_i^y) are the respective velocity components.

$$\begin{pmatrix} x_i(t+\Delta t) \\ y_i(t+\Delta t) \end{pmatrix} = \begin{pmatrix} x_i(t) \\ y_i(t) \end{pmatrix} + \Delta t \begin{pmatrix} v_i^x(t) \\ v_i^y(t) \end{pmatrix}. \tag{24.25}$$

However, there is an additional constraint. *The magnitude of velocity, $v = \sqrt{\left(v_i^x\right)^2 + \left(v_i^y\right)^2}$, is kept the same for all the particles and only the direction of the velocity (θ) is allowed to change* according to the following equation:

$$\boxed{\theta_i(t+\Delta t) = \langle \theta(t) \rangle_R + \Delta \theta}. \tag{24.26}$$

Here $\langle \theta(t) \rangle_R$ is the average orientation of particles around the ith particle within a circle (or a sphere in 3D) of radius R and $\Delta\theta$ is a random noise term that is chosen from a uniform probability distribution between $-\eta/2$ to $+\eta/2$. The circle of radius R is called the *circle of influence*. We provide a schematic diagram to explain the model in Figure 24.2.

As the system is out of thermal equilibrium, there is no concept of temperature. Nevertheless, the magnitude of η (intrinsic noise) is related to the magnitude of thermal fluctuations in the system. This is often considered as a *temperature like variable*.

By simulating Eq. (24.25) and (24.26) one can see a new kind of phase transition. The system makes a transition from a disordered state (at $t = 0$) to a dynamically ordered state after some time. In Figure 24.3, we show the snapshots from the Vicsek model simulations for systems with various densities and thermal noise. It is found that in a moderately dense system with a high value of noise, extremely weak dynamical ordering develops. However, for low number densities and a low value of noise one can see local ordering. The collective flocking develops when one looks into high-density systems with a low value of noise.

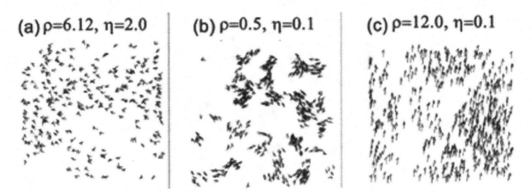

FIGURE 24.3 Snapshots of three different dynamical systems that obey the Vicsek model. (a) System with moderate density (ρ) and high value of noise (η) shows weak dynamical ordering. (b) System with low density and a low value of noise shows region wise dynamical ordering. (c) System with high density and low noise shows collective flocking behavior.

Source: Taken from [8].

   As we know, in order to study any kind of phase transition we need to define an order parameter. Here the most suitable order parameter is the absolute value of the average normalized velocity [Eq. (24.27)].

$$v_a = \frac{1}{Nv}\left|\sum_{i=1}^{N}\mathbf{v}_i\right|.$$ (24.27)

v_a shows a sharp transition with increasing η, which denotes the transition from a dynamical ordered phase to a random chaotic phase. If one assumes that in the thermodynamic limit, the Vicsek model exhibits a kinetic phase transition analogous to the continuous phase transition in equilibrium systems, the one can write Eq. (24.28) where η_c is the crucial noise and β is the exponent, which is found to be ~0.45.

$$v_a \sim \left[\eta - \eta_c\right]^{\beta}.$$ (24.28)

   A similar behavior of v_a can be observed with respect to the number density of the system. The Vicsek model is a historically important and simple model that can successfully describe the flocking or herd behavior. Next, we focus on some of the rigorous theoretical analyses, namely, the hydrodynamic model and the field theory approach.

24.5 HYDRODYNAMIC CONSIDERATION

The hydrodynamic approach provides a semi-quantitative description of the emergence and stability of directional flow in self-propelled objects. These theories serve to provide a macroscopic explanation of the results obtained in the Vicsek model. *At the core of these theories is the hydrodynamic interaction between moving objects described as the advective term. This advective term is a sum of forces propagated by the Oseen tensor due to hydrodynamic flows of surrounding objects.* When all the particles in the system are self-propelled the direction is determined by its orientation (a body-fixed axis*). The dependence of the advective force on the velocity of all the other particles is the reason for the emergence of the coherent motion, as further illustrated below.*

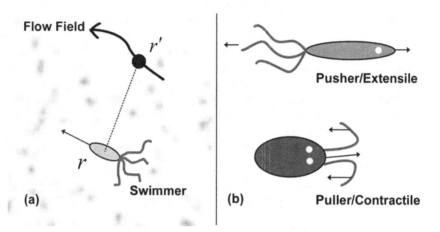

FIGURE 24.4 (a) Schematic illustration of the origin of the advective term in the Navier-Stokes equation. A swimmer a position r creates a flow field at r'. (a) Two different kinds of active swimmers, namely, pusher/extensile and puller/contractile swimmers.

The advective term arises because a swimmer at **r** creates a hydrodynamic flow at **r'** with a certain lag time [Figure 24.4(a)]. The range of influence and the lag time would depend on the viscosity of the medium. We note that, the self-propelling particles do not possess monopole moment, as the mutual forces of the swimmer get canceled by the force exerted by the fluid medium. Hence, one needs to consider a self-propelled particle with a permanent force dipole with strength W. The value of W is positive for extensile swimmers (pushers) and negative for contractile swimmers (pullers) [Figure 24. 4(b)].

Let the velocity of the ith particle at position r_i and time t be $v_i\left(r_i, t\right)$. Let the self-propelled force acting on the ith particle is F_i. There are additional forces on the ith particle from the other j^{th} particle. In a dissipative environment, we can write the velocity vector as

$$\mathbf{v}_i\left(r_i, t\right) = \left[\frac{1}{\zeta_i}\mathbf{F}_i\left(t\right) + \sum_{j \neq i} A_{ij}\right] \tag{24.29}$$

where ζ_i is the friction on the ith particle and A_{ij} is given by the Oseen tensor (Happel and Brenner)

$$A_{ij} = \frac{F_j}{8\pi\eta r_{ij}}\left[\mathbf{n}_i.\mathbf{n}_j + \left(\mathbf{n}_i.\mathbf{r}_{ij}\right)\left(\mathbf{n}_i.\mathbf{r}_{ij}\right)\right] \tag{24.30}$$

where $\mathbf{n}_i$ is the vector in the direction of the flow of particle i. In Eq. (24.30) the second term is the advective force term, which corresponds to the additive or the cooperative term in the Vicsek model. We can now decompose the velocity of the ith particle in parallel and perpendicular components to the driven force direction,

$$v_{\parallel}^i = \frac{F_i}{\zeta_i} + \sum_{j \neq i}\frac{F_j}{8\pi\eta r_{ij}}\left\{\mathbf{n}_i.\mathbf{n}_j + \left(\mathbf{n}_i.\mathbf{r}_{ij}\right)\left(\mathbf{r}_{ij}.\mathbf{n}_j\right)\right\} \tag{24.31}$$

$$v_{\perp}^i = \sum\frac{F_j}{8\pi\eta r_{ij}}\left\{\mathbf{n}_i \times \mathbf{n}_j + \left(\mathbf{n}_i \times \mathbf{r}_{ij}\right)\left(\mathbf{r}_{ij}.\mathbf{n}_j\right)\right\}. \tag{24.32}$$

Here ζ_i is the friction on the ith particle. There is a simple and clear physical interpretation within hydrodynamics of the emergence of a coherent motion arising from hydrodynamic interactions between velocities. Due to the hydrodynamic interactions, the motion in the particular direction, call it the parallel direction that comes from the active force term, gets reinforced, but the same does not happen in the perpendicular direction. Moreover, the motion in the parallel direction can get further reinforced beyond a certain concentration. This hydrodynamic picture is similar to the elegant proof of the emergence of coherent collective motion provided by Ramaswamy and coworkers. The latter authors demonstrated that the coherent motion could emerge in the long wavelength limit, which means that all the active particles (that is, a large number) interact with each other. This is also similar to the Vicsek model.

24.6   HYDRODYNAMIC FIELD THEORY APPROACH

Tu and Toner identified the density field (ρ) and the velocity field ($\mathbf{v}$) as the slow variables in the Vicsek model and developed a coarse-grained dynamical description for the active matter. Therefore, one can write the following equation:

$$\frac{\partial}{\partial t}\mathbf{v} + \lambda\left(\mathbf{v}.\nabla\right)\mathbf{v} = \left(\alpha - \beta\mathbf{v}.\mathbf{v}\right)\mathbf{v} + \Gamma\nabla\nabla\mathbf{v} - \nabla P(\rho) + f. \tag{24.33}$$

Here α and β are two control parameters, f is a Gaussian random noise, Γ is a tensor that controls the elastic restoring forces and $P(\rho)$ is a function of the number density that obeys the equation of continuity.

$$\frac{\partial}{\partial t}\rho + \nabla.\rho\mathbf{v} = 0. \tag{24.34}$$

The Navier-Stokes like λ-term in Eq. (24.33) says that the distortions in the velocity field, $\mathbf{v}$, are advected by $\mathbf{v}$ itself, as already discussed in the last section. The nonequilibrium character arises from this advective term in Eq. (24.33) and the current term in Eq. (24.34). The right-hand side of Eq. (24.33) can also be written as the sum of the derivative of a free-energy functional ($F[\mathbf{v}]$) and the noise. $F[\mathbf{v}]$ is given by Eq. (24.35) [*cf.* Ginzburg-Landau theory]

$$F[\mathbf{v}] = \int d^d r \left[-\frac{\alpha}{2}\mathbf{v}.\mathbf{v} + \frac{\beta}{4}\left(\mathbf{v}.\mathbf{v}\right)^2 + \frac{\Gamma}{2}\nabla\mathbf{v}\nabla\mathbf{v} + \mathbf{v}.\nabla P(\rho)\right]. \tag{24.35}$$

According to Tu and Toner, the dynamics of a flock is governed by the above three equations, that is, Eqs. (24.33), (24.34), and (24.35). The isotropic (or disordered) state is characterized by $\mathbf{v} = 0$ and an ordered flock by $|\langle\mathbf{v}\rangle| \approx \sqrt{\alpha/\beta}$ if the gradients are negligible in Eq. (24.33). If α is tuned from a negative to a high positive value, one can observe a spontaneous breaking of the rotational symmetry.

24.7   ACTIVE BROWNIAN PARTICLES WITH INTERMOLECULAR
INTERACTIONS

Intermolecular interactions can influence dynamics of active matter and has been studied theoretically. In the Vicsek model this feature has been taken partly into account by placing the particles on a lattice. In a continuous system, in the presence of liquid-like disorder, intermolecular interactions like collisions can influence dynamical properties, such as the emergence of coherence.

Let us consider N active self-propelling Brownian particles/swimmers in overdamped condition with positions $\{\mathbf{r}_i\}$ and orientation $\{\theta_i\}$. The latter is defined by the instantaneous direction of the self-propulsion force. One can write a Langevin equation in $\mathbf{r}$ and θ for the ith particle as follows:

$$m\frac{\partial^2 \mathbf{r}_i}{\partial t^2} = -\zeta_{tr}\frac{\partial \mathbf{r}_i}{\partial t} + \mathbf{F}\left(\mathbf{r}^{(N)},\boldsymbol{\theta}^{(N)}\right) + \delta\mathbf{F}_{tr}(t) \tag{24.36}$$

$$I\frac{\partial^2 \theta_i}{\partial t^2} = -\zeta_{rot}\frac{\partial \theta_i}{\partial t} + \boldsymbol{\Gamma}\left(\mathbf{r}^{(N)},\boldsymbol{\theta}^{(N)}\right) + \delta\mathbf{F}_{rot}(t) \tag{24.37}$$

Here m is the mass of a particle, I is the moment of inertia, ζ_{tr} and ζ_{rot} are respectively the translational and rotational friction, $\mathbf{F}$ and $\boldsymbol{\Gamma}$ respectively denote forces and torques, other than the drag (such as *intermolecular interactions*), and $\delta\mathbf{F}$ is the random noise. In the overdamped limit, the left-hand sides of Eqs. (24.36) and (24.37) can be ignored. Therefore, one can write the equations of motion as

$$\frac{\partial \mathbf{r}_i}{\partial t} = v_o\mathbf{f}\left(\theta_i\right) + \beta D_{tr}\mathbf{F}^{ext}\left(\mathbf{r}_i,\theta_i,t\right) + \beta D_{tr}\sum_j \mathbf{F}_{ij}\left(\mathbf{r}_{ij},t\right) + \sqrt{2D_{tr}}\,\eta_i(t) \tag{24.38}$$

$$\frac{\partial \theta_i}{\partial t} = \beta D_{rot}\text{“}^{ext}\left(\mathbf{r}_i,\theta_i,t\right) + \beta D_{rot}\sum_j \text{“}_{ij}\left(\mathbf{r}_{ij},\theta_i,\theta_j,t\right) + \sqrt{2D_{rot}}\,\xi_i(t). \tag{24.39}$$

Here v_0 is the speed of the particle and $v_0\mathbf{f}(\theta_i)$ is the translational self-propulsion force (also known as *motor force*) F^{ext} is the external forced and Γ^{ext} is the external torque. β is the inverse of $(k_B T)$. In this formalism, intermolecular interactions enter through $\mathbf{F}_{ij}$ and $\boldsymbol{\Gamma}_{ij}$ terms. $\eta_i(t)$ and $\xi_i(t)$ are the Brownian noise terms that satisfy two properties – (a) the first moment is zero [Eq. (24.40)], and (b) they are delta-correlated in time [Eq. (24.41)].

$$\langle\eta_i(t)\rangle = 0 = \langle\xi_i(t)\rangle \tag{24.40}$$

$$\begin{aligned}\langle\eta_i(t)\eta_j(t')\rangle &= 2\zeta_{tr}k_B T\delta_{ij}\delta(t-t')\\ \langle\xi_i(t)\xi_j(t')\rangle &= 2\zeta_{rot}k_B T\delta_{ij}\delta(t-t').\end{aligned} \tag{24.41}$$

One can simulate Eqs. (24.38) and (24.39) to obtain a trajectory of the system in terms of $\mathbf{r}$ and θ. This trajectory contains all the influence of the interactions that can reproduce the emergence of coherent behavior. We do not pursue this topic any further but refer to [12] to [14].

24.8 SUMMARY

The field of irreversible thermodynamics is still under development. It is a difficult area whose inception lies in the work of Clausius and Boltzmann. Onsager's work is restricted because of the assumptions that the nonequilibrium system must be close to an equilibrium state. This is because of the assumption of microscopic reversibility.

While progress has been made in the calculation of the free energy change through several fluctuation theorems, little progress has been made in understanding the stability and the decay of fluctuations of a steady state far from equilibrium. The latter is an important problem in the study of stability of biological systems. The emergence of active matter as a subfield has motivated progress. In particular, the Vicsek model captures a model of dynamic cooperativity, which is both appealing and general.

BIBLIOGRAPHY AND SUGGESTED READING

1. Boltzmann, L. 1872. Weitere studien über das wärmegleichgewicht unter gasmolekülen, *Sitzungsberichte Akademie der Wissenschaften*, *66*, 275–370. [English translation: Boltzmann, L. 2003. Further studies on the thermal equilibrium of gas molecules. *History of Modern Physical Sciences*, *1*, 262–349.]
2. Onsager, L. 1931. Reciprocal relations in irreversible processes. I, *Phys. Rev.*, *37*(no. 4), 405.
3. Onsager, L. 1931. Reciprocal relations in irreversible processes. II, *Phys. Rev.*, *38*(no. 12), 2265.
4. Groot, S. R. De., and P. Mazur. 2013, *Non-Equilibrium Thermodynamics*. Chelmsford: Courier Corporation.
5. Jarzynski, C. 1997. Nonequilibrium equality for free energy differences. *Physical Review Letters*, *78*(no. 14), 2690.
6. Crooks, G. E., 1998. Nonequilibrium measurements of free energy differences for microscopically reversible Markovian systems. *J. Statistical Physics*, *90*(nos. 5–6), 1481.
7. Crooks, G. E. 1999, Entropy production fluctuation theorem and the nonequilibrium work relation for free energy differences. *Physical Review E*, *60*(no. 3), 2721.
8. Vicsek, T., A. Czirók, E. Ben-Jacob, I. Cohen, and O. Shochet. 1995. Novel type of phase transition in a system of self-driven particles. *Physical Review Letters*, *75*(no. 6), 1226.
9. Toner, J., and Y. Tu, 1995. Long-range order in a two-dimensional dynamical XY model: how birds fly together. *Physical Review Letters*, *75*(no. 23), 4326.
10. Marchetti, M. C., J. F. Joanny, J. F., S. Ramaswamy, T. B. Liverpool., J. Prost, J., M. Rao, and R. A. Simha, 2013. Hydrodynamics of soft active matter. *Reviews of Modern Physics*, *85*(no. 3), 1143.
11. Ramaswamy, S. 2010. The mechanics and statistics of active matter. *Annual Reviews of Condensed Matter Physics*, *1*(no. 1), 323–345.
12. Del Junco, C., L. Tociu., and S. Vaikuntanathan. 2018. Energy dissipation and fluctuations in a driven liquid. *Proceedings of the National Academy of Sciences*, *115*(no. 14), 3569–3574.
13. Shaebani, M. R., A. Wysocki, R. G. Winkler, G. Gompper, and H. Rieger. 2020. Computational models for active matter. *Nature Reviews Physics*, *2*(no. 4), 181–199.
14. D'Orsogna, M. R., Y. L. Chuang, A. L. Bertozzi, and L. S. Chayes. 2006. Self-propelled particles with soft-core interactions: patterns, stability, and collapse. *Physical Review Letters*, *96*(10), 104302.

25 Rate of Rare Events

OVERVIEW

As discussed in Chapter 20, traditional methods for calculation of the rate of a chemical reaction require two ingredients: (i) a quantitative knowledge of the free energy barrier, and (ii) a method to obtain the rate of transit. The traditional computational approaches are found inapplicable to complex chemical or physical transformations (like protein folding or protein association) because the free energy barrier is just not accessible by standard methods of MD simulations due to a large separation of time scales between the fast molecular motions (typically of the order of *fs* or *ps*) and the slow rate of the reaction. This large separation prevents us from accessing the free energy barrier. Such slow processes are collectively termed as "rare events". An additional complexity faced in the development of a reduced description is that often the reaction energy surface is multidimensional but the selection or the choice of coordinates is not obvious. For example, it might involve both rotation and translation (like in some cis-trans isomerization reactions). Another example is protein dimer dissociation (as in insulin dimer) that involves both the monomer-monomer distance separation and the number of native contacts, as two order parameters. In the latter case in particular, there could be a multitude of alternative paths. This challenge has led to the development of new techniques that involve clever use of probabilistic techniques in computer simulations. There are now several such techniques: umbrella sampling, metadynamics, transition matrix Monte Carlo, to name a few. These techniques approach the problem in different ways, and each prove to be effective in a class of problems. In this chapter, we provide an initial understanding for the students of the relevant concepts and techniques, with a view from a reaction dynamics perspective.

25.1 INTRODUCTION

It is a real wonder and puzzle that many molecular motions that occur on very short time scales, like femtoseconds or so, and that appear chaotic, and but these movements ultimately coordinate to give rise to slow, well-defined macroscopically observable processes that occur on the time scale of seconds or minutes and hours. This is due to the fact that at the microscopic level, the behavior of individual particles is governed by the laws of mechanics, which can be highly complex and difficult to understand. However, at the macroscopic level, these individual motions can be averaged out using laws of probability and statistics, resulting in more predictable and well-defined behavior.

We can describe the driving force for microscopic changes in terms of the free energy differences. Linear response theory (LRT) provides us with a theoretical method that allows us to relate the

DOI: 10.1201/9781003157601-29

rate of change, like transport properties, to the natural molecular motions, through time correlation functions. Thus, LRT helps us to formulate theories of reaction rates, as discussed in Chapters 11 and 20. However, in complex processes, like protein folding or protein association-dissociation reactions, we cannot easily apply LRT in a straightforward fashion because the reaction path is not known. We do not know the reaction free energy surface *a priori*, and a large unknown barrier in a complex energy surface separates the reactant from the product.

Many of the processes involved in natural changes are called rare processes as these are slow on the time scale of molecular motions. Thus, on the time scale of molecular motion, a system appears to reside in the reactant state, and it is hard to understand the rare transitions happening on a time scale 10 or 12 orders of magnitudes longer than molecular motions, from studying those much faster motions. The conventional simulation techniques, like molecular dynamics simulations, can access time scales to a maximum of milliseconds, that too with state-of-the-art computational facilities. Hence, the events which take place in time scales much greater than these, become inaccessible to such simulation techniques. These are known as *rare events*.

Let us consider a realistic example. In an aqueous solution of a protein, the various relaxation events involved are the movement of the protein backbone, side-chain fluctuations, dynamics of the counter-ions that neutralize the protein, fluctuations in hydration water and the bulk, to name a few. The time scales of these events can spread over a wide range: from femtoseconds/picoseconds (water vibrational motion, hydrogen bond relaxation etc.) to hundreds of ps. Nevertheless, these same motions together lead to protein folding, which can occur on a time scale of milliseconds, even seconds. Direct simulations fail to capture these slow rare processes.

In physical chemistry/chemical physics courses/discussions on chemical kinetics, one routinely uses a quantity called reaction coordinate (RC). We have discussed examples in Chapter 20. RC is used to understand the pathway and the mechanism of the reaction and is needed to formulate a theory. For example, in barrierless chemical reactions, the lack of temperature dependence and strong viscosity dependence on the isomerization of TPM dyes (like malachite green) gave clues about the RC and then the free energy surface. Rare events that we are interested are often more complex. Most of the complex processes require multiple reaction coordinates, which are termed either collective variables or order parameters – the last terminology is borrowed from phase transition literature.

A reactive process broadly involves the following steps. The process starts in a reactant well, goes over an energy barrier to yield the product. A schematic free energy profile for an imaginary chemical reaction is shown in Figure 25.1.

While we have discussed such cases in Chapter 20, we did not discuss any theoretical method to calculate the free energy barrier. In some specific cases, quantum chemical calculation might prove useful. But for complex chemical and biological processes in solution quantum calculations are not feasible. Thus, one needed to evolve new simulation methods to sample the barrier region. There are two strategies that are employed. First, one can "pull" the system out of the reactant minimum, step-by-step towards the barrier. This is achieved by adding an appropriate term to the Hamiltonian. A second method is to "push" the system away from the minimum by superimposing a distribution with a maximum at the reactant minimum where the equilibrium distribution by itself is a maximum. The former (the method of "pull") is called umbrella sampling (US) while the latter (the method of "push") is called metadynamics (MetaD). In both the cases, the effect of the extra terms can be removed exactly by using methods of statistical mechanics. We show below a cartoon (Figure 25.2) that brings out the essence of the two techniques.

Although US and MetaD are the two most popular methods, there are several other techniques to evaluate the free energy surface across the barrier.

In the following we discuss some of these techniques, along with relevant concepts.

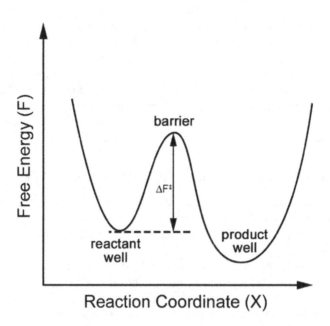

FIGURE 25.1 Schematic free energy (F) profile of a model chemical reaction. The total landscape involves a reactant well and a product well, which are separated by a free energy barrier of $\Delta F^{\ddagger}$. The reaction will only take place if the system is supplied with enough energy that is greater than the barrier. X is the reaction coordinate that determines the progress of the process.

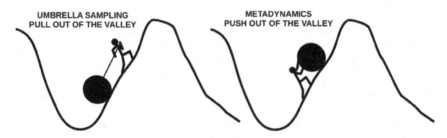

FIGURE 25.2 An illustration by cartoon (for beginners) to demonstrate the difference between umbrella sampling and metadynamics enhanced sampling techniques.

25.2 THE RATE OF A RARE PROCESS

The rate (k) of any chemical reaction is primarily determined by this free energy barrier. For example, from the transition state theory, we get the relation between reaction rate and free energy barrier given elegantly by the Eyring equation [Eq. 25.1].

$$k = \frac{k_B T}{h} \exp\left(-\frac{\Delta G^{\ddagger}}{k_B T} \right)$$

(25.1)

Here, k_B is the Boltzmann constant, h is Planck's constant, and T is the temperature of the system. Use of this expression just needs the activation barrier. Detailed discussion on activated barrier crossing and chemical reaction dynamics is given in Chapter 20.

The above simple expression of Eyring is inadequate in complex situations that we have in mind here. Most importantly, we need to consider multidimensional energy surfaces. In Chapter 20 we discussed multidimensional rate theories, such as that of Langer. In this context we just want to point out *that a calculation of rate requires not just the barrier height but also the frequencies at the reactant well and also at the barrier top.* In addition, we need the friction. The magnitude of the friction on the reactive mode and also on the nonreactive modes could be important in rate determination. However, these frictions are hard to determine. An elaborate procedure has been developed to determine the frictions, and we discussed a few of those in Chapter 20.

25.3   COMPUTATION OF REACTION FREE ENERGY SURFACE

With this basic theory in mind, let us get back to the discussion of *rare events*. As mentioned before, these are ultraslow processes with long timescales. This implies that the free energy barrier involved in these processes is very large. To put this into perspective let us look into some approximate numbers given in Table 25.1.

Table 25.1 demonstrates that an increase of 5 kcal mol^{-1} in the free energy barrier of a process results in an increase in the time scale by several (more than 3) orders of magnitude. Clearly, the rates of these reactions are very small. Naturally, traditional computer simulations are unable to capture these rare events. Hence, we need advanced sampling techniques to navigate the barrier region. In such techniques, the Hamiltonian of the system is intentionally altered so that it can sample greater extents of the phase space. But before entering the discussion of such techniques, we need to clarify a few basic concepts of use here.

25.3.1   Selection of Collective Variables (or Order Parameters or Reaction Coordinates)

Description of the change from the reactant to the product is quantified by a set of parameters. These parameters are often termed order parameters or collective variables. In chemical kinetics, it is known as a reaction coordinate. In the study of phase transition, they are called order parameters. In the study of rare events, they are termed collective variables.

The choice of collective variables in a complex reaction is not always obvious and might require certain trial and error.

From equilibrium statistical mechanics, we know that the free energy (A) of a system in an NVT (canonical) ensemble is given by the logarithm of the canonical partition function (Q)

$$A = -k_B T \ln Q = -k_B T \ln \int d\mathbf{r}^N \exp\left[-\frac{U(\mathbf{r}^N)}{k_B T} \right].$$

(25.2)

TABLE 25.1
Approximate Time Scales of Events for Given Free Energy Barriers

Barrier (kcal mol^{-1})	Time scale (T=300K)
5	< 1 ns
10	> 1 μs
15	> 10 ms
20	~ 1 min
25	~ 3 days
30	~ 40 years

TABLE 25.2
Examples of Some Possible Collective Variables/Order Parameters for Some Well-Studied Processes

Process	Collective variable(s)
Bond dissociation in a chemical reaction. $A\text{-}B + C \rightarrow A + B\text{-}C$	1. Distance between A and B atoms. 2. Distance between B and C atoms.
Nucleation	Diameter of the nucleus.
Water to ice phase transition	1. Tetrahedral order parameter that depends on the orientation of nearest-neighbor water molecules with respect to each other. 2. The number of hydrogen bonds. 3. Energy per water molecule.
Vapor-liquid phase transition	The density of the system
Protein folding/unfolding	1. Radius of gyration of the protein. 2. Solvent accessible surface area.
Protein dimer to monomer dissociation	1. Distance between monomer centers of mass. 2. The number of native contacts between the monomers.
Conformational changes in small polypeptides	Backbone dihedral angles (Ramachandran plot).

Now, the partition function Q spans a $3N$-dimensional space occupied by N particles. Here we have ignored the kinetic energy term from momentum, which is (in classical systems) simply a multiplicative factor in Q and an additive term is free energy, which is the ideal gas contribution.

Practically, it is impossible to calculate the $3N$-dimensional integral for a system with a large number of interacting particles (with a potential U). Fortunately, we have a way around this. A chemical/physical/biological process can, in practice, be expressed by only a few generalized coordinates of the system, rather than all the $3N$ degrees of freedom. This reduction in the number of degrees of freedom is not only beneficial from a computational point of view but is also a more convenient way to monitor the progress of a reaction. This set of generalized coordinates is often referred to as *collective variables* or *order parameters* or *reaction coordinates* depending on the system or process under investigation. In Figure 25.1 the reaction coordinate of the model reaction is given by X.

Let us look into a few examples of such collective variables with respect to some well-known processes as given in Table 25.2.

From Table 25.2 it becomes clear that the choice of collective variable to monitor a process is completely dependent on the nature of the process. These serve as sufficient degrees of freedom that almost completely define the process. Hence, we may rewrite Eq. (25.2) as

$$A(X) = -k_B T \ln \int d\mathbf{r}^N \delta\left(f(\mathbf{r}^N) - X\right) \exp\left[-\beta U\left(\mathbf{r}^N\right)\right] \qquad (25.3)$$

where the free energy A is now a function of the reaction coordinate X, and the function $f(\mathbf{r}^N)$ describes the dependence of the reaction coordinate X on the position coordinates of all the particles. However, this equation is hard to visualize for all practical purposes, which leads us to express the free energy in simpler terms.

We know that the entropy (S) of a system is given by the logarithm of the number of microstate arrangements (Ω) possible in the system with respect to the collective variable X.

$$S = k_B \ln \Omega. \qquad (25.4)$$

Hence, we get

$$\Omega = \exp\left(\frac{S}{k_B}\right). \tag{25.5}$$

Now, the probability that the system is in a given state with respect to X is given by the product of Ω and the Boltzmann factor. Therefore, we have

$$
\begin{aligned}
P(X) &= \Omega \exp\left(-\frac{E(X)}{k_B T}\right) \\
&= \exp\left(\frac{S}{k_B}\right)\exp\left(-\frac{E(X)}{k_B T}\right) \\
&= \exp\left(-\frac{E(X)-TS}{k_B T}\right) \\
&= \exp\left(-\frac{A(X)}{k_B T}\right).
\end{aligned}
\tag{25.6}
$$

Hence, we finally obtain

$$\boxed{A(X) = -k_B T \ln P(X)}. \tag{25.7}$$

The relation in Eq. (25.7) is illustrated in Figure 25.3. Therefore, the free energy (also known as the *potential of mean force*) is given by the logarithm of the probability with respect to the considered collective variable. Now, from Figure 25.1 and Eq. . for calculating the rate of a reaction, or studying the differences between the product and the reaction states, we are more interested in the free energy difference, rather than its absolute values in the different states. Hence, if we can simply obtain the probability distribution of the collective variable X, we can get the underlying free energy profile from Eq. (25.7).

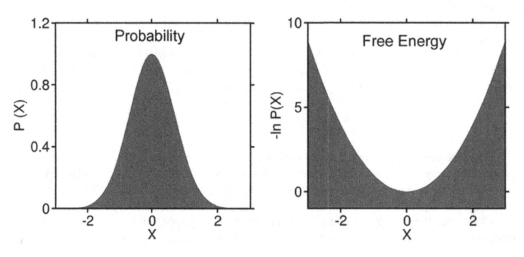

FIGURE 25.3 The probability distribution $P(X)$ (left) and the corresponding free energy profile (in $k_B T$ unit) (right) for a representative order parameter X.

25.3.2 THERMODYNAMIC INTEGRATION (TI) TO FREE ENERGY

As discussed before, we are interested in the free energy difference between two states, rather than their absolute free energies, for all practical purposes. From experiments and simulations, one can easily derive the values of free energy derivatives (like pressure, specific heat, etc.). However, obtaining free energy difference is nontrivial. One way to compute this from computer simulations (molecular dynamics or Monte Carlo) is the *thermodynamic integration* method.

Let us consider a system with two states A and B, with potential energies U_A and U_B respectively. The total energy of the system is given by Eq. (25.8).

$$U\left(\mathbf{r}^N,\lambda\right)=\left(1-\lambda\right)U_A\left(\mathbf{r}^N\right)+\lambda U_B\left(\mathbf{r}^N\right).$$

(25.8)

Here λ is a parameter (known as Kirkwood's coupling parameter) that reversibly connects the two states A and B. The total potential energy is a continuous function of λ. Hence, the partition function of the system follows suit.

$$Q\left(\lambda\right)=\frac{1}{\Lambda^{3N}N!}\int d\mathbf{r}^N e^{-\beta U(\lambda)}.$$

(25.9)

Now, the Helmholtz free energy in a canonical ensemble is given by Eq. (25.10)

$$A\left(\lambda\right)=-k_B T \ln Q\left(\lambda\right).$$

(25.10)

Considering free energy to be a continuous function of λ, we can express the free energy difference between state A ($\lambda = 0$) and state B ($\lambda = 1$) as

$$\begin{aligned}
\Delta A &= \int_0^1 \partial\lambda \frac{\partial A\left(\lambda\right)}{\partial\lambda} \\
&= -k_B T \int_0^1 \partial\lambda \frac{\partial \ln Q\left(\lambda\right)}{\partial\lambda} \\
&= -k_B T \int_0^1 \partial\lambda \frac{1}{Q\left(\lambda\right)} \frac{\partial Q\left(\lambda\right)}{\partial\lambda}.
\end{aligned}$$

(25.11)

Hence, following Eq. (25.9), we can write

$$\frac{\partial A}{\partial\lambda}=\int d\mathbf{r}^N \frac{\partial U(\mathbf{r}^N,\lambda)}{\partial\lambda} \frac{e^{-U(\mathbf{r}^N,\lambda)/k_B T}}{Q(\lambda)},$$

(25.12)

which can be translated in the following form [Eq. (25.13)] to get the free energy difference between the two states:

$$\boxed{\Delta A = \int_0^1 d\lambda \left\langle \frac{\partial U(\mathbf{r}^N,\lambda)}{\partial\lambda} \right\rangle_\lambda.}$$

(25.13)

$\langle...\rangle$ denotes an ensemble average. In an NPT ensemble, one can derive the Gibbs free energy difference in a similar way.

25.3.3 Free Energy Perturbation (FEP) Method

Kirkwood introduced a powerful technique to calculate properties of a real system in terms of a reference system. This method is known as "charging". The philosophy is simple and as follows. We know the detail properties of the reference system (for example, a system of ideal gas, or hard spheres). If we now tune the reference system continuously and gradually, we would be able to calculate the incremental change incurred in the property that we are interested in, by the method of perturbation. *The important point here is a slow gradual change towards the final interaction potential can allow us to access the properties of the final state when the charging is complete, because we can calculate properties of the intermediate states by successive charging.*

We have discussed this topic in great detail in our *Equilibrium Statistical Mechanics* text book, cited below.

This method was developed into a powerful scheme (when combined with modern day computers) by Zwanzig in the early 1950s, for computation of the free energy difference between two states. This is now known generally as the free energy perturbation technique. Let us consider an NVT system, such that the free energy difference between the two states A and B is given by

$$
\begin{aligned}
\Delta A &= A_B - A_A \\
&= -k_B T \ln \frac{Q_B}{Q_A} \\
&= -k_B T \ln \frac{\int d\mathbf{r}^N \exp\left[-\beta U_B\left(\mathbf{r}^N\right)\right]}{\int d\mathbf{r}^N \exp\left[-\beta U_A\left(\mathbf{r}^N\right)\right]} \\
&= -k_B T \ln \frac{\int d\mathbf{r}^N \exp\left[-\beta U_B\left(\mathbf{r}^N\right)\right]\exp\left[-\beta U_A\left(\mathbf{r}^N\right)\right]}{\int d\mathbf{r}^N \exp\left[-\beta U_A\left(\mathbf{r}^N\right)\right]}
\end{aligned}
\tag{25.14}
$$

which gives us

$$
\Delta A = -k_B T \ln \left\langle \exp\left(-\beta \Delta U\right)\right\rangle_{U_A}
\tag{25.15}
$$

The ensemble average $\left\langle \exp\left(-\beta \Delta U\right)\right\rangle_{U_A}$ is computed over the configurations generated in the potential energy surface of state A. Although the fundamental philosophies between the thermodynamic integration (TI) and the FEP techniques are the same, the latter is more elegant and exact. In FEP, it is sufficient to take the ensemble average over only one state, irrespective of the potential energy of the other state.

25.3.4 Umbrella Sampling

Umbrella sampling is an advanced simulation technique that allows a system to enter and sample those regions in the phase space which are not accessible from unbiased simulations. This technique was developed in the late 1970s by Torrie and Valleau and has been extensively used since then.

As discussed before, rare events are associated with a high free energy barrier that localizes the system in an energy valley, which may not be the global minimum. This results in an inefficient

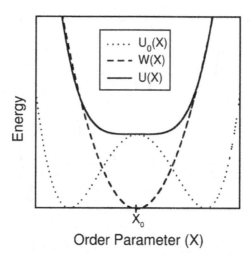

FIGURE 25.4 A schematic representation of the effect of bias potential [$W(X)$] on the original potential energy profile [$U_0(X)$]. The latter is a bistable potential given by the dotted line. On addition of the bias potential (dashed line) the resultant potential [$U(X)$] profile assumes the form given by the solid line.

sampling of the phase space culminating to a nonergodicity in the system. Hence, the obtained free energy surface is erroneous. Thus, if we modify the Hamiltonian of the system, such that it naturally comes out of the trapped state, we will be able to obtain better sampling of the phase space.

Let us consider that the original energy of the system is $U_0(X)$ with a high energy barrier at $X = X_0$. If we add a bias of the form $W(X) = k(X - X_0)^2$, then the new biased potential of the system becomes $U(X)$ as in Eq. (25.16). The bias may not necessarily be harmonic, we, however, continue here with a harmonic constrain for simplicity.

$$U(X) = U_0(X) + k(X - X_0)^2. \tag{25.16}$$

Redefined $U(X)$ has a minimum at $X = X_0$. Consequently, the system is now forced to sample the barrier region, which was inaccessible previously. Performing simulations with this Hamiltonian, we are hence able to gather sufficient statistics of this region. Figure 25.4 shows the effects of biasing potential on the starting original potential energy surface.

After sufficient sampling is done, it is necessary to recover the probability distribution for the unbiased trajectory, which is described by the unperturbed potential energy $U_0(X)$. This can be done in the following way. The average of an observable A in the unbiased trajectory is given by

$$\langle A \rangle_{U_0} = \frac{\int d\mathbf{r}\, A(\mathbf{r}) \exp\left(-\beta U_0(\mathbf{r})\right)}{\int d\mathbf{r} \exp\left(-\beta U_0(\mathbf{r})\right)}. \tag{25.17}$$

This is obviously an average with respect to the unbiased potential ($U_0(\mathbf{r})$), which can now be expressed in terms of the biased potential as follows:

$$U_0(X) = -U(X) + k(X - X_0)^2. \tag{25.18}$$

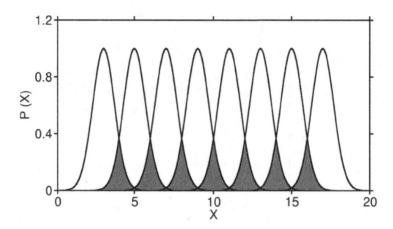

FIGURE 25.5 Probability distributions of the order parameter X at each umbrella. Note that to obtain a correct free energy landscape, there must be significant overlap between these distributions (shaded regions).

Hence, Eq. (25.17) can be rewritten as,

$$
\begin{aligned}
\langle A \rangle_{U_0} &= \frac{\int d\mathbf{r}\, A(\mathbf{r}) \exp(\beta W) \exp(-\beta U(\mathbf{r}))}{\int d\mathbf{r}\, \exp(\beta W) \exp(-\beta U(\mathbf{r}))} \\
&= \frac{\langle A \exp(\beta W) \rangle_U}{\langle \exp(\beta W) \rangle_U}.
\end{aligned}
\tag{25.19}
$$

Hence, we represent the mean observable with respect to the unbiased Hamiltonian in terms of the biased trajectory. Hence, by efficiently modifying the bias, we can sample the desired region of the phase space. Here, we have used an umbrella like potential to bias the simulation in a particular region; hence the name.

There are two crucial points that one needs to remember in umbrella sampling. Firstly, it is necessary to set up several umbrella potentials to completely sample the total stretch of the collective variable. Secondly, the umbrellas should be taken such that the probability distributions thus obtained should have a considerable amount of overlap as shown in Figure 25.5. The overlap between the probability distributions is also affected by the harmonic constant K in the bias potential $W(X)$. A very large value of K would lead to very narrow distribution as the harmonic constraint would be very strong. These narrow distributions would also lead to poor overlap. On the other hand, if we set the constant to a very low value, the system might not be able to exit the valley as the restrain would not be strong enough to pull the system out the metastable basin. Thus, one needs to be careful about choosing the window centers and the harmonic spring constant K. Good overlap in the distributions assures that the new distribution is not too far from the earlier, preceding, one. This helps in minimizing the cumulative error from the individual simulations.

Following this method, we get PMFs for each window, which are generated from individual simulations with different bias potentials. Hence, the obvious next task is to stitch these PMFs together to obtain the total free energy landscape. This is accomplished by the weighted histogram analysis method (WHAM). A detailed discussion of this technique is out of the scope of this book. Interested readers are directed to references.

A common collective variable used in umbrella sampling is the distance between two entities. For example, protein-ligand dissociation, ion-pair separation, solute-solvent interaction etc. However, several other order parameters can also be used, depending on the system of interest. Umbrella sampling can also be easily extended for a multiparameter description of the free energy surface.

Sampling techniques are actively being developed and are the active field of research. The major point of concern is the form of the bias potential $W(X)$. There have been successful attempts to develop adaptive techniques, where the bias is automatically generated during the sampling process. The most popular among these techniques is metadynamics, which we discuss in the next section.

25.3.5 METADYNAMICS

The biased sampling scheme discussed above uses an "attractive localizing" bias potential to constraint the system in the desired part of the phase space. But there is another class of biased sampling technique, where a "repulsive" bias potential is added to force the system out of the local minima. Metadynamics is one of the most popular techniques in this class. Here the bias potential is history-dependent, and has the following form:

$$W(X,t) = w\sum_{t_i \leq t}\exp\left(-\frac{\left(X - X(t_i)\right)^2}{2\sigma^2}\right)$$

$$= w\left[\exp\left(-\frac{\left(X - X(t_0)\right)^2}{2\sigma^2}\right) + \exp\left(-\frac{\left(X - X(t_1)\right)^2}{2\sigma^2}\right) + \ldots\ldots\ldots\right] \qquad (25.20)$$

where $W(X, t)$ is the history-dependent (hence, time-dependent) bias potential, which is given by a sequential sum of a Gaussian function of the reaction coordinate X. The parameters w and σ denote the height and width of the Gaussian bias function, respectively.

Let us consider that at the beginning of the simulation trajectory ($t = t_0$) the system may start at a local minimum ($X = X(t_0)$). At this point the bias potential will have a single term: $w\exp\left(-\frac{\left(X - X(t_0)\right)^2}{2\left(\delta X\right)^2}\right)$.

Thus, the bias potential is added at that particular value of the reaction coordinate where the system resides at that instant. Since this is a repulsive potential, the system does not find itself in a minimum any longer, so it relaxes to the next available local minimum. So if we wait long enough before adding the next Gaussian, it will be added to the next available minimum of the current effective energy landscape. As time evolves, and more such Gaussians are added, the free energy minimum is filled up, and the system escapes to the next available minimum. You might find this approach analogous to the story of the intelligent crow that was dropping pebbles in a water jar in order to bring the water to the top of the jar. Basically we keep adding Gaussian bias potentials wherever the system gets too comfortable to push it out of its comfort zone and we keep pushing it to explore newer regions of the phase space. Figure 25.6 illustrates the resultant time evolution owing to the addition of a one dimensional model potential.

The following three parameters determine W:

(i)   The Gaussian height w.
(ii) The Gaussian width σ.
(iii) The frequency at which the Gaussians are added.

These parameters influence the accuracy and efficiency of the free energy calculation. If the Gaussian potentials are too large or added too quickly, the free energy surface would get explored in too short a time while the reconstructed profile could be affected by large errors. Instead, if the Gaussians

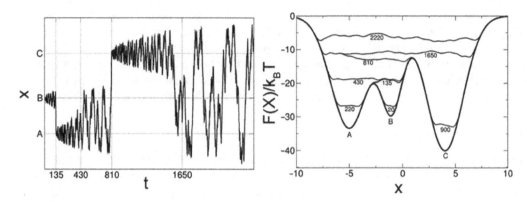

FIGURE 25.6 The time evolution of the sum of a one-dimensional model potential and the accumulating Gaussian terms of Eq. (25.20). The dynamic evolution (thin lines) is labeled by the number of terms added so far. The starting potential (thick line) has three minima and the dynamics is initiated in the second minimum.

Source: Taken with permission from [11].

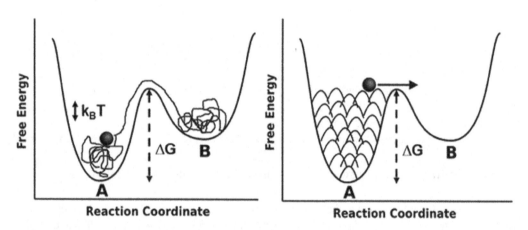

FIGURE 25.7 A schematic illustration of the metadynamics technique to demonstrate how the accumulation of the bias functions fills up one valley (local minimum) of the free energy surface so that it "spills over" to the next available minimum.

are small enough, and added slow enough so that system gets sufficient time to relax to the next available minimum, the reconstructed free energy surface would be accurate, but it would certainly require longer time.

For example, in the schematic diagram given in Figure 25.7, the system starts at the minimum in the middle. After adding 20 small Gaussians it completely fills up the second minimum and then spills over to the first minimum. Again after adding 160 Gaussians, the system finds the global minimum (third). After all the minima are filled up (320 Gaussians), the whole effective underlying energy surface becomes flat. At this point, the original free energy surface is nothing but the negative sum of all the Gaussians added so far. A point of caveat is that we must allow enough time between adding two Gaussians so that the system can adjust itself to the new effective energy landscape and reach the new minimum. Otherwise, we might end up adding too many Gaussians at the same spot of the reaction coordinate space leading to an incorrect free energy surface.

The strength of this method over umbrella sampling lies in the fact that one does not need to know the locations to add the bias potentials *a priori*. The system learns to escape the minima and discover new ones. Thus, in the context of biomolecular systems with complex energy landscapes, this method can help us discover new structural entities, which can be metastable in nature. Also, one is not limited to using only one reaction coordinate. The above treatment can be directly extended to multiple reaction coordinates, and we can obtain a multidimensional free energy surface as well.

25.3.6 REPLICA EXCHANGE MOLECULAR DYNAMICS

The techniques like umbrella sampling or metadynamics are routinely used to sample the unfavorable regions of the phase space or to accelerate barrier crossing rare events. But a practical challenge associated with these methods lies in the identification of suitable reaction coordinates or order parameters. Any practitioner in this field of research needs to rely on chemical intuition to come up with a limited set of physical parameters that can faithfully capture the key events. But often this task is not trivial *a priori* (before running the simulations and understanding the nature of the transitions). Another popular method called replica exchange molecular dynamics (REMD) somewhat alleviates this problem as discussed below.

Instead of biasing the sampling in a specific lower-dimensional space, we may use temperature as the driving force to enhance sampling of the higher energy states. When one encounters a multidimensional energy landscape with several deep minima, the system tends to get stuck in some of these minima as discussed before. But if the same system is simulated at a higher temperature, it would now sample higher energy regions of the landscape, and hopefully, cross the barriers of the order of $\sim k_B T$. Subsequently, the temperature can be brought down and the system relaxes to another local/global minimum. This technique is called simulated annealing and this is a nonequilibrium simulation protocol to search for newer local minima or global minimum.

REMD is an equilibrium and parallel variant of the simulated annealing technique (also called parallel tempering), where multiple replicas of the same system are simulated at different temperatures (normally ranging from ambient temperature to much higher values), where each replica exists in its own thermal equilibrium. As the simulation of each replica progresses, configurations between two different replicas can be exchanged with a certain probability such that the Boltzmann distribution is preserved. Suppose we attempt to exchange a configuration with energy U_1 from a replica with temperature T_1 and a configuration with energy U_2 from a replica with temperature T_2. From the Boltzmann distribution, it can be easily shown that the probability of the configuration exchange: (U_1 at T_1 and U_2 at T_2) to (U_2 at T_1 and U_1 at T_2) is given by

$$\frac{\exp\left[-\left(U_2/k_B T_1\right)-\left(U_1/k_B T_2\right)\right]}{\exp\left[-\left(U_1/k_B T_1\right)-\left(U_2/k_B T_2\right)\right]} = \exp\left[\left(U_1 - U_2\right)\left(1/k_B T_1 - 1/k_B T_2\right)\right]. \qquad (25.21)$$

A schematic description of the exchanges on an energy landscape is shown in Figure 25.8.

The above protocol maintains the thermodynamic equilibrium at each replica while the exchanges with higher (and lower) temperature replica bring in configurations from other distant regions of the phase space. Thus, the system may sample configurations from two different local minima simultaneously at a given temperature. The strength of this method lies in the fact that we did not have to resort to any specific reaction coordinate or order parameter to bias the simulation *a priori*. The REMD protocol itself ensures acceleration of barrier crossing and enhanced sampling. We can post-process the simulation trajectory, and extract PMF along any suitable reaction coordinate of choice by using Eq. (33.7) (Boltzmann inversion of the probability distribution) after the REMD simulations are completed.

Of course, there are practical problems that at very high temperatures the system might undergo substantial undesired changes such as the system would sample irrelevant regions of the phase space

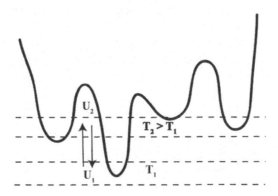

FIGURE 25.8 A schematic representation of replica exchange molecular dynamics simulation. The dashed lines indicate the typical ranges sampled in respective temperature/replica. The arrows indicate the exchange attempt between two configurations from the different replicas.

(e.g., protein unfolding instead of specific local conformational transitions). Moreover, rare events related to the entropic barriers (as discussed in the next section) cannot be overcome by temperature-assisted enhanced sampling methods.

25.4 SUMMARY

The well-established, time-honored theoretical methods to obtain rate of a transformation fail to provide results in many complex situations, not only because that the free energy surface is not available, which is indeed a pre-requisite, but also, even when the surface is available, the calculation of the rate is not always possible. The methods presented here strive to solve the first problem, the evaluation of the free energy surface. The second half of the problem, that is the evaluation of the rate in a complex free energy landscape, still remains partly unsolved, and has been discussed in Chapter 20. In a certain philosophical sense, the techniques evolved to treat the rare events that symbolize man's quest to conquer uncharted territories. The techniques developed have a certain "brute force" character in them in the sense that one drives the system to or pulls the system into an unfavorable part of the free energy surface otherwise not visited in the course of the natural motion of the system. The basic principles are fairly simple but the implementation requires considerable skill, which can be developed by practice. These techniques have now become necessary and routine tools of nonequilibrium statistical mechanics. The theories of reaction dynamics also remained useful because once the reaction free energy surface becomes available, we call upon these theories to obtain the reaction rates. In fact, the availability of multidimensional reaction free energy surface has provided an impetus to the study of reaction dynamics on multidimensional reaction energy surfaces, with new challenges. Nevertheless, the primary goal of the techniques described here is to obtain an accurate estimate of the free energy barrier.

All these techniques have their individual strength and weaknesses, and sometimes one needs to combine the two techniques. For example, metadynamics could face difficulty when the free energy surface is flat, while umbrella sampling could find difficulty with multiple barriers separated by minima.

An important issue is the separation of the computed free energy surface into entropy and enthalpy components. Such a separation would be useful in rate calculation. However, this advanced topic is beyond the scope of this chapter.

BIBLIOGRAPHY

1. Zwanzig, R. W. 1954. High-Temperature Equation of State by a Perturbation Method. I. Nonpolar Gases, *Journal of Chemical Physics, 22*, 1420–26.
2. Sugitaa, Y., and Y. Okamotoab. 1999. Replica-exchange molecular dynamics method for protein folding, *Chemical Physics Letters, 314*, 141–51.
3. Zwanzig, R. W. 1983. Effective diffusion coefficient for a Brownian particle in a two-dimensional periodic channel, *Physica A, 117*, 277–80.
4. Machta, J., and R. W. Zwanzig. 1983. Diffusion in a periodic Lorentz gas, *Physical Review Letter, 50*, 1959–62.
5. Frenkel, D., and B. Smit. 2002. *Understanding Molecular Simulation: From Algorithms to Applications.* California: Academic Press.
6. Allen, M. P., and D. J. Tildesley. 1987. *Computer Simulation of Liquids.* Oxford: Oxford Univ. Press.
7. Torrie, G. M., and J. P. Valleau. 1977. Nonphysical sampling distributions in Monte Carlo free-energy estimation: umbrella sampling, *Computational Physics, 23*, 187–89.
8. Ng, K.-C., J. P. Valleau, G. M. Torrie, and G. N. Patey. 1979. Liquid-vapour co-existence of dipolar hard spheres, *Molecular Physics, 28*, 781–88.
9. Laio, A., and F. L. Gervasio. 2008. Metadynamics: a method to simulate rare events and reconstruct the free energy in biophysics, chemistry and material science, *Reports on Progress in Physics, 71*, 126601.
10. Laio, A., and M. Parrinello. 2002. Escaping free-energy minima, *Proceeding of the National Academy of Sciences USA*, 99, 12562–566.
11. Barducci, A., M. Bonomi, and M. Parrinello. 2011. Metadynamics. *Wiley Interdisciplinary Reviews: Computational Molecular Science, 1*, 826–43.

Epilogue

The first and foremost motivation for writing this book was to expose students to the main concepts of time-dependent statistical mechanics and relaxation phenomena while at the same time providing them with the necessary tools. I found that chemistry students in particular are not exposed to, and often find it difficult to absorb, such concepts as time-dependent distribution, linear response, and correlation function. It is frustrating, and even sad, that the beautiful (although formidable) work of Boltzmann lies largely neglected these days, although so much of our understanding evolved from this work.

As research in this area becomes increasingly specialized and dependent on computer simulations, the understanding of the basics has become more important than ever before. For example, such theories like linear response, and the concepts that evolve from them, form the basis of many of our more advanced applications. Similarly, time correlation function formalism remains the key to understand transport properties and line shapes measured in a number of experiments. We also wanted to bring home the point that Langevin's equation can be derived from first principles, as was demonstrated by Zwanzig and by Caldeira and Leggett. This is a common knowledge to advanced researchers, but I clearly remember that when I was a student, this bridge between the fundamentals and phenomenology was not made clear. This actually is a source of considerable relief that such a bridge indeed exists.

There are a few chapters that are unique to this book. For example, Chapter 3 on the relationship between theory and experiment, Chapter 20 on the rate theory, Chapter 23 on the mode-coupling theory and Chapter 24 on irreversible thermodynamics. These chapters have been included to familiarize students to the huge breadth and depth of nonequilibrium statistical mechanics.

Despite my effort to keep it short, the book did become somewhat long. This is because of the many chapters (some mentioned above) that I thought need to be included to make a comprehensive and exhaustive textbook at the intermediate level that can serve multiple purposes.

There are clearly many limitations of the book, the foremost being the absence of any discussion on the more sophisticated techniques, not just in computer simulations but also in theoretical methods. However, as we experienced in the chapter on the mode-coupling theory, inclusion of such details would rob the book of the introductory appeal. We really wanted the whole book to be accessible to students of MSc and first- or second-year Ph.D. students.

We hope that the book would come to be of use by students and researchers in this vast and rapidly evolving field. Even in this era of specialization, some beautiful concepts and elegant techniques retain their universal appeal and utility.

DOI: 10.1201/9781003157601-30

Index

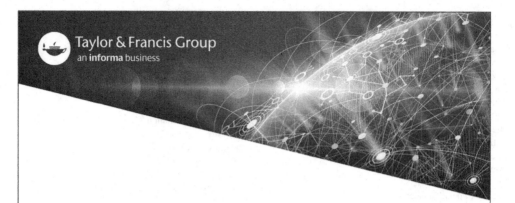